Mycoviruses

Mycoviruses

Special Issue Editor

Ioly Kotta-Loizou

MDPI • Basel • Beijing • Wuhan • Barcelona • Belgrade

MDPI

Special Issue Editor
Ioly Kotta-Loizou
Department of Life Sciences
Faculty of Natural Sciences
Imperial College London
UK

Editorial Office
MDPI
St. Alban-Anlage 66
4052 Basel, Switzerland

This is a reprint of articles from the Special Issue published online in the open access journal *Viruses* (ISSN 1999-4915) from 2018 to 2019 (available at: https://www.mdpi.com/journal/viruses/special_issues/mycoviruses)

For citation purposes, cite each article independently as indicated on the article page online and as indicated below:

LastName, A.A.; LastName, B.B.; LastName, C.C. Article Title. *Journal Name* **Year**, *Article Number*, Page Range.

ISBN 978-3-03897-996-8 (Pbk)
ISBN 978-3-03897-997-5 (PDF)

Cover image courtesy of Ioly Kotta-Loizou.

Contents

About the Special Issue Editor

Ioly Kotta-Loizou is currently a Research Associate at the Department of Life Sciences, Imperial College London, UK. She holds a BSc in Biology, a MSc in Bioinformatics and a PhD in Molecular Virology from the University of Athens, Greece. She has a keen interest in microorganisms, especially viruses, and her studies focus on microbial life cycles and pathogenicity, together with potential applications in medicine and agriculture. Currently, she explores how mycoviruses found in pathogenic fungi targeting humans, plants or insects can affect fungus–host interactions and potentially be used in biological control. In parallel, she investigates RNA expression, damage and repair in bacteria and how these processes can be linked to antibiotic resistance. She teaches mycovirology and next-generation sequencing to undergraduate and postgraduate students. She is an Editor for the scientific journals *Archives of Virology* and *Viruses* and a Deputy Chair of the International Committee for the Taxonomy of Viruses.

viruses

MDPI

Editorial

Mycoviruses: Past, Present, and Future

Ioly Kotta-Loizou

Department of Life Sciences, Imperial College London, London SW7 2AZ, UK; i.kotta-loizou13@imperial.ac.uk

Received: 18 April 2019; Accepted: 19 April 2019; Published: 19 April 2019

Approximately a year ago, when I accepted the offer to act as a Guest Editor for the Special Issue 'Mycoviruses' organised by the MDPI journal *Viruses*, I dared not expect that 'Mycoviruses' would include such a large number of manuscripts. Therefore, it is with great delight that I can count today a total of twenty-three, high-quality publications and I would like to take this opportunity to thank deeply all the contributing authors who chose 'Mycoviruses' as a vehicle for sharing their fascinating work with the mycovirology community.

'Mycoviruses' consists of three timely reviews on virus structure [1] and viruses of significant fungal pathogens causing damage to forests [2] and crops [3], together with twenty original research articles covering a range of relevant topics. These include the discovery and characterisation of novel members in the families *Chrysoviridae, Endornaviridae, Hypoviridae, Mymonaviridae, Narnaviridae, Partitiviridae, Totiviridae, Quadriviridae*, and others [4–16]; the study of the distribution, transmission, evolution, and dynamics of viruses in fungal populations [5,9,10,14,16–18] and the development of new techniques and methods to be used in the field of mycovirology [19,20]. Of particular interest are the investigations regarding the effects of viruses on their fungal hosts, most prominently on fungal morphology, spore production, growth, virulence [4,8,9,13–15,17,21–23], and, in the case of killer yeast systems, toxin production [6,20]. Understanding how these effects are mediated is crucial and applications of high-throughput next-generation sequencing technologies such as transcriptome and small RNA profiling provide insight into the molecular mechanisms underpinning the observed phenotypes [21,22], in addition to increasing sensitivity of virus detection [12,15,16]. Furthermore, the link between antiviral RNA silencing and infection is undoubtedly a significant one, given its implications for virus-induced hypovirulence, hypervirulence, and other phenotypic alterations [22,23].

As a relatively young mycovirologist with hopefully many active years to look forward to and a scientist with some first-hand knowledge in animal and plant virology, I frequently question how our work on mycoviruses conforms with the rapidly developing research in other fields and which future directions will lead to mycovirology being rightfully accepted as a mainstream research area and not merely esoteric, as it is often regarded. Mycovirology is not as advanced as human, animal, or even plant virology; overall our understanding of mycoviruses is not as detailed and in depth while the methodology we use to study them is not at the cutting edge. Since mycoviruses are not causative agents of significant diseases, they receive much less attention than, for instance, life-threatening human pathogens. Their 'moment of glory' was the use of *Cryphonectria parasitica* hypoviruses to control chestnut blight in Europe during the last century and today the major interest stems from the potential of mycoviruses as biological control agents in the context of integrative pest management programs. Based on this observation, a way forward would be to genetically engineer mycovirus-mediated hypovirulence or hypervirulence instead of merely hoping for a fortunate discovery. To this end, both the understanding of the molecular mechanisms underpinning these phenotypes and the development of reverse genetics systems for mycoviruses is an essential prerequisite.

Conflicts of Interest: The author declares no conflict of interest.

References

1. Luque, D.; Mata, C.P.; Suzuki, N.; Ghabrial, S.A.; Castón, J.R. Capsid structure of dsRNA fungal viruses. *Viruses* **2018**, *10*, 481. [CrossRef]
2. Botella, L.; Hantula, J. Description, distribution, and relevance of viruses of the forest pathogen *Gremmeniella abietina*. *Viruses* **2018**, *10*, 654. [CrossRef]
3. Moriyama, H.; Urayama, S.I.; Higashiura, T.; Le, T.M.; Komatsu, K. Chrysoviruses in *Magnaporthe oryzae*. *Viruses* **2018**, *10*, 697. [CrossRef]
4. Hao, F.; Ding, T.; Wu, M.; Zhang, J.; Yang, L.; Chen, W.; Li, G. Two novel hypovirulence-associated mycoviruses in the phytopathogenic fungus *Botrytis cinerea*: Molecular characterization and suppression of infection cushion formation. *Viruses* **2018**, *10*, 254. [CrossRef]
5. Hao, F.; Wu, M.; Li, G. Molecular characterization and geographic distribution of a mymonavirus in the population of *Botrytis cinerea*. *Viruses* **2018**, *10*, 432. [CrossRef] [PubMed]
6. Vepštaitė-Monstavičė, I.; Lukša, J.; Konovalovas, A.; Ežerskytė, D.; Stanevičienė, R.; Strazdaitė-Žielienė, Ž.; Serva, S.; Servienė, E. *Saccharomyces paradoxus* K66 killer system evidences expanded assortment of helper and satellite viruses. *Viruses* **2018**, *10*, 564. [CrossRef] [PubMed]
7. Chun, J.; Yang, H.E.; Kim, D.H. Identification and molecular characterization of a novel partitivirus from *Trichoderma atroviride* NFCF394. *Viruses* **2018**, *10*, 578. [CrossRef]
8. Mizutani, Y.; Abraham, A.; Uesaka, K.; Kondo, H.; Suga, H.; Suzuki, N.; Chiba, S. Novel mitoviruses and a unique tymo-like virus in hypovirulent and virulent strains of the Fusarium head blight fungus, *Fusarium boothii*. *Viruses* **2018**, *10*, 584. [CrossRef] [PubMed]
9. Yang, D.; Wu, M.; Zhang, J.; Chen, W.; Li, G.; Yang, L. Sclerotinia minor endornavirus 1, a novel pathogenicity debilitation-associated mycovirus with a wide spectrum of horizontal transmissibility. *Viruses* **2018**, *10*, 589. [CrossRef] [PubMed]
10. Schoebel, C.N.; Prospero, S.; Gross, A.; Rigling, D. Detection of a conspecific mycovirus in two closely related native and introduced fungal hosts and evidence for interspecific virus transmission. *Viruses* **2018**, *10*, 628. [CrossRef] [PubMed]
11. Liu, C.; Zeng, M.; Zhang, M.; Shu, C.; Zhou, E. Complete nucleotide sequence of a partitivirus from *Rhizoctonia solani* AG-1 IA strain C24. *Viruses* **2018**, *10*, 703. [CrossRef] [PubMed]
12. Neupane, A.; Feng, C.; Feng, J.; Kafle, A.; Bücking, H.; Lee Marzano, S.Y. Metatranscriptomic analysis and *in silico* approach identified mycoviruses in the arbuscular mycorrhizal fungus *Rhizophagus* spp. *Viruses* **2018**, *10*, 707. [CrossRef]
13. Shah, U.A.; Kotta-Loizou, I.; Fitt, B.D.L.; Coutts, R.H.A. Identification, molecular characterization, and biology of a novel quadrivirus infecting the phytopathogenic fungus *Leptosphaeria biglobosa*. *Viruses* **2019**, *11*, 9. [CrossRef]
14. Kamaruzzaman, M.; He, G.; Wu, M.; Zhang, J.; Yang, L.; Chen, W.; Li, G. A novel partitivirus in the hypovirulent isolate QT5-19 of the plant pathogenic fungus *Botrytis cinerea*. *Viruses* **2019**, *11*, 24. [CrossRef]
15. Tran, T.T.; Li, H.; Nguyen, D.Q.; Jones, M.G.K.; Wylie, S.J. Co-infection with three mycoviruses stimulates growth of a *Monilinia fructicola* isolate on nutrient medium, but does not induce hypervirulence in a natural host. *Viruses* **2019**, *11*, 89. [CrossRef] [PubMed]
16. Nibert, M.L.; Debat, H.J.; Manny, A.R.; Grigoriev, I.V.; De Fine Licht, H.H. Mitovirus and mitochondrial coding sequences from basal fungus *Entomophthora muscae*. *Viruses* **2019**, *11*, 351. [CrossRef]
17. Filippou, C.; Garrido-Jurado, I.; Meyling, N.V.; Quesada-Moraga, E.; Coutts, R.H.A.; Kotta-Loizou, I. Mycoviral population dynamics in Spanish isolates of the entomopathogenic fungus *Beauveria bassiana*. *Viruses* **2018**, *10*, 665. [CrossRef] [PubMed]
18. Rigling, D.; Borst, N.; Cornejo, C.; Supatashvili, A.; Prospero, S. Genetic and phenotypic characterization of Cryphonectria hypovirus 1 from Eurasian Georgia. *Viruses* **2018**, *10*, 687. [CrossRef]
19. Özkan-Kotiloğlu, S.; Coutts, R.H.A. Multiplex detection of *Aspergillus fumigatus* mycoviruses. *Viruses* **2018**, *10*, 247. [CrossRef] [PubMed]
20. Crabtree, A.M.; Kizer, E.A.; Hunter, S.S.; Van Leuven, J.T.; New, D.D.; Fagnan, M.W.; Rowley, P.A. A rapid method for sequencing double-stranded RNAs purified from yeasts and the identification of a potent K1 killer toxin isolated from *Saccharomyces cerevisiae*. *Viruses* **2019**, *11*, 70. [CrossRef]

21. Ejmal, M.A.; Holland, D.J.; MacDiarmid, R.M.; Pearson, M.N. The effect of Aspergillus thermomutatus chrysovirus 1 on the biology of three *Aspergillus* species. *Viruses* **2018**, *10*, 539. [CrossRef] [PubMed]

22. Lee Marzano, S.Y.; Neupane, A.; Domier, L. Transcriptional and small RNA responses of the white mold fungus *Sclerotinia sclerotiorum* to infection by a virulence-attenuating hypovirus. *Viruses* **2018**, *10*, 713. [CrossRef] [PubMed]

23. Mochama, P.; Jadhav, P.; Neupane, A.; Lee Marzano, S.Y. Mycoviruses as triggers and targets of RNA silencing in white mold fungus *Sclerotinia sclerotiorum*. *Viruses* **2018**, *10*, 214. [CrossRef]

viruses

MDPI

Review

Capsid Structure of dsRNA Fungal Viruses

Daniel Luque [1,2], Carlos P. Mata [1,†], Nobuhiro Suzuki [3], Said A. Ghabrial [4] and José R. Castón [1,*]

[1] Department of Structure of Macromolecules, Centro Nacional de Biotecnología (CNB-CSIC), Campus Cantoblanco, 28049 Madrid, Spain; dluque@isciii.es (D.L.); cpmata@mrc-lmb.cam.ac.uk (C.P.M.)
[2] Centro Nacional de Microbiología/ISCIII, Majadahonda, 28220 Madrid, Spain
[3] Institute of Plant Science and Resources, Okayama University, Kurashiki 710-0046, Japan; nsuzuki@okayama-u.ac.jp
[4] Department of Plant Pathology, University of Kentucky, Lexington, KY 40546, USA; saghab00@uky.edu
* Correspondence: jrcaston@cnb.csic.es; Tel.: +34-91585-4506
† Current Adress: Department of Medicine, University of Cambridge, MRC Laboratory of Molecular Biology, Cambridge CB2 0QH, UK.

Received: 21 August 2018; Accepted: 5 September 2018; Published: 7 September 2018

Abstract: Most fungal, double-stranded (ds) RNA viruses lack an extracellular life cycle stage and are transmitted by cytoplasmic interchange. dsRNA mycovirus capsids are based on a 120-subunit T = 1 capsid, with a dimer as the asymmetric unit. These capsids, which remain structurally undisturbed throughout the viral cycle, nevertheless, are dynamic particles involved in the organization of the viral genome and the viral polymerase necessary for RNA synthesis. The atomic structure of the T = 1 capsids of four mycoviruses was resolved: the L-A virus of *Saccharomyces cerevisiae* (ScV-L-A), *Penicillium chrysogenum* virus (PcV), *Penicillium stoloniferum* virus F (PsV-F), and *Rosellinia necatrix* quadrivirus 1 (RnQV1). These capsids show structural variations of the same framework, with 60 asymmetric or symmetric homodimers for ScV-L-A and PsV-F, respectively, monomers with a duplicated similar domain for PcV, and heterodimers of two different proteins for RnQV1. Mycovirus capsid proteins (CP) share a conserved α-helical domain, although the latter may carry different peptides inserted at preferential hotspots. Insertions in the CP outer surface are likely associated with enzymatic activities. Within the capsid, fungal dsRNA viruses show a low degree of genome compaction compared to reoviruses, and contain one to two copies of the RNA-polymerase complex per virion.

Keywords: dsRNA virus; mycovirus; capsid protein; capsid structure; virus evolution; viral lineage; ScV-L-A; PcV; PsV-F; RnQV1

1. Introduction

Double-stranded RNA (dsRNA) viruses infect a diversity of host organisms, from bacteria to unicellular and simple eukaryotes (fungi and protozoa), through to plants and animals. No archaea-infecting dsRNA viruses have yet been reported [1]. Although dsRNA viruses are a rather diverse group, they share general architectural principles and numerous functional features. The complexity of the capsid ranges from a single shell [2,3] to a multilayered and concentric capsid [4–7]. Whereas the outer shell has a protective role and is involved in cell entry, the innermost capsid (or inner core), which all these viruses possess, is devoted to the organization of the viral genome and viral polymerase. As a whole this specialized capsid consists of 120 protein subunits arranged in a T = 1 icosahedral shell, i.e., a capsid protein (CP) dimer is the asymmetric unit. This T = 1 capsid is also referred as a "T = 2 layer"—an exception to the quasi-equivalence theory proposed by Caspar and Klug [8].

The T = 1 capsids of dsRNA viruses are known to be critical for genome replication (minus-strand synthesis) and transcription (plus-strand synthesis), with the viral RNA-dependent RNA polymerase(s) (RdRp) frequently packaged as an integral component of the capsid. T = 1 capsids also function as molecular sieves, allowing the exit of single-stranded (ss) RNA transcripts for translation in the host cytoplasm, and the entrance of nucleotides for intra-capsid RNA synthesis. The pores are presumably small enough to exclude potentially degradative enzymes.

T = 1 capsids remain structurally undisturbed throughout the viral cycle [9], isolating dsRNA molecules and any replicative intermediates, thus preventing the triggering of dsRNA sensor-mediated antiviral host defense mechanisms, such as RNA silencing, interferon synthesis, and apoptosis [10–12].

The totiviruses ScV-L-A and UmV-P4, which infect the yeast *Saccharomyces cerevisiae* and the smut fungus *Ustilago maydis* respectively, were the first unambiguously described viruses with a T = 1 capsid formed by 12 decamers rather than 12 pentamers [13]. The conservation of this stoichiometry and architecture is probably related to the stringent requirements of capsid RNA metabolism-associated activity [14,15] as the capsid organizes the packaged genome and the replicative complex(es). RdRp is incorporated as a replicative complex at the pentameric vertex (as in rotavirus capsids [16,17]), as a fusion protein with the CP (as in the totivirus ScV-L-A [3]), or as a separate, non-fused protein (as in the victorivirus HvV190SV [18]).

The 120-subunit T = 1 capsids have been described for members of the *Reo-* [19–23] and *Picobirnaviridae* [24], which mostly infect higher eukaryotic organisms. They have also been described for members of the family *Cystoviridae*—bacteriophages that infect the prokaryote *Pseudomonas syringae* [25,26]. Members of the Toti- [27–29], Partiti- [30,31], Megabirna- [32], Chryso- [33–35], and Quadriviridae [36,37] families, which infect unicellular and simple eukaryotes, such as fungi, protozoa, but also some plants, also have these capsids. Members of the family Birnaviridae, which infect vertebrates, mollusks, insects, and rotifers, are exceptions, since they lack the T = 1 core of 60 CP dimers [38,39]. Rather, these have a single T = 13 shell that encapsidates a polyploid dsRNA genome organized as ribonucleoprotein complexes [40,41].

To date, 14 T = 1 capsid proteins have been resolved at the atomic level (Figure 1): VP3 of orbivirus bluetongue virus (BTV) [4], λ1 of reovirus (genus *Orthoreovirus*) [42], P3 of rice dwarf virus (RDV) [43], VP1 of cytoplasmic polyhedrosis virus (CPV) [20], VP2 of rotavirus [44], and VP3 of grass carp reovirus (GCRV) [23] (these last six viruses are all members of Reoviridae), CP of picobirnavirus (PBV) [24], φ6 P1 [45] and φ8 P1 [46] of the family *Cystoviridae*, Gag of the yeast virus ScV-L-A [27] (family Totiviridae), CP of *Penicillium chrysogenum* virus (PcV, family Chrysoviridae) [47], CP of *Penicillium stoloniferum* virus F (PsV-F, family Partitiviridae) [30], and the heterodimer P2–P4 of *Rosellinia necatrix* quadrivirus 1 (RnQV1, family Quadriviridae) [37]. The amino acid sequences of the above 14 CPs are quite different. All 14 proteins are, however, predominantly α-helical. The T = 1 CPs of *Reoviridae* members share an overall conformation [48], and those of PsV-F and PBV are based on similar folding. However, most CPs of mycoviruses have a tertiary structure that bears little resemblance to the reovirus structure. Rather, the 120-subunit capsids of fungal dsRNA viruses have a corrugated outer surface with protuberances rising above the continuous protein shell. Notably, the average thickness of a 120-subunit T = 1 CP is 15–30 Å in mammalian dsRNA viruses, but those of mycoviruses are thicker.

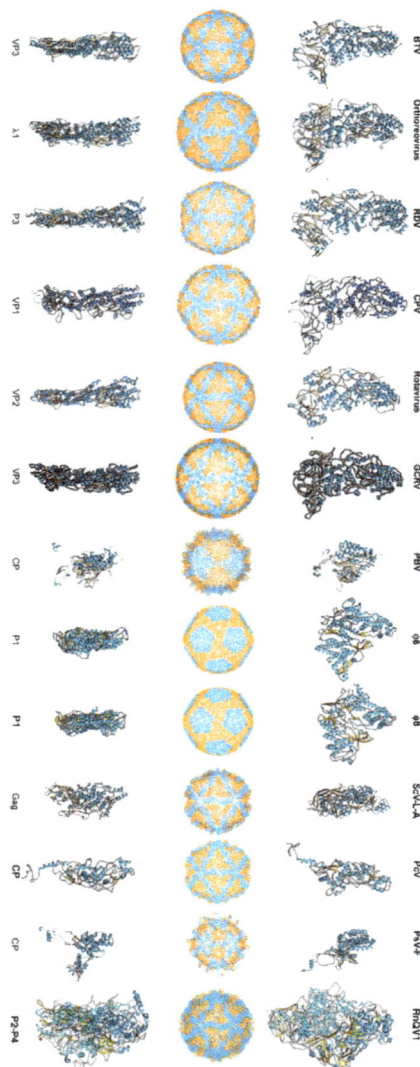

Figure 1. T = 1 capsid protein X-ray- and cryo-EM-based structures. Top row: T = 1 capsids of bluetongue virus (BTV), orthoreovirus, rice dwarf virus (RDV), cytoplasmic polyhedrosis virus (CPV), rotavirus, grass carp reovirus (GCRV), picobirnavirus (PBV), φ6 phage and φ8 phage, L-A virus of *Saccharomyces cerevisiae* (ScV-L-A), *Penicillium chrysogenum* virus (PcV), *Penicillium stoloniferum* virus F (PsV-F), and *Rosellinia necatrix* quadrivirus 1 (RnQV1), viewed along a two-fold axis of icosahedral symmetry (center row). BTV VP3 [2] (PDB accession number 2btv; 901 residues), λ1 [41] (1ej6; 1275 residues), P3 [42] (1uf2; 1019 residues), VP1 [19] (3cnf; 1333 residues), VP2 [43] (3kz4; 880 residues), GCRV VP3 [22] (3k1q; 1027 residues), PBV CP [23] (2vf1; 590 residues), φ6 P1 [44] (4btq; 769 residues) and φ8 P1 [45] (4btp; 792 residues), Gag [26] (1m1c; 680 residues), PcV CP domain A [46] (3j3i; 498 residues; complete CP 982 residues), PsV-F CP [29] (3es5; 420 residues) and RnQV1-W1118 P2 and P4 [36] (5nd1; 972 and 1005 residues, respectively), shown from top view. Bottom row: side views of the same structures (T = 1 shell exterior at right).

Unlike their bacteria- and higher eukaryote-infecting counterparts, most mycoviruses are transmitted by cytoplasmic interchange; they never leave the host, and indeed have no strategy for entering host cells [49]. Recent studies of fungal and protozoan dsRNA viruses identified functional and structural features unlike those recorded for members of the family *Reoviridae*, as well as evolutionary relationships among T = 1 capsid structural proteins. Whereas T = 1 capsids of reoviruses and cystoviruses share the same structural pattern, i.e., a 120-subunit capsid, most dsRNA mycoviruses exhibit high structural variability. ScV-L-A is built from 60 copies of a dimer of chemically identical subunits (as reo- and cystovirus), but the PcV T = 1 capsid is a variant of the 120-subunit capsid, as the CP has two motifs with the same fold, and the RnQV1 T = 1 capsid is composed of 60 dimers of two different proteins with a similar fold. In addition, the close relationship between the fungal dsRNA virus and its host probably place many constraints on the virus that it overcame by increasing CP complexity. In contrast to the plate-like protein found in reovirus and cystovirus T = 1 capsids, the 120-subunit capsid of fungal dsRNA viruses share a corrugated outer surface with domains rising above the continuous protein shell. In ScV-L-A virus, the CP has an extra domain with decapping activity, and the PcV and RnQV1 CP showed similar extra domains on the outer capsid surface with unknown functions. The present review focuses on the structure of dsRNA mycoviruses, and discusses how the lack of an extracellular phase has had unanticipated functional effects in their life cycles.

2. Structure of dsRNA Virus Capsids

2.1. Totiviruses

The L-A virus of the yeast *Saccharomyces cerevisiae* (ScV-L-A) is the type species of the genus *Totivirus* (family Totiviridae) [50,51]. The ScV-L-A genome is a 4.6 kb, single-segment dsRNA molecule that encodes a major capsid protein (Gag; 680 residues, 76 kDa) and viral polymerase (Pol; 868 residues, 94 kDa), as a Gag-Pol fusion protein generated by −1 ribosomal frameshifting [52,53]. Gag is bound covalently to the inside of the particle wall.

The structure of ScV-L-A was first examined by three-dimensional cryo-electron microscopy (3D cryo-EM) and later by X-ray crystallography (resolution 3.4 Å) [27]. Dark-field scanning transmission electron microscopy (STEM) was used to determine the virus stoichiometry [3,13]. The rough, icosahedral, ~400 nm diameter T = 1 lattice of ScV-L-A has 120 copies of Gag, of which one or two are fused to the Pol moiety [54] (Figure 2A). The protein shell is 56 Å thick. The structural unit is an asymmetric Gag dimer. Each Gag monomer can adopt one of two conformations, termed subunits A and B, with notable structural differences in specific surface regions and with entirely different bonding environments (non-equivalent contacts) (Figure 2B). These subunits are arranged in two sets of five: five A subunits directly surrounding the icosahedral five-fold axis, leaving an 18 Å diameter channel as a portal for the entry of nucleotide triphosphates and the exit of viral mRNA; and five B subunits intercalated between the A subunits, forming a decamer. This quaternary organization is similar to the 120-subunit T = 1 inner core of reoviruses [4,20,23,42–44] and cystoviruses [45,46], in which subunits A and B are arranged in nearly parallel positions (Figure 2C).

Gag functions as an enzyme and has a major role in the sophisticated interaction between ScV-L-A and the host cell. The Gag segment Gln139-Ser182, in which His154 is the active site, contributes to the rough outer surface of the capsid, and is responsible for the cellular RNA decapping activity that transfers the 7-methyl-GMP (m^7GMP) cap from the 5′ end of the cellular mRNA to the 5′ end of the viral RNA [55,56] (Figure 2D). L-A counters a host exoribonuclease that targets uncapped RNAs (such as viral mRNA), allowing the latter to compete with host mRNA for use of the translation machinery.

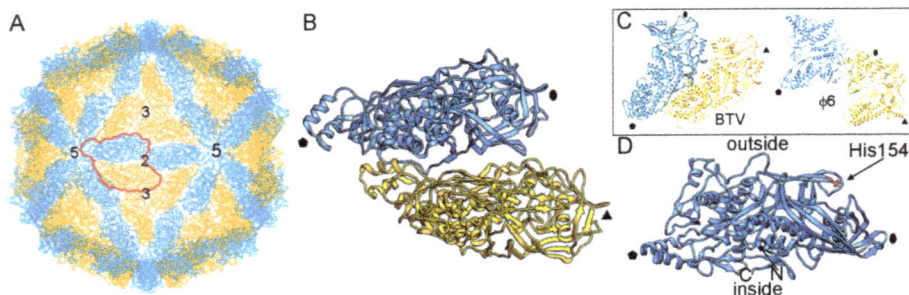

Figure 2. ScV-L-A T = 1 capsid protein; X-ray-based structure. (**A**) T = 1 capsid of ScV-L-A viewed along a two-fold axis of icosahedral symmetry, showing the Gag subunits A (blue) and B (yellow), with the boundaries of the asymmetric unit outlined in red. Numbers indicate icosahedral symmetry axes. (**B**) Atomic model of a Gag dimer (1m1c; 680 residues). Icosahedral symmetry five- (pentagon), three- (triangle), and two-fold (oval) axes are indicated in black. (**C**) T = 1 CP dimers of BTV (2btv) and φ6 (4btq). The structural unit is an asymmetric dimer in which subunits A and B are oriented in parallel with numerous side contacts. Symbols indicate icosahedral symmetry axes. (**D**) Side view of a Gag monomer (T = 1 shell exterior top). His154, the active site for decapping activity is indicated.

The *Helminthosporium victoriae* virus 190S (HvV190S), a prototype of the genus *Victorivirus*, family Totiviridae, infects the filamentous fungus *H. victoriae*, and has a similar capsid organization to that of ScV-L-A [18,57]. The smooth HvV190S capsid (average thickness 35 Å) is composed of 120 CP monomers, with RdRp incorporated as a separate, non-fused protein synthesized by a stop/reinitiation mechanism [58,59]. The RdRp is either non-covalently associated with the underside of the capsid, as in reoviruses, free in the capsid interior, or non-covalently bound to the genome [18,57]. *Trichomonas vaginalis* virus 1 (TTV1), a totivirus of the genus *Trichomonasvirus* that infects a human-hosted protozoan, has its RdRp fused to the CP, as in ScV-L-A, but by −2 ribosomal frameshifting [29]. Notably, both the protozoan-infecting *Giardia lamblia* virus (GLV, genus *Giardiavirus*) [28] and the metazoan-infecting myonecrosis virus (IMNV, a tentative member of the family Totiviridae) [60,61] share the 120-subunit T = 1 capsid organization, but can be transmitted extracellularly.

2.2. Chrysoviruses

Chrysoviruses are isometric virions characterized by a multipartite genome [62]. *Penicillium chrysogenum* virus (PcV) is the prototype of the Chrysoviridae, a family of typically symptomless mycoviruses with a genome consisting of four monocistronic dsRNA segments (genome size 2.4–3.6 kbp). Each segment is encapsidated separately in a similar particle [63,64], i.e., chrysoviruses are multi-segmented and multi-particulate virions. dsRNA-1 (3.6 kbp) encodes the RdRp (1117 amino acid residues with a molecular mass of 128.5 kDa; one or two copies per virion), dsRNA-2 (3.2 kbp) encodes the CP (982 amino acid residues, 109 kDa), and dsRNA-3 and -4 (3 and 2.9 kbp) code for virion-associated proteins of unknown function (912 amino acid residues and 101 kDa, and 847 amino acid residues and 95 kDa, respectively).

So far, the 3D structures of the capsids of two chrysoviruses have been determined by cryo-EM analysis, that of PcV at atomic resolution [47], and that of *Cryphonectria nitschkei* chrysovirus virus 1 (CnCV1) at subnanometer resolution [35]. Analytical ultracentrifugation analysis has shown that PcV and CnCV1 virions are exceptions to the most-extended tendency among dsRNA viruses—a T = 1 core with 60 equivalent dimers—since they have an authentic T = 1 capsid formed by 60 copies of a single monomer [33,34]. The capsid diameter is 400 Å and the protein shell is 48 Å thick (Figure 3A). Similar to ScV-L-A, the outer capsid surface of PcV is relatively uneven with 12 outwardly protruding pentons, each containing five copies of the CP; this contrasts with the smooth outer surface of reoviruses, in which the CP has a plate-like structure. The 982-residue CP of PcV is formed by duplication of an α-helical domain; this is indicative of gene duplication despite negligible sequence similarity between the two roughly parallel α-helical domains (Figure 3B). The N-terminal A domain (residues 1–498) and the C-terminal B domain (residues 516–982) are connected by a 16-residue linker (Ala499-Ile515), accessible from the capsid outer surface. These domains are arranged in two sets of five: five A domains directly surround the icosahedral fivefold axis and five B domains intercalated between them, forming a pseudodecamer. This organization is clearly reminiscent of the 120-subunit T = 1 lattice of totivirus and megabirnavirus (as well as reovirus and cystovirus) capsids, in which the two asymmetrical dimer components are arranged in near-parallel fashion. The structural details of the PcV capsid reinforce the idea that a T = 1 layer with a dimer as the asymmetric unit provides an optimal framework for managing dsRNA metabolism.

Superimposition of the PcV A and B αhelical domains identifies a single "hotspot" on the outer capsid surface where variation is introduced by insertion of 50–100 residue segments (Figure 3C,D). A preferential insertion site would allow the acquisition of new functions while preserving basic CP folding. It is plausible that, in addition to its structural role, chrysovirus CP also has enzymatic activity.

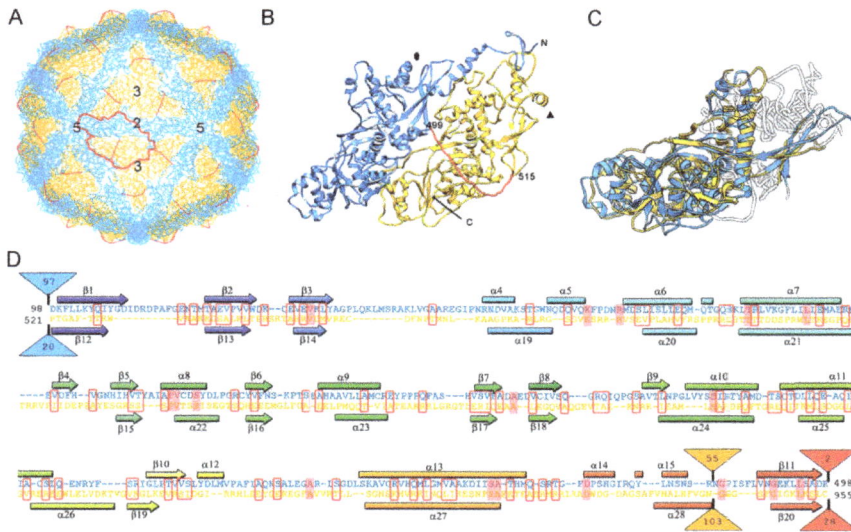

Figure 3. PcV T = 1 capsid protein; cryo-EM-based structure. (**A**) T = 1 capsid of PcV viewed along a two-fold axis of icosahedral symmetry, showing the N-terminal domain A (1–498, blue), the linker segment (499–515, red), and the C-terminal domain B (516–982, yellow), with the boundaries of an asymmetric unit outlined in red. Numbers indicate icosahedral symmetry axes. (**B**) Top view of the atomic of the PcV CP (3j3i; 982 residues). Symbols indicate icosahedral symmetry axes. (**C**) PcV capsid protein is a structural duplication. Superimposed A and B domains (white segments indicate non-superimposed regions for both domains). (**D**) Sequence alignment of domains A (blue) and B (yellow) resulting from Dali structural alignment. α-helices (rectangles) and β-strands (arrows) are rainbow-colored from blue (N terminus) to red (C terminus) for each domain. Triangles represent non-aligned segments (sizes indicated): the orange triangle indicates the single "hotspot" on the outer capsid surface. Strictly conserved residues are on a red background and partially conserved residues are in a red rectangle.

2.3. Partitiviruses

Members of the family Partitiviridae have bisegmented, 1.4–2.4 kbp-long genomes. Each segment is encapsidated separately in a similar virus particle. dsRNA1 encodes RdRp (one copy per virion), whereas dsRNA2 encodes the CP. The partitiviruses that infect fungi are grouped into three genera:

alpha-, beta-, and gamma-partitiviruses [65,66]. Alpha- and beta-partitiviruses infect plants and filamentous fungi, whereas gamma-partitiviruses infect only the latter. In general, partitivirus infections are largely symptomless.

Four fungal partitivirus structures have been resolved by 3D cryo-EM, including those of the gamma-partitiviruses *Penicillium stoloniferum* virus S (PsV-S) [67,68] and *Penicillium stoloniferum* virus F (PsV-F) (by X-ray crystallography at 3.3 Å resolution) [30], and of the beta-partitiviruses *Fusarium poae* virus 1 (FpV1) [31] and *Sclerotinia sclerotiorum* partitivirus 1 (SsPV1) [69].

The single-layered, 120-subunit capsids of these viruses are 35–42 nm in diameter and distinct in that they have "arch-like" surface features that protrude above the continuous capsid shell (Figure 4A). These T = 1 capsids have a different quaternary organization, their CP dimer having almost perfect local two-fold symmetry (Figure 4B). The quasi-symmetric CP dimer is stabilized by domain swapping within the shell region of the A and B subunits, as well as by intradimeric interactions between equivalent protruding arch domains on the particle surface (Figure 4C). A similar organization has been found in a picobirnavirus [24]—a bisegmented dsRNA virus that infects humans and other vertebrates. This might represent convergent evolution. Brome mosaic virus (BMV) and cowpea chlorotic mottle virus (CCMV), two plant positive-sense (+) ssRNA viruses with a T = 3 capsid, have a CP (with a β-barrel domain) that assembles into a 120-subunit capsid with a quaternary organization similar to that of PsV-F and PBV [70,71] (Figure 4D).

Based on their capsid organization, partiti- and picobirnaviruses appear to be assembled from dimers of CP dimers (i.e., tetramers). In contrast, the proposed assembly pathway for the 120-subunit capsids of Totiviridae and Reoviridae members is based on pentamers of CP dimers (i.e., decamers). Notably, the capsid protein P1 of bacteriophage ϕ8 (a cystovirus) appears as a soluble tetramer in an in vitro assembly system [72].

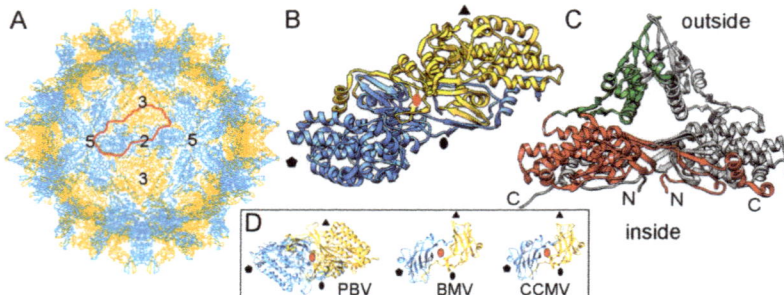

Figure 4. PsV-F T = 1 capsid protein X-ray-based structure. (**A**) T = 1 capsid of PsV-F viewed along a two-fold axis of icosahedral symmetry, showing the CP subunits A (blue) and B (yellow), with the boundaries of an asymmetric unit (a quasi-symmetric A-B dimer) outlined in red. Numbers indicate icosahedral symmetry axes. (**B**) Top view of the atomic model of a PsV-F CP dimer (3es5; 420 residues). Symbols indicate icosahedral symmetry axes; red oval indicates a local two-fold symmetry axis. (**C**) Side view of a PsV-F CP dimer. The arch (green) and shell domains (red) are indicated. For clarity, the second subunit is shown in grey. (**D**) T = 1 CP dimers of PBV (2vf1), BMV, brome mosaic virus (1js9), and cowpea chlorotic mottle virus (CCMV; 1cwp). Dimers are related by a local quasi-two-fold symmetry axis (red oval), and show molecular swapping. BMV and CCMV are plant ssRNA viruses with T = 3 capsids, but their CP (with a β-barrel domain) can assemble into 120-subunit capsids that show a quaternary organization similar to that of PsV-F and PBV.

2.4. Quadriviruses

Rosellinia necatrix quadrivirus 1 (RnQV1) is the type species of the genus *Quadrivirus* in the family Quadriviridae [73,74]. The filamentous ascomycete *Rosellinia necatrix*, a pathogen of many plants, can be infected by dsRNA viruses belonging to at least six families [75,76]. RnQV1 is associated with latent infections (i.e., it causes no apparent slowing of host growth), and has a multipartite genome consisting of four monocistronic dsRNA segments (as in chrysoviruses) with genome sizes ranging from 3.7 to 4.9 kbp. DsRNA-1 (4.9 kbp) codes for a protein of unknown function (1602 amino acid residues), dsRNA-2 (4.3 kbp) encodes the P2 CP (1356 amino acids), dsRNA-3 (4 kbp) codes for RdRp (1117 amino acids), and dsRNA-4 (3,7 kbp) codes for the P4 CP (1061 amino acids).

RnQV1 virus strains W1075 and W1118, isolated from different locations in Japan, have been analyzed by 3D cryo-EM and analytical ultracentrifugation [36]. Their P2 and P4 proteins co-assemble into isometric virus particles ~45 nm in diameter, which each package either one or two of the four genome segments. Whereas most dsRNA virus capsids are based on dimers of a single protein, RnQV1 has a single-shelled T = 1 capsid formed by 60 P2 and P4 protein heterodimers (Figure 5A). Whereas P2 and P4 of RnQV1 strain W1118 remain nearly intact, in strain W1075, both proteins are cleaved into discrete polypeptides, apparently without altering capsid structural integrity. The atomic structure of the RnQV1 W1118 capsid at 3.7 Å resolution shows that P2–P4 heterodimers are organized into a quaternary structure similar to that of the homodimers of reoviruses, chrysoviruses, and totiviruses [37] (Figure 5B,C). Although the RnQV1 capsid, and that of PcV, is an exception to the rule that all dsRNA viruses have a T = 1 capsid with a CP homodimer as the asymmetric unit, it follows the architectural principle that a 120-subunit capsid is a conserved assembly that supports dsRNA replication and organization.

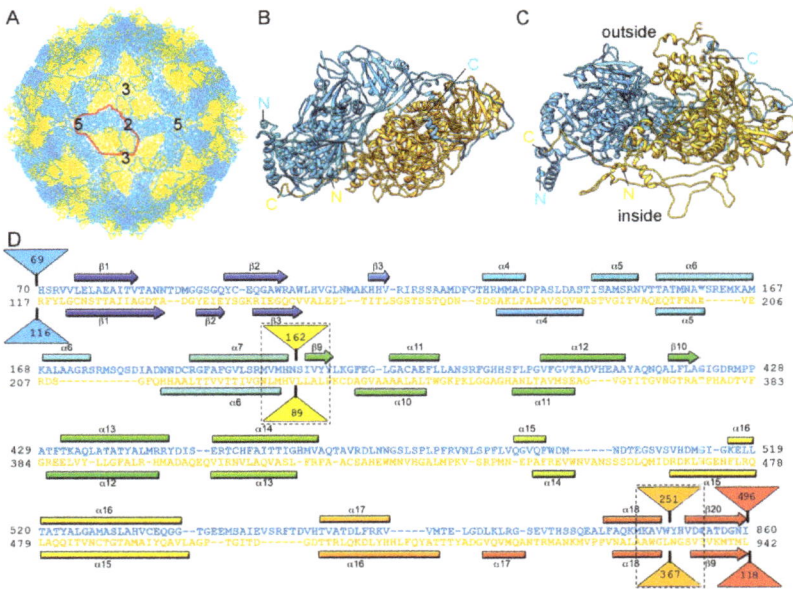

Figure 5. RnQV1 T = 1 capsid cryo-EM-based structure. (**A**) T = 1 capsid of RnQV1 viewed along a two-fold axis of icosahedral symmetry, showing P2 (blue) and P4 (yellow). Boundaries for an asymmetric unit are outlined in red. Numbers indicate icosahedral symmetry axes. (**B**) Top and (**C**) side views of the atomic models of P2 (blue; 972 residues) and P4 (yellow; 1005 residues) (5nd1). The last visible P2 C-terminal residue is located on a P4 surface crevice. (**D**) Sequence alignment of P2 (blue) and P4 (yellow) resulting from Dali structural alignment. α-helices (rectangles) and β-strands (arrows) are rainbow-colored from blue (N terminus) to red (C terminus) for each protein. Dashed rectangles indicate favorable insertion sites, triangles represent non-aligned segments (sizes indicated).

Despite their low sequence similarity, the superimposition of P2 and P4 revealed their having a common α-helical domain (Figure 5D). As described for the PcV capsid, P2 and P4 have also acquired new functions through the insertion of complex domains at preferential insertion sites on the capsid outer surface. These are also probably related to enzyme activity. The P2 insertion has a fold similar to that of gelsolin and profilin, two actin-binding proteins with a function in cytoskeleton metabolism; whereas the P4 insertion suggests a protease activity involved in cleavage of the P2 383-residue C-terminal region (absent in the mature viral particle). This P2 C-terminal segment might represent an external scaffolding domain [37].

3. Evolutionary Relationships Based on Structural Comparisons

Structural comparisons of CPs have been used to establish relatedness when sequence conservation is limited [77–79], and have detected relationships among viruses that infect organisms that, in evolutionary terms, are widely separated [78,80–82]. Icosahedral viruses are grouped into four lineages [80]: the dsDNA viruses with an upright double β-barrel CP (the prototypes are phage PRD1 and adenoviruses), the head-tailed phages and herpesviruses that share the Hong Kong 97 (HK97)-like CP fold (also termed the Johnson fold), the picornavirus-like superfamily with a single β-barrel as the CP fold, and the dsRNA or bluetongue virus (BTV)-like viruses. The PRD1- and HK97-like lineages include archaea-, bacteria-, and eukaryote-infecting viruses, suggesting that their last common ancestral hosts were infected by the progenitors of the current viral lineages before the host organisms diverged [82–84]. Although many viruses are not included in these four lineages, the number of folds that satisfy the assembly constraints for a viable viral shell is thought to be limited.

The similarity of the A and B α-helical domains of PcV CP, which have many well-matching secondary structural elements, indicate a common fold in both domains [47]. Gene duplication (or joined folds) has been a recurrent evolutionary event in other viral lineages, for example, involving the trimeric capsomeres of adenoviruses [85], *Paramecium bursaria* chlorella virus 1 (PBCV1) [86], and bacteriophage PRD1 [87], and the large subunit of comoviruses ([+] ssRNA viruses that infect plants) [88]. The conserved ~350 residue-long PcV fold is also preserved in the Gag of ScV-L-A [33,47] (Figure 6). This basic α-helical domain shares many secondary structural elements with L-A Gag, in particular those regions involved in interactions at the five-, three-, and two-fold symmetry axes. The preserved fold in Gag has three peptide insertion sites facing the outer capsid surface, one of which colocalizes with the single-insertion hotspots of the PcV CP domains. This colocalization suggests that these preferential insertion sites are ancient, and provide a means for the acquisition of new functions without altering the structural and functional motifs of the dsRNA virus CP.

P2 and P4 of RnQV1 also have a common fold some 300 residues long, with two preferential insertion sites on the outer surface [37]. Both coincide with the ScV-L-A Gag insertion sites, and one with the single-insertion site of the PcV A and B α-helical CP domains. Notably, the conserved folds of PcV and ScV-L-A CP are similar to the common fold of P2 and P4, indicating that this fold may have evolved from a common ancestral domain of the dsRNA virus lineage (Figure 6).

Despite their size and overall shape differences, the preserved ~300-residue PcV domains can be compared with the 1000–1300-residue reovirus T = 1 CP through the use of robust structural alignment methods for highly diverged CP structures [46,89]. There are discernible similarities in the arrangement of the secondary structural elements that place φ8 CP as an intermediate between reovirus CP and mycovirus CP [46], i.e., at the furthest distance within the structure-based phylogenetic tree. The preserved α-helical domain of mycoviruses is broken by much longer insertions in reovirus CP, resulting in basic structural motifs or subdomains.

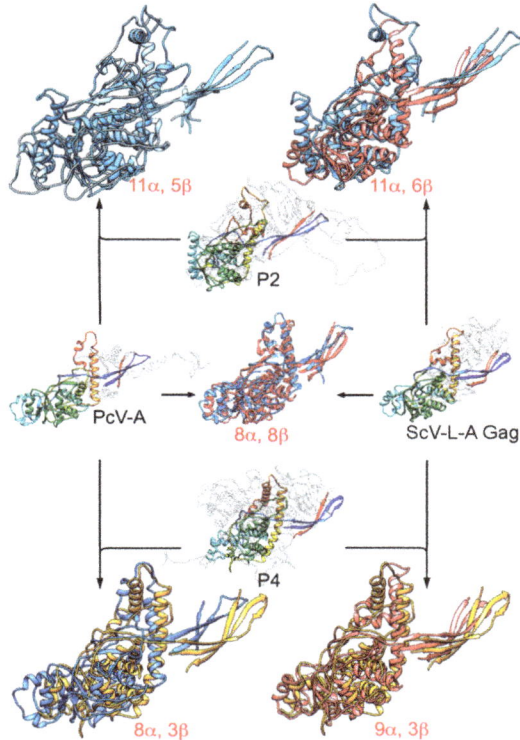

Figure 6. Structural homology of mycovirus T = 1 CP. The PcV CP A domain (PcV-A, left, center) was structurally aligned with ScV-L-A Gag (right, center), and P2 (top, center) and P4 (bottom, center) with PcV-A and ScV-L-A Gag. Center rainbow-colored structures indicate conserved secondary structure elements within the dsRNA viruses. PcV-A is aligned with ScV-L-A Gag (blue and pink, center). P2 is aligned with PcV-A (blue and light blue, top left) and with ScV-L-A Gag (blue and pink, top right). P4 is aligned with PcV-A (yellow and blue, bottom left) and P4 with ScV-L-AL-A Gag (yellow and pink). Total numbers of secondary structural elements with close relative spatial locations are indicated.

Duplication of an ancestral gene for a CP with the BTV-like fold might have resulted in two separate (as in quadriviruses) or covalently joined folds (as in chrysoviruses). This event could direct the assembly of a T = 1 capsid with 120 subunits or domains with a dimer as the asymmetric unit—a necessary arrangement for dsRNA replication/transcription. Separate and joined folds are found in the CP of other virus families, such as picornaviruses [79] and comoviruses [88], respectively. Once the 120-subunit capsid was well-established, later divergent evolutionary events would have introduced additional changes in each copy, or even the complete removal of one of them, producing a CP that assembles as a dimer of unfused identical monomers. Alternatively, the ancestral CP could have initially acquired dimer assembly ability, followed by gene duplication.

The CP of many-tailed dsDNA phages with the HK97-like fold has additional domains with specific functions related to capsomere and/or capsid stability (reviewed in Suhanovsky, M.M.. et al. [90]). Human cytomegalovirus (HCMV), a herpesvirus, has a 1370-residue CP folded into seven domains [91], with the Johnson fold or floor domain in the shell, and a six-domain protruding tower. The Johnson fold has a five-stranded β-core that acts as the organizational hub of the CP; the additional domains in the Johnson fold are considered modular insertions into the peripheral loops [91]. In this context, tailed dsDNA phages and herpesviruses share some similarities with dsRNA mycoviruses. Conserved α-helices and/or the β-sheet structure preserved in the dsRNA virus basic fold might form a similar functional center for domain insertion.

4. RdRp and dsRNA Organization within Mycovirus Capsids

Reovirus T = 1 cores have 10–12 RdRp complexes per virion, around which the dsRNA is densely coiled [92,93]. RdRp complexes are non-covalently anchored to the capsid inner surface near the icosahedral 5-fold axes [19,94,95], as presumably they are in mycoreoviruses. In addition to RdRp molecules, reovirus replicase complexes include a few minor core proteins with ATPase- and/or RNA-binding abilities. For members of the Toti-, Chryso- Partiti-, and Quadriviridae families, the RdRp molecules are incorporated into one or two copies per virion, and show more variability than reovirus. For chryso-, partiti-, and quadriviruses, the RdRp is expressed as a physically separate protein from a discrete genome segment, and is incorporated into virions via non-covalent interactions with the capsid and/or genome. The same is true for victoriviruses (genus *Victorivirus*, family Totiviridae), such as HvV190S, except that the RdRp is expressed from the single genome segment of those viruses via a coupled termination-reinitiation mechanism [58,59,96]. For totiviruses such as ScV-L-A (genus *Totivirus*), in contrast, the RdRp is expressed as a C-terminal fusion product with the CP (i.e., as a Gag-Pol). As a result, in ScV-L-A, the one or two RdRp domains per virion are covalently tethered to the capsid via the fused CP domain, which occupy one or two subunit positions in the capsid.

The anchoring of RdRp at the five-fold axes on the reovirus capsid inner surface seems likely to occur in toti-, chryso-, partiti-, and quadriviruses too, with important consequences for the channeling of freshly synthesized transcripts into an exit pore.

The mycovirus T = 1 capsid wall is perforated by many pores and channels, but none is large enough to pass an A-form 23-Å-diameter duplex (Figure 7A,B). Whereas the largest pores (15–20 Å diameter and usually located near the five-fold axis) would allow the passage of nascent mRNA into the host cytoplasm, the smallest holes (5–10 Å in diameter and usually located at the three-fold axis) could be used for nucleotide substrate or pyrophosphate byproduct diffusion. In non-transcribing T = 1 capsids, the pores are very narrow, but the N- or C-termini or the side chains of residues that face the channel wall might adopt alternative conformations to allow the exit of viral transcripts.

With the exception of totiviruses, which have a single genomic segment, many fungal dsRNA viruses, including chryso-, partiti-, and quadriviruses, have multisegmented dsRNA genomes. In addition, the multisegmented viruses appear to be multiparticulate, i.e., segments are encapsidated separately [97]. Fungal dsRNA viruses have spacious capsids in comparison with the inner cores of complex eukaryotic dsRNA viruses (Table 1). Whereas reoviruses have 9–12 genome dsRNA segments packed into liquid crystalline arrays at high density (~40 bp/100 nm^3, a spacing between dsRNA strands of 25–30 Å) [6,98–100], fungal virus capsids (including ScV-L-A, PcV, PsV-F, and RnQV1) contain a single loosely packed dsRNA molecule (~20 bp/100 nm^3, an interstrand spacing of ~40–45 Å) [34,49,63]. In reoviruses, individual genome segments must be transported through the active sites of the RdRp complexes at the five-fold axes, and template motion could be a limiting factor. ScV-L-A is a simplified version of these viruses, with a single-segment genome. The looser packing of the dsRNA would probably improve template motion in the more spacious transcriptional and replicative active particles, minimizing electrostatic repulsion between dsRNA strands.

Table 1. Genome packaging densities in double-stranded ribonucleic acid (dsRNA) viruses.

Virus Family	dsRNA Features			Capsid Features		
	N° Segments	Size (kbp)	MW [c] (MDa)	CP (residue)	Φ [d]/ir [e] (nm)	dsRNA Density (bp/100 nm³) [f]
HSV [a]	1	~152	103.7	1374	~130/43	46
Reoviridae						
Orthoreovirus	10	~23.5	16	1275	~60/24.5	38
Rotavirus	11	~18.5	12.6	880	~52/23.5	34
Orbivirus, BTV	10	~19.2	13.1	901	~52/22	43
Aquareovirus, GCRV	11	~23.6	16	1027	~60/23	46
Phytoreovirus, RDV	12	~25.7	17.5	1019	~57/26	35
Cypovirus, CPV	10	~31.4	21.4	1333	~58/24	54
Mycoreovirus, MyRV1	11	23.4	16			
Picobirnaviridae	2	~4.2	2.9	590	~35/14	
Cystoviridae, phage φ6	3	~13.4	9.1	769	~50/20	40
Totiviridae, ScV-L-A	1	~4.6	3.1	680	~43/17	22
Partitiviridae, PsV-S	1 (2) [b]	~1.7 (3.3)	1.2 (2.2)	420	~35/12	23
Chrysoviridae, PcV	1 (4) [b]	~3.2 (12.6)	2.2 (8.6)	109	~40/16	19
Megabirnaviridae, RnMBV1	1 (2) [b]	~8.1 (16.2)	5.5 (11)	135	~52/19	28
Quadriviridae, RnQV1	1–2 (4)	~4.3 (17.1)	2.9 (11.7)	1356 + 1061	~47/16	25 (50) [g]

[a] Herpes simplex virus, a dsDNA virus with liquid-crystalline packing of the encapsidated, B-form dsDNA. [b] For PsV-S, PcV, and RnMBV1 dsRNA, the genome is formed by two, four, or two dsRNA molecules, respectively, but a mean value was calculated for one dsRNA molecule/particle in each column. [c] Molecular weights (MW) were calculated assuming a mass of 682 Da/bp. HSV dsDNA is assumed to have a B-form. [d] Outer diameter. [e] Inner radius. [f] Densities when volume of a perfect sphere is assumed and any other internal components are ignored. [g] 25 if there is one dsRNA molecule/particle; 50 if there are two dsRNA molecules/particle.

Most mycovirus T = 1 capsids are negatively charged on their inner surface, a feature common to many such capsids of dsRNA viruses [37]. This might facilitate the movement of template and/or product RNA molecules by repulsion, maintaining the RNA layer at ~25 Å from the capsid surface (Figure 7B,C). The PcV capsid is an exception. It has positively charged regions on the inner surface (Figure 7A) and has numerous interactions with the underlying genome, which is ordered in the outermost RNA layer [33]. As a result, there is almost no space between the latter layer and the inner capsid surface. These contacts have been defined at the atomic level in PcV and PsV-F virions [30,47]. The lower density of the central region and the associated slight increase in dsRNA mobility might be necessary for maximum RdRp activity in the context of a non-fused RdRp complex.

Comparative analysis of dsRNA packing densities in dsRNA virions have revealed two major tendencies among T = 1 capsids of dsRNA viruses: (1) those with 9–12 dsRNA segments densely packaged within the same particle and containing 9–12 RdRp complexes, as seen in reoviruses, and (2) those with a single-genomic dsRNA segment with less internal order and one or two copies of the RdRp complex per particle, as seen in mycoviruses.

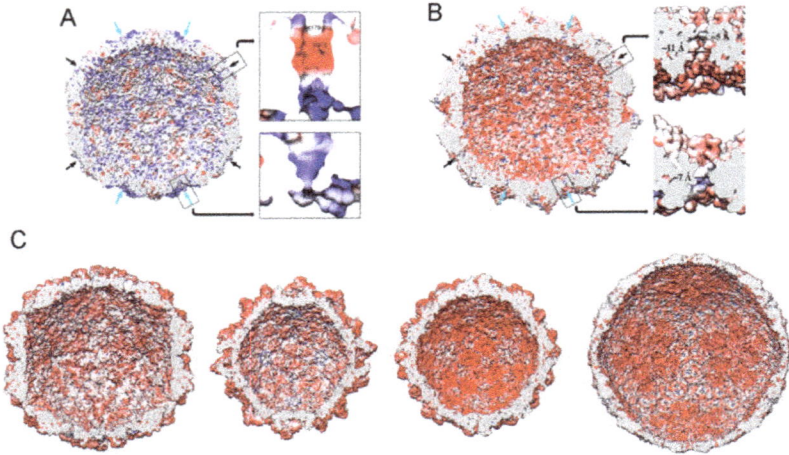

Figure 7. dsRNA virus T = 1 capsid inner surfaces with electrostatic potentials. (**A**) PcV and (**B**) RnQV1 capsid inner surfaces viewed along a two-fold axis of icosahedral symmetry. The inner surface charge representations of these capsids show the distribution of negative (red) and positive (blue) charges. Note the numerous electropositive areas in PcV. Arrows indicate the capsid pores at the five-fold (black) and three-fold (blue) axes. Boxes: magnified views of the five- (top) and three-fold (bottom) pores showing charge distribution on the channel walls. (**C**) T = 1 capsids of ScV-L-A, PsV-Fe, rabbit picobirnavirus, and rotavirus (from left to right), viewed along a two-fold axis of icosahedral symmetry. dsRNA and packaged proteins (such as RNA polymerases) were removed computationally.

5. Concluding Remarks and Future Perspectives

Structural studies of a limited number of fungal viruses have revealed them to conform to the basic concepts of dsRNA viruses, but also to have unexpected features that have contributed to a better understanding of their structure, function, and evolution. dsRNA mycovirus capsids, exemplified by ScV-L-A, PcV, PsV-F, and RnQV1, show structural variations of the same framework optimized for RNA metabolism; they possess 60 asymmetric or symmetric dimers of a single protein (ScV-L-A and PsV-F, respectively), dimers of similar domains (PcV), or dimers of two different proteins (RnQV1). Since mycoviruses are transmitted by cytoplasmic interchange and commonly confined to their hosts, their capsids incorporate polypeptides and domains on their outer surfaces for the acquisition of new functions without altering the structure and function of the CP. Such acquisitions would eventually lead to optimal viral-host interactions.

Despite recent advances in understanding the structure of dsRNA mycoviruses, many aspects of several fungal (and protozoan) viruses remain unknown. Recent work has identified a positive-sense ssRNA virus—the yado-kari virus 1 (YkV1)—that hijacks the CP of a dsRNA virus that resembles totivirus—the yado-nushi virus 1 (YnV1) [101,102]. There are several papers reporting yadokari-like viruses with sequence similarity to YkV1 [102–105], but their possible mutualism with potential partners has yet to be elucidated. Another notable example includes Aspergillus fumigatus tetramycovirus 1 (AfuTmV1) [106], Colletotrichum camelliae filamentous virus 1 (CcFV-1) [107], and related viruses. Despite similarity in genome organization and sequence, these viruses seemingly utilize different genome packaging strategies; namely, the genomic dsRNAs are associated with a virally encoded protein in a colloidal form (AfuTmV1) or packaged in filamentous particles (CcFV-1), for which infectivity as purified dsRNA has also been demonstrated. Future structural studies should focus on the asymmetric substructures and components of their capsids [108,109]—such as their RdRp (isolated or packaged inside virions)—and their packaged dsRNA genome.

Author Contributions: D.L., C.P.M., N.S., S.A.G. and J.R.C. contributed to the writing, editing and content of this manuscript.

Funding: This work was supported by grants from the Spanish Ministry of Economy and Competitivity (BFU2017-88736-R to J.R.C), and the *Comunidad Autónoma de Madrid* (S2013/MIT-2807 to J.R.C), and Grants-in-Aid for Scientific Research on Innovative Areas from the Japanese Ministry of Education, Culture, Sports, Science and Technology (KAKENHI 25252011 and 16H06436, 16H06429 and 16K21723 to N.S).

Acknowledgments: The authors thank Javier M. Rodríguez (CSIC) for critical reading of the manuscript. C.P.M. was a fellow of the La Caixa Foundation International Fellowship Program (La Caixa/CNB).

Conflicts of Interest: The authors declare no conflict of interest.

References

1. Patton, J.T. *Segmented double-stranded RNA viruses. Structure and Molecular Biology*; Caister Academic Press: Norfolk, UK, 2008.
2. Hill, C.L.; Booth, T.F.; Prasad, B.V.; Grimes, J.M.; Mertens, P.P.; Sutton, G.C.; Stuart, D.I. The structure of a cypovirus and the functional organization of dsRNA viruses. *Nat. Struct. Biol.* **1999**, *6*, 565–568. [PubMed]
3. Castón, J.R.; Trus, B.L.; Booy, F.P.; Wickner, R.B.; Wall, J.S.; Steven, A.C. Structure of L-A virus: A specialized compartment for the transcription and replication of double-stranded RNA. *J. Cell Biol.* **1997**, *138*, 975–985. [CrossRef] [PubMed]
4. Grimes, J.M.; Burroughs, J.N.; Gouet, P.; Diprose, J.M.; Malby, R.; Zientara, S.; Mertens, P.P.; Stuart, D.I. The atomic structure of the bluetongue virus core. *Nature* **1998**, *395*, 470–478. [CrossRef] [PubMed]
5. Prasad, B.V.; Rothnagel, R.; Zeng, C.Q.; Jakana, J.; Lawton, J.A.; Chiu, W.; Estes, M.K. Visualization of ordered genomic RNA and localization of transcriptional complexes in rotavirus. *Nature* **1996**, *382*, 471–473. [CrossRef] [PubMed]
6. Shaw, A.L.; Samal, S.K.; Subramanian, K.; Prasad, B.V. The structure of aquareovirus shows how the different geometries of the two layers of the capsid are reconciled to provide symmetrical interactions and stabilization. *Structure* **1996**, *4*, 957–967. [CrossRef]
7. Lu, G.; Zhou, Z.H.; Baker, M.L.; Jakana, J.; Cai, D.; Wei, X.; Chen, S.; Gu, X.; Chiu, W. Structure of double-shelled rice dwarf virus. *J. Virol.* **1998**, *72*, 8541–8549. [PubMed]
8. Caspar, D.L.D.; Klug, A. Physical principles in the construction of regular viruses. *Cold Spring Harbor. Symp. Quant. Biol.* **1962**, *27*, 1–24. [CrossRef] [PubMed]
9. Harrison, S.C. Principles of virus structure. In *Fields Virology*, 5th ed.; Knipe, D.M., Howley, P.M., Griffin, D.E., Lamb, R.A., Martin, M.A., Roizman, B., Strauss, S.E., Eds.; Lippincott Williams & Wilkins: Philadelphia, PA, USA, 2007; Volume 1, pp. 59–98.
10. Mertens, P. The dsRNA viruses. *Virus Res.* **2004**, *101*, 3–13. [CrossRef] [PubMed]
11. Arnold, M.M.; Sen, A.; Greenberg, H.B.; Patton, J.T. The battle between rotavirus and its host for control of the interferon signaling pathway. *PLoS Pathog.* **2013**, *9*, e1003064. [CrossRef] [PubMed]
12. Nuss, D.L. Mycoviruses, RNA silencing, and viral RNA recombination. *Adv. Virus Res.* **2011**, *80*, 25–48. [PubMed]
13. Cheng, R.H.; Castón, J.R.; Wang, G.J.; Gu, F.; Smith, T.J.; Baker, T.S.; Bozarth, R.F.; Trus, B.L.; Cheng, N.; Wickner, R.B.; et al. Fungal virus capsids, cytoplasmic compartments for the replication of double-stranded RNA, formed as icosahedral shells of asymmetric gag dimers. *J. Mol. Biol.* **1994**, *244*, 255–258. [CrossRef] [PubMed]
14. Lawton, J.A.; Estes, M.K.; Prasad, B.V. Mechanism of genome transcription in segmented dsRNA viruses. *Adv. Virus Res.* **2000**, *55*, 185–229. [PubMed]
15. Patton, J.T.; Spencer, E. Genome replication and packaging of segmented double-stranded RNA viruses. *Virology* **2000**, *277*, 217–225. [CrossRef] [PubMed]
16. Li, Z.; Baker, M.L.; Jiang, W.; Estes, M.K.; Prasad, B.V. Rotavirus architecture at subnanometer resolution. *J. Virol.* **2009**, *83*, 1754–1766. [CrossRef] [PubMed]
17. Trask, S.D.; McDonald, S.M.; Patton, J.T. Structural insights into the coupling of virion assembly and rotavirus replication. *Nat. Rev. Microbiol.* **2012**, *10*, 165–177. [CrossRef] [PubMed]

18. Castón, J.R.; Luque, D.; Trus, B.L.; Rivas, G.; Alfonso, C.; González, J.M.; Carrascosa, J.L.; Annamalai, P.; Ghabrial, S.A. Three-dimensional structure and stoichiometry of helmintosporium victoriae190s totivirus. *Virology* **2006**, *347*, 323–332. [CrossRef] [PubMed]

19. Zhang, X.; Walker, S.B.; Chipman, P.R.; Nibert, M.L.; Baker, T.S. Reovirus polymerase lambda 3 localized by cryo-electron microscopy of virions at a resolution of 7.6 A. *Nat. Struct. Biol.* **2003**, *10*, 1011–1018. [CrossRef] [PubMed]

20. Yu, X.; Jin, L.; Zhou, Z.H. 3.88 A structure of cytoplasmic polyhedrosis virus by cryo-electron microscopy. *Nature* **2008**, *453*, 415–419. [CrossRef] [PubMed]

21. Lawton, J.A.; Estes, M.K.; Prasad, B.V. Three-dimensional visualization of mRNA release from actively transcribing rotavirus particles. *Nat. Struct. Biol.* **1997**, *4*, 118–121. [CrossRef] [PubMed]

22. McClain, B.; Settembre, E.; Temple, B.R.; Bellamy, A.R.; Harrison, S.C. X-ray crystal structure of the rotavirus inner capsid particle at 3.8 A resolution. *J. Mol. Biol.* **2010**, *397*, 587–599. [CrossRef] [PubMed]

23. Cheng, L.; Zhu, J.; Hui, W.H.; Zhang, X.; Honig, B.; Fang, Q.; Zhou, Z.H. Backbone model of an aquareovirus virion by cryo-electron microscopy and bioinformatics. *J. Mol. Biol.* **2010**, *397*, 852–863. [CrossRef] [PubMed]

24. Duquerroy, S.; Da Costa, B.; Henry, C.; Vigouroux, A.; Libersou, S.; Lepault, J.; Navaza, J.; Delmas, B.; Rey, F.A. The picobirnavirus crystal structure provides functional insights into virion assembly and cell entry. *EMBO J.* **2009**, *28*, 1655–1665. [CrossRef] [PubMed]

25. Jaalinoja, H.T.; Huiskonen, J.T.; Butcher, S.J. Electron cryomicroscopy comparison of the architectures of the enveloped bacteriophages phi6 and phi8. *Structure* **2007**, *15*, 157–167. [CrossRef] [PubMed]

26. Huiskonen, J.T.; de Haas, F.; Bubeck, D.; Bamford, D.H.; Fuller, S.D.; Butcher, S.J. Structure of the bacteriophage phi6 nucleocapsid suggests a mechanism for sequential RNA packaging. *Structure* **2006**, *14*, 1039–1048. [CrossRef] [PubMed]

27. Naitow, H.; Tang, J.; Canady, M.; Wickner, R.B.; Johnson, J.E. L-A virus at 3.4 A resolution reveals particle architecture and mRNA decapping mechanism. *Nat. Struct. Biol.* **2002**, *9*, 725–728. [CrossRef] [PubMed]

28. Janssen, M.E.; Takagi, Y.; Parent, K.N.; Cardone, G.; Nibert, M.L.; Baker, T.S. Three-dimensional structure of a protozoal double-stranded RNA virus that infects the enteric pathogen giardia lamblia. *J. Virol.* **2015**, *89*, 1182–1194. [CrossRef] [PubMed]

29. Parent, K.N.; Takagi, Y.; Cardone, G.; Olson, N.H.; Ericsson, M.; Yang, M.; Lee, Y.; Asara, J.M.; Fichorova, R.N.; Baker, T.S.; et al. Structure of a protozoan virus from the human genitourinary parasite trichomonas vaginalis. *MBio* **2013**, *4*, e00056–013. [CrossRef] [PubMed]

30. Pan, J.; Dong, L.; Lin, L.; Ochoa, W.F.; Sinkovits, R.S.; Havens, W.M.; Nibert, M.L.; Baker, T.S.; Ghabrial, S.A.; Tao, Y.J. Atomic structure reveals the unique capsid organization of a dsRNA virus. *Proc. Natl. Acad. Sci. USA* **2009**, *106*, 4225–4230. [CrossRef] [PubMed]

31. Tang, J.; Ochoa, W.F.; Li, H.; Havens, W.M.; Nibert, M.L.; Ghabrial, S.A.; Baker, T.S. Structure of fusarium poae virus 1 shows conserved and variable elements of partitivirus capsids and evolutionary relationships to picobirnavirus. *J. Struct. Biol.* **2010**, *172*, 363–371. [CrossRef] [PubMed]

32. Miyazaki, N.; Salaipeth, L.; Kanematsu, S.; Iwasaki, K.; Suzuki, N. Megabirnavirus structure reveals a putative 120-subunit capsid formed by asymmetrical dimers with distinctive large protrusions. *J. Gen. Virol.* **2015**, *96*, 2435–2441. [CrossRef] [PubMed]

33. Luque, D.; González, J.M.; Garriga, D.; Ghabrial, S.A.; Havens, W.M.; Trus, B.; Verdaguer, N.; Carrascosa, J.L.; Castón, J.R. The T = 1 capsid protein of penicillium chrysogenum virus is formed by a repeated helix-rich core indicative of gene duplication. *J. Virol.* **2010**, *84*, 7256–7266. [CrossRef] [PubMed]

34. Castón, J.R.; Luque, D.; Gómez-Blanco, J.; Ghabrial, S.A. Chrysovirus structure: Repeated helical core as evidence of gene duplication. *Adv. Virus Res.* **2013**, *86*, 87–108. [PubMed]

35. Gómez-Blanco, J.; Luque, D.; Gonzalez, J.M.; Carrascosa, J.L.; Alfonso, C.; Trus, B.; Havens, W.M.; Ghabrial, S.A.; Castón, J.R. Cryphonectria nitschkei virus 1 structure shows that the capsid protein of chrysoviruses is a duplicated helix-rich fold conserved in fungal double-stranded RNA viruses. *J. Virol.* **2012**, *86*, 8314–8318. [CrossRef] [PubMed]

36. Luque, D.; Mata, C.P.; Gonzalez-Camacho, F.; González, J.M.; Gómez-Blanco, J.; Alfonso, C.; Rivas, G.; Havens, W.M.; Kanematsu, S.; Suzuki, N.; et al. Heterodimers as the structural unit of the T = 1 capsid of the fungal double-stranded RNA rosellinia necatrix quadrivirus 1. *J. Virol.* **2016**, *90*, 11220–11230. [CrossRef] [PubMed]

37. Mata, C.P.; Luque, D.; Gómez-Blanco, J.; Rodríguez, J.M.; González, J.M.; Suzuki, N.; Ghabrial, S.A.; Carrascosa, J.L.; Trus, B.L.; Castón, J.R. Acquisition of functions on the outer capsid surface during evolution of double-stranded RNA fungal viruses. *PLoS Pathog.* **2017**, *13*, e1006755. [CrossRef] [PubMed]
38. Coulibaly, F.; Chevalier, C.; Gutsche, I.; Pous, J.; Navaza, J.; Bressanelli, S.; Delmas, B.; Rey, F.A. The birnavirus crystal structure reveals structural relationships among icosahedral viruses. *Cell* **2005**, *120*, 761–772. [CrossRef] [PubMed]
39. Castón, J.R.; Martínez-Torrecuadrada, J.L.; Maraver, A.; Lombardo, E.; Rodríguez, J.F.; Casal, J.I.; Carrascosa, J.L. C terminus of infectious bursal disease virus major capsid protein VP2 is involved in definition of the t number for capsid assembly. *J. Virol.* **2001**, *75*, 10815–10828. [CrossRef] [PubMed]
40. Luque, D.; Rivas, G.; Alfonso, C.; Carrascosa, J.L.; Rodríguez, J.F.; Castón, J.R. Infectious bursal disease virus is an icosahedral polyploid dsrna virus. *Proc. Natl. Acad. Sci. USA* **2009**, *106*, 2148–2152. [CrossRef] [PubMed]
41. Luque, D.; Saugar, I.; Rejas, M.T.; Carrascosa, J.L.; Rodríguez, J.F.; Castón, J.R. Infectious bursal disease virus: Ribonucleoprotein complexes of a double-stranded RNA virus. *J. Mol. Biol.* **2009**, *386*, 891–901. [CrossRef] [PubMed]
42. Reinisch, K.M.; Nibert, M.L.; Harrison, S.C. Structure of the reovirus core at 3.6 Å resolution. *Nature* **2000**, *404*, 960–967. [CrossRef] [PubMed]
43. Nakagawa, A.; Miyazaki, N.; Taka, J.; Naitow, H.; Ogawa, A.; Fujimoto, Z.; Mizuno, H.; Higashi, T.; Watanabe, Y.; Omura, T.; et al. The atomic structure of rice dwarf virus reveals the self-assembly mechanism of component proteins. *Structure* **2003**, *11*, 1227–1238. [CrossRef] [PubMed]
44. Settembre, E.C.; Chen, J.Z.; Dormitzer, P.R.; Grigorieff, N.; Harrison, S.C. Atomic model of an infectious rotavirus particle. *Embo J.* **2011**, *30*, 408–416. [CrossRef] [PubMed]
45. Nemecek, D.; Boura, E.; Wu, W.; Cheng, N.; Plevka, P.; Qiao, J.; Mindich, L.; Heymann, J.B.; Hurley, J.H.; Steven, A.C. Subunit folds and maturation pathway of a dsrna virus capsid. *Structure* **2013**, *21*, 1374–1383. [CrossRef] [PubMed]
46. El Omari, K.; Sutton, G.; Ravantti, J.J.; Zhang, H.; Walter, T.S.; Grimes, J.M.; Bamford, D.H.; Stuart, D.I.; Mancini, E.J. Plate tectonics of virus shell assembly and reorganization in phage phi8. a distant relative of mammalian reoviruses. *Structure* **2013**, *21*, 1384–1395. [CrossRef] [PubMed]
47. Luque, D.; Gómez-Blanco, J.; Garriga, D.; Brilot, A.F.; González, J.M.; Havens, W.M.; Carrascosa, J.L.; Trus, B.L.; Verdaguer, N.; Ghabrial, S.A.; et al. Cryo-EM near-atomic structure of a dsRNA fungal virus shows ancient structural motifs preserved in the dsRNA viral lineage. *Proc. Natl. Acad. Sci. USA* **2014**, *111*, 7641–7646. [CrossRef] [PubMed]
48. Reinisch, K.M. The dsRNA viridae and their catalytic capsids. *Nat. Struct. Biol.* **2002**, *9*, 714–716. [CrossRef] [PubMed]
49. Ghabrial, S.A.; Castón, J.R.; Jiang, D.; Nibert, M.L.; Suzuki, N. 50-plus years of fungal viruses. *Virology* **2015**, *479–480*, 356–368. [CrossRef] [PubMed]
50. Wickner, R.B. Viruses of yeast, fungi, and parasitic microorganisms. In *Fields. Virology*, 4nd ed.; Knipe, D.M., Howley, P.M., Griffin, D.E., Martin, M.A., Lamb, R.A., Roizman, B., Strauss, S.E., Eds.; Lippincott Williams & Wilkins: Philadelphia, PA, USA, 2001; Volume 1, pp. 629–658.
51. Wickner, R.B.; Fujimura, T.; Esteban, R. Viruses and prions of saccharomyces cerevisiae *Adv. Virus Res.* **2013**, *86*, 1–36. [PubMed]
52. Dinman, J.D.; Icho, T.; Wickner, R.B. A-1 ribosomal frameshift in a double-stranded RNA virus of yeast forms a gag-pol fusion protein. *Proc. Natl. Acad. Sci. USA* **1991**, *88*, 174–178. [CrossRef] [PubMed]
53. Ribas, J.C.; Wickner, R.B. The gag domain of the gag-pol fusion protein directs incorporation into the l-a double-stranded rna viral particles in saccharomyces cerevisiae. *J. Biol. Chem.* **1998**, *273*, 9306–9311. [CrossRef] [PubMed]
54. Dinman, J.D.; Wickner, R.B. Ribosomal frameshifting efficiency and gag/gag-pol ratio are critical for yeast M1 double-stranded RNA virus propagation. *J. Virol.* **1992**, *66*, 3669–3676. [PubMed]
55. Tang, J.; Naitow, H.; Gardner, N.A.; Kolesar, A.; Tang, L.; Wickner, R.B.; Johnson, J.E. The structural basis of recognition and removal of cellular mRNA 7-methyl G 'caps' by a viral capsid protein: A unique viral response to host defense. *J. Mol. Recognit.* **2005**, *18*, 158–168. [CrossRef] [PubMed]
56. Fujimura, T.; Esteban, R. Cap-snatching mechanism in yeast L-A double-stranded RNA virus. *Proc. Natl. Acad. Sci. USA* **2011**, *108*, 17667–17671. [CrossRef] [PubMed]

57. Dunn, S.E.; Li, H.; Cardone, G.; Nibert, M.L.; Ghabrial, S.A.; Baker, T.S. Three-dimensional structure of victorivirus HvV190s suggests coat proteins in most totiviruses share a conserved core. *PLoS Pathog.* **2013**, *9*, e1003225. [CrossRef] [PubMed]

58. Huang, S.; Ghabrial, S.A. Organization and expression of the double-stranded RNA genome of helminthosporium victoriae 190s virus, a totivirus infecting a plant pathogenic filamentous fungus. *Proc. Natl. Acad. Sci. USA* **1996**, *93*, 12541–12546. [CrossRef] [PubMed]

59. Soldevila, A.I.; Ghabrial, S.A. Expression of the totivirus helminthosporium victoriae 190s virus RNA-dependent rna polymerase from its downstream open reading frame in dicistronic constructs. *J. Virol.* **2000**, *74*, 997–1003. [CrossRef] [PubMed]

60. Tang, J.; Ochoa, W.F.; Sinkovits, R.S.; Poulos, B.T.; Ghabrial, S.A.; Lightner, D.V.; Baker, T.S.; Nibert, M.L. Infectious myonecrosis virus has a totivirus-like, 120-subunit capsid, but with fiber complexes at the fivefold axes. *Proc. Natl. Acad. Sci. USA* **2008**, *105*, 17526–17531. [CrossRef] [PubMed]

61. Nibert, M.L.; Takagi, Y. Fibers come and go: Differences in cell-entry components among related dsRNA viruses. *Curr. Opin. Virol.* **2013**, *3*, 20–26. [CrossRef] [PubMed]

62. Ghabrial, S.A.; Castón, J.R.; Coutts, R.H.A.; Hillman, B.I.; Jiang, D.; Kim, D.H.; Moriyama, H.; Ictv Report, C. ICTV virus taxonomy profile: Chrysoviridae. *J. Gen. Virol.* **2018**, *99*, 19–20. [CrossRef] [PubMed]

63. Castón, J.R.; Ghabrial, S.A.; Jiang, D.; Rivas, G.; Alfonso, C.; Roca, R.; Luque, D.; Carrascosa, J.L. Three-dimensional structure of penicillium chrysogenum virus: A double-stranded RNA virus with a genuine T = 1 capsid. *J. Mol. Biol.* **2003**, *331*, 417–431. [CrossRef]

64. Jiang, D.; Ghabrial, S.A. Molecular characterization of penicillium chrysogenum virus: Reconsideration of the taxonomy of the genus chrysovirus. *J. Gen. Virol.* **2004**, *85*, 2111–2121. [CrossRef] [PubMed]

65. Nibert, M.L.; Ghabrial, S.A.; Maiss, E.; Lesker, T.; Vainio, E.J.; Jiang, D.; Suzuki, N. Taxonomic reorganization of family partitiviridae and other recent progress in partitivirus research. *Virus Res.* **2014**, *188*, 128–141. [CrossRef] [PubMed]

66. Vainio, E.J.; Chiba, S.; Ghabrial, S.A.; Maiss, E.; Roossinck, M.; Sabanadzovic, S.; Suzuki, N.; Xie, J.; Nibert, M.; Ictv Report Consortium. ICTV virus taxonomy profile: Partitiviridae. *J. Gen. Virol.* **2018**, *99*, 17–18. [CrossRef] [PubMed]

67. Tang, J.; Pan, J.; Havens, W.M.; Ochoa, W.F.; Guu, T.S.; Ghabrial, S.A.; Nibert, M.L.; Tao, Y.J.; Baker, T.S. Backbone trace of partitivirus capsid protein from electron cryomicroscopy and homology modeling. *Biophys. J.* **2010**, *99*, 685–694. [CrossRef] [PubMed]

68. Ochoa, W.F.; Havens, W.M.; Sinkovits, R.S.; Nibert, M.L.; Ghabrial, S.A.; Baker, T.S. Partitivirus structure reveals a 120-subunit, helix-rich capsid with distinctive surface arches formed by quasisymmetric coat-protein dimers. *Structure* **2008**, *16*, 776–786. [CrossRef] [PubMed]

69. Xiao, X.; Cheng, J.; Tang, J.; Fu, Y.; Jiang, D.; Baker, T.S.; Ghabrial, S.A.; Xie, J. A novel partitivirus that confers hypovirulence on plant pathogenic fungi. *J. Virol.* **2014**, *88*, 10120–10133. [CrossRef] [PubMed]

70. Krol, M.A.; Olson, N.H.; Tate, J.; Johnson, J.E.; Baker, T.S.; Ahlquist, P. RNA-controlled polymorphism in the in vivo assembly of 180-subunit and 120-subunit virions from a single capsid protein. *Proc. Natl. Acad. Sci. USA* **1999**, *96*, 13650–13655. [CrossRef] [PubMed]

71. Tang, J.; Johnson, J.M.; Dryden, K.A.; Young, M.J.; Zlotnick, A.; Johnson, J.E. The role of subunit hinges and molecular "switches" in the control of viral capsid polymorphism. *J. Struct. Biol.* **2006**, *154*, 59–67. [CrossRef] [PubMed]

72. Kainov, D.E.; Butcher, S.J.; Bamford, D.H.; Tuma, R. Conserved intermediates on the assembly pathway of double-stranded RNA bacteriophages. *J. Mol. Biol.* **2003**, *328*, 791–804. [CrossRef]

73. Lin, Y.H.; Chiba, S.; Tani, A.; Kondo, H.; Sasaki, A.; Kanematsu, S.; Suzuki, N. A novel quadripartite dsRNA virus isolated from a phytopathogenic filamentous fungus, rosellinia necatrix. *Virology* **2012**, *426*, 42–50. [CrossRef] [PubMed]

74. Lin, Y.H.; Hisano, S.; Yaegashi, H.; Kanematsu, S.; Suzuki, N. A second quadrivirus strain from the phytopathogenic filamentous fungus rosellinia necatrix. *Arch. Virol.* **2013**, *158*, 1093–1098. [CrossRef] [PubMed]

75. Kondo, H.; Kanematsu, S.; Suzuki, N. Viruses of the white root rot fungus, rosellinia necatrix. *Adv. Virus Res.* **2013**, *86*, 177–214. [PubMed]

76. Arjona-Lopez, J.M.; Telengech, P.; Jamal, A.; Hisano, S.; Kondo, H.; Yelin, M.D; Arjona-Girona, I.; Kanematsu, S.; López-Herrera, C.J.; Suzuki, N. Novel, diverse RNA viruses from mediterranean isolates of the phytopathogenic fungus, rosellinia necatrix: Insights into evolutionary biology of fungal viruses. *Environ. Microbiol.* **2018**, *20*, 1464–1483. [CrossRef] [PubMed]

77. Baker, M.L.; Jiang, W.; Rixon, F.J.; Chiu, W. Common ancestry of herpesviruses and tailed DNA bacteriophages. *J. Virol.* **2005**, *79*, 14967–14970. [CrossRef] [PubMed]

78. Bamford, D.H.; Grimes, J.M.; Stuart, D.I. What does structure tell us about virus evolution? *Curr. Opin. Struct. Biol.* **2005**, *15*, 655–663. [CrossRef] [PubMed]

79. Rossmann, M.; Johnson, J. Icosahedral rna virus structure. *Annu. Rev. Biochem.* **1989**, *58*, 533–573. [CrossRef] [PubMed]

80. Abrescia, N.G.; Bamford, D.H.; Grimes, J.M.; Stuart, D.I. Structure unifies the viral universe. *Annu. Rev. Biochem.* **2012**, *81*, 795–822. [CrossRef] [PubMed]

81. Benson, S.D.; Bamford, J.K.; Bamford, D.H.; Burnett, R.M. Does common architecture reveal a viral lineage spanning all three domains of life? *Mol. Cell* **2004**, *16*, 673–685. [CrossRef] [PubMed]

82. Krupovic, M.; Bamford, D.H. Virus evolution: How far does the double beta-barrel viral lineage extend? *Nat. Rev. Microbiol.* **2008**, *6*, 941–948. [CrossRef] [PubMed]

83. Rissanen, I.; Grimes, J.M.; Pawlowski, A.; Mantynen, S.; Harlos, K.; Bamford, J.K.; Stuart, D.I. Bacteriophage P23-77 capsid protein structures reveal the archetype of an ancient branch from a major virus lineage. *Structure* **2013**, *21*, 718–726. [CrossRef] [PubMed]

84. Pietila, M.K.; Laurinmaki, P.; Russell, D.A.; Ko, C.C.; Jacobs-Sera, D.; Hendrix, R.W.; Bamford, D.H.; Butcher, S.J. Structure of the archaeal head-tailed virus HSTV-1 completes the HK97 fold story. *Proc. Natl. Acad. Sci. USA* **2013**, *110*, 10604–10609. [CrossRef] [PubMed]

85. Roberts, M.M.; White, J.L.; Grutter, M.G.; Burnett, R.M. Three-dimensional structure of the adenovirus major coat protein hexon. *Science* **1986**, *232*, 1148–1151. [CrossRef] [PubMed]

86. Nandhagopal, N.; Simpson, A.A.; Gurnon, J.R.; Yan, X.; Baker, T.S.; Graves, M.V.; Van Etten, J.L.; Rossmann, M.G. The structure and evolution of the major capsid protein of a large, lipid-containing DNA virus. *Proc. Natl. Acad. Sci. USA* **2002**, *99*, 14758–14763. [CrossRef] [PubMed]

87. Abrescia, N.G.; Cockburn, J.J.; Grimes, J.M.; Sutton, G.C.; Diprose, J.M.; Butcher, S.J.; Fuller, S.D.; San Martin, C.; Burnett, R.M.; Stuart, D.I.; et al. Insights into assembly from structural analysis of bacteriophage PRD1. *Nature* **2004**, *432*, 68–74. [CrossRef] [PubMed]

88. Lomonossoff, G.P.; Johnson, J.E. The synthesis and structure of comovirus capsids. *Prog. Biophys. Mol. Biol.* **1991**, *55*, 107–137. [CrossRef]

89. Ravantti, J.; Bamford, D.; Stuart, D.I. Automatic comparison and classification of protein structures. *J. Struct. Biol.* **2013**, *183*, 47–56. [CrossRef] [PubMed]

90. Suhanovsky, M.M.; Teschke, C.M. Nature's favorite building block: Deciphering folding and capsid assembly of proteins with the HK97-fold. *Virology* **2015**, *479–480*, 487–497. [CrossRef] [PubMed]

91. Yu, X.; Jih, J.; Jiang, J.; Zhou, Z.H. Atomic structure of the human cytomegalovirus capsid with its securing tegument layer of pp150. *Science* **2017**, *356*. [CrossRef] [PubMed]

92. Liu, H.; Cheng, L. Cryo-em shows the polymerase structures and a nonspooled genome within a dsrna virus. *Science* **2015**, *349*, 1347–1350. [CrossRef] [PubMed]

93. Zhang, X.; Ding, K.; Yu, X.; Chang, W.; Sun, J.; Zhou, Z.H. In situ structures of the segmented genome and RNA polymerase complex inside a dsRNA virus. *Nature* **2015**, *527*, 531–534. [CrossRef] [PubMed]

94. Li, X.; Zhou, N.; Chen, W.; Zhu, B.; Wang, X.; Xu, B.; Wang, J.; Liu, H.; Cheng, L. Near-atomic resolution structure determination of a cypovirus capsid and polymerase complex using cryo-EM at 200kV. *J. Mol. Biol.* **2017**, *429*, 79–87. [CrossRef] [PubMed]

95. Estrozi, L.F.; Settembre, E.C.; Goret, G.; McClain, B.; Zhang, X.; Chen, J.Z.; Grigorieff, N.; Harrison, S.C. Location of the dsRNA-dependent polymerase, VP1, in rotavirus particles. *J. Mol. Biol.* **2013**, *425*, 124–132. [CrossRef] [PubMed]

96. Li, H.; Havens, W.M.; Nibert, M.L.; Ghabrial, S.A. An RNA cassette from helminthosporium victoriae virus 190s necessary and sufficient for stop/restart translation. *Virology* **2015**, *474*, 131–143. [CrossRef] [PubMed]

97. Sato, Y.; Castón, J.R.; Suzuki, N. The biological attributes, genome architecture and packaging of diverse multi-component fungal viruses. *Curr. Opin. Virol.* **2018**, *33*, 55–65. [CrossRef] [PubMed]

98. Dryden, K.; Wang, G.; Yeager, M.; Nibert, M.; Coombs, K.; Furlong, D.; Fields, B.; Baker, T. Early steps in reovirus infection are associated with dramatic changes in supramolecular structure and protein conformation: Analysis of virions and subviral particles by cryoelectron microscopy and image reconstruction. *J. Cell Biol.* **1993**, *122*, 1023–1041. [CrossRef] [PubMed]

99. Gouet, P.; Diprose, J.M.; Grimes, J.M.; Malby, R.; Burroughs, J.N.; Zientara, S.; Stuart, D.I.; Mertens, P.P. The highly ordered double-stranded RNA genome of bluetongue virus revealed by crystallography. *Cell* **1999**, *97*, 481–490. [CrossRef]

100. Pesavento, J.B.; Lawton, J.A.; Estes, M.E.; Venkataram Prasad, B.V. The reversible condensation and expansion of the rotavirus genome. *Proc. Natl. Acad. Sci. USA* **2001**, *98*, 1381–1386. [CrossRef] [PubMed]

101. Zhang, R.; Hisano, S.; Tani, A.; Kondo, H.; Kanematsu, S.; Suzuki, N. A capsidless ssrna virus hosted by an unrelated dsRNA virus. *Nat. Microbiol.* **2016**, *1*, 15001. [CrossRef] [PubMed]

102. Hisano, S.; Zhang, R.; Faruk, M.I.; Kondo, H.; Suzuki, N. A neo-virus lifestyle exhibited by a (+)ssRNA virus hosted in an unrelated dsrna virus: Taxonomic and evolutionary considerations. *Virus Res.* **2018**, *244*, 75–83. [CrossRef] [PubMed]

103. Kozlakidis, Z.; Herrero, N.; Coutts, R.H. The complete nucleotide sequence of a totivirus from aspergillus foetidus. *Arch. Virol.* **2013**, *158*, 263–266. [CrossRef] [PubMed]

104. Nerva, L.; Ciuffo, M.; Vallino, M.; Margaria, P.; Varese, G.C.; Gnavi, G.; Turina, M. Multiple approaches for the detection and characterization of viral and plasmid symbionts from a collection of marine fungi. *Virus Res.* **2016**, *219*, 22–38. [CrossRef] [PubMed]

105. Osaki, H.; Sasaki, A.; Nomiyama, K.; Tomioka, K. Multiple virus infection in a single strain of fusarium poae shown by deep sequencing. *Virus Genes* **2016**, *52*, 835–847. [CrossRef] [PubMed]

106. Kanhayuwa, L.; Kotta-Loizou, I.; Ozkan, S.; Gunning, A.P.; Coutts, R.H. A novel mycovirus from aspergillus fumigatus contains four unique dsrnas as its genome and is infectious as dsRNA. *Proc. Natl. Acad. Sci. USA* **2015**, *112*, 9100–9105. [CrossRef] [PubMed]

107. Jia, H.; Dong, K.; Zhou, L.; Wang, G.; Hong, N.; Jiang, D.; Xu, W. A dsRNA virus with filamentous viral particles. *Nat. Commun.* **2017**, *8*, 168. [CrossRef] [PubMed]

108. Dai, X.; Li, Z.; Lai, M.; Shu, S.; Du, Y.; Zhou, Z.H.; Sun, R. In situ structures of the genome and genome-delivery apparatus in a single-stranded RNA virus. *Nature* **2017**, *541*, 112–116. [CrossRef] [PubMed]

109. Huiskonen, J.T. Image processing for cryogenic transmission electron microscopy of symmetry-mismatched complexes. *Biosci. Rep.* **2018**. [CrossRef] [PubMed]

viruses

MDPI

Article

A Rapid Method for Sequencing Double-Stranded RNAs Purified from Yeasts and the Identification of a Potent K1 Killer Toxin Isolated from *Saccharomyces cerevisiae*

Angela M. Crabtree [1], Emily A. Kizer [1], Samuel S. Hunter [2], James T. Van Leuven [1], Daniel D. New [2], Matthew W. Fagnan [2] and Paul A. Rowley [1,*]

[1] Department of Biological Sciences, University of Idaho, Moscow, ID 83844, USA; amcrabtree@uidaho.edu (A.M.C.); mrs@jkizer.com (E.A.K.); jvanleuven@uidaho.edu (J.T.V.L.)
[2] IBEST Genomics Core, University of Idaho, Moscow, ID 83843, USA; shunter@uidaho.edu (S.S.H.); dnew@uidaho.edu (D.D.N.); mfagnan@uidaho.edu (M.W.F.)
[*] Correspondence: prowley@uidaho.edu

Received: 31 October 2018; Accepted: 11 January 2019; Published: 16 January 2019

Abstract: Mycoviruses infect a large number of diverse fungal species, but considering their prevalence, relatively few high-quality genome sequences have been determined. Many mycoviruses have linear double-stranded RNA genomes, which makes it technically challenging to ascertain their nucleotide sequence using conventional sequencing methods. Different specialist methodologies have been developed for the extraction of double-stranded RNAs from fungi and the subsequent synthesis of cDNAs for cloning and sequencing. However, these methods are often labor-intensive, time-consuming, and can require several days to produce cDNAs from double-stranded RNAs. Here, we describe a comprehensive method for the rapid extraction and sequencing of dsRNAs derived from yeasts, using short-read next generation sequencing. This method optimizes the extraction of high-quality double-stranded RNAs from yeasts and 3' polyadenylation for the initiation of cDNA synthesis for next-generation sequencing. We have used this method to determine the sequence of two mycoviruses and a double-stranded RNA satellite present within a single strain of the model yeast *Saccharomyces cerevisiae*. The quality and depth of coverage was sufficient to detect fixed and polymorphic mutations within viral populations extracted from a clonal yeast population. This method was also able to identify two fixed mutations within the alpha-domain of a variant K1 killer toxin encoded on a satellite double-stranded RNA. Relative to the canonical K1 toxin, these newly reported mutations increased the cytotoxicity of the K1 toxin against a specific species of yeast.

Keywords: mycovirus; dsRNA; sequencing; killer toxin; totivirus

1. Introduction

Double-stranded RNAs (dsRNAs) found within fungi are the hallmark of infection by mycoviruses and their associated satellites. The majority of mycoviruses do not cause overt pathology in their host fungi, therefore, the direct extraction of dsRNAs and the visualization of viral particles using electron microscopy are the best methods to identify the presence of mycoviruses in fungal cultures. Surveys of pure fungal cultures indicate that dsRNA mycoviruses are abundant and present within every major group of fungi [1–3]. Although most mycoviruses appear to be avirulent, they can still lead to phenotypic changes in their fungal hosts, including changes in pigmentation, growth rate, and sporulation efficiency, and can improve stress tolerance, cause hypo- or hypervirulence in pathogens, or enable the production of extracellular antifungal toxins [4–6]. To better understand the contribution of mycoviruses to fungal ecology and pathogenicity, there is a need to improve the existing dsRNA purification and sequencing methods to enable the exploration of mycovirus diversity within fungi.

The purification of dsRNAs from cell culture, tissues, and environmental samples has been achieved by applying a variety of techniques, including selective precipitation, solvent extraction, size exclusion chromatography, gel electrophoresis, affinity purification using dsRNA-binding proteins or cellulose, selective degradation of non-dsRNA nucleic acids, or a combination of these methods [7–11]. DNAs can be synthesized from a template of purified dsRNAs using reverse transcriptase, a reaction that requires priming to initiate synthesis in a 5' to 3' direction. The 3' polyadenylation of cellular mRNAs can be used to prime cDNA synthesis using oligo(dT) primers, but the 3' termini of linear mycoviral dsRNAs lack these terminal homopolymeric sequences. However, homopolymeric adenine-rich tracts that are known to be encoded internally by satellite dsRNAs found within yeasts have been successfully used to prime cDNA synthesis [12–14]. To initiate reverse transcription from an RNA template of unknown sequence, primer binding sites can be added to the 3' termini, either by enzymatic polyadenylation [15–19] or the ligation of short oligonucleotides [11,20,21]. This enables the synthesis of cDNAs that are representative of the full-length dsRNAs. Alternatively, random hexamers can be used to generate cDNAs that are annealed and repaired to form double-stranded DNAs for cloning and sequencing [22,23].

Cloning of dsRNA-derived cDNA products into plasmids has been widely used to determine their nucleic acid sequence via Sanger sequencing. However, there has been a shift towards using next-generation sequencing (NGS) methods to determine the genetic sequence of dsRNAs and random amplification of cDNA ends (RACE) to resolve the terminal ends of dsRNAs. The use of NGS has also enabled the interrogation of pooled environmental samples, plant tissues, and animal-derived samples to identify dsRNA viruses, including those associated with disease [21,23–26]. However, there remain opportunities to improve the efficiency of dsRNA sequencing techniques by optimizing dsRNA purification, reverse transcription, and 3' end tailing to increase the quality of cDNAs used for NGS sequencing. Here we report a comprehensive approach that leverages NGS technologies to determine the genetic sequence of dsRNAs purified from yeasts. Specifically, we combine rapid dsRNA extraction and cDNA synthesis protocols to create high-quality cDNAs for downstream NGS library preparation using an affordable 'tagmentation' procedure [27]. We demonstrate the optimization of 3' polyadenylation of viral dsRNAs for 'anchored' oligo(dT) priming, which simplifies the reverse-transcription, amplification, and sequencing of cDNAs by NGS. This method has allowed the description of the diversity of dsRNAs found within a single strain of *Saccharomyces cerevisiae*.

2. Materials and Methods

2.1. Double-Stranded RNA Extraction

Double-stranded RNAs were extracted from both live yeast cultures inoculated in yeast peptone dextrose (YPD) broth and from commercial packets of dried yeasts (all yeast strains used in this study are described in Table S2). YPD cultures were grown overnight at 30 °C and washed once with sterile water. Dried yeasts were rehydrated with 500 µL of sterile water and vortexed until homogenized. All cultures were then centrifuged for 5 min at 8000× g and the supernatant aspirated. For each extraction, approximately 0.04 g of wet biomass (approximately 1×10^9 yeast cells) from YPD cultures and 0.06 g dry weight of dried yeasts was used for dsRNA extractions. The following protocol was modified from the dsRNA extraction method previously published by Okada et. al [28]. Cellulose columns were prepared by puncturing the bottom of a 0.6 mL tube with a hot 20-gauge needle and nesting it in a 2.0 mL tube. Approximately 0.06 g of cellulose powder D (Advantec, Japan) was added to the 0.6 mL tube, followed by 500 µL of wash buffer (1× STE (100 mM NaCl; 10 mM Tris–HCl, pH 8.0; 1 mM EDTA, pH 8.0) containing 16% (v/v) ethanol). Wash buffer was removed by a 10 s centrifugation, just before use. To extract dsRNAs, 450 µL of 2× LTE (500 mM LiCl; 20 mM Tris-HCl, pH 8.0; 30 mM EDTA, pH 8.0) containing 0.1% (v/v) beta-mercaptoethanol (14.3 M) (Amresco) was added to the harvested yeast cells. The cell mixture was vortexed for 3 min at 3000 rpm (Disruptor Genie, Scientific Industries, Bohemia, NY, USA). Fifty microliters of 10% (w/v) SDS solution and 500 μL of phenol–chloroform–isoamyl alcohol [25:24:1] pH 8.0 were added to the crude cell extracts and vortexed until homogenous. Samples were centrifuged at 20,000× g for 5 min and the supernatant was transferred to a clean tube and a second 500 μL of phenol–chloroform–isoamyl alcohol extraction was performed. A 0.2× volume of oligo d(T)$_{25}$ magnetic beads (New England Biolabs, Ipswich, MA, USA) was added to the recovered supernatant before the sample was vortexed, agitated at 250 rpm at ambient temperature for 10 min, and then allowed to stand on a magnetic rack for 5 min. The supernatant was transferred to a clean tube whereupon a one-fifth volume of ethanol was added to precipitate the nucleic acids from solution. Tubes were centrifuged at 20,000× g for 3 min to remove precipitates and the supernatant was transferred to the pre-prepared cellulose spin column. The column was centrifuged at 10,000× g for 10 s, and the flow-through was discarded. Four hundred microliters of wash buffer was added to the columns, centrifuged at 10,000× g for 10 s and the flow-through was discarded. This step was repeated twice, for a total of three washes. After the last wash, the columns were dried by centrifugation at 10,000× g for 10 s. Cellulose columns were transferred to clean tubes, 400 μL of 1× STE was added, and columns were centrifuged at 10,000× g for 10 s to collect the eluate. Forty microliters of 3 M aqueous sodium acetate, pH 5.2, and 1 mL of absolute ethanol were added to the eluate, which was inverted to mix, and then centrifuged at 20,000× g for 5 min to precipitate the dsRNAs. The ethanol mix was aspirated, and dsRNA pellets were allowed to air-dry, before being suspended in 11 μL of nuclease-free water. To remove any remaining DNAs from the dsRNA-enriched sample, 0.5 μL of *E. coli* DNase I enzyme (New England Biolabs) was added with 1.2 μL of NEB Buffer 2.1, 0.5 μL of 10 mM CaCl$_2$, and incubated at 37 °C, for 10 min. DMSO was added to a final concentration of 15% (v/v) and the sample was incubated at 95 °C, for 10 min to deactivate the DNase I and denature the dsRNAs, prior to cDNA synthesis. Samples were rapidly cooled in an ice bath to reduce the annealing of dsRNAs. A more rapid variation of this method for screening yeasts for the presence of dsRNAs was also used and involved only a single phenol:chloroform extraction, no oligo d(T)$_{25}$ magnetic beads, and no DNase digestion. This rapid protocol gives higher yields and a clear visualization of the dsRNAs by agarose gel electrophoresis.

2.2. Sequencing Sample Preparation

Poly(A) polymerase (New England Biolabs) was used to synthesize a poly(A) tail at the 3′ termini of all denatured dsRNAs. To 12.5 μL of purified dsRNAs, the following was added: 1.5 μL 10× poly(A) polymerase reaction buffer, 1.5 μL adenosine 5′ triphosphate [10 mM], 0.5 μL of poly(A) polymerase (diluted 1:32 in nuclease-free water), and 0.5 μL murine RNAse inhibitor. Samples were incubated at 37 °C for 30 min, 65 °C for 20 min, 98 °C for 5 min, and then immediately placed in a wet ice slurry. Superscript IV (Invitrogen, Carlsbad, CA, USA) with an "anchored" NV(dT)$_{20}$ primer (Invitrogen) was used to reverse transcribe the poly(A)-tailed single-stranded RNAs (ssRNAs) into cDNAs according to the manufacturer's protocol. Murine RNase Inhibitor (New England Biolabs) was used in place of the RNaseOUT™ RNase Inhibitor. Each sample was digested with 1 μL of RNase H (New England Biolabs) and incubated at 37 °C, for 20 min to remove ssRNAs. cDNAs were annealed at 65 °C, for 2 h. To fully extend cDNA overhangs, 1 μL of *E. coli* DNA Polymerase I enzyme (New England Biolabs) was added to 3.5 μL of NEB Buffer 2.0 and 0.5 μL of 10 mM dNTPs and was incubated at 37 °C, for 30 min. DMSO was then added to a final concentration of 15% (v/v) and the reaction was incubated at 75 °C, for 20 min, to deactivate the polymerase. Five microliters of cDNAs were used as a template for PCR amplification, using 25 μL of Phusion Master Mix with HF Buffer (New England Biolabs), 1 μL of anchored oligo(dT) primer (0.7 ug/μL), and 1.5 μL of DMSO, to a final reaction volume of 50 μL. Reactions were subjected to the following parameters on a thermal cycler: (1) 72 °C for 10 min, (2) 98 °C for 30 s, (3) 98 °C for 5 s, (4) 50 °C for 10 s, and 72 °C for 45 s, (5) go to step 3 for 30 cycles, (6) 72 °C for 5 min. Six 50 μL PCR reactions were pooled and concentrated using HighPrep™ PCR reagent with magnetic beads, following the manufacturer's protocol, using 0.5× sample volume of the reagent and five times the specified volume of ethanol wash (MagBio, Gaithersburg, MD, USA). Samples were eluted from the beads using 30 µL of nuclease-free water and subjected to fragment analysis (Fragment Analyzer, Advanced Analytical), prior to Illumina library preparation and NGS.

2.3. Illumina Library Preparation Using a Modified Nextera Protocol

All cDNA samples were normalized to 2.5 ng/µL for the desired final average library insert size of 550 bp. Fluorometric quantification was performed with SpectraMax Gemini XPS plate reader (Molecular Devices, San Jose, CA, USA) and PicoGreen (Invitrogen). For the fluorometric quantification, 2 µL of cDNA was diluted in 98 µL 1× TE buffer (10 mM Tris-HCl, 1 mM EDTA, pH 7.5), and mixed with 100 µL of PicoGreen (diluted 1:200 in TE). Standards were prepared as per the manufacturer's protocol and by scaling the volumes to one-tenth of that stated. Samples and standards were incubated at ambient temperature, in the dark, for 5 min, before analysis. Tagmentation, PCR (Applied Biosystems thermal cycler, Hercules, CA, USA), PCR-mediated adapter addition and library amplification were performed according to Baym et. al [27], with the post-tagmentation PCR using the following thermal cycling parameters: (1) 72 °C for 3 min, (2) 98 °C for 5 min, (3) 98 °C for 10 s, (4) 63 °C for 1 min, 72 °C for 30 s, (5) go to step 3 for 13 cycles, (6) 72 °C 5 min. For magnetic bead purification, 0.8× sample volume of HighPrep™ PCR reagent was used while following the manufacturer's protocol. Samples were suspended in 50 µL of nuclease-free water and a two-sided size selection was performed to further narrow the insert size distribution. Then, 0.4× sample volume of HighPrep reagent was added to the sample with magnetic beads, and after an incubation at ambient temperature, for 5 min, the beads were discarded; 0.6× sample volume of HighPrep reagent was then added to the sample with magnetic beads and after incubation at ambient temperature for 5 min, the supernatant was removed. DNAs were then eluted from the magnetic beads and suspended in 50 µL of nuclease-free water. Samples were then quantified with a fluorometer and pooled by mass proportionally to the desired read distribution in the downstream sequencing run. Library-distribution, size-weighted fragment length, and nucleic acid concentration were determined by fragment analysis (Fragment Analyzer, Agilent Technologies Inc, La Jolla, CA, USA).

2.4. Sequencing

The prepared DNA libraries were sequenced by the IBEST Genomics Resources Core at the University of Idaho, using an Illumina MiSeq sequencing platform and Micro v2 300 cycle reagent kit. Base calling and demultiplexing was performed using the Illumina bcl2fastq v2.17.1.14 software tool (Illumina, San Diego, CA, USA).

2.5. Bioinformatics Analysis

Bioinformatic analysis was done in two stages. First, to determine the approximate percentage of viral sequence, reads were mapped against a collection of previously published viral sequences using bowtie2 v 2.3.4.1 run with "–local" parameter [29]. Of the 471,742 reads sequenced for this sample, 97.84% could be mapped against viral sequences (NCBI GenBank accession numbers: ScV-L-A1, M28353.1; ScV-L-BC, NC_001641.1; ScV-M1, NC_001782.1). The resulting BAM file was further analyzed using SAMtools v1.5 to confirm the mapping depth across the full length of the viral reference sequence [30].

After confirming that the majority of sequenced reads were viral in origin, we performed a de novo assembly of reads in order to confirm the applicability of this method for the discovery of novel dsRNA viruses. Prior to assembly, HTStream (https://github.com/ibest/HTStream) was used to clean the reads. Due to the extremely high coverage, stringent cleaning parameters were used in order to retain the highest quality reads. Cleaning was done using the following steps and parameters: (1) PCR duplicates were identified and removed using hts_SuperDeduper; (2) reads were screened with hts_SeqScreener to remove PhiX control sequences, which were spiked in following Illumina protocols; (3) sequencing adapters were trimmed using hts_AdapterTrimmer; (4) reads were screened against a database of known sequencing adapters, using hts_SeqScreener and a collection of known adapter sequences to remove reads containing adapters that could not be trimmed during step 3; and (5) reads were quality trimmed by using a minimum q-score of 25 and retaining reads at least 148 bp in length using hts_QWindowTrim.

Cleaned reads were assembled de novo using the SPAdes assembler v3.11.1, with default parameters [31]. To assess the assembly quality and mapping depth, the contigs produced for each sample were used to build a bowtie2 index, and the cleaned reads from the respective sample were mapped. The resulting BAM files were visualized using Geneious 8.1 (https://www.geneious.com), which was also used to align the assembled contigs against previously published sequences for comparison. The read qualities were visualized in R using seqTools (R package version 1.14.0). Sequence reads were deposited to the NCBI Sequence Read Archive with the accession number: SAMN10274163.

Polymorphic and fixed mutations were identified within the mapped reads, using Geneious v11.1.4. Significant mutations were selected from the output, using cutoffs for the minimum variant frequency (5 %) and minimum coverage (50 reads). Mutations with more than or equal to 95% variant frequency were specified as fixed, while the remaining mutations were considered polymorphic.

2.6. Cloning of dsRNAs

K1 toxin-encoding inducible plasmids were constructed by cloning reverse transcriptase PCR-derived K1 genes into pCR8 by TOPO-TA cloning (Thermo Fisher) using the primers PRUI1 and PRUI2 (Table S1). The nucleic acid sequence of all cloned K1 genes was confirmed by Sanger sequencing. Utilizing Gateway™ technology (Thermo Fisher), K1 genes were sub-cloned into the destination vector pAG426-GAL-ccdB to create the high copy number, galactose-inducible plasmids pEK005 (reference K1 sequence) and pEK006 (K1 BJH001) [32]. To amplify and clone a putative polymorphic frameshift region from ScV-LA1, we used reverse transcriptase-PCR with primers PRUI132 and PRUI133. Amplified cDNAs were cloned into pCR8 by TOPO-TA cloning (Thermo Fisher) and the nucleic acid sequence was confirmed by Sanger sequencing.

2.7. Killer Toxin Assays

To test yeast strains for the production of killer toxins, single colonies were inoculated in YPD broth and grown at ambient temperature for 24 h. Putative killer yeasts were spotted at high cell density onto YPD dextrose 'killer assay' agar plates (0.003% w/v methylene blue, pH 4.6), seeded with a killer toxin-susceptible yeast strain. Plates were visually inspected for evidence of killer toxin production after incubation at ambient temperature, for 3 days. Toxin production by a strain of yeast was identified by either a zone of growth inhibition or methylene blue-staining of the yeasts that were spread as a lawn. To quantitatively compare the antifungal activities of the different K1 toxins, single colonies of *S. cerevisiae*, transformed with the plasmids pEK005 or pEK006, were inoculated in 1 mL of complete liquid media lacking uracil with 2% galactose. These cultures were incubated at ambient temperature, for 48 h, with shaking at 250 rpm. K1 toxin-susceptible yeasts were inoculated in 1 mL of YPD and incubated at ambient temperature for 48 h with shaking at 250 rpm. 6×10^5 K1 toxin-susceptible yeast cells were spread onto YPD galactose killer assay agar plates (10% w/v galactose, 0.003% w/v methylene blue, pH 4.6). Five microliters with 6×10^6 cells of K1-expressing yeast were spotted onto the inoculated plates and incubated at ambient temperature for 4 days. Areas of growth inhibition were determined by measuring the diameter of the growth inhibition zones.

2.8. Verifying the Presence of DsRNA Elements in S. cerevisiae BJH001 by Reverse Transcriptase-PCR

DsRNAs extracted from *S. cerevisiae* BJH001 were used as templates for Superscript IV two-step reverse transcriptase-PCR, according to the manufacturer's protocol, with primers specific for ScV-L-A1, ScV-L-BC, and ScV-M1 (Table S1).

3. Results

3.1. Extraction of High-Quality dsRNAs from Saccharomyces Yeasts

The presence of dsRNA mycoviruses in *S. cerevisiae* is often correlated with the production of antifungal proteins (killer toxins). We used two yeast strains that have been previously reported as killer yeasts (*S. cerevisiae* BJH001 [33] and *S. paradoxus* Y8.5 [34,35]), one non-killer yeast (*S. paradoxus* CBS12357), and several commercially-available dried yeasts to assay the effectiveness of a modified protocol based on a dsRNA extraction method previously optimized for filamentous fungi and plant material (Figure 1) [28]. Approximately 0.04 g of biomass (~1×10^9 yeast cells) were used as input for the extraction of dsRNAs. Cells were first subjected to homogenization in LTE buffer, followed by two rounds of phenol:chloroform:isoamyl alcohol extraction. The resulting aqueous phase was then incubated with oligo(dT) beads to deplete cellular polyadenylated single-stranded RNAs (ssRNAs), before loading onto a cellulose D spin column. Eluted material from the cellulose D had a higher concentration of dsRNAs compared to a rapid method using guanidinium thiocyanate and phenol that we previously described (Figure S1) [33]. To remove residual DNAs, samples were incubated with DNase I. This protocol was used to identify the dsRNA content of several strains of *S. cerevisiae* "killer yeasts" that produce killer toxins (Figure 1A), which is often dependent on the presence of dsRNA totiviruses and associated satellite dsRNAs (Figure 1B) [5,36]. After extracting dsRNAs directly from rehydrated commercial dried yeasts or from yeasts grown in a laboratory culture, we were able to resolve dsRNAs in killer and non-killer yeasts that correspond to the presence of mycoviruses and satellite dsRNAs (Figure 1B).

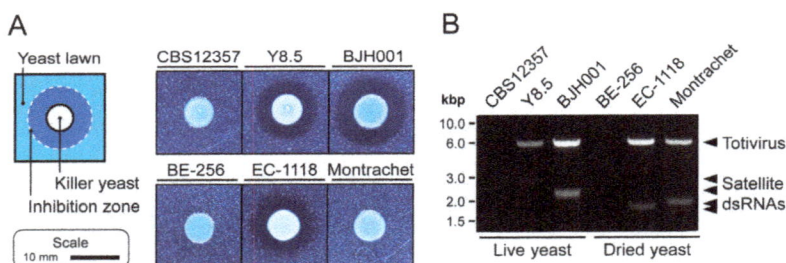

Figure 1. The extraction of dsRNAs from dried and actively growing cultures of killer and non-killer Saccharomyces yeasts. (**A**) Killer toxin production by different strains of *Saccharomyces* yeasts. The ability of yeasts to inhibit the growth of the lawn strain *Saccharomyces bayanus* CBS7001 indicate killer toxin production by strains Y8.5, BJH001, and EC-1118. (**B**) dsRNAs were separated by 0.8% agarose gel electrophoresis and stained with ethidium bromide. Molecular weight standards are DNA-specific and provide an approximate size of dsRNAs. Larger dsRNAs represent putative totiviruses, whereas the smaller heterogenous dsRNAs are strain-specific satellite dsRNAs.

3.2. Next Generation Sequencing of cDNAs Derived from dsRNAs

To initiate reverse transcription and create full-length cDNAs from a purified mixture of dsRNAs, we used poly(A) polymerase to polyadenylate the 3′ end of denatured dsRNAs (Figure 2A). Analysis of dsRNAs, before and after poly(A) polymerase incubation by fragment analysis, revealed a significant increase in the molecular weight of the treated RNAs (Figure 2B). Reverse transcription was primed using an anchored oligo(dT) primer (sequence: NV(T_{20})) to minimize priming within the poly(A) tail. The resultant cDNAs were annealed and repaired by *E. coli* polymerase I and amplified with anchored oligo(dT) primers, using Phusion polymerase. After magnetic bead purification, the size distribution and quantity of cDNAs was determined by fragment analysis. Total cDNA yields ranged from 420–810 ng and had a broad size distribution (Figure 2C). The small size of mycovirus dsRNAs means that many different cDNAs can be analyzed using a fraction of the reads available during the

NGS, therefore, the cost of conventional library preparation becomes a limiting factor for the sequencing of large numbers of fungal dsRNAs. To reduce the amount of time and resources required for the NGS library preparation, we applied a previously described inexpensive transposon-based 'tagmentation' method for preparing fragmented and tagged DNA libraries [27]. The resulting cDNA libraries were sequenced with an Illumina sequencing platform, using the MiSeq Sequencing v2 (Micro 300) package. Reads were cleaned, deduplicated, and trimmed, as described in the materials and methods.

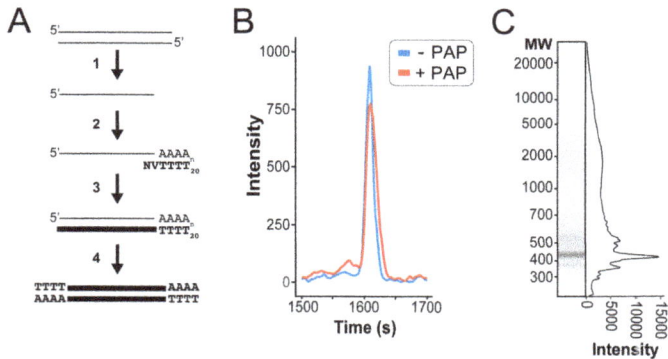

Figure 2. Overview of cDNAs synthesis from dsRNAs using poly(A) polymerase for next-generation sequencing. (**A**) (**1**) Purified dsRNAs are denatured and rapidly cooled to separate RNA strands. (**2**) ssRNA is 3′ polyadenylated by poly(A) polymerase and anchored oligo(dT) primers are annealed (N = any nucleotide, V = A, G, C). (**3**) Reverse transcription is initiated from anchored oligo(dT) primers to create cDNAs that are complementary to both the positive and negative strand of the dsRNAs. (**4**) RNAs are removed by RNase H digestion followed by the annealing and repairing of cDNAs. (**B**) The increase in the molecular weight of dsRNAs by the 3′ addition of adenine nucleotides by poly(A) polymerase (PAP) as visualized by intensity trace. (**C**) High molecular weight cDNA synthesis from dsRNAs extracted from *S. cerevisiae* by anchored oligo(dT) priming and PCR amplification visualized by intensity trace.

We found that high concentrations of poly(A) polymerase reduced the number of high-quality reads of viral origin after NGS and resulted in a large percentage of homopolymeric reads (Table 1). Titration of poly(A) polymerase was able to increase the overall number and quality of reads (25 U; 23,000 reads, 1.25 U; 28,000 reads, and 0.5 U; 42,000 reads), but we observed that using 0.02 U of poly(A) polymerase with an anchored oligo(dT) primer increased read count by 21-fold and reduced homopolymers by more than a 100-fold, relative to 25 U and a homopolymeric oligo(dT) primer (Table 1). In concert with the reduction in homopolymeric reads, we also observed an improvement in sequenced read quality. This improvement was caused by an increased base diversity and we were able to assemble long contigs of viral origin with a mean coverage of 610 (Table 1) (Figure 3A). This demonstrated that the enzymatic addition of the 3′ poly(A) tracts, which was previously used for the direct cloning of dsRNAs, is a feasible and rapid approach for the creation and NGS of dsRNA-derived cDNAs.

Table 1. Sequencing and assembly statistics for libraries made from dsRNAs polyadenylated with varying poly(A) polymerase concentrations.

Criteria	25 Units Poly(A) Polymerase	0.02 Units Poly(A) Polymerase	Fold Increase
Number of reads	22,521	471,742	21
Homopolymeric reads	68%	0.6%	0.01
De novo assembled contigs > 100nt	21	98	5
Length of largest contig	925	5022	5
Mean coverage of viral contigs	1	610	610

Using the SPAdes assembler, a de novo assembly of high-quality sequence reads derived from the dsRNAs extracted from *S. cerevisiae* BJH001, produced four long contigs with a high sequence coverage (Figure 3B). The most significant hits from the BLAST analysis of these contigs revealed that BJH001 harbors three distinct dsRNA species - two totiviruses (Saccharomyces cerevisiae virus L-A1 (ScV-L-A1) and Saccharomyces cerevisiae virus L-BC (ScV-L-BC)) and one satellite dsRNA (Saccharomyces cerevisiae satellite M1 (ScV-M1)) (Figure 3B). We have previously described the presence of ScV-L-A1 and ScV-M1 within this strain but were unaware of the totivirus ScV-L-BC [33]. Reverse transcriptase PCR was used to confirm the presence of these dsRNAs within the strain BJH001 (Figure 3B, inset).

The assembly of the sequence reads onto the published reference sequences of these dsRNA viruses and satellite demonstrated that 99.9%, 99.5%, and 89.2% of the ScV-L-A1, ScV-L-BC, and ScV-M1 dsRNAs, were sequenced to a read depth greater than 50, respectively (Figure 3C). Median read depth for all dsRNAs was greater than 2500 (Figure 3C). The 3' terminal ends of the dsRNAs were also resolved but with a low coverage (< 50 reads), especially for the terminal nucleotide (Figure S2). The 5' terminal end of ScV-L-A1 and ScV-L-BC were also resolved at a low coverage, but we were unable to resolve the 5' terminal nucleotide of the ScV-M1 (Figure S2). The only other region that was not well resolved was the low complexity ~200 bp homopolymeric adenine-rich tract contained within the ScV-M1 satellite dsRNA, which was masked prior to the read mapping. The increased coverage of the 5' half of ScV-M1 was likely due to the initiation of the reverse transcription and PCR from this internal adenine-rich tract (Figure 3C) [14]. The overall high-quality and deep coverage of the dsRNAs present within the *S. cerevisiae* strain BJH001 using short-read Illumina sequencing, demonstrated the utility of the described method for the future discovery and characterization of novel mycoviruses.

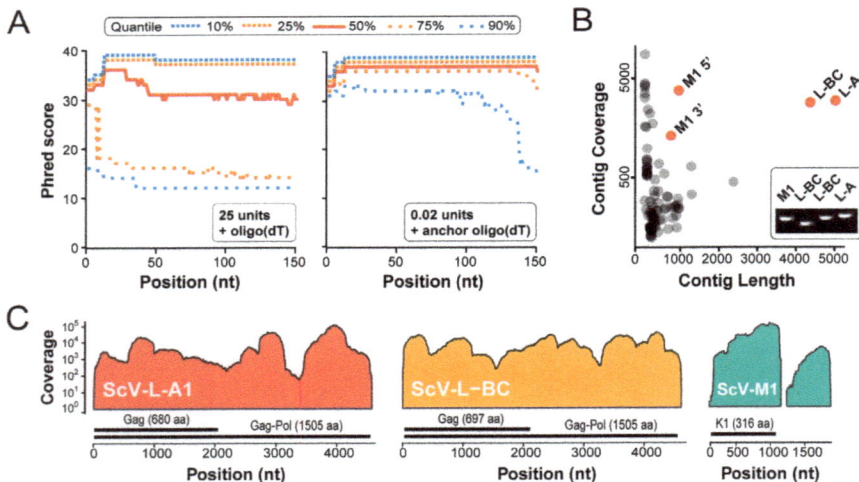

Figure 3. Next-generation sequencing of totiviruses and dsRNA satellites from *Saccharomyces cerevisiae* strain BJH001. (**A**) Read quality (phred score) after Illumina QC shown along the length of the sequencing reads when using different concentrations of poly(A) polymerase and oligo(dT) or anchored oligo(dT) primers; 10%, 25%, 75%, and 90% quantile and median (50% quantile) read quality at each position along the reads are shown. (**B**) Sequence contigs after de novo assembled represented by contig coverage and contig length. BLAST analysis of the four contigs with the longest length and deepest coverage enabled their identification as totiviruses (ScV-L-A1 and ScV-L-BC) and a dsRNA satellite (ScV-M1), the latter was assembled as two separate contigs. *Inset* reverse transcriptase-PCR was used to confirm the presence of each type of dsRNA. Two primer pairs were used to amplify the ScV-L-BC. (**C**) Read depth coverage across the reference-assembled ScV-L-A, ScV-L-BC, and ScV-M1 contigs. Open reading frames present within each dsRNA are shown above the nucleotide position.

3.3. Sequence Variation in dsRNAs Identified by NGS

The high median read depth of our NGS datasets enabled the detection of fixed synonymous and non-synonymous mutations and indels within the dsRNAs extracted from the strain BJH001 (Figure 4) (File S2). Even though the dsRNAs isolated from the strain were extracted from a clonal population, single nucleotide polymorphisms and polymorphic indels were detected within both the ScV-L-A1 and ScV-L-BC contigs (Table 2). No polymorphic nucleotides were found in assembled contigs for the ScV-M1 dsRNA. Two polymorphic indels that are present together in 21% of the ScV-L-A1 dsRNAs caused a +1 frameshift followed, after 55 base pairs, by a -1 frameshift. However, we were unable to confirm these by reverse transcriptase-PCR, cloning, and Sanger sequencing, meaning that they could have appeared due to replication errors during sample preparation. The proximity of the observed mutations to the secondary structure of the frameshift region could account for the observed discrepancy. However, two fixed indels (one nucleotide insertion and one deletion) that were observed in all sequence reads of the ScV-L-BC, resulted in a small 4 amino acid frameshift within the C-terminus of the Gag-Pol fusion protein (Figure 4B).

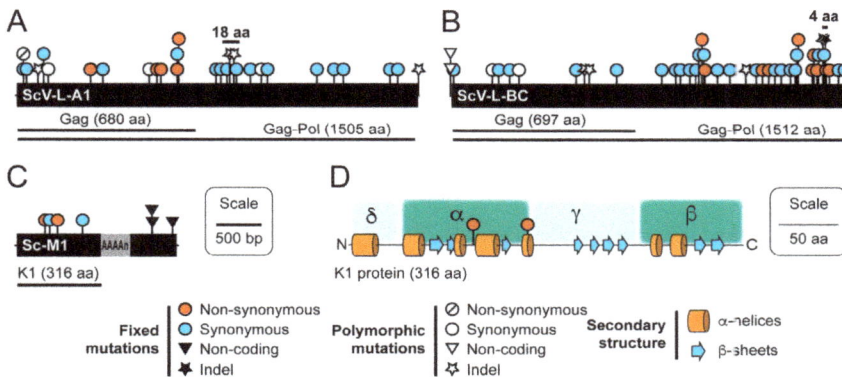

Figure 4. Natural variation in dsRNAs isolated from *S. cerevisiae* detected by NGS. A linear representation of the three dsRNAs isolated from the strain BJH001 showing the position of ORFs, relative to observed mutations (**A**) ScV-L-A1; 4575 bp, (**B**) ScV-L-BC; 4633 bp, and (**C**) ScV-M1; ~1700 bp. Bars above the indels represent the indel pairs that result in frameshifts of 18 and 4 amino acids in the Gag-Pol fusion protein. (**D**) Secondary structure prediction of the K1 killer toxin showing the position of the non-synonymous mutations in the strain BJH001.

Table 2. Polymorphic sites in endogenous dsRNA virus genomes from a clonal isolate of the *S. cerevisiae* strain BJH001.

dsRNA Element	Start Position (bp)	Polymorphism Type	Mutation	Coding Change	Coverage	Freq. (%)
ScV-L-A1	89	SNP (transition)	U to C	Non-synonymous	1002	5.8
ScV-L-A1	269	Indel	Insertion U	Gag truncation	1332	5.4
ScV-L-A1	407	SNP (transition)	C to U	Synonymous	848	31.8
ScV-L-A1	1526	SNP (transition)	U to C	Synonymous	1297	7.9
ScV-L-A1	2362	Indel	Insertion A	+1 frameshift	583	20.8
ScV-L-A1	2417	Indel	Deletion U	−1 frameshift	618	20.9
ScV-L-A1	2860	SNP (transition)	G to A	Synonymous	43,300	68.4
ScV-L-A1	4557	Indel	Insertion A	n/a	140	7.1
ScV-L-BC	10	SNP (transition)	G to A	n/a	245	81.6
ScV-L-BC	10	SNP (transversion)	G to U	n/a	245	8.6
ScV-L-BC	455	SNP (transition)	G to A	Synonymous	1882	8.3
ScV-L-BC	884	SNP (transition)	A to G	Synonymous	13,398	8.9
ScV-L-BC	1551	Indel	Insertion U	Gag truncation	207	8.9
ScV-L-BC	1569	Indel	Deletion U	Gag truncation	203	13.3
ScV-L-BC	3424	Indel	Insertion U	Pol truncation	8674	5.4

The *S. cerevisiae* strain BJH001 expresses a potent K1 killer toxin, which we have found to be capable of inhibiting the growth in a variety of different strains and species of yeast, unlike the non-killer *S. cerevisiae* strain BY4741 (Figure S3). Our NGS data suggests that the BJH001 K1 killer toxin differs from the canonical killer toxin gene sequence by two synonymous and two non-synonymous mutations (Figure 4C,D). To confirm the presence of the four mutations identified by NGS and test the functional significance of these mutations, we used reverse transcriptase PCR to directly amplify the K1 gene from the dsRNAs isolated from the strain BJH001. As a positive control, we also amplified the canonical K1 gene from the plasmid pM1TF (+) GAL [37]. The PCR products were cloned using TOPO-TA and Gateway™ methods into a galactose inducible yeast expression vector [32]. Importantly, the four mutations identified within the BJH001 K1 gene by Illumina NGS were confirmed by Sanger sequencing. The K1 expression vectors were used to transform the non-killer *S. cerevisiae* strain BY4741. To compare the biological activities of the two cloned K1 toxins, 6×10^6 cells of each isogenic K1-expressing strain were spotted, in triplicate, onto galactose-containing agar plates seeded with various K1-sensitive yeasts. Qualitative comparison of the specificity of the BJH001 K1 toxin expressed from a plasmid or the dsRNA satellite demonstrated that the ectopic expression does not alter its specificity toward the toxin-sensitive yeasts (Figures S3 and S4). However, measurement of the area of growth inhibition revealed that the BJH001 K1 toxin produces significantly larger zones of growth inhibition on the K1-sensitive yeast *Kazachstania africana*, compared to the canonical K1 reference toxin (T-test, two-tailed, $p < 0.01$) (Figure 5). This zone of growth inhibition was 28% larger than the K1 reference toxin (Figure 5). For the other seven K1-sensitive strains tested, the differences between the two killer toxins did not significantly alter the area of the zone of growth inhibition, which suggests that the mutations in the K1 toxin from the strain BJH001 did not affect the amount of killer toxin produced or the rate of diffusion through the agar (Figure S4). These data demonstrated that mutations in K1 killer toxin can alter their toxicity to specific species of yeasts.

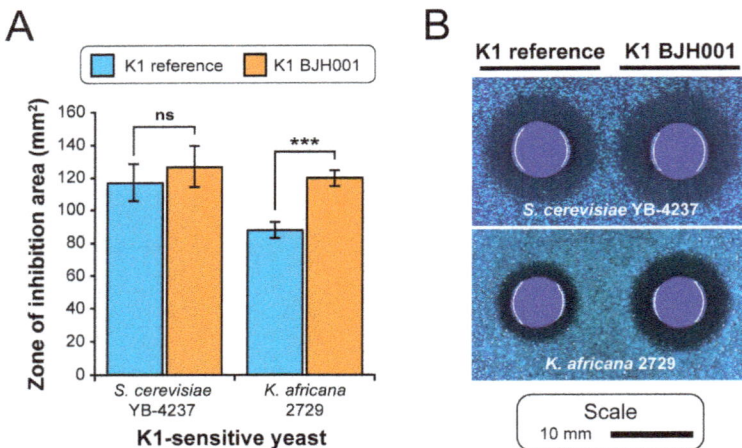

Figure 5. Mutations within the K1 gene increases the ability of the killer toxin to inhibit the growth of the yeast *Kazachstania africana in vitro*. (**A**) The change in the area of growth inhibition around K1-expressing *S. cerevisiae* challenged with different strains of K1-sensitive yeasts measured in mm². Asterisks are indicative of a significant difference in the mean zone of the inhibition area (T-test, two-tailed, *** $p < 0.01$, ns indicates no significant difference). Error bars represent standard error of three independent repeats. (**B**) Representative images of the isogenic non-killer yeast strains expressing different K1 killer toxins (derived from the K1 reference sequence or K1 from *S. cerevisiae* BJH001), on agar seeded with yeasts known to be sensitive to K1 killer toxins.

4. Discussion

The methods that we describe constitute a broadly applicable approach to the sequencing of dsRNA purified from fungi using Illumina NGS. We have successfully applied this approach to determine the nucleotide sequence of dsRNAs purified from the yeast *S. cerevisiae*. We also show the feasibility of extracting high-quality dsRNAs from commercial dried yeasts as well as laboratory grown cultures. Polyadenylation of dsRNAs was one of the first methods used to modify the 3' termini of RNAs to enable cDNA synthesis for Sanger sequencing [15,16,18]. However, these methods were limited in their ability to clone full-length cDNAs derived from viral dsRNAs [9,18,38]. Most recent methods have focused on using 3' oligo ligation or random priming to initiate cDNA synthesis from unknown dsRNAs, prior to cloning or NGS, and have been successful in determining the genetic sequence of many viral and satellite dsRNAs [11–13,19–21,35]. However, these previously described methods often involve labor- and time-intensive steps during dsRNA purification, 3' oligomer ligation, and NGS library preparation. We have evaluated these methods to develop a protocol that is rapid and feasible for sequencing large numbers of small dsRNA molecules extracted from fungi. From cells to sequencer-ready libraries the described protocol takes 17 h; 4 h for dsRNA extraction, 8 h for cDNA synthesis, and 5 h for library creation by tagmentation [27]. CDNA synthesis and library construction using the described method takes 13 h, which compares well to contemporary commercial kits using mRNAs that take 12 h (TruSeq RNA sequencing kit; Illumina) and is faster than methods that require a long (up to 18 h) incubation for the efficient 3' ligation of oligonucleotides, prior to cDNA synthesis [11,20,21]. To the best of our knowledge, the combination of 3' polyadenylation, anchored oligo(dT) priming, and tagmentation for NGS library preparation has never been applied as a technique for the rapid synthesis of high-quality cDNAs from dsRNAs for Illumina NGS.

Cloning of the dsRNA-derived cDNAs and 5' or 3' RACE enable the efficient resolution of the dsRNA terminal ends. NGS methods alone have been largely unsuccessful in the resolution of the terminal ends of cDNAs [21,39]. However, there are some notable exceptions that have leveraged a combination of commercial kits and homopolymeric primers to completely sequence the dsRNAs isolated from yeasts [12,13].

Our NGS method was also able to resolve most terminal ends when mapping to a reference sequence, but coverage appeared to be dependent on the terminal sequence of the dsRNAs. Moreover, we were able to resolve the 3' termini of all dsRNAs within *S. cerevisiae* BJH001 but the 5' termini had a reduced coverage (Figure S2). ScV-M1 and ScV-L-A1 have A/U-rich 5' termini that may have resulted in ambiguities during sequence mapping, causing a poor sequence coverage. Except for the low coverage of the terminal nucleotides, we were able to assemble long viral contigs from the mixtures of different dsRNAs extracted from a single strain of yeast, independent of a reference sequence (Figure 3B). For example, we have previously studied the mycoviruses and dsRNAs present within *S. cerevisiae* BJH001 by agarose gel electrophoresis but were unaware of the presence of a variant ScV-L-BC within this strain because of its similar electrophoretic mobility to ScV-L-A1 [33]. By applying our NGS method we were able to identify the presence of ScV-L-BC and assemble a large contig with a high similarity to the reference sequence of this totivirus (Figure 3). The high coverage of the majority of the dsRNAs allowed the identification of fixed and polymorphic mutations within the populations of dsRNAs. We also observed an indel that resulted in a small, but dramatic change to the amino acid sequence of the polymerase domain in ScV-L-BC. This frameshifted region is peripheral to the conserved motifs of the catalytic core of the RNA-dependent RNA polymerase [40], which would suggest that these mutations might not disrupt the polymerase function. Furthermore, because ScV-L-BC is stably associated with *S. cerevisiae* BJH001, we do not expect that the fixed 4 amino acid frameshift to significantly affect viral replication and persistence. Frameshift mutations most often result in premature stop codons and defective proteins that are truncated, with the most prominent examples of frameshift mutations being those that cause human disease [41–43]. Moreover, mutant polymerases could be incorporated into viral capsids to form defective interfering particles with likely negative consequences for viral replication. Alternatively, frameshift mutations can result in novel protein functions, although they are less frequently reported [44]. The co-occurrence of a +1 and -1 frameshift indel suggests that there has been selection to maintain the reading frame of the polymerase gene, but any functional consequence to the polymerase enzyme and on the replication and persistence of ScV-L-BC remains unexplored.

ScV-M1 was found to contain four fixed mutations in the K1 killer toxin gene that we were able to confirm by cloning and Sanger sequencing. The two non-synonymous mutations (I103S and T146I) map to the K1 alpha-domain that is important for the cytotoxicity of K1 and are positioned close to known mutations that are defective in cell wall binding and toxicity (D101R and D140R) (Figure 4D) [45]. Relative to the cloned canonical K1 toxin, the mutations I103S and T146I significantly increased the toxicity of K1 to *K. africana* but not to other strains of *Saccharomyces* yeasts that were challenged by the toxin (Figure 5). Previously, two different K1 variants have been described by reverse transcriptase PCR and Sanger sequencing within different species of *Saccharomyces* yeasts [34]. Expression of these variant K1 toxins in *S. cerevisiae* and *S. paradoxus* appeared to show that a single gain-of-function mutation in the K1 beta-domain (L251F) can increase the cytotoxicity of the K1 toxin, but the results were not quantified to assess the statistical significance [34]. Furthermore, a separate study failed to identify these K1 variants in the same strains of yeast [35]. We anticipate that our NGS method could be applied to rapidly elucidate the genetic sequence of satellite dsRNAs to investigate the effect of genetic variation on killer toxin activity. The large number of killer yeasts with unique antifungal activities discovered since the 1970s suggests that killer toxins are numerous and diverse [46–52]. Indeed, this is highlighted by the recent description of several novel satellite dsRNAs and associated killer toxins within *Saccharomyces* yeasts [35]. Ultimately, a better understanding of the relationship between killer toxin genotype and phenotype will clarify their contribution to fungal ecology with broad significance to human health and agriculture.

Supplementary Materials: The following are available online at http://www.mdpi.com/1999-4915/11/1/70/s1, Figure S1: Comparing methods for dsRNA extraction from the *Saccharomyces* yeasts. Figure S2: Resolution of the 5′ and 3′ termini of dsRNAs using NGS. Figure S3: *S. cerevisiae* BJH001 produces a killer toxin that can inhibit the growth of different yeast strains and species. Figure S4: Mutations within the K1 gene increase the ability of the K1 killer toxin to inhibit the growth of *K. africana* in vitro. Table S1: Primers used in this study. Table S2: Yeast strains and species used in this study. File S1: The DNA sequences of the plasmids used in this study. File S2: Mutations identified within dsRNAs extracted and sequenced in this study. File S3: Raw images of agarose gels presented in this study.

Author Contributions: Conceptualization, P.A.R.; Data curation, A.M.C., S.S.H., J.T.V.L., and P.A.R.; Formal analysis, A.M.C. and P.A.R.; Funding acquisition, P.A.R.; Investigation, A.M.C., E.A.K., D.D.N., and M.W.F.; Methodology, A.M.C., S.S.H., D.D.N., M.W.F., and P.A.R.; Project administration, P.A.R.; Supervision, P.A.R.; Writing—original draft, A.M.C., and P.A.R.; Writing—review and editing, A.M.C, E.A.K., S.S.H., and P.A.R.

Funding: This work was supported by the Center for Modeling Complex Interactions at the University of Idaho (NIH grant P20 GM104420), Idaho INBRE Program Core Technology Access Grant (NIH Grant Nos. P20 GM103408 and P20 GM109095, National Science Foundation Grant Nos., 1818368, 0619793, and 0923535; the MJ Murdock Charitable Trust; and the Idaho State Board of Education). Undergraduate researchers were supported by the INBRE Program, NIH grant P20 GM103408, the University of Idaho, Office of Undergraduate Research, and the Department of Biological Sciences. Publication of this article was funded by the University of Idaho Open Access Publishing Fund.

Acknowledgments: Douglas Cole for critical comments on the manuscript. Library preparation, sequencing, and bioinformatic analysis was performed at the IBEST Genomics Core (University of Idaho, Moscow).

Conflicts of Interest: The authors declare no conflict of interest.

References

1. Ghabrial, S.A.; Suzuki, N. Viruses of plant pathogenic fungi. *Annu. Rev. Phytopathol.* **2009**, *47*, 353–384. [CrossRef] [PubMed]

2. Bozarth, R.F. Mycoviruses: A new dimension in microbiology. *Environ. Health Perspect.* **1972**, *2*, 23–39. [CrossRef]

3. Milgroom, M.G.; Hillman, B.I. *Studies in Viral Ecology: Microbial and Botanical Host Systems*, 1st ed.; Hurst, C.J., Ed.; John Wiley & Sons, Inc.: Hoboken, NJ, USA, 2011; Volume 1, pp. 217–253.

4. Nuss, D.L. Hypovirulence: Mycoviruses at the fungal–plant interface. *Nat. Rev. Microbiol.* **2005**, *3*, 632–642. [CrossRef] [PubMed]

5. Schmitt, M.J.; Breinig, F. Yeast viral killer toxins: Lethality and self-protection. *Nat. Rev. Microbiol.* **2006**, *4*, 212–221. [CrossRef] [PubMed]

6. Márquez, L.M.; Redman, R.S.; Rodriguez, R.J.; Roossinck, M.J. A virus in a fungus in a plant: Three-way symbiosis required for thermal tolerance. *Science* **2007**, *315*, 513–515. [CrossRef] [PubMed]

7. Franklin, R.M. Purification and properties of the replicative intermediate of the RNA bacteriophage R17. *Proc. Natl. Acad. Sci. USA* **1966**, *55*, 1504–1511. [CrossRef]

8. Kobayashi, K.; Tomita, R.; Sakamoto, M. Recombinant plant dsRNA-binding protein as an effective tool for the isolation of viral replicative form dsRNA and universal detection of RNA viruses. *J. Gen. Plant Pathol.* **2009**, *75*, 87–91. [CrossRef]

9. Cashdollar, L.W.; Esparza, J.; Hudson, G.R.; Chmelo, R.; Lee, P.W.; Joklik, W.K. Cloning the double-stranded RNA genes of reovirus: Sequence of the cloned S2 gene. *Proc. Natl. Acad. Sci. USA* **1982**, *79*, 7644–7648. [CrossRef]

10. Fried, H.M.; Fink, G.R. Electron microscopic heteroduplex analysis of "killer" double-stranded RNA species from yeast. *Proc. Natl. Acad. Sci. USA* **1978**, *75*, 4224–4228. [CrossRef]

11. Lambden, P.R.; Cooke, S.J.; Caul, E.O.; Clarke, I.N. Cloning of noncultivatable human rotavirus by single primer amplification. *J. Virol.* **1992**, *66*, 1817–1822.

12. Ramìrez, M.; Velázquez, R.; Maqueda, M.; López-Piñeiro, A.; Ribas, J.C. A new wine *Torulaspora delbrueckii* killer strain with broad antifungal activity and its toxin-encoding double-stranded RNA virus. *Front. Microbiol.* **2015**, *6*, 403–412. [CrossRef] [PubMed]

13. Ramìrez, M.; Velázquez, R.; López-Piñeiro, A.; Naranjo, B.; Roig, F.; Llorens, C. New Insights into the Genome Organization of Yeast Killer Viruses Based on "Atypical" Killer Strains Characterized by High-Throughput Sequencing. *Toxins* **2017**, *9*, 292. [CrossRef] [PubMed]

14. Brizzard, B.L.; de Kloet, S.R. Reverse transcription of yeast double-stranded RNA and ribosomal RNA using synthetic oligonucleotide primers. *Biochim. Biophys. Acta* **1983**, *739*, 122–131. [CrossRef]

15. Hagenbüchle, O.; Santer, M.; Steitz, J.A.; Mans, R.J. Conservation of the primary structure at the 3′ end of 18S rRNA from eucaryotic cells. *Cell* **1978**, *13*, 551–563. [CrossRef]

16. Taniguchi, T.; Palmieri, M.; Weissmann, C. QB DNA-containing hybrid plasmids giving rise to QB phage formation in the bacterial host. *Nature* **1978**, *274*, 223–228. [CrossRef] [PubMed]

17. Both, G.W.; Bellamy, A.R.; Street, J.E.; Siegman, L.J. A general strategy for cloning double-stranded RNA: Nucleotide sequence of the Simian-11 rotavirus gene 8. *Nucleic Acids Res.* **1982**, *10*, 7075–7088. [CrossRef] [PubMed]

18. Ghibelli, L.; Usala, S.J.; Mukhopadhyay, R.; Haselkorn, R. Polyadenylation and reverse transcription of bacteriophage φ6 double-stranded RNA. *Virology* **1982**, *120*, 318–328. [CrossRef]

19. Grybchuk, D.; Akopyants, N.S.; Kostygov, A.Y.; Konovalovas, A.; Lye, L.-F.; Dobson, D.E.; Zangger, H.; Fasel, N.; Butenko, A.; Frolov, A.O.; et al. Viral discovery and diversity in trypanosomatid protozoa with a focus on relatives of the human parasite Leishmania. *Proc. Natl. Acad. Sci. USA* **2018**, *115*, E506–E515. [CrossRef]

20. Imai, M.; Richardson, M.A.; Ikegami, N.; Shatkin, A.J.; Furuichi, Y. Molecular cloning of double-stranded RNA virus genomes. *Proc. Natl. Acad. Sci. USA* **1983**, *80*, 373–377. [CrossRef]

21. Potgieter, A.C.; Page, N.A.; Liebenberg, J.; Wright, I.M.; Landt, O.; van Dijk, A.A. Improved strategies for sequence-independent amplification and sequencing of viral double-stranded RNA genomes. *J. Gen. Virol.* **2009**, *90*, 1423–1432. [CrossRef]

22. Bobek, L.A.; Bruenn, J.A.; Field, L.J.; Gross, K.W. Cloning of cDNA to a yeast viral double-stranded RNA and comparison of three viral RNAs. *Gene* **1982**, *19*, 225–230. [CrossRef]

23. Roossinck, M.J.; Saha, P.; Wiley, G.B.; Quan, J.; White, J.D.; Lai, H.; Chavarría, F.; Shen, G.; Roe, B.A. Ecogenomics: Using massively parallel pyrosequencing to understand virus ecology. *Mol. Ecol.* **2010**, *19*, 81–88. [CrossRef] [PubMed]

24. Decker, C.J.; Parker, R. Analysis of double-stranded RNA from microbial communities identifies double-stranded RNA virus-like elements. *Cell Rep.* **2014**, *7*, 898–906. [CrossRef] [PubMed]

25. Coetzee, B.; Freeborough, M.-J.; Maree, H.J.; Celton, J.-M.; Rees, D.J.G.; Burger, J.T. Deep sequencing analysis of viruses infecting grapevines: Virome of a vineyard. *Virology* **2010**, *400*, 157–163. [CrossRef] [PubMed]

26. Adams, I.P.; Glover, R.H.; Monger, W.A.; Mumford, R.; Jackevicien, E.; Navalinskiene, M.; Samuitiene, M.; Boonham, N. Next-generation sequencing and metagenomic analysis: A universal diagnostic tool in plant virology. *Mol. Plant Pathol.* **2009**, *10*, 537–545. [CrossRef] [PubMed]

27. Baym, M.; Kryazhimskiy, S.; Lieberman, T.D.; Chung, H.; Desai, M.M.; Kishony, R. Inexpensive multiplexed library preparation for megabase-sized genomes. *PLoS ONE* **2015**, *10*, e0128036. [CrossRef]

28. Okada, R.; Kiyota, E.; Moriyama, H.; Fukuhara, T.; Natsuaki, T. A simple and rapid method to purify viral dsRNA from plant and fungal tissue. *J. Gen. Plant Pathol.* **2015**, *81*, 103–107. [CrossRef]

29. Langmead, B.; Salzberg, S.L. Fast gapped-read alignment with Bowtie 2. *Nat. Methods* **2012**, *9*, 357–359. [CrossRef]

30. Li, H.; Handsaker, B.; Wysoker, A.; Fennell, T.; Ruan, J.; Homer, N.; Marth, G.; Abecasis, G.; Durbin, R. 1000 Genome Project Data Processing Subgroup. The Sequence Alignment/Map format and SAMtools. *Bioinformatics* **2009**, *25*, 2078–2079. [CrossRef]

31. Bankevich, A.; Nurk, S.; Antipov, D.; Gurevich, A.A.; Dvorkin, M.; Kulikov, A.S.; Lesin, V.M.; Nikolenko, S.I.; Pham, S.; Prjibelski, A.D.; et al. SPAdes: A new genome assembly algorithm and its applications to single-cell sequencing. *J. Comput. Biol.* **2012**, *19*, 455–477. [CrossRef]

32. Alberti, S.; Gitler, A.D.; Lindquist, S. A suite of Gateway cloning vectors for high-throughput genetic analysis in *Saccharomyces cerevisiae*. *Yeast* **2007**, *24*, 913–919. [CrossRef] [PubMed]

33. Rowley, P.A.; Ho, B.; Bushong, S.; Johnson, A.; Sawyer, S.L. *XRN1* Is a Species-Specific Virus Restriction Factor in Yeasts. *PLoS Pathog.* **2016**, *12*, e1005890. [CrossRef]

34. Chang, S.-L.; Leu, J.-Y.; Chang, T.-H. A Population Study of Killer Viruses Reveals Different Evolutionary Histories of Two Closely Related *Saccharomyces sensu stricto* Yeasts. *Mol. Ecol.* **2015**, *24*, 4312–4322. [CrossRef]

35. Rodríguez-Cousiño, N.; Gómez, P.; Esteban, R. Variation and Distribution of L-A Helper Totiviruses in *Saccharomyces sensu stricto* Yeasts Producing Different Killer Toxins. *Toxins* **2017**, *9*, 313. [CrossRef] [PubMed]

36. Rowley, P.A. The frenemies within: Viruses, retrotransposons and plasmids that naturally infect *Saccharomyces* yeasts. *Yeast* **2017**, *49*, 111–292. [CrossRef] [PubMed]

37. Russell, P.J.; Bennett, A.M.; Love, Z.; Baggott, D.M. Cloning, sequencing and expression of a full-length cDNA copy of the M1 double-stranded RNA virus from the yeast, Saccharomyces cerevisiae. *Yeast* **1997**, *13*, 829–836. [CrossRef]

38. Shapouri, M.R.; Kane, M.; Letarte, M.; Bergeron, J.; Arella, M.; Silim, A. Cloning, sequencing and expression of the S1 gene of avian reovirus. *J. Gen. Virol.* **1995**, *76 Pt 6*, 1515–1520. [CrossRef]

39. Rodríguez-Cousiño, N.; Esteban, R. Relationships and Evolution of Double-Stranded RNA Totiviruses of Yeasts Inferred from Analysis of L-A-2 and L-BC Variants in Wine Yeast Strain Populations. *Appl. Environ. Microbiol.* **2017**, *83*, e02991-16. [CrossRef]

40. Bruenn, J.A. A structural and primary sequence comparison of the viral RNA-dependent RNA polymerases. *Nucleic Acids Res.* **2003**, *31*, 1821–1829. [CrossRef]

41. Liu, R.; Paxton, W.A.; Choe, S.; Ceradini, D.; Martin, S.R.; Horuk, R.; MacDonald, M.E.; Stuhlmann, H.; Koup, R.A.; Landau, N.R. Homozygous defect in HIV-1 coreceptor accounts for resistance of some multiply-exposed individuals to HIV-1 infection. *Cell* **1996**, *86*, 367–377. [CrossRef]

42. Lau, M.M.; Neufeld, E.F. A frameshift mutation in a patient with Tay-Sachs disease causes premature termination and defective intracellular transport of the alpha-subunit of beta-hexosaminidase. *J. Biol. Chem.* **1989**, *264*, 21376–21380. [PubMed]

43. White, M.B.; Amos, J.; Hsu, J.; Gerrard, B.; Finn, P.; Dean, M. A Frame-Shift Mutation in the Cystic-Fibrosis Gene. *Nature* **1990**, *344*, 665–667. [CrossRef] [PubMed]

44. Raes, J.; Van de Peer, Y. Functional divergence of proteins through frameshift mutations. *Trends Genet.* **2005**, *21*, 428–431. [CrossRef] [PubMed]

45. Zhu, H.; Bussey, H. Mutational analysis of the functional domains of yeast K1 killer toxin. *Mol. Cell. Biol.* **1991**, *11*, 175–181. [CrossRef] [PubMed]

46. Rosini, G. The Occurrence of Killer Characters in Yeasts. *Can. J. Microbiol.* **1983**, *29*, 1462–1464. [CrossRef] [PubMed]

47. Young, T.W.; Yagiu, M. A comparison of the killer character in different yeasts and its classification. *Antonie Van Leeuwenhoek* **1978**, *44*, 59–77. [CrossRef]

48. Kandel, J.S.; Stern, T.A. Killer Phenomenon in Pathogenic Yeast. *Antimicrob. Agents Chemother.* **1979**, *15*, 568–571. [CrossRef]

49. Philliskirk, G.; Young, T.W. The occurrence of killer character in yeasts of various genera. *Antonie Van Leeuwenhoek* **1975**, *41*, 147–151. [CrossRef]

50. Starmer, W.T.; Ganter, P.F.; Aberdeen, V. Geographic distribution and genetics of killer phenotypes for the yeast *Pichia kluyveri* across the United States. *Appl. Environ. Microbiol.* **1992**, *58*, 990–997.

51. Stumm, C.; Hermans, J.; Middelbeek, E.J.; Croes, A.F.; de Vries, G. Killer-Sensitive Relationships in Yeasts From Natural Habitats. *Antonie Van Leeuwenhoek* **1977**, *43*, 125–128. [CrossRef]

52. Pieczynska, M.D.; de Visser, J.A.G.M.; Korona, R. Incidence of symbiotic dsRNA "killer" viruses in wild and domesticated yeast. *FEMS Yeast Res.* **2013**, *13*, 856–859. [CrossRef] [PubMed]

MDPI

Article

Saccharomyces paradoxus K66 Killer System Evidences Expanded Assortment of Helper and Satellite Viruses

Iglė Vepštaitė-Monstavičė [1], Juliana Lukša [1], Aleksandras Konovalovas [2], Dovilė Ežerskytė [1], Ramunė Stanevičienė [1], Živilė Strazdaitė-Žielienė [1], Saulius Serva [2,3,*] and Elena Servienė [1,3,*]

[1] Laboratory of Genetics, Institute of Botany, Nature Research Centre, LT-08412 Vilnius, Lithuania; igle.vepstaite-monstavice@gamtc.lt (I.V.-M.); juluksa@gmail.com (J.L.); ezerskytedovile@gmail.com (D.E.); ramune.staneviciene@gamtc.lt (R.S.); zivile.strazdaite-zieliene@gamtc.lt (Ž.S.-Ž.)

[2] Department of Biochemistry and Molecular Biology, Institute of Biosciences, Vilnius University, LT-10257 Vilnius, Lithuania; aleksandras.konovalovas@gf.vu.lt

[3] Department of Chemistry and Bioengineering, Vilnius Gediminas Technical University, LT-10223 Vilnius, Lithuania

* Correspondence: saulius.serva@gf.vu.lt (S.S.); elena.serviene@gamtc.lt (E.S.); Tel.: +370-52-72-9363 (S.S.); Tel.: +370-52-39-8244 (E.S.); Fax: +370-52-39-8231 (S.S.); Fax: +370-52-72-9352 (E.S.)

Received: 28 August 2018; Accepted: 15 October 2018; Published: 16 October 2018

Abstract: The *Saccharomycetaceae* yeast family recently became recognized for expanding of the repertoire of different dsRNA-based viruses, highlighting the need for understanding of their cross-dependence. We isolated the *Saccharomyces paradoxus* AML-15-66 killer strain from spontaneous fermentation of serviceberries and identified helper and satellite viruses of the family *Totiviridae*, which are responsible for the killing phenotype. The corresponding full dsRNA genomes of viruses have been cloned and sequenced. Sequence analysis of SpV-LA-66 identified it to be most similar to *S. paradoxus* LA-28 type viruses, while SpV-M66 was mostly similar to the SpV-M21 virus. Sequence and functional analysis revealed significant differences between the K66 and the K28 toxins. The structural organization of the K66 protein resembled those of the K1/K2 type toxins. The AML-15-66 strain possesses the most expressed killing property towards the K28 toxin-producing strain. A genetic screen performed on *S. cerevisiae* YKO library strains revealed 125 gene products important for the functioning of the *S. paradoxus* K66 toxin, with 85% of the discovered modulators shared with *S. cerevisiae* K2 or K1 toxins. Investigation of the K66 protein binding to cells and different polysaccharides implies the β-1,6 glucans to be the primary receptors of *S. paradoxus* K66 toxin. For the first time, we demonstrated the coherent habitation of different types of helper and satellite viruses in a wild-type *S. paradoxus* strain.

Keywords: *Saccharomyces paradoxus*; *Totiviridae*; dsRNA virus; killer system

---

1. Introduction

Yeasts constitute a large group of microorganisms characterized by the ability to grow and survive in stressful conditions and to colonize a wide range of environmental ecosystems [1]. The secretion of yeast killer toxins confers a competitive edge to the producer strain by excluding other yeasts from shared habitat without direct cell-to-cell contact [2]. Rather than providing immediate advantages for the respective host, killer toxin-immunity systems also have to be considered as important players in the autoselection system [3]. Recently, the mutual incompatibility of double-stranded RNA virus-based killer systems and the RNA interference mechanism in yeast has been demonstrated [4].

Double-stranded RNA-based killer systems have been described in different yeast species, such as *Saccharomyces cerevisiae*, *S. paradoxus*, *S. uvarum*, *Ustilago maydis*, *Zygosaccharomyces bailii*, *Hanseniaspora uvarum*, and *Torulaspora delbrueckii* [5–7]. The mycoviruses of the *Totiviridae* family are incapsulated into virus-like particles (VLPs) and stably persist in the host cell without causing cell lysis; they are transmitted by vegetative cell division or through sexual fusion [5,8–10]. Most killer toxins are encoded by dsRNA viruses called M satellites, which depend for their propagation and maintenance on an L-A helper virus [8,11]. Among the representatives of the genus *Saccharomyces*, the killer phenomenon has been studied most extensively in *S. cerevisiae*, where four different viral-originated killer toxins (K1, K2, K28, and Klus) have been described [12–15]. For effective functioning of the killer system, well organized communication between the functionally distinct ScV-LA and ScV-M viruses is required. The L-A virus has a 4.6 kb segment that encodes the major structural capsid protein Gag, encapsulating either L or M virus dsRNA in icosahedral structures, and the Gag-Pol fusion protein responsible for replication and encapsidation [11,16,17]. There are several variants of L-A (ScV-LA-1, ScV-LA-2, ScV-LA-28 and ScV-LA-lus), with an average of 74% identity in nucleotide sequences, associated with different M viruses and displaying distinct phenotypic properties [14,18]. Four different *S. cerevisiae* M dsRNA viruses have been described (ScV-M1, ScV-M2, ScV-M28 and ScV-Mlus) so far [15]. Certain relationships between L-A and M viruses have been observed (LA-1 and M1, LA-2 and M2, LA-lus and Mlus, LA-28 and M28) [14,18,19] and the possible role of the toxin-producing M viruses in selecting the L-A variants to support them has been proposed [19]. A certain heterogeneity in L-A itself has also been reported with functional phenotypic variants that exhibited differences in maintaining the K1 and K2 phenotypes with the involvement of MKT genes [20]. The observed exceptions of specificity were essentially limited to laboratory strains or hybrids, as well as strains featuring significantly elevated amounts of L-A dsRNA or proteins encoded by this virus [19,20]. The association of distinct L-As with different M viruses suggests their co-evolution, leading to the propensity of a particular L-A virus to maintain a certain type of M virus [14,19].

The M viral genome is 1.6–2.4 kb in size and encodes a specific preprotoxin, which is subsequently processed into a mature protein (K1, K2, K28, and Klus). The M dsRNA viruses show no sequence homology to each other, though the organization of their genomes is strikingly similar. The positive strand contains an open reading frame (ORF) in the 5′-terminal region that encodes for the toxin precursor, followed by a unique internal AU-rich region, and a 3′-terminal non-coding region of variable length possessing *cis* signals for encapsidation and replication by the viral RNA polymerase [6]. Different killer toxins are secreted glycoproteins lacking amino acid sequence conservation and adopting diverse cell killing mechanisms. The proposed mechanism of the action for the K1 and K2 toxins is a two-step process, whereby the killer protein first binds to the primary cell wall receptor-β-1,6-glucan [21,22], then, at the second step, the toxins approach a plasma membrane receptor and form lethal cation-selective ion channels [15,23–25]. In contrast, the K28 toxin binds to α-1,3-mannoproteins positioned in the cell wall, then interacts with a plasma membrane receptor Erd2 and enters the cell by endocytosis. In the cell, the K28 toxin travels to the nucleus by retrograde passage and blocks DNA synthesis causing G1/S cell cycle arrest [26,27]. The lethal mechanism of the Klus toxin is yet to be uncovered [14,28].

Many host genes affect the maintenance of *S. cerevisiae* L-A and M viruses [11], as well as the performance of viral killer toxins [29–31]. The SKI family genes block expression of the non-polyadenylated viral mRNA [32]. Species-specific virus restriction factor Xrn1p (encoded by SKI1 gene) appears to co-evolve with totiviruses to control viral propagation in *Saccharomyces* yeasts [33]. MAK family gene products are necessary for L-A and M propagation [12,34,35]. The exact function and interplay of these genes in virus replication and maintenance is not fully understood [36,37]. On the other hand, the presence of dsRNA viruses impacts the expression of numerous host genes [38,39], many of them tightly integrated into cellular metabolism. The phylogenetic analysis uncovers that at least some mycoviruses co-evolve with their hosts, suggesting a close interaction between participants [40,41]. Viruses and their hosts exist in a constant state of genetic conflict, where an advantage for one party is often a disadvantage for the other [33]. The stable persistence of L-A virus in some 20% of wild *S. cerevisiae* [42] cells suggests a generally detrimental impact to the host, probably because of consumption of energy and material resources [11]. In addition, M dsRNAs are rather rare in wild strains [42,43].

Saccharomyces paradoxus is the closest relative of the domesticated yeast *S. cerevisiae*, mainly found in the wild [6]. Killer toxins of *S. paradoxus* have been considered as chromosome-coded for a long time and it is only relatively recently that dsRNA viruses (L and M) in such yeast have been discovered [44]. The sizes of L and M dsRNAs genomes are similar to those of *S. cerevisiae* strains. The degree of nucleotide variation in different types of L-A viruses ranged from 73% to 90%, depending on the geographical location of *S. paradoxus* strains [6,9]. The essential features of *S. cerevisiae* L-A viruses-frameshift region and encapsidation signal-remain conserved in all *S. paradoxus* L-A variants investigated thus far. The encoded Gag-Pol proteins demonstrate 85% to 98% amino acid identity. At least five different *S. paradoxus* killer toxin-producing viruses (SpV-M21, SpV-M28, SpV-M45, SpV-M62, and SpV-M74) have been identified. They encode toxins differing in sequences from any of those previously known, while structural-functional characterization has not been accomplished yet [6].

In this study, we performed cloning and structure-functional analysis of the *S. paradoxus* SpV-LA-66 and SpV-M66 viruses, recently isolated from the natural environment. Comparison of the SpV-LA-66 sequence and phylogenetic analysis demonstrated its close relationship to SpV-LA-28 and SpV-LA-21 viruses. The SpV-M66 virus-encoded preprotoxin sequence analysis allowed prediction of the structural elements of the active K66 killer protein. The *S. paradoxus* K66 toxin was isolated and the essential parameters of protein activity were investigated. For the first time, we have identified genetic factors, involved in the functioning of the *S. paradoxus* virus-originated K66 toxin, and those important for the susceptibility of the target cell. The gene products, connected to cell wall organization and biogenesis, as well as involved in the regulation of response to osmotic stress, were demonstrated as significantly enriched. By performing in vivo and in vitro toxin binding assays, we demonstrated that β-1,6 glucans could play the role of primary receptors for the *S. paradoxus* K66 toxin. We concluded that SpV-LA-66 and SpV-M66 represent a previously undescribed combination of helper and satellite viruses in the wild-type *S. paradoxus* AML-15-66 strain.

2. Materials and Methods

2.1. Strains and Media

The killer strain employed in this study (*Saccharomyces paradoxus* AML-15-66) was originally isolated from spontaneous fermentation of serviceberries (*Amelanchier ovalis* Medik.). The following yeast strains were used for the killer assay: *S. cerevisiae* α'1 (*MATα leu2-2 (KIL-0)*) [45], M437 (*wt, HM/HM (KIL-K2)*) [46], K7 (*MATα arg9 (KIL-K1)*) [47], MS300 (*MATα leu2 ura 3-52 (KIL-K28)*) [13], SRB-15-4 (wt, HM/HM (KIL-Klus) (laboratory collection), BY4741 (*MATa; his3 D1; leu2Δ0; met15Δ0; ura3Δ0 (KIL-0)*) (Thermo Scientific Molecular Biology, Lafayette, CO, USA), and *S. paradoxus* T21.4 strain (kindly provided by Dr. G. Liti, Université Côte d'Azur, CNRS, INSERM, IRCAN, Nice, France). The curing of yeast strains from dsRNA viruses was accomplished as described in [39].

For identification of yeast, the regions between the 18S rRNA and 28S rRNA genes containing two non-coding spacers (ITS-A and ITS-B) separated by the 5.8S rRNA gene were PCR-amplified using ITS1 (5′-TCCGTAGGTGAACCTGCGG-3′) and ITS4 (5′-TCCTCCGCTTATTGATATGC-3′) primers [48], and sequenced at Base Clear (Leiden, ZH, The Netherlands). The obtained sequences were compared with those found in the FASTA network service of the EMBL-EBI database (http://www.ebi.ac.uk/Tools/sss/fasta/nucleotide.html). Screening for genetic factors modulating the K66 toxin activity was performed with a *S. cerevisiae* single ORFs deletion strains (BY4741 background, *MATa; his3 D1; leu2Δ0; met15Δ0; ura3Δ0*) (Thermo Scientific Molecular Biology, Lafayette, CO, USA).

Yeast strains were grown in standard YEPD medium (1% yeast extract, 2% peptone, 2% dextrose, 2% agar). For the killing assay, MBA medium (0.5% yeast extract, 0.5% peptone, 2% dextrose) was used, adjusted to appropriate pH 3.2–6.0 with the 75 mM phosphate-citrate buffer and supplemented with 0.002% methylene blue dye. For toxin preparation, liquid synthetic medium SC (2% dextrose,

0.2% K_2HPO_4, 0.1% $MgSO_4 \times 7H_2O$, 0.1% $(NH_4)_2SO_4$, 1.29% citric acid, 2.76% $Na_2HPO_4 \times 12H_2O$) was used, containing 5% glycerol, adjusted to appropriate pH 3.2–6.0.

2.2. Assay for Killing/Resistance Phenotypes

For detection of killing phenotype, the tested *S. paradoxus* strain was spotted on the MBA agar plates seeded with a lawn of the sensitive *S. cerevisiae* strain BY4741 (2×10^6 cells/plate) or strains of different yeast species. After incubation of the plates at 25 °C for 3 days, clear zones of growth inhibition surrounding the killer cells were evaluated and interpreted as a killer activity.

The sensitivity/resistance tests were performed by spotting *S. cerevisiae* killer strains onto the MBA plates with an overlay of the *S. paradoxus* strain AML-15-66. The absence of lysis zones indicates a resistant phenotype, while non-growth zones around the different types of killer toxins producing colonies were attributed to the sensitive phenotype [49].

2.3. Viral dsRNA Isolation From Yeast

Total extraction of nucleic acids from yeast was based on the previously described method [50] with modifications. *S. paradoxus* culture was grown in YEPD media overnight at 30 °C. Cells were collected by centrifugation for 5 min at $5000 \times g$ at 20 °C and washed with 1/10 part of the starting volume of the culture media supplemented by 50 mM EDTA. Cells were collected and re-suspended in 1/10 part of the starting volume of TB buffer (50 mM Tris-HCl pH 9.3; 1% β-mercaptoethanol) and incubated for 15 min at the room temperature. Cells were pelleted and re-suspended in 2/10 part of the starting volume of TES buffer (10 mM Tris-HCl pH 8.0; 100 mM NaCl; 10 mM EDTA; 0.2% (*w/v*) SDS). Subsequently, an equal volume of phenol (pH 5.2) preheated to 80 °C was added, and the suspension was vigorously shaken for 45 min at room temperature. Afterwards, an equal volume of chloroform was added and mixed thoroughly. The mix was subjected to centrifugation at $18,000 \times g$ for 45 min at 4 °C and separated by pipetting. Nucleic acids were pelleted from an aqueous fraction by adding 1 volume of isopropanol supplemented with 1/10 volume of 3 M sodium acetate (pH 5.2) and subjected to centrifugation at $18,000 \times g$ for 10 min at 4 °C. The resulting pellets were washed with 75% ethanol and dissolved in DEPC-treated water. Re-suspended nucleic acids were separated in 1% agarose gels and visualized by staining with ethidium bromide.

For double-stranded RNA (dsRNA) preparation, an isolated 700 µg of the total nucleic acid fraction was incubated in 2.8 M LiCl overnight at 4 °C [51]. The single-stranded nucleic acids were removed by centrifugation at $18,000 \times g$ for 45 min at 4 °C. The aqueous phase was substituted with 1/10 volume of 3 M NaCl and 1 volume of isopropanol, dsRNA pelleted by centrifugation at $18,000 \times g$ for 10 min at room temperature, washed with 75% ethanol and dissolved in DEPC treated water. The resulting L and M dsRNAs were visualized by gel electrophoresis and gel-purified using GeneJet Gel Extraction Kit (Thermo Fisher Scientific, Vilnius, Lithuania).

2.4. cDNA Synthesis, Amplification, and Cloning

Viral dsRNA cDNA synthesis and amplification were performed as described [52], with modifications. PC3-T7 loop primer (5′-GGATCCCGGGAATTCGGTAATACGACTCACTAT ATTTTTATAGTGAGTCGTATTA-3′) was ligated to gel-extracted dsRNA following the primer:dsRNA molar ratio of 250:1. The ligation reaction was performed by T4 RNA ligase (Thermo Fisher Scientific, Vilnius, Lithuania) in the supplier's buffer with additional adding of up to 20% PEG-6000, 10% DMSO, 0.01% BSA and 20U RiboLock RNase Inhibitor (Thermo Fisher Scientific, Vilnius, Lithuania) overnight at 37 °C. DsRNA with ligated primers was purified using a GeneJet PCR Purification Kit (Thermo Fisher Scientific, Vilnius, Lithuania), before cDNA synthesis step was denatured by adding of DMSO to a final concentration of 15% (*v/v*), heating at 95 °C for 2 min, and immediately transferred onto the ice for 5 min. Prepared dsRNA was reverse-transcribed using Maxima Reverse Transcriptase (Thermo Fisher Scientific, Vilnius, Lithuania). Alkaline hydrolysis of residual RNA was performed and cDNA strands re-annealed at 65 °C for at least 90 min followed by gradual cooling to 4 °C for 2 h

The cDNA was amplified using Phusion High-Fidelity DNA Polymerase (Thermo Fisher Scientific, Vilnius, Lithuania) with PC2 primer (5′-CCGAATTCCCGGGATCC-3′) by using the manufacturer's recommended cycling conditions, with an initial step carried at 72 °C for 2 min added. PCR products were cloned into the pUC19 vector (Thermo Fisher Scientific, Vilnius, Lithuania), sequenced at Base Clear (Leiden, ZH, The Netherlands) and the obtained sequences blasted against known sequences in the NCBI database.

2.5. Sequence Analysis

A maximum likelihood phylogenetic tree was constructed using the IQ-Tree v1.6.3 [53] with automatic selection of best-fit amino acid substitution and site heterogeneity models. LG + R3 proved to be the best-fit model. Edge support was estimated with bootstrap test (1000 replicates). Phylogenetic tree visualization was performed using Fig-Tree v1.4.3 program (http://tree.bio.ed.ac.uk/software/figtree/). Phobius server http://phobius.sbc.su.se/ [54] was employed for transmembrane topology identification. Sites of N-glycosylation in the protein sequences were identified using NetNGlyc web server (http://www.cbs.dtu.dk/services/NetNGlyc/). Conservative domains of the protein were determined using Pfam database [55]. DIANNA server (http://clavius.bc.edu/~clotelab/DiANNA) [56] was employed for disulfide bond connectivity prediction.

2.6. Analysis of the Partially Purified K66 Toxin for Thermal and pH Activity

S. paradoxus strain AML-15-66 was grown in synthetic SC-medium at various pH in the range of 3.2–6.0 for 4 to 6 days at 18 °C until reaching comparable cell density (OD600: 0.6–0.8). Yeast cells were separated by centrifugation at $3000 \times g$ for 10 min and filtration of supernatant through a 0.22 μm sterile PVDF membrane (Millipore, Bedford, MA, USA). The supernatant was then filtered using pressure-based Amicon system (membrane MWCO 10 kDa, Sigma-Aldrich, St. Louis, MO, USA) and serial centrifugations followed through Amicon ultra centrifugal filters with different cut-offs (10 and 30 kDa). The preparation of partially purified K66 toxin was used for assessment of optimal toxin activity and screening for modulators.

To determine K66 toxin activity at different pH values, MBA plates adjusted to pH values between 3.2 and 6 were seeded with the sensitive *S. cerevisiae* strain α′1 (2×10^6 cells/plate) and incubated at 25 °C in the presence of aliquots of the toxin (100 μL) extracted from killer protein-producing strain grown at appropriate pH and concentrated 100-fold. The inhibition zones were determined on triplicate plates after 2–5 days of incubation and the mean of remaining toxin activity was expressed in percent. Toxin preparation obtained at pH 4.8 was used for temperature activity measurement following the method described in [57].

2.7. Screening for Modulators of K66 Activity and Bioinformatic Analysis

The sensitivity was tested by either depositing 100 mL of concentrated 100-fold K66 toxin into 10 mm diameter "punched-wells" in the agar plate or spotting K66-producing cells onto the MBA medium overlaid with the yeast strain of interest (2×10^6 cells/plate). Plates were incubated for 2 days at 25 °C, and the diameter of the lysis zones measured and compared with those formed on BY4741 overlay. The screening was repeated 3 times.

The GO-term analysis was performed using the BiNGO 3.0.3 plug-in embedded into the Cytoscape 3.6.1. platform [58]. Significance *p* values were calculated with the hypergeometric test, using the Benjamini and Hochberg false discovery rate (FDR) correction for the enrichment of each GO term. Fold enrichment (F.E.) was determined by dividing the frequency of specific gene cluster to the total frequency for each GO term.

Network diagrams were generated using STRING web resource (version 10.5, http://string-db.org) [59]. Our created network uses the "confidence view" option of the program, where stronger associations are represented by thicker lines. The experiments-based active prediction method was used, and the medium confidence score (0.400) was utilized.

2.8. Evaluation of K66 Toxin Binding to the Yeast Mutants and Different Polysaccharides

S. cerevisiae BY4741 as control and mutant strains from *S. cerevisiae* deletion library were cultivated at 30 °C in YEPD medium overnight. 2×10^6 cells were sedimented by centrifugation at $3000 \times g$ for 10 min and washed with 1 mL SC medium, pH 4.8 and incubated with 500 µL of 100-fold concentrated K66 toxin at 4 °C for 1 h. The supernatant was collected by centrifugation 1 min $10,000 \times g$ and tested by the well-test. After 2 days incubation at 25 °C the diameter of lysis zones was measured.

For analysis of K66 binding to different polysaccharides, 9 mg of chitin, laminarin, pullulan, or pustulan were mixed with 100 µL of 100-fold concentrated K66 toxin. Samples were incubated for 1 h at 25 °C. After centrifugation (1 min $10,000 \times g$), 100 µL of supernatant was transferred into the wells on MBA medium (pH 4.8) containing α'1 cells. Lysis zones were analyzed after 2 days of incubation at 25 °C.

2.9. GenBank Accession Numbers

The SpV-L-A66 and SpV-M66 cDNA nucleotide sequences appear in NCBI/GenBank under GenBank accession no. MH784501 and MH784500, respectively.

3. Results

3.1. Characterization of the S. paradoxus Killer Strain

The yeast strain AML-15-66 was isolated from spontaneous fermentation of serviceberries. Based on the ITS region sequencing data and RFLP-PCR profiles, the strain was identified as *S. paradoxus* (Figure S1). We observed that AML-15-66 exhibits killing activity against *S. cerevisiae* non-killer and different types of killer virus-possessing strains at pH ranging from 3.6 to 5.6 (Table 1).

Table 1. Killing phenotype of *S. paradoxus* AML-15-66 strain. Diameter of zone of inhibition in mm: +++ (3–2.5), +++/- (2.5–2), ++ (2–1.5), ++/- (1.5–1), + (1–0.5), +/- (0.5–0).

Target Strain (Killer Type)	Killing Phenotype of *S. Paradoxus* AML-15-66							
	pH							
	3.2	3.6	4.0	4.4	4.8	5.2	5.6	6.0
S. cerevisiae α 1 (K0)	-	+/-	+	++	+++	++/-	-	-
S. cerevisiae BY4741 (K0)	-	+/-	+	++	+++/-	+	-	-
S. cerevisiae K7 (K1)	-	-	+/-	+	+	+/-	-	-
S. cerevisiae K7 [L-M-] (K0)	-	-	+/-	+	++/-	+	-	-
S. cerevisiae M437 (K2)	-	-	+/-	+/-	+/-	+/-	-	-
S. cerevisiae M437 [L-M-] (K0)	-	-	-	+/-	+	+/-	-	-
S. cerevisiae CRB-15-4 (Klus)	-	-	-	+/-	+	+	+/-	-
S. cerevisiae CRB-15-4 [L-M-] (K0)	-	-	-	+/-	+	+	+/-	-
S. cerevisiae MS300 (K28)	-	-	+	+	++	+	-	-
S. cerevisiae MS300 [L-M-] (K0)	-	-	+	++/-	++	++/-	+/-	-
S. paradoxus AML-15-66 (K66)	-	-	-	-	-	-	-	-
S. paradoxus AML-15-66 [L-M-] (K0)	-	-	+/-	+	++/-	+/-	-	-
S. paradoxus T.21.4 (K21)	-	-	-	-	-	-	-	-

The strongest killing activity was determined at pH 4.4–4.8 against non-killer *S. cerevisiae* strains α'1 and BY4741. The killing phenotype against various types of killer toxin-producing strains was weaker, compared to that of the killer-free strains. *S. paradoxus* AML-15-66 strain demonstrated the lowest activity against *S. cerevisiae* M437 cells maintaining ScV-M2 virus (Table 1) and was not active against tested *Pichia*, *Hanseniaspora*, *Candida*, and *Torulaspora* spp. *S. paradoxus* AML-15-66 strain was found to be resistant to the action of *S. cerevisiae* K1, K28, and Klus mycotoxins as well as to the K21 toxin, produced by *S. paradoxus* T21.4 strain, while susceptible to the action of *S. cerevisiae* K2 toxin produced by strain M437 only (Figure S2A).

3.2. Double-Stranded RNA Viruses from the S. Paradoxus Strain

To delve into the nature of the killing phenotype, we extracted dsRNAs from AML-15-66 and performed electrophoretic analysis (Figure S3). The size of observed L and M dsRNAs was compared to that of the dsRNAs isolated from reference strains of *S. cerevisiae* K7 (LA-1, M1), M437 (LA-lus, M2), MS300 (LA-28, M28), and SRB-15-4 (LA-lus, Mlus). The size of the L fraction was about 4.6 kb and thus highly similar to all L dsRNAs, while M dsRNA was about 1.6 kb and thus close to M2 dsRNA. We named these viruses SpV-LA-66 and SpV-M66, respectively. Purified dsRNA of the SpV-LA-66 virus was used as a substrate for primer ligation, subsequent reverse transcription, and cDNA amplification. In total, the genome of the SpV-LA-66 virus was found to possess 4580 nucleotides. Like other known L-A viruses, SpV-LA-66 genome features two overlapping open reading frames, which encode capsid protein Gag and RNA dependent RNA polymerase Gag-pol, formed by ribosomal frameshift. All features inherent for L-A viruses, such as conservative frameshift region, packing and replication signals, and catalytic histidine residue required for cap-snatching were present in the SpV-LA-66 genome sequence. Tentative ORFs coding for the Gag-pol and Gag proteins were compared with corresponding fragments of *S. cerevisiae* and *S. paradoxus* dsRNA sequences. At the nucleotide level, all entries display 74 to 92% identity (Figure 1).

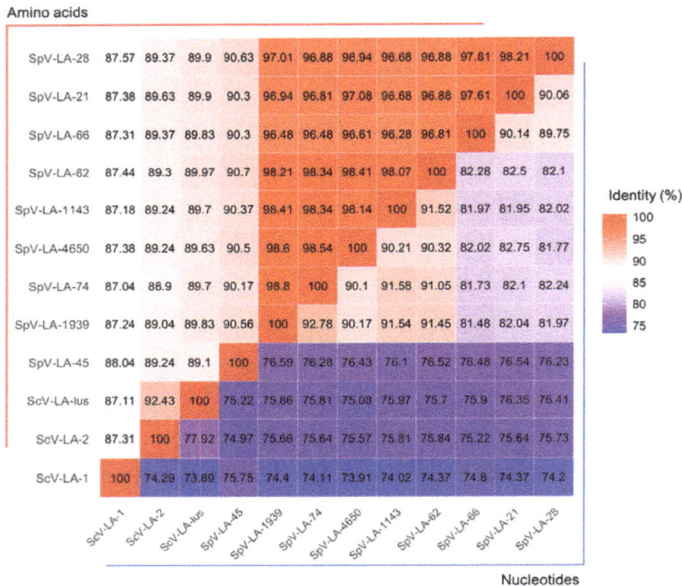

Figure 1. The similarities of *S. cerevisiae* and *S. paradoxus* dsRNA L-A virus-encoded Gag-pol proteins. ORFs coding for the Gag-pol proteins were compared with corresponding fragments of *S. cerevisiae* and *S. paradoxus* dsRNA sequences, namely: GenBank entry J04692 for the ScV-LA-1 virus, KC677754 for ScV-LA-2, and JN819511 for ScV-LA-lus, KU845301 for SpV-LA-28 (formerly attributed to *S. cerevisiae*), KY489962 for SpV-LA-21, KY489963 for SpV-LA-45, KY489964 for SpV-LA-74, KY489965 for SpV-LA-4650, KY489966 for SpV-LA-1939, KY489967 for SpV-LA-1143, KY489968 for SpV-LA-62. Identity at nucleotide level is represented in the lower right triangle, amino acid level–in the upper left triangle, framed by corresponding blue and red lines.

Similarity at the amino acid level is higher: coat proteins (Gag) are 88 to 99% homologous (Figure S4), while RNA polymerases (Gag-pol) are 87 to 98% homologous (Figure 1). Proteins originating from SpV-LA-66, SpV-LA-28 (initially reported as ScV-LA-28, origin recently updated by [6]) and SpV-LA-21 are the most closely related and comprise a separate cluster in relation to other L-A viruses (Figure 2). The remaining L-A viruses from *S. paradoxus* comprise another cluster. Altogether, *S. paradoxus* L-A viruses are significantly more homogenous than those from *S. cerevisiae*.

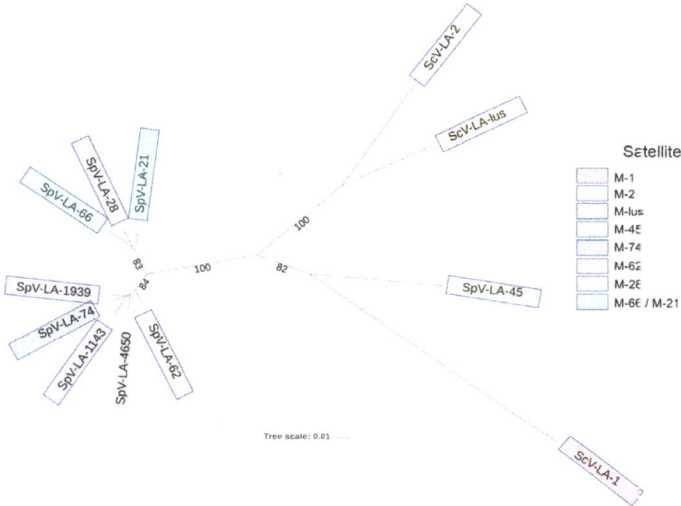

Figure 2. The phylogenetic tree of dsRNA-encoded Gag-pol proteins from *S. cerevisiae* and *S. paradoxus* yeasts.

In the same AML-15-66 strain, we discovered and cloned a 1553 bp long M dsRNA, named as SpV-M66. Sequence analysis of M66 satellite shows 86% identity in nucleotides to *S. paradoxus* SpV-M21 virus (GenBank MF358732) (Figure S5). The SpV-M66 genome consists of 5′-end 4 bp non-translating region, single ORF of 1,038 bp, which has 92% aa identity to K21 killer preprotoxin (GenBank ATN38270), about 60 bp polyA region and 330 bp non-translating region located at 3′-end (Figure S6). We expressed the complete ORF sequence from SpV-M66 in *S. cerevisiae* BY4741 strain and confirmed that the encoded protein confers the host with the killer activity (Figure S2B). Sequence analysis of preprotoxin revealed three potential recognition sites for Kex2 protease ending at amino acids Arg59, Arg181, and Arg239 (Figure 3, Figure S5); prediction of disulfide bridge formation sites remains ambiguous. Up to five putative protein N-glycosylation sites were found in the sequence of the K66 protein, four of them overlapping with those for K21. The amino acid at 141 position has the highest probability to be modified. A hydrophobicity profile reveals three transmembrane domains (from 21 to 42 aa, 62 to 85 aa, and 97 to 119 aa) in the K66 protein (Figure 3). In the C-proximal part of K66 precursor, a conservative Pfam family domain DUF5341 of presumably unknown function has been identified.

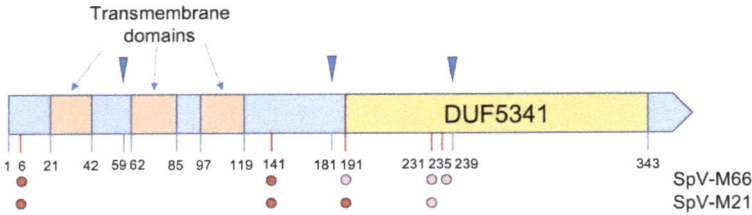

Figure 3. Features of the protein coded by SpV-M66. Blue triangles above the picture mark Kex2 sites. Predicted transmembrane domains and DUF5341 conservative Pfam family domain are marked within the picture. Dots below the picture mark putative glycosylation sites for K66 and K21 proteins, color intensity corresponds to the value of reliability index of the given position. Feature-linked positions of amino acids are indicated.

3.3. Effect of pH and Temperature on the Action of the S. Paradoxus K66 Toxin

The activity of partially purified *S. paradoxus* K66 viral protein was assayed on a lawn of *S. cerevisiae* strain α'1 with adjusted pH value from 3.2 to 6. K66 toxin exhibits killing activity in a narrow pH range between 3.6 and 5.2, with an activity peak at pH 4.4–4.8 (Figure 4A).

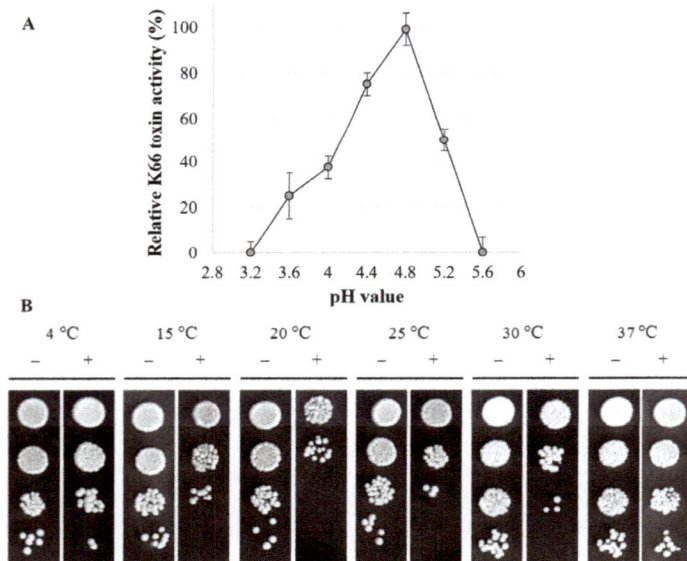

Figure 4. Impact of pH and temperature on K66 toxin functionality. (**A**) Sensitive *S. cerevisiae* strain a'1 was seeded into MBA medium (2×10^6 cells/plate, adjusted to pH values between 3.2 and 6) and 100 μL of the concentrated K66 toxin poured into 10 mm wide wells, cut in the agar layer. Plates were incubated for 2 days at 25 °C, non-growth zones around the wells measured and expressed as mean of three independent experiments in percent ± SD. (**B**) Sensitive yeast a'1 cells (5×10^5 cells) were mixed with 500 μL of 100-fold concentrated K66 toxin or the same volume of heat-inactivated K66 toxin and incubated for 24 h at different temperatures. Yeast cells were then serially diluted and spotted onto YPD-agar plates following for 2 days incubation at 25 °C.

More acidic pH values of 4.0–3.6 result in a reduction of toxin activity up to 38% and 25% respectively, and more basic than optimal pH 5.2 results in about 50% of toxin activity remained (Figure 4A). By analyzing yeast cells that survived the treatment by K66 protein at different temperatures (from 4 °C to 37 °C), we found that extracted viral protein is active at temperatures between 15 °C and 30 °C, with optimal temperature of 20 °C (Figure 4B).

3.4. Genetic Factors Modulating the Functionality of the Viral K66 Toxin

To determine the genetic factors important for the action of *S. paradoxus* K66 toxin and involved in the formation of cellular resistance to the viral agent, we screened 526 *S. cerevisiae* single-gene deletion mutants, previously demonstrated to alter the functioning of *S. cerevisiae* K1, K2, or K28 toxins [29–31].

We identified 125 *S. cerevisiae* YKO library mutants demonstrating different degrees of phenotypic response to the K66 toxin (Table S1), of which 73 were more resistant than the control strain BY4741 and 52 more sensitive to the toxin treatment. We manually annotated groups of all identified mutants resistant or susceptible to K66 toxin. The largest groups contain genes associated with cell wall organization and biogenesis (15), membrane formation/secretion/transport (16), chromatin organization/gene expression (18), and translation (13) (Table S1, Figure 5). Deletions of 19 genes were common in all four screens performed and cause resistance/sensitivity alterations in the cells, not depending on the toxin type. In 13 mutants, different responses to K66 and some *S. cerevisiae* toxins were recorded (Table S1, Figure S2C). 85% of modulators identified in *S. paradoxus* K66 screen (106 gene products) were also identified in screens of *S. cerevisiae* K2 or K1 toxins.

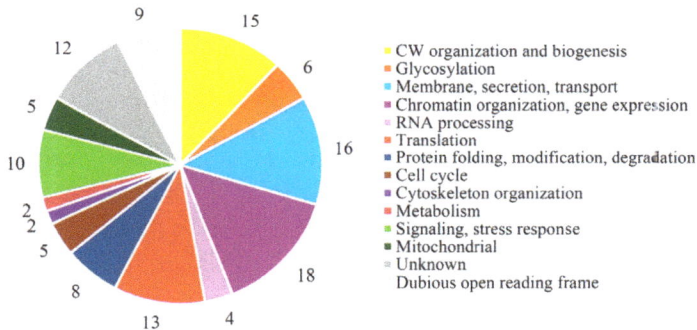

Figure 5. Distribution of cellular processes and cellular components involved in the action of K66 toxin. The number of genes identified in each class is indicated.

The GO-term analysis ("biological process") reveals a statistically significant enrichment in genes involved in cell wall organization and biogenesis (F.E. (fold enrichment) of 4.1, $p < 5.4 \times 10^{-6}$), response to osmotic stress (F.E. of 5.6, $p < 2.7 \times 10^{-3}$), and signaling pathway (F.E. of 3.3, $p < 1.3 \times 10^{-3}$) (Table S2).

Based on published high-throughput datasets, we built a protein-interconnection network and documented that the majority of all identified genetic factors were a part of one main functional cluster, at medium confidence level (0.4) (Figure 6). Most of the observed proteins are involved in stress response and signaling processes, cell wall organization and biogenesis, belong to ribosomal components or translation machinery, and are connected to membranes or protein transport.

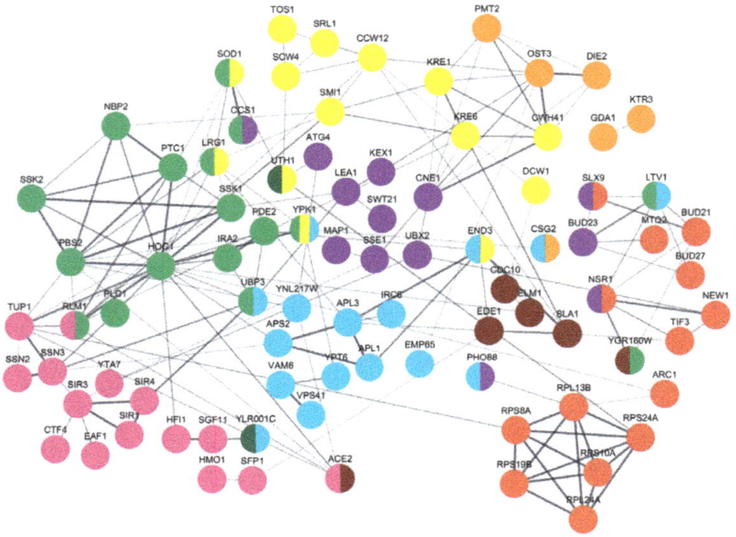

Figure 6. Interconnections of gene products involved in the modulation of susceptibility to K66 toxin. An integrated functional interaction network is obtained from STRING database. Subnetworks of proteins associated with ribosomes/translation (red), signaling and stress response (green), chromatin organization and gene expression (purple), glycosylation (orange), CW organization/biogenesis (yellow), cell cycle (brown), membrane and transport (light blue), RNA and protein modification (dark blue), and mitochondrial (dark green) are represented.

3.5. Targeting of the Viral K66 Killer Protein to the Cell Wall

To correlate the K66 toxin binding with the cell wall composition, we investigated the binding properties of the viral protein to mutant cells with altered levels of β-1,3 and β-1,6 glucans [21,29]. The level of glucan content was based on that reported in [21,29]. During this study, we determined that mutants with decreased levels of β-1,6 glucans Δ*aim26*, Δ*smi1*, and Δ*kre1* bind from 35% to 42% less of the K66 toxin molecules than control cells (BY4741) (Figure 7A). The killing activity of K66 is independent of β-1,3 glucan concentration in the cell wall (β-1,3 glucan amount in Δ*aim26* is as in *wt*, Δ*smi1*—50% decreased and in Δ*kre1*—10% increased). When mutant cells have increased amount of β-1,6-linkages at the cell wall, as in Δ*bud27*, Δ*map1*, and Δ*end3* mutants, K66 binding is boosted at about 30% over the control cells level. The defects in cell wall structure resulted in a major impact on the efficiency of K66 toxin binding to the cells, as in the case of K2 and K1 killer toxins [21,29,31,60]. We observed a good correlation between the genes whose deletion led to decreased K66 toxin binding and increased resistance to the toxin. Similarly, genes whose deletion led to increased toxin binding correlated with increased sensitivity towards the K66 toxin (Figure 7B).

To confirm the in vivo data on the importance of β-1,6 glucans as a binding target of K66, the ability of toxin to directly complex the polysaccharides bearing different glucan linkages was evaluated. After incubation with either laminarin (consisting of β-1,3 and β-1,6 linkages), pustulan (β-1,6 linkage), pullulan (α-1,4 and α-1,6 linkages), or chitin (β-1,4 linkage), residual toxin activity was tested by well test on the sensitive *S. cerevisiae* strain α'1 (Figure 7C). Unbound toxin forms clear lysis zones around the well. The competitive inhibition of the action of viral protein demonstrated that the β-1,6-glucan exclusively present in pustulan provides binding sites for the K66 toxin. K66 toxin activity was completely abolished by pustulan only.

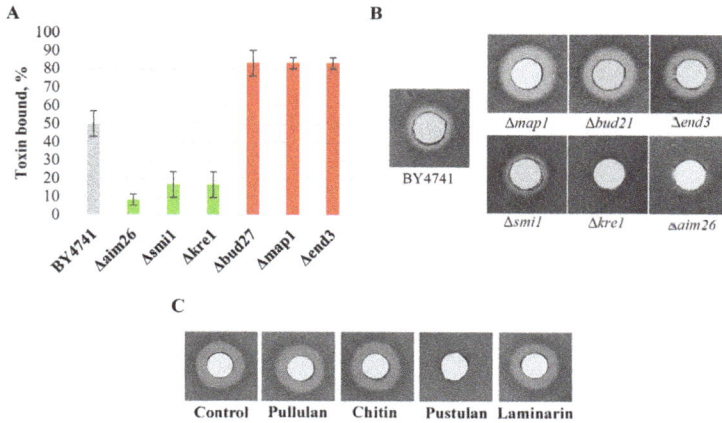

Figure 7. K66 toxin binding to yeast mutants with altered levels of β-glucans and different polysaccharides. (**A**) Cells of different yeast mutants (2×10^6 each) were incubated with 500 μL of concentrated K66 toxin, the remaining toxin activity was measured by the well assay. After incubation with the indicated yeast strain, the unbound K66 toxin is able to kill sensitive tester strain α′1, seeded in the MBA plates. The size of the formed lysis zones was converted to the relative toxin activity and subtracted from the total activity to calculate the binding efficiency. The data are averages ± standard deviations (SD) (n = 3). (**B**) Responses of deletion strains to the action of K66 toxin were measured in the well assay using BY4741 strain in a lawn. (**C**) Nine milligrams of each polysaccharide (chitin, laminarin, pullulan, or pustulan) in 100 μL of concentrated K66 toxin preparation was incubated for 1 h at 25 °C and residual toxin activity was analyzed in the well assay using sensitive α′1 strain in the lawn.

None of the other polysaccharides tested exhibited K66 protein binding effects, forming lysis zones equal to that of control sample without polysaccharide added. The in vitro and in vivo approaches used here to access the binding specificity suggest that type β-1,6-glucans can play the role of primary cell surface receptor for the K66 toxin.

4. Discussion

Our work for the first time provides deep insight into the composition and functioning of the *S. paradoxus* viral killer system. The study stemmed from the comprehensive analysis of the viral sequences, analyzed the genetic factors of the host and targets of the virus-encoded killer toxin.

The high similarity between *S. paradoxus* L-A viruses has been reported recently [6]. This observation is in line with the high relatedness observed previously between *S. paradoxus* dsRNA virus-possessing strains based on the distance matrix of PCR profiles with the microsatellite primer $(GTG)_5$ [61]. Here, we demonstrate that L-A viruses from *S. paradoxus* discovered so far are significantly more homogenous than those from *S. cerevisiae*. Within *S. paradoxus* L-A viruses, two clades can be confidently separated: LA-28 type, including SpV-LA-66, SpV-LA-21 and SpV-LA-28, and the rest of *S. paradoxus* L-A viruses, except for the SpV-LA-45.

No sequence homology between different *Saccharomyces sensu stricto* yeast dsRNA M viruses is detected, except for the high similarity of *S. paradoxus* SpV-M66 and SpV-M21 viruses. Even though SpV-LA-66 and SpV-LA-28 are highly related, no homology between SpV-M66 and SpV-M28 in nucleotide or in amino acid level can be documented. Genome organization of SpV-M66 resembles all so far described *S. cerevisiae* and *S. paradoxus* dsRNA M viruses [19,28,62]. The coding region is located at the 5′ terminus, followed by an A-rich sequence and non-coding region with secondary stem-loop structure important for encapsidation at the 3′ terminus. The ORF encoding for a K66 preprotoxin features three potential recognition sites of Kex2 protease, acting in the late Golgi compartment.

Thus, the maturation process of K66 toxin may proceed by two alternative scenarios: first, by removing pre-pro-sequence and potential γ-peptide (from 182 till 239 aa) and forming a disulfide-bonded α/β heterodimer. A similar structural organization is typical for almost all known *S. cerevisiae* killer toxins, except for the K2 killer protein lacking γ-subunit [15,62]. In the second scenario, the Kex2 protease may not cleave at the position 239 and, similar to the K2 killer protein, γ-peptide is not released. In this case, the β subunit will start from the amino acid position 182, retaining the integral DUF5341 domain. Future studies are needed to fully understand the role of DUF5341 in the killing phenotype of K66, as this domain is found in numerous proteins of *Ascomycota*, including the N-terminal part of KHS killer toxin. Of special interest is the presence of the DUF5341 domain identified within C-terminal part of the K2 toxin [63]. However, the patterns of DUF5341 sequence similarity to K66 and K2 do not match, making direct sequence comparison not possible. Therefore, K66 and K2 toxins appear to share the same DUF5341 domain core. Three transmembrane helixes detected in the K66 predict this protein to form an ion channel after reaching the plasma membrane, similar to *S. cerevisiae* K2 and K1 toxins [15,25]. The possibility of the K66 toxin acting in monomeric form could not be excluded, as the probability level of disulfide bond prediction is rather low. ER retention motif, typical for K28 protein and essential for its activity [37] was not found in the structure of K66, separating the organization and therefore the modes of action of these toxins further on.

BLAST search revealed K66 homologues in different yeast strains. We found that all ORFs coding homologues of K66 in *S. cerevisiae* are at the telomeric region of chromosome 5 and code for YER187W-like proteins. In some strains, YER187W ORF contains an in-frame stop codon (for example, in *S. cerevisiae* BY4741). YER187W ORF neighboring YER188W is homologous to Kbarr-1 killer toxin (GenBank KT429819), encoded by *Torulaspora delbrueckii* dsRNA Mbarr-1 killer virus. An identity of some regions of *T. delbrueckii* Mbarr-1 genome with the putative replication and packaging signals of most of the M-virus RNAs was observed, suggesting the evolutionary relationship [7]. Several chromosomal ORFs with homology to *S. cerevisiae* Klus, K1, or K2 preprotoxins have been observed in yeasts before, suggesting that the M virus might have originated from the host messenger RNAs, been encapsidated and replicated by the L-A virus-encoded RNA polymerase after acquiring sequences needed for both events, probably from the genome of the L-A virus itself [14]. In addition, it has been demonstrated that genes of dsRNA viruses from *Totiviridae* and *Partitiviridae* have widespread homologues in the nuclear genomes of eukaryotic organisms, such as plants, arthropods, fungi, nematodes, and protozoa, suggesting that viral genes might have been transferred horizontally from viral to eukaryotic genomes [8,64].

Despite the striking similarity in virus genome organization and toxin maturation, the modes of action of the viral toxins are clearly distinct [15]. There are differences between toxins with respect to killing, interaction with the host cells, and immunity formation mechanisms. Mycovirus-originated killer proteins are usually active between pH 4.0 and 5.4 and at a temperature below 30 °C [1,65]. The favorable conditions for the establishment of *Saccharomyces* killer species have been found on fruits where the pH is moderately low [49,66,67], growing best in a natural environment with optimal temperature range of 20 °C to 30 °C [57]. These conditions strongly correlate with activity and stability profile of most secreted killer toxins and are in line with our data pointing to the optimal activity of *S. paradoxus* K66 toxin at a temperature of 20 °C at pH about 4.8. The *S. cerevisiae* killer toxins have narrow target range, inhibiting only strains or species within the same genus [68], except for the Klus toxin, which executes the activity against broader spectrum of yeast [28]. *S. paradoxus* killer strain AML-15-66 exhibits similar features towards the majority of *S. cerevisiae* yeast, demonstrating killing activity against *S. cerevisiae* cells but being not active against other yeast genera tested. The AML-15-66 strain possessed the lowest activity against *S. cerevisiae* K2 toxin-producing strain M437, suggesting relation of both toxins, while K28-bearing *S. cerevisiae* strain M300 was killed the most efficiently among all killer strains tested. At the same time, AML-15-66 was sensitive to the action of *S. cerevisiae* K2 toxin, probably due to the high activity of this toxin [69,70], and completely resistant to other known killer types of *S. cerevisiae* K1, K28, and Klus.

The genetic screen performed on *S. cerevisiae* single-gene deletion strains revealed 125 gene products important for the functioning of the *S. paradoxus* K66 toxin and involved in the target cell susceptibility. Previous yeast genome-wide screens performed with all known dsRNA virus-originated *S. cerevisiae* killer toxins revealed a 753 gene set, contributing to the functioning of viral agents: 268 for the K1, 332 for the K2, and 365 for the K28 [29–31]. Page and colleagues [29] determined that the resistance to the *S. cerevisiae* K1 toxin is mostly conferred by gene products linked to the synthesis of cell wall components, secretion pathway, and cell surface signal transduction. Mutant cells with an increased amount of β-1,6-glucans in the cell wall, those unable to grow at high osmolarity, and bearing compromised stress response pathways, exhibited increased sensitivity to the K1 toxin [29]. The K2 toxin executed a similar mechanism of action as K1; therefore, changes in the cell wall structure and proper functioning of mitochondria remain crucial for the killing phenotype. The cells defective in HOG and CWI signaling pathways, with affected maintenance of pH and ion homeostasis, demonstrated hypersensitivity to the K2 toxin [31]. Even for the action of the different-by-mechanism K28 toxin, interrupted cell wall biogenesis process and lipid organization led to increased resistance, while most genes related to hypersensitivity to K28 toxin are those involved in stress-activated signaling and protein degradation [30]. In this study, based on manual annotation, the largest groups of K66 modulators have been identified as those connected to cell wall organization and biogenesis, membrane formation and transport, chromatin organization, and gene expression. GO analysis and protein interconnection networks highlighted the importance of cell wall structure for the functioning of *S. paradoxus* K66 toxin and allowed us to speculate that K66 acts via the disruption of ion homeostasis, since genes involved in regulation of osmotic stress response were highly represented in the screen. Importantly, 85% of modulators identified in the *S. paradoxus* K66 screen (106 gene products) were involved in the functioning of *S. cerevisiae* K1 or K2 toxins as well, pointing to a high similarity of their action. At the same time, only 19 unique modulators common for both *S. paradoxus* K66 and *S. cerevisiae* K28 were found. Functions of several of those gene products are connected to endocytosis and, therefore, remain unclear in the context of *S. paradoxus* K66 toxin predicted function. Until the relevant experiments are carried out, the possibility that the K66 toxin possesses yet another unique mode of the action cannot be excluded.

Yeast cell wall is a primary target for cytotoxic activity of most mycotoxins, and different components of the cell wall could play the role of receptors [71]. Mutants with altered cell wall biogenesis process demonstrated the most resistant phenotype to the action of *S. paradoxus* viral protein K66. This is expected, because the toxin primarily interacts with the cell wall and components of the plasma membrane. By investigating the efficiency of the K66 toxin binding to *S. cerevisiae* mutant cells with altered levels of β-1,3 and β-1,6 glucans in vivo and performing competition experiments with the polysaccharides, bearing different glucan linkages, we observed a good correlation between the level of β-1,6 glucans and binding ability. This highlights the potential of β-1,6 glucans to act as primary targets on the cell wall. The similar genetic factors related to cell wall organization and biogenesis processes modulate the functioning of *S. cerevisiae* K1 and K2 toxins [29,31]. Even more, β-1,6-glucan was originally proposed to be a cell wall receptor for *S. cerevisiae* K1 and K2 toxins [21,22]. Therefore, we propose that β-1,6 glucans could play the role of primary receptors of *S. paradoxus* K66 toxin, further relating its action to that of *S. cerevisiae* K1 and K2 toxins.

Our finding that the majority of L-A viruses from *S. paradoxus* are significantly more homogenous than those from *S. cerevisiae*, might substantiate an important evolutionary crossroad between wild *S. paradoxus* and domesticated *S. cerevisiae* yeast. It remains to be discovered the reason or driving force for the extreme level of homology of L-A viruses from *S. paradoxus*, if any: genomes of *S. paradoxus* were shown to be at least several times more diverse than *S. cerevisiae* genomes [72]. One can envision the evolutionary pressure from the satellite M dsRNA virus; however, many of *S. paradoxus* killer systems are rather diverse in terms of toxin specificity [9,19]. At the same time, *S. paradoxus* SpV-LA-45 is more similar to *S. cerevisiae* counterparts, than to L-As from *S. paradoxus*, extending further limits of L-A virus variability in host species. *S. cerevisiae* features at least four killer systems, maintained by corresponding L-A variants [15,37], therefore highlighting a route for possible diversification of L-A viruses within species. Here, fitness pressure determined by the domestication of *S. cerevisiae* should not be overlooked.

Ample specific combinations of *S. paradoxus* killer systems cast doubts on homogeneity within killer-helper duos of *S. paradoxus*. The specificity paradigm in maintenance of M satellite virus by corresponding L-A virus only has been challenged by the *S. cerevisiae* K2 killer system, where two distinct L-A type viruses were found to support the ScV-M2 virus in wild strains [14,28]. This paradigm is further refused by the *S. paradoxus* K66 killer system, consisting of the previously unreported combination of LA-28 type SpV-LA-66 virus and M1/M2 type SpV-M66 virus. To the best of our knowledge, this is the first report of LA-28 type helper virus maintaining other than the M28 type satellite virus in a wild-type strain. The presented uniqueness of the K66 killer system raises questions of the true limits and the factors behind helper-satellite compatibility and distribution. Genetic background is essential for the consolidating of any virus in a cell; the diversity of yeast killer systems should obey the cellular context, unique for each and every host species. Results from this study extend our knowledge on the *Totiviridae* viruses in *Saccharomyces sensu stricto* yeasts and functioning of killer systems, urging exploration of new horizons of their diversity.

Supplementary Materials: The following are available online at http://www.mdpi.com/1999-4915/10/10/564/s1. Figure S1: Identification of *S. paradoxus* AML-15-66 strain, Figure S2: Characterization of killing and immunity features of K66 toxin-producing *S. paradoxus* strain, Figure S3: Electrophoretic analysis of dsRNAs extracted from different killer yeast, Figure S4: The similarities of *S. cerevisiae* and *S. paradoxus* dsRNA L-A virus-encoded Gag proteins, Figure S5: Sequence alignment of SpV-M66 and SpV-M21 genomes, Figure S6: Nucleotide sequence of the SpV-M66 genome and amino acid sequence of the putative ORF of K66 toxin, Table S1: Mutant strains with an altered *S. paradoxus* K66 killer toxin phenotype, Table S2: GO terms in biological process for genes involved in the functioning of *S. paradoxus* K66 toxin.

Author Contributions: E.S. and S.S. conceived and designed the experiments; I.V.-M., J.L., A.K., R.S., D.E. and Ž.S.-Ž. performed the experiments; I.V.-M., J.L., A.K. and E.S. analyzed the data; E.S. contributed to analysis tools; J.L., S.S. and E.S. wrote the paper.

Funding: This research was funded by a grant from the Lithuanian Research Council (No. SIT-7/2015).

Acknowledgments: We thank Antanas Žilakauskis and Vytautas Balnionis for technical assistance. Authors would like to thank Vyacheslav Yurchenko for critical reading of the manuscript and Jonathan Robert Stratford for the English language review.

Conflicts of Interest: The authors declare no conflict of interest. The founding sponsors had no role in the design of the study; in the collection, analyses, or interpretation of data; in the writing of the manuscript, and in the decision to publish the results.

References

1. Muccilli, S.; Restuccia, C. Bioprotective role of yeasts. *Microorganisms* **2015**, *3*, 588–611. [CrossRef] [PubMed]
2. Yap, N.A.; De Barros Lopes, M.; Langridge, P.; Henschke, P.A. The incidence of killer activity of non-Saccharomyces yeasts towards indigenous yeast species of grape must: Potential application in wine fermentation. *J. Appl. Microbiol.* **2000**, *89*, 381–389. [CrossRef] [PubMed]
3. Kast, A.; Voges, R.; Schroth, M.; Schaffrath, R.; Klassen, R.; Meinhardt, F. Autoselection of cytoplasmic yeast virus like elements encoding toxin/antitoxin systems involves a nuclear barrier for immunity gene expression. *PLoS Genet.* **2015**, *11*, e1005005. [CrossRef] [PubMed]
4. Drinnenberg, I.A.; Fink, G.R.; Bartel, D.P. Compatibility with killer explains the rise of RNAi-deficient fungi. *Science* **2011**, *333*, 1592. [CrossRef] [PubMed]
5. Golubev, V.I. Wine yeast races maintained in the All-Russia Collection of Microorganisms (VKM IBPM RAS). *Prikl. Biokhim. Mikrobiol.* **2005**, *41*, 592–595. [PubMed]
6. Rodríguez-Cousiño, N.; Gómez, P.; Esteban, R. Variation and distribution of L-A helper totiviruses in *Saccharomyces sensu stricto* yeasts producing different killer toxins. *Toxins (Basel)* **2017**, *9*, 313. [CrossRef] [PubMed]

7. Ramírez, M.; Velázquez, R.; López-Piñeiro, A.; Naranjo, B.; Roig, F.; Llorens, C. New insights into the genome organization of yeast killer viruses based on "atypical" killer strains characterized by high-throughput sequencing. *Toxins (Basel)* **2017**, *9*, 292. [CrossRef] [PubMed]

8. Wickner, R.B. Prions and RNA viruses of *Saccharomyces cerevisiae*. *Annu. Rev. Genet.* **1996**, *30*, 109–139. [CrossRef] [PubMed]

9. Chang, S.L.; Leu, J.Y.; Chang, T.H. A population study of killer viruses reveals different evolutionary histories of two closely related *Saccharomyces sensu stricto* yeasts. *Mol. Ecol.* **2015**, *24*, 4312–4322. [CrossRef] [PubMed]

10. Ghabrial, S.A.; Caston, J.R.; Jiang, D.; Nibert, M.L.; Suzuki, N. 50-plus years of fungal viruses. *Virology* **2015**, *479–480*, 356–368. [CrossRef] [PubMed]

11. Wickner, R.B.; Fujimura, T.; Esteban, R. Viruses and prions of *Saccharomyces cerevisiae*. *Adv. Virus Res.* **2013**, *86*, 1–36. [CrossRef] [PubMed]

12. Wickner, R.B. Double-stranded RNA viruses of *Saccharomyces cerevisiae*. *Microbiol. Rev.* **1996**, *60*, 250–265. [CrossRef] [PubMed]

13. Schmitt, M.J.; Tipper, D.J. K28, a unique double-stranded RNA killer virus of *Saccharomyces cerevisiae*. *Mol. Cell. Biol.* **1990**, *10*, 4807–4815. [CrossRef] [PubMed]

14. Rodríguez-Cousiño, N.; Gómez, P.; Esteban, R. L-A-lus, a new variant of the L-A totivirus found in wine yeasts with klus killer toxin-encoding mlus double-stranded RNA: Possible role of killer toxin-encoding satellite RNAs in the evolution of their helper viruses. *Appl. Environ. Microbiol.* **2013**, *79*, 4661–4674. [CrossRef] [PubMed]

15. Schaffrath, R.; Meinhardt, F.; Klassen, R. Yeast killer toxins: Fundamentals and applications. In *Physiology and Genetics*, 2nd ed.; Anke, T., Schüffler, A., Eds.; Springer: Cham, Switzerland, 2018; pp. 87–118, ISBN 978-3-319-71739-5.

16. Icho, T.; Wickner, R.B. The double-stranded RNA genome of yeast virus L-A encodes its own putative RNA polymerase by fusing two open reading frames. *J. Biol. Chem.* **1989**, *264*, 6716–6723. [PubMed]

17. Dinman, J.D.; Icho, T.; Wickner, R.B. A-1 ribosomal frameshift in a double-stranded RNA virus of yeast forms a fag-pol fusion protein. *Proc. Natl. Acad. Sci. USA* **1991**, *88*, 174–178. [CrossRef] [PubMed]

18. Konovalovas, A.; Serviené, E.; Serva, S. Genome sequence of *Saccharomyces cerevisiae* double-stranded RNA virus L-A-28. *Genome Announc.* **2016**, *4*, e00549-16. [CrossRef] [PubMed]

19. Rodríguez-Cousiño, N.; Esteban, R. Relationships and evolution of double stranded RNA totiviruses of yeasts inferred from analysis of L-A-2 and L-BC variants in wine yeast strain populations. *Appl. Environ. Microbiol.* **2017**, *83*, 1–18. [CrossRef] [PubMed]

20. Wickner, R.B. Killer systems in *Saccharomyces cerevisiae*: Three distinct modes of exclusion of M2 double-stranded RNA by three species of double-stranded RNA, M1, L-A-E, and L-A-HN. *Mol. Cell. Biol.* **1983**, *3*, 654–661. [CrossRef] [PubMed]

21. Luksa, J.; Podoliankaite, M.; Vepstaite, I.; Strazdaite-Zieliene, Z.; Urbonavicius, J.; Serviene, E. Yeast beta-1,6-glucan is a primary target for the *Saccharomyces cerevisiae* K2 toxin. *Eukaryot. Cell* **2015**, *14*, 406–414. [CrossRef] [PubMed]

22. Hutchins, K.; Bussey, H. Cell wall receptor for yeast killer toxin: Involvement of (1 leads to 6)-beta-D-glucan. *J. Bacteriol.* **1983**, *154*, 161–169. [PubMed]

23. Breinig, F.; Tipper, D.J.; Schmitt, M.J. Kre1p, the plasma membrane receptor for the yeast K1 viral toxin. *Cell* **2002**, *108*, 395–405. [CrossRef]

24. Orentaite, I.; Poranen, M.M.; Oksanen, H.M.; Daugelavicius, R.; Bamford, D.H. K2 killer toxin-induced physiological changes in the yeast *Saccharomyces cerevisiae*. *FEMS Yeast Res.* **2016**, *16*. [CrossRef] [PubMed]

25. Martinac, B.; Zhu, H.; Kubalski, A.; Zhou, X.L.; Culbertson, M.; Bussey, H.; Kung, C. Yeast K1 killer toxin forms ion channels in sensitive yeast spheroplasts and in artificial liposomes. *Proc. Natl. Acad. Sci. USA* **1990**, *87*, 6228–6232. [CrossRef] [PubMed]

26. Schmitt, M.J.; Klavehn, P.; Wang, J.; Schonig, I.; Tipper, D.J. Cell cycle studies on the mode of action of yeast K28 killer toxin. *Microbiology* **1996**, *142*, 2655–2662. [CrossRef] [PubMed]

27. Reiter, J.; Herker, E.; Madeo, F.; Schmitt, M.J. Viral killer toxins induce caspase-mediated apoptosis in yeast. *J. Cell Biol.* **2005**, *168*, 353–358. [CrossRef] [PubMed]

28. Rodríguez-Cousiño, N.; Maqueda, M.; Ambrona, J.; Zamora, E.; Esteban, R.; Ramírez, M. A new wine *Saccharomyces cerevisiae* killer toxin (Klus), encoded by a double-stranded RNA virus, with broad antifungal

activity is evolutionarily related to a chromosomal host gene. *Appl. Environ. Microbiol.* **2011**, *77*, 1822–1832. [CrossRef] [PubMed]

29. Page, N.; Gerard-Vincent, M.; Menard, P.; Beaulieu, M.; Azuma, M.; Dijkgraaf, G.J.P.; Li, H.; Marcoux, J.; Nguyen, T.; Dowse, T.; et al. A *Saccharomyces cerevisiae* genome-wide mutant screen for altered sensitivity to K1 killer toxin. *Genetics* **2003**, *163*, 875–894. [PubMed]

30. Carroll, S.Y.; Stirling, P.C.; Stimpson, H.E.M.; Giesselmann, E.; Schmitt, M.J.; Drubin, D.G. A yeast killer toxin screen provides insights into a/b toxin entry, trafficking, and killing mechanisms. *Dev. Cell* **2009**, *17*, 552–560. [CrossRef] [PubMed]

31. Serviene, E.; Luksa, J.; Orentaite, I.; Lafontaine, D.L.J.; Urbonavicius, J. Screening the budding yeast genome reveals unique factors affecting K2 toxin susceptibility. *PLoS ONE* **2012**, *7*, e50779. [CrossRef] [PubMed]

32. Masison, D.C.; Blanc, A.; Ribas, J.C.; Carroll, K.; Sonenberg, N.; Wickner, R.B. Decoying the cap- mRNA degradation system by a double-stranded RNA virus and poly(A)- mRNA surveillance by a yeast antiviral system. *Mol. Cell. Biol.* **1995**, *15*, 2763–2771. [CrossRef] [PubMed]

33. Rowley, P.A.; Ho, B.; Bushong, S.; Johnson, A.; Sawyer, S.L. XRN1 is a species-specific virus restriction factor in yeasts. *PLoS Pathog.* **2016**, *12*, e1005890. [CrossRef] [PubMed]

34. Tercero, J.C.; Wickner, R.B. MAK3 encodes an N-acetyltransferase whose modification of the L-A gag NH2 terminus is necessary for virus particle assembly. *J. Biol. Chem.* **1992**, *267*, 20277–20281. [PubMed]

35. Tercero, J.C.; Riles, L.E.; Wickner, R.B. Localized mutagenesis and evidence for post-transcriptional regulation of MAK3. A putative N-acetyltransferase required for double-stranded RNA virus propagation in *Saccharomyces cerevisiae*. *J. Biol. Chem.* **1992**, *267*, 20270–20276. [PubMed]

36. Ohtake, Y.; Wickner, R.B. Yeast virus propagation depends critically on free 60S ribosomal subunit concentration. *Mol. Cell. Biol.* **1995**, *15*, 2772–2781. [CrossRef] [PubMed]

37. Becker, B.; Schmitt, M.J. Yeast killer toxin K28: Biology and unique strategy of host cell intoxication and killing. *Toxins* **2017**, *9*, 333. [CrossRef] [PubMed]

38. McBride, R.C.; Boucher, N.; Park, D.S.; Turner, P.E.; Townsend, J.P. Yeast response to LA virus indicates coadapted global gene expression during mycoviral infection. *FEMS Yeast Res.* **2013**, *13*, 162–179. [CrossRef] [PubMed]

39. Lukša, J.; Ravoitytė, B.; Konovalovas, A.; Aitmanaitė, L.; Butenko, A.; Yurchenko, V.; Serva, S.; Servienė, E. Different metabolic pathways are involved in response of *Saccharomyces cerevisiae* to L-A and M viruses. *Toxins* **2017**, *9*, 233. [CrossRef] [PubMed]

40. Göker, M.; Scheuner, C.; Klenk, H.-P.; Stielow, J.B.; Menzel, W. Codivergence of mycoviruses with their hosts. *PLoS ONE* **2011**, *6*, e22252. [CrossRef] [PubMed]

41. Liu, H.; Fu, Y.; Xie, J.; Cheng, J.; Ghabrial, S.A.; Li, G.; Yi, X.; Jiang, D. Discovery of novel dsRNA viral sequences by in silico cloning and implications for viral diversity, host range and evolution. *PLoS ONE* **2012**, *7*, e42147. [CrossRef] [PubMed]

42. Nakayashiki, T.; Kurtzman, C.P.; Edskes, H.K.; Wickner, R.B. Yeast prions [URE3] and [PSI+] are diseases. *Proc. Natl. Acad. Sci. USA* **2005**, *102*, 10575–10580. [CrossRef] [PubMed]

43. Young, T.W.; Yagiu, M. A comparison of the killer character in different yeasts and its classification. *Antonie Van Leeuwenhoek* **1978**, *44*, 59–77. [CrossRef] [PubMed]

44. Naumov, G.I.; Ivannikoiva, I.V.; Naumova, E.S. Molecular polymorphism of viral dsRNA of yeast *Saccharomyces paradoxus*. *Mol. Gen. Mikrobiol. Virusol.* **2005**, *1*, 38–40.

45. Čitavičius, D.; Inge-Večtomov, S.G. Množestvennye mutanty u drožžej *Saccharomyces cerevisiae*-I. Polučenie i obščaja kharakteristika. *Genetika* **1972**, *1*, 95–102.

46. Naumova, T.I.; Naumov, G.I. Sravnitel'naja genetika drožžej. Soobščenie XII. Izučenie antagonističeskikh otnošenij u drožžej roda Saccharomyces. *Genetika* **1973**, *9*, 85–90.

47. Somers, J.M.; Bevan, E.A. The inheritance of the killer character in yeast. *Genet. Res.* **1969**, *13*, 71–83. [CrossRef] [PubMed]

48. Vepstaite-Monstavice, I.; Luksa, J.; Staneviciene, R.; Strazdaite-Zieliene, Z.; Yurchenko, V.; Serva, S.; Serviene, E. Distribution of apple and blackcurrant microbiota in Lithuania and the Czech Republic. *Microbiol. Res.* **2018**, *206*, 1–8. [CrossRef] [PubMed]

49. Gulbiniene, G.; Kondratiene, L.; Jokantaite, T.; Serviene, E.; Melvydas, V.; Petkuniene, G. Occurrence of killer yeast strains in fruit and berry wine yeast populations. *Food Technol. Biotechnol.* **2004**, *42*, 159–163.

50. Fried, H.M.; Fink, G.R. Electron microscopic heteroduplex analysis of "killer" double-stranded RNA species from yeast. *Proc. Natl. Acad. Sci. USA* **1978**, *75*, 4224–4228. [CrossRef] [PubMed]
51. Grybchuk, D.; Akopyants, N.S.; Kostygov, A.Y.; Konovalovas, A.; Lye, L.F.; Dobson, D.E.; Zangger, H.; Fasel, N.; Butenko, A.; Frolov, A.O.; et al. Viral discovery and diversity in trypanosomatid protozoa with a focus on relatives of the human parasite *Leishmania*. *Proc. Natl. Acad. Sci. USA* **2018**, *115*, E506–E515. [CrossRef] [PubMed]
52. Potgieter, A.C.; Page, N.A.; Liebenberg, J.; Wright, I.M.; Landt, O.; van Dijk, A.A. Improved strategies for sequence-independent amplification and sequencing of viral double-stranded RNA genomes. *J. Gen. Virol.* **2009**, *90*, 1423–1432. [CrossRef] [PubMed]
53. Nguyen, L.T.; Schmidt, H.A.; von Haeseler, A.; Minh, B.Q. IQ-TREE: A fast and effective stochastic algorithm for estimating maximum-likelihood phylogenies. *Mol. Biol. Evol.* **2015**, *32*, 268–274. [CrossRef] [PubMed]
54. Kall, L.; Krogh, A.; Sonnhammer, E.L.L. Advantages of combined transmembrane topology and signal peptide prediction–the Phobius web server. *Nucleic Acids Res.* **2007**, *35*, W429–W432. [CrossRef] [PubMed]
55. Finn, R.D.; Coggill, P.; Eberhardt, R.Y.; Eddy, S.R.; Mistry, J.; Mitchell, A.L.; Potter, S.C.; Punta, M.; Qureshi, M.; Sangrador-Vegas, A.; et al. The Pfam protein families database: Towards a more sustainable future. *Nucleic Acids Res.* **2016**, *44*, D279–D285. [CrossRef] [PubMed]
56. Ferre, F.; Clote, P. DiANNA: A web server for disulfide connectivity prediction. *Nucleic Acids Res.* **2005**, *33*, W230–W232. [CrossRef] [PubMed]
57. Lukša, J.; Serva, S.; Serviene, E. *Saccharomyces cerevisiae* K2 toxin requires acidic environment for unidirectional folding into active state. *Mycoscience* **2016**, *57*, 51–57. [CrossRef]
58. Maere, S.; Heymans, K.; Kuiper, M. BiNGO: A Cytoscape plugin to assess overrepresentation of gene ontology categories in biological networks. *Bioinformatics* **2005**, *21*, 3448–3449. [CrossRef] [PubMed]
59. Szklarczyk, D.; Franceschini, A.; Kuhn, M.; Simonovic, M.; Roth, A.; Minguez, P.; Doerks, T.; Stark, M.; Muller, J.; Bork, P.; et al. The STRING database in 2011: Functional interaction networks of proteins, globally integrated and scored. *Nucleic Acids Res.* **2011**, *39*, D561–D568. [CrossRef] [PubMed]
60. Novotna, D.; Flegelova, H.; Janderova, B. Different action of killer toxins K1 and K2 on the plasma membrane and the cell wall of *Saccharomyces cerevisiae*. *FEMS Yeast Res.* **2004**, *4*, 803–813. [CrossRef] [PubMed]
61. Naumov, G.I.; Ivannikova, I.V.; Chernov, I.I.; Naumova, E.S. Genetic polymorphism of double stranded RNA of *Saccharomyces* plasmids. *Mikrobiologiia* **2009**, *78*, 242–247. [PubMed]
62. Dignard, D.; Whiteway, M.; Germain, D.; Tessier, D.; Thomas, D.Y. Expression in yeast of a cDNA copy of the K2 killer toxin gene. *Production* **1991**, *7*, 127–136. [CrossRef]
63. Frank, A.C.; Wolfe, K.H. Evolutionary capture of viral and plasmid DNA by yeast nuclear chromosomes. *Eukaryot. Cell* **2009**, *8*, 1521–1531. [CrossRef] [PubMed]
64. Liu, H.; Fu, Y.; Jiang, D.; Li, G.; Xie, J.; Cheng, J.; Peng, Y.; Ghabrial, S.A.; Yi, X. Widespread horizontal gene transfer from double-stranded RNA viruses to eukaryotic nuclear genomes. *J. Virol.* **2010**, *84*, 11876–11887. [CrossRef] [PubMed]
65. Meinhardt, F.; Klassen, R. Yeast killer toxins: Fundamentals and applications. In *Physiology and Genetics. The Mycota (a Comprehensive Treatise on Fungi as Experimental Systems for Basic and Applied Research)*; Anke, T., Weber, D., Eds.; Springer: Berlin/Heidelberg, Germany, 2009; Volume 15, pp. 107–130, ISBN 978-3-642-00286-1.
66. Sun, H.; Ma, H.; Hao, M.; Pretorius, I.; Chen, S. Identification of yeast population dynamics of spontaneous fermentation in Beijing wine region, China. *Ann. Mikrobiol.* **2009**, *59*, 69–76. [CrossRef]
67. Barata, A.; Malfeito-Ferreira, M.; Loureiro, V. The microbial ecology of wine grape berries. *Int. J. Food Microbiol.* **2012**, *153*, 243–259. [CrossRef] [PubMed]
68. Mannazzu, I.; Clementi, F.; Ciani, M. Strategies and criteria for the isolation and selection of autochthonous starters. In *Biodiversity and Biotechnology of Wine Yeasts*; Ciani, M., Ed.; Research Signpost: Kerala, India, 2002; pp. 19–35, ISBN 978-8-177-36120.
69. Podoliankaite, M.; Luksa, J.; Vysniauskis, G.; Sereikaite, J.; Melvydas, V.; Serva, S.; Serviene, E. High-yield expression in *Escherichia coli*, purification and application of budding yeast K2 killer protein. *Mol. Biotechnol.* **2014**, *56*, 644–652. [CrossRef] [PubMed]
70. Lebionka, A.; Servienė, E.; Melvydas, V. Isolation and purification of yeast *Saccharomyces cerevisiae* K2 killer toxin. *Biologija* **2002**, *4*, 2–4.

71. Liu, G.L.; Chi, Z.; Wang, G.Y.; Wang, Z.P.; Li, Y.; Chi, Z.M. Yeast killer toxins, molecular mechanisms of their action and their applications. *Crit. Rev. Biotechnol.* **2015**, *35*, 222–234. [CrossRef] [PubMed]

72. Liti, G.; Carter, D.M.; Moses, A.M.; Warringer, J.; Parts, L.; James, S.A.; Davey, R.P.; Roberts, I.N.; Burt, A.; Koufopanou, V.; et al. Population genomics of domestic and wild yeasts. *Nature* **2009**, *458*, 337–341. [CrossRef] [PubMed]

Communication

Multiplex Detection of
Aspergillus fumigatus Mycoviruses

Selin Özkan-Kotiloğlu [1,2,*] and Robert H. A. Coutts [2,3]

1 Department of Molecular Biology and Genetics, Ahi Evran University, Kırşehir 40100, Turkey
2 Department of Life Sciences, Faculty of Natural Sciences, Imperial College London, London SW7 2AZ, UK; r.coutts@herts.ac.uk
3 Department of Biological and Environmental Sciences, University of Hertfordshire, Hatfield AL10 9AB, UK
* Correspondence: selin.ozkan@ahievran.edu.tr, slnozkan@gmail.com; Tel.: +90-533-265-5386

Received: 5 April 2018; Accepted: 6 May 2018; Published: 8 May 2018

Abstract: Mycoviruses are viruses that naturally infect and replicate in fungi. They are widespread in all major fungal groups including plant and animal pathogenic fungi. Several dsRNA mycoviruses have been reported in *Aspergillus fumigatus*. Multiplex polymerase chain reaction (PCR) amplification is a version of PCR that enables amplification of different targets simultaneously. This technique has been widely used for detection and differentiation of viruses especially plant viruses such as those which infect tobacco, potato and garlic. For rapid detection, multiplex RT-PCR was developed to screen new isolates for the presence of *A. fumigatus* mycoviruses. Aspergillus fumigatus chrysovirus (AfuCV), Aspergillus fumigatus partitivirus (AfuPV-1), and Aspergillus fumigatus tetramycovirus-1 (AfuTmV-1) dsRNAs were amplified in separate reactions using a mixture of multiplex primer pairs. It was demonstrated that in the presence of a single infection, primer pair mixtures only amplify the corresponding single virus infection. Mixed infections using dual or triple combinations of dsRNA viruses were also amplified simultaneously using multiplex RT-PCR. Up until now, methods for the rapid detection of Aspergillus mycoviruses have been restricted to small scale dsRNA extraction approaches which are laborious and for large numbers of samples not as sensitive as RT-PCR. The multiplex RT-PCR assay developed here will be useful for studies on determining the incidence of *A. fumigatus* mycoviruses. This is the first report on multiplex detection of *A. fumigatus* mycoviruses.

Keywords: dsRNA mycoviruses; multiplex PCR; Aspergillus fumigatus chrysovirus; Aspergillus fumigatus partitivirus-1; Aspergillus fumigatus tetramycovirus-1

---

1. Introduction

Aspergillus is a genus in the phylum Ascomycota, kingdom Fungi. *Aspergillus* species (spp.) are universal and ubiquitous in nature, and are usually found as saprophytes in soil, decaying vegetation, seed and grains [1]. They are important both medically and commercially. *Aspergillus fumigatus* is an important pathogen with the capability of infecting humans, invading lungs and leading to opportunistic infections in immunocompromised hosts along with the allergic reactions in healthy individuals [1,2].

Mycoviruses were first reported in *Aspergillus foetidus* and *Aspergillus niger* [3]. Research on dsRNA elements showed that various *Aspergillus* spp. were infected by mycoviruses [4] (Varga et al., 1998). Previously, research on 61 isolates of *A. fumigatus* showed that none of the strains contained dsRNA elements [4]. However, after screening more than 360 *A. fumigatus* isolates from environmental and clinical sources for the presence of dsRNA elements, three different dsRNA profiles were reported with a 6.6% incidence in the UK isolates [5]. Recently, different dsRNA mycoviruses have been reported in *A. fumigatus* with an 18.6% incidence in Holland [6]. Those previous studies reporting variable mycovirus incidence in *A. fumigatus* have employed common nucleic acid extraction methods [4–6].

The impacts of the viruses on the host were also investigated in *A. fumigatus* using murine and *Galleria mellonella* models and those viruses are known to affect host virulence [7,8].

Multiplex detection technologies where simultaneous detection of multiple viruses in a single assay can be achieved, enables a reduction in cost, time and labor as compared to other detection methods [9]. Various methods such as enzyme-linked immunoabsorbant assay (ELISA), real-time polymerase chain reaction (RT-PCR), loop-mediated isothermal amplification (LAMP), Luminex bead arrays and next generation sequencing (NGS) have all been used for virus detection [9–12]. Multiplex PCR is a version of PCR, which facilitates amplification of several different target sequences simultaneously. In this technique, more than one pair of primers is used to amplify different targets in one reaction. Multiplex PCR is a useful tool to detect various viruses infecting the same sample [10,13]. It is called multiplex reverse transcription PCR (RT-PCR) when the starting material is RNA instead of DNA. This technique has been widely used for the detection and differentiation of the viruses especially plant viruses such as tobacco, potato and garlic viruses [10,13–16].

The main aim of this study was to optimize a rapid detection method for three characterized *A. fumigatus* mycoviruses namely Aspergillus fumigatus chrysovirus (AfuCV) [17], Aspergillus fumigatus partitivirus (AfuPV-1) [18], and Aspergillus fumigatus tetramycovirus-1 (AfuTmV-1) [19], which have been reviewed recently by Kotta-Loizou and Coutts [20]. For rapid detection, multiplex PCR was developed to screen new isolates in terms of presence of *A. fumigatus* mycoviruses.

2. Materials and Methods

2.1. Fungal Growth Conditions and Mycoviruses

A. fumigatus strains used in this study were isolates A56, A88 and Af293 for Aspergillus fumigatus chrysovirus (AfuCV), Aspergillus fumigatus partitivirus-1 (AfuPV-1) and Aspergillus fumigatus tetramycovirus-1 (AfuTmV-1), respectively [5]. All *A. fumigatus* strains were cultured from glycerol stocks on Aspergillus complete medium (ACM) agar at 37 °C for 3–4 days and spores were harvested in order to inoculate ACM broth which was incubated at 37 °C with shaking at 130 rpm. After 5 days, mycelia were harvested using sterile Miracloth (Merck Millipore, MA, USA), rapidly frozen in liquid N_2 and kept at -80 °C until processing. *A. fumigatus* isolates were selected from the ones screened previously by Bhatti et al., 2012 [5].

2.2. Primer Design

Sequence-specific oligonucleotide primers were designed for amplification of conserved areas within the *RdRP* genes of AfuCV, AfuPV-1 and AfuTmV-1. The *RdRP* gene sequences of each mycovirus were obtained from the National Center for Biotechnology Information (NCBI). Specificity of the primers was checked by BLAST analysis of mycoviral genomes and the *A. fumigatus* genome. In order to facilitate correct multiplex PCR amplification, all primers were designed with similar annealing temperatures. The sequences of the oligonucleotide primers used are shown in Table 1.

Table 1. Oligonucleotide primers used for multiplex polymerase chain reaction (PCR) to detect virus infection in *A. fumigatus*. AfuCV: Aspergillus fumigatus chrysovirus; AfuPV-1: Aspergillus fumigatus partitivirus-1; AfuTmV-1: Aspergillus fumigatus tetramycovirus-1.

Virus	Primer ID	Sequence 5'→3'	Tm (°C)	Amplicon Size and Positions in Genomes	GenBank No
AfuCV	MCV1F MCV1R	TCGACACAGAAGGCGATATG CGCCGTTGATAAAAGTCCAT	63.863.6	592 bp (1114–1705)	FN178512.1
AfuPV-1	MPV1F MPV1R	TCAGCTGGAGCCACCTTTAT CTCCACTTCTGAGCATCACG	63.763.7	497 bp (546–1042)	FN376847.3
AfuTmV-1	MTmV1F MTmV1R	AACCAGGACGTCGTTTCCTTC GAACAGTGTATTGAGGGTGTC	64.763.3	328 bp (953–1280)	HG975302

2.3. RNA Extraction and Reverse Transcription

In order to perform multiplex PCR for mycovirus-infected *A. fumigatus* isolates, total fungal RNA was extracted using the RNeasy Plant Mini kit (Qiagen, Hilden, Germany) from 100 mg of grounded mycelium, quantified by using NanoDrop 2000C spectrophotometer (Thermo Fischer, Waltham, MA, USA) and cDNA was synthesized using Superscript-III first-strand synthesis system (Invitrogen, Carlsbad, CA, USA) according to the manufacturers' protocol as follows; 5 µg of RNA, 10 mM of dNTP mix and 250 ng of random primers along with 6.5 µL of DEPC-treated water were incubated at 65 °C for 5 min and snap cooled on ice for at least 2 min. Then 4 µL of 5× first strand buffer, 0.1 M DTT, 40 U of RNasin RNase inhibitor (Promega, Madison, WI, USA) and 200 U of Superscript-III reverse transcriptase were added to the reaction after a brief centrifugation. The 20 µL reaction mixture was then subjected to the following cycling regime of incubation at 25 °C, 50 °C and 70 °C for 5 min, 1 h and 15 min, respectively in a DNA Engine DYAD thermocycler. In an attempt to confirm if there was direct amplification from DNA or not, cDNA synthesis was performed without the reverse transcriptase enzyme. Additionally, viral dsRNAs were extracted using LiCl extraction as described previously [21] and used as a positive control in order to check the efficiency of the kit at extracting viral dsRNAs.

2.4. Specificity and Sensitivity Testing of Single PCR and Multiplex PCR

Single PCR amplification was performed prior to multiplex PCR in order to check the primers and amplicon sizes. A no template control was included along with the specificity tests consisting of (i) a mixture of the three primer pairs and cDNA template from each virus, and (ii) a mixture of the three primer pairs and the mixture of three viruses (2 ng from each mycovirus). The PCR master mix contained 10 µL of 5× GoTaq Buffer (Promega, Fitchburg, WI, USA), 1 µL of 100 mM dNTP mix (Promega), 3 µL of forward primer mixture (10 µM of each primer), 3 µL of reverse primer mixture (10 µM of each primer), 2 U of GoTaq polymerase (Promega) and 5 ng DNA or cDNA template in 50 µL reaction.

Multiplex PCR was performed using a program consisting of an initial activation at 95 °C for 3 min followed by 32 cycles of 3 steps including 95 °C for 30 s, 62 °C for 45 s and 72 °C for 30 s and a final extension at 72 °C for 3 min. The resulting PCR amplicons were analyzed by electrophoresis in 2% agarose gel containing ethidium bromide as before [20].

3. Results

3.1. The Efficiency of the RNA Extraction Method

The multiplex RT-PCR amplification assay was used to detect known and characterized *A. fumigatus* mycoviruses, namely AfuCV, AfuPV-1 and AfuTmV-1. The effect of the dsRNA extraction method on the efficiency of PCR was tested using LiCl extraction and the RNeasy Plant Mini kit (Qiagen). It was found that both gave the same amplicon with similar efficiency (Figure 1). We demonstrated that targets of interest can be amplified with the same efficiency from viral dsRNAs obtained using different procedures such as virus purification or an RNA extraction kit.

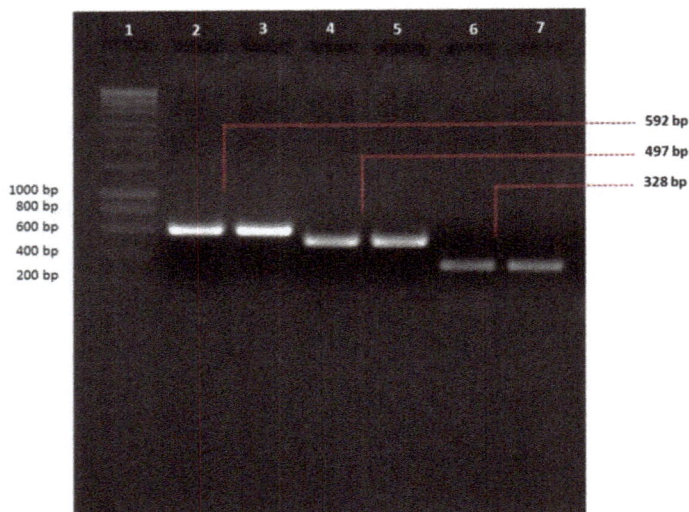

Figure 1. Conventional PCR to check amplicon size prior to multiplex PCR. Amplicon sizes were checked on two percent agarose gel prior to performing multiplex PCR for *A. fumigatus* dsRNA mycoviruses. AfuCV, AfuPV-1 and AfuTmV-1 dsRNAs were extracted using LiCl extraction (Lanes 2, 4 and 6, respectively) and RNeasy Plant Mini kit (Lanes 3, 5 and 7, respectively) and used as templates for amplification with AfuCV, AfuPV-1 and AfuTmV-1 primers (2–3, 4–5, 6–7, respectively). Hyperladder-I was used as a marker to estimate the size of the amplicons (Lane 1).

3.2. Specificity and Sensitivity of the Multiplex PCR

After establishing the dsRNA extraction method, primers were examined in terms of specificity under multiplex conditions. AfuCV, AfuPV-1 and AfuTmV-1 dsRNAs were amplified in separate reactions using a mixture of multiplex primer pairs. It was determined that in the presence of single infection, mixtures of primer pairs only amplify the virus corresponding single infection (Figure 2). The specificity of the mixture of three primer pairs used in the present study was tested and shown in Figure 2. As positive controls oligonucleotide primer pairs were used for RT-PCR amplification of amplicons 592, 497 and 328 bp in size from respectively AfuCV, AfuPV-1 and AfuTmV-1. No cross-reaction with non-targets was identified.

Mixed infections using dual or triple combinations of dsRNA viruses were also amplified simultaneously using multiplex RT-PCR (Figure 3). Combinations of AfuCV and AfuPV-1, AfuCV and AfuTmV-1, AfuPV-1 and AfuTmV-1 and AfuCV, AfuPV and AfuTmV-1 were used as template and multiplex RT-PCR was performed using all three primer pairs. In all cases, amplicons were only generated from their respective mycovirus template RNAs.

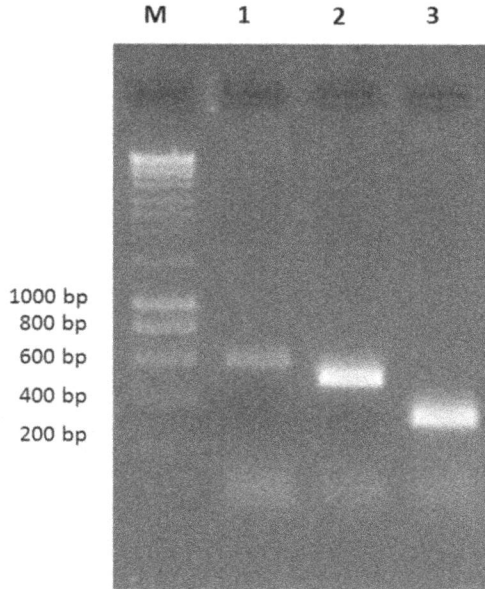

Figure 2. Specificity testing of multiplex RT-PCR with a mixture of three primer pairs. Multiplex RT-PCR was performed using the mixture of three primer pairs in a reaction including single template. Lanes 1, 2, 3 indicate the amplicons from AfuCV, AfuPV-1 and AfuTmV-1 templates in the presence of mixed primers, respectively. Hyperladder-I was used as a marker to estimate the size of the amplicons (Lane M).

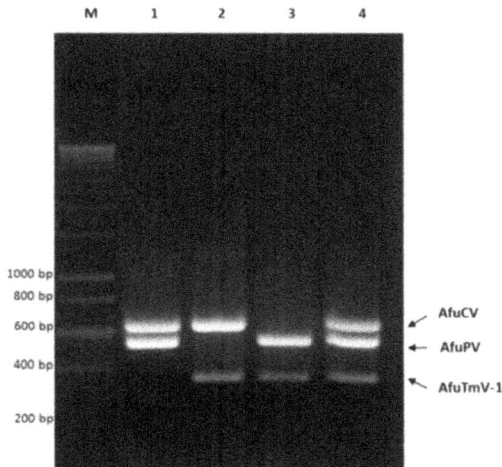

Figure 3. Simultaneous detection of *A. fumigatus* mycoviruses using multiplex RT-PCR. Mixed dsRNA combinations were used to perform multiplex RT-PCR. In lanes 1, 2 and 3, combinations of AfuCV and AfuPV-1, AfuCV and AfuTmV-1 and AfuPV-1 and AfuTmV-1, were used as template, respectively. In lane 4, dsRNAs belonging to three *A. fumigatus* mycoviruses were mixed and multiplex RT-PCR was performed using all three primer pairs. Hyperladder-I was used as a marker to estimate the size of the amplicon (Lane M).

The efficiency of the multiplex PCR amplification assay was also tested using a naturally mixed infected *A. fumigatus* isolate A80 [5], which was naturally infected with a combination of AfuCV and AfuTmV-1 (Figure 4). Extracted RNA, as described above, was used as template for single PCRs including AfuCV, AfuPV-1 and AfuTmV-1 primers separately first. It was found that A80 gave amplicons of the correct sizes when AfuCV and AfuTmV-1 primer sets were used. However, as anticipated the A80 isolate did not give any amplicon when the AfuPV-1 primer set was used. When the multiplex PCR was conducted with all three primer sets the A80 isolate only gave amplicons with primers specific for AfuCV and AfuTmV-1. However, the amplicon generated from AfuTmV-1 was small in quantity as compared to that generated from AfuCV suggesting the former was present in lesser amounts in the mixedly infected A80 isolate (Figure 4, lane 4).

In addition, all positive amplicons from multiplex PCR were sequenced in order to confirm the specificity of PCR products. The sequences were blasted to the deposited GenBank sequences to confirm the specificity of the PCR amplified DNA fragments. Sequences of amplified products were identical to the *RdRP* gene sequences of respectively AfuCV, AfuPV-1 and AfuTmV-1.

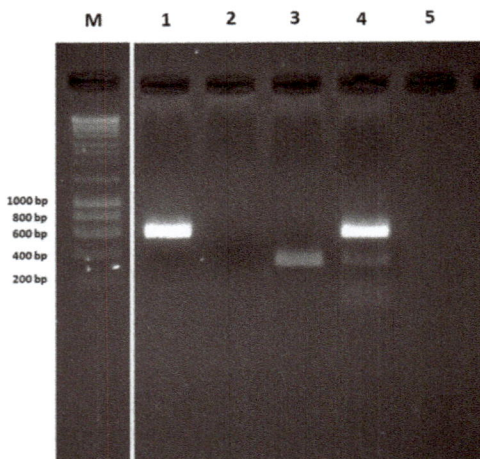

Figure 4. Simultaneous detection of mycoviruses in an *A. fumigatus* isolate (A80) which was infected naturally by a mixture of AfuCV and AfuTmV-1. In lanes 1, 2 and 3, amplicons were produced using AfuCV, AfuPV-1 or AfuTmV-1 primers, respectively. In lane 4, RNA isolated from the mixedly infected A80 isolate was used as template and multiplex RT-PCR was performed using all three primer pairs. Lane 5 showed the negative template control. Hyperladder-I was used as a marker to estimate the size of the amplicon (Lane M).

4. Discussion

It is essential to develop a rapid and reliable method to detect *A. fumigatus* mycoviruses. This study developed a multiplex PCR assay for rapid detection of three *A. fumigatus* mycoviruses (AfuCV, AfuPV-1 and AfuTmV-1) using specific primers for each mycovirus.

There are various studies on *Aspergillus* mycoviruses, especially ones infecting the medically important human pathogen *A. fumigatus*. In these studies, variable mycovirus incidence was reported in different *A. fumigatus* populations using common nucleic acid extraction methods [4–6]. Recently, a simple and rapid method has been developed to purify viral dsRNA from plant and fungal tissues using cellulose powder [22]. There is no study reported on rapid detection of *Aspergillus* mycoviruses except those using small-scale dsRNA extraction approaches [23]. Even though viruses can be detected with the small-scale dsRNA extraction method, it can be laborious where large numbers of samples are to be screened and when the virus is in low titre and sensitivity is a problem. Conversely, multiplex RT-PCR amplification provides sensitivity and specificity in diagnosis of virus infection along with the other advantages such as being easy to perform, rapid and cost-efficient [10]. Recently, a number of novel techniques to detect mycoviruses, record their incidence and characterize them have been developed. These include RT-LAMP [12] and next generation sequencing-mediated virus detection and characterization [24,25], and it is conceivable that such techniques could be combined with multiplex-PCR technology in virus detection and characterization in the future.

Additionally, the method of RNA extraction is of great importance as is one of the factors determining the efficiency of multiplex PCR. The method developed in this study uses the RNA template obtained using a rapid kit extraction method which also makes the procedure quicker and reproducible.

To our knowledge, this is the first study reporting multiplex RT-PCR optimized for mycoviruses. However, it is only applicable to characterized viruses that have a known genomic sequence. It is also possible that novel mycoviruses in *A. fumigatus* can be missed as this method has only been developed based on sequence information of known *A. fumigatus* mycoviruses. In order to overcome this problem, a primer pair could be designed based on the conserved sequence (if there is any) of known *A. fumigatus* mycoviruses assuming that the undiscovered *A. fumigatus* mycoviruses also have this conserved sequence. This primer pair could be added to the multiplex PCR in order not to miss any potential novel mycoviruses. Since it is known that all viruses, especially those with RNA as their genome, evolve and mutate, it is entirely feasible that the multiplex PCR amplification protocol as described might have limited application in the future. However, we have been careful in our design of oligonucleotides used in the protocol such that they are in conserved regions of the genomes, mutations to which are likely lethal for virus replication. Multiple detection by deep-sequencing is also a very useful approach to screen and furthermore elucidate the viromes [25,26]. In conclusion, a novel multiplex RT-PCR assay which is rapid, reliable and cost-effective was developed to detect AfuCV, AfuPV-1 and AfuTmV-1 infection of *A. fumigatus* isolates. The multiplex RT-PCR assay developed here would be useful for the studies on determining the incidence of known *A. fumigatus* mycoviruses and to establish a system for a comprehensive survey of them. Moreover, it could be useful to detect mycovirus infection rapidly in order to design treatment to fungal infection as mycoviruses can alter fungal pathogenicity.

Author Contributions: S.Ö.-K. and R.H.A.C. conceived and designed the experiments; S.Ö.-K. performed the experiments; S.Ö.-K. and R.H.A.C. analyzed the data; S.Ö.-K. and R.H.A.C. wrote the paper.

Acknowledgments: We thank Catherine Townsend for working on this project during her summer placement. Robert Coutts wishes to acknowledge the support of the Leverhulme Trust through a Leverhulme Emeritus Research Fellowship. Selin Özkan thanks the Turkish Government Higher Education program for supporting her PhD investigations.

Conflicts of Interest: The authors declare no conflict of interest. The founding sponsors had no role in the design of the study; in the collection, analyses, or interpretation of data; in the writing of the manuscript, and in the decision to publish the results.

References

1. Pitt, J.I. The current role of *Aspergillus* and *Penicillium* in human and animal health. *J. Med. Vet. Mycol.* **1994**, *32* (Suppl. 1), 17–32. [CrossRef] [PubMed]

2. Latge, J.P. *Aspergillus fumigatus* and aspergillosis. *Clin. Microbiol. Rev.* **1999**, *12*, 310–350. [PubMed]

3. Banks, G.T.; Buck, K.W.; Chain, E.B.; Darbyshire, J.E.; Himmelweit, F.; Ratti, G.; Sharpe. T.J.; Planterose, D.N. Antiviral activity of double stranded RNA from a virus isolated from *Aspergillus foetidus*. *Nature* **1970**, *227*, 505–507. [CrossRef] [PubMed]

4. Varga, J.; Rinyu, E.; Kevei, E.; Toth, B.; Kozakiewicz, Z. Double-stranded RNA mycoviruses in species of *Aspergillus* sections *Circumdati* and *Fumigati*. *Can. J. Microbiol.* **1998**, *44*, 569–574. [CrossRef] [PubMed]

5. Bhatti, M.F.; Jamal, A.; Bignell, E.M.; Petrou, M.A.; Coutts, R.H.A. Incidence of dsRNA mycoviruses in a collection of *Aspergillus fumigatus* isolates. *Mycopathologia* **2012**, *174*, 323–326. [CrossRef] [PubMed]

6. Refos, J.M.; Vonk, A.G.; Eadie, K.; Lo-Ten-Foe, J.R.; Verbrugh, H.A.; van Diepeningen, A.D.; van de Sande, W.W. Double-stranded RNA mycovirus infection of *Aspergillus fumigatus* is not dependent on the genetic make-up of the host. *PLoS ONE* **2013**, *8*, e77381. [CrossRef] [PubMed]

7. Bhatti, M.F.; Jamal, A.; Petrou, M.A.; Cairns, T.C.; Bignell, E.M.; Coutts, R.H. The effects of dsRNA mycoviruses on growth and murine virulence of *Aspergillus fumigatus*. *Fungal Genet. Biol.* **2011**, *48*, 1071–1075. [CrossRef] [PubMed]

8. Özkan, S.; Coutts, R.H. Aspergillus fumigatus mycovirus causes mild hypervirulent effect on pathogenicity when tested on *Galleria mellonella*. *Fungal Genet. Biol.* **2015**, *76*, 20–26. [CrossRef] [PubMed]

9. Boonham, N.; Kreuze, J.; Winter, S.; van der Vlugt, R.; Bergervoet, J.; Tomlinson, J.; Mumford, R. Methods in virus diagnostics: From ELISA to next generation sequencing. *Virus Res.* **2014**, *186*, 20–31. [CrossRef] [PubMed]

10. Dai, J.; Cheng, J.; Huang, T.; Zheng, X.; Wu, Y. A multiplex reverse transcription PCR assay for simultaneous detection of five tobacco viruses in tobacco plants. *J. Virol. Methods* **2012**, *183*, 57–62. [CrossRef] [PubMed]

11. Uehara-Ichiki, T.; Shiba, T.; Matsukura, K.; Ueno, T.; Hirae, M.; Sasaya, T. Detection and diagnosis of rice-infecting viruses. *Front. Microbiol.* **2013**, *4*, 289. [CrossRef] [PubMed]

12. Komatsu, K.; Urayama, S.; Katoh, Y.; Fuji, S.; Hase, S.; Fukuhara, T.; Arie, T.; Teraoka, T.; Moriyama, H. Detection of Magnaporthe oryzae chrysovirus 1 in Japan and establishment of a rapid, sensitive and direct diagnostic method based on reverse transcription look-mediated isothermal amplification. *Arch. Virol.* **2016**, *161*, 317–326. [CrossRef] [PubMed]

13. Meunier, A.; Schmit, J.-F.; Stas, A.; Kutluk, N.; Bragard, C. Multiplex reverse transcription-PCR for simultaneous detection of *Beet necrotic yellow vein virus*, *Beet soilborne virus*, and *Beet virus* Q and their vector *Polymyxa betae* KESKIN on sugar beet. *Appl. Environ. Microbiol.* **2003**, *69*, 2356–2360. [CrossRef] [PubMed]

14. Lorenzen, J.H.; Piche, L.M.; Gudmestad, N.C.; Meacham, T.; Shiel, P. A multiplex PCR assay to characterize *Potato virus Y* isolates and identify strain mixtures. *Plant Dis.* **2006**, *90*, 935–940. [CrossRef]

15. Kumar, S.; Udaya Shankar, A.C.; Nayaka, S.C.; Lund, O.S.; Prakash, H.S. Detection of *Tobacco mosaic virus* and *Tomato mosaic virus* in pepper and tomato by multiplex RT-PCR. *Lett. Appl. Microbiol.* **2011**, *53*, 359–363. [CrossRef] [PubMed]

16. Majumder, S.; Baranwal, V.K. Simultaneous detection of four garlic viruses by multiplex reverse transcription PCR and their distribution in Indian garlic accessions. *J. Virol. Methods* **2014**, *202*, 34–38. [CrossRef] [PubMed]

17. Jamal, A.; Bignell, E.M.; Coutts, R.H.A. Complete nucleotide sequences of four dsRNAs associated with a new chrysovirus infecting *Aspergillus fumigatus*. *Virus Res.* **2010**, *153*, 64–70. [CrossRef] [PubMed]

18. Bhatti, M.F.; Bignell, E.M.; Coutts, R.H.A. Complete nucleotide sequences of two dsRNAs associated with a new partitivirus infecting *Aspergillus fumigatus*. *Arch. Virol.* **2011**, *156*, 1677–1680. [CrossRef] [PubMed]

19. Kanhayuwa, L.; Kotta-Loizou, I.; Özkan, S.; Gunning, A.P.; Coutts, R.H.A. A novel mycovirus from *Aspergillus fumigatus* contains four unique dsRNAs as its genome and is infectious as dsRNA. *Proc. Natl. Acad. Sci. USA* **2015**, *112*, 9100–9105. [CrossRef] [PubMed]

20. Kotta-Loizou, I.; Coutts, R.H.A. Mycoviruses in Aspergilli: A Comprehensive Review. *Front. Microbiol.* **2017**, *8*, 1699. [CrossRef] [PubMed]

21. Diaz-Ruiz, J.R.; Kaper, J.M. Isolation of viral double-stranded RNAs using a LiCl fractionation procedure. *Prep. Biochem.* **1978**, *8*, 1–17. [CrossRef] [PubMed]

22. Okada, R.; Kiyota, E.; Moriyama, H.; Fukuhara, T.; Natsuaki, T. A simple and rapid method to purify viral dsRNA from plant and fungal tissue. *J. Gen. Plant Pathol.* **2015**, *81*, 103–107. [CrossRef]

23. Coenen, A.; Kevei, F.; Hoekstra, R.F. Factors affecting the spread of double-stranded RNA viruses in *Aspergillus nidulans*. *Genet. Res.* **1997**, *69*, 1–10. [CrossRef] [PubMed]

24. Marzano, S.L.; Nelson, B.D.; Ajayi-Oyetunde, O.; Bradley, C.A.; Hughes, T.J.; Hartman, G.L.; Eastburn, D.M.; Domier, L.L. Identification of diverse mycoviruses through metatranscriptomics characterization of the viromes of five major fungal plant pathogens. *J. Virol.* **2016**, *90*, 6846–6863. [CrossRef] [PubMed]

25. Nerva, L.; Ciuffo, M.; Vallino, M.; Margaria, P.; Varese, G.C.; Gnavi, G.; Turina, M. Multiple approaches for the detection and characterization of viral and plasmid symbionts from a collection of marine fungi. *Virus Res.* **2016**, *219*, 22–38. [CrossRef] [PubMed]

26. Osaki, H.; Sasaki, A.; Nomiyama, K.; Tomioka, K. Multiple virus infection in a single strain of *Fusarium poae* shown by deep sequencing. *Virus Genes* **2016**, *52*, 835–847. [CrossRef] [PubMed]

viruses

MDPI

Article

The Effect of Aspergillus thermomutatus Chrysovirus 1 on the Biology of Three *Aspergillus* Species

Mahjoub A. Ejmal [1], David J. Holland [2], Robin M. MacDiarmid [1,3] and Michael N. Pearson [1,*]

[1] School of Biological Sciences, the University of Auckland, Auckland 1142, New Zealand;
    mejm158@aucklanduni.ac.nz (M.A.E.); robin.macdiarmid@plantandfood.co.nz (R.M.M.)
[2] Infectious Diseases Unit, Division of Medicine, Middlemore Hospital, Auckland 1640, New Zealand;
    David.Holland@middlemore.co.nz
[3] Plant and Food Research, Auckland 1142, New Zealand
* Correspondence: m.pearson@auckland.ac.nz

Received: 4 September 2018; Accepted: 29 September 2018; Published: 2 October 2018

Abstract: This study determined the effects of *Aspergillus thermomutatus* chrysovirus 1 (AthCV1), isolated from *Aspergillus thermomutatus*, on *A. fumigatus*, *A. nidulans* and *A. niger*. Protoplasts of virus-free isolates of *A. fumigatus*, *A. nidulans* and *A. niger* were transfected with purified AthCV1 particles and the phenotype, growth and sporulation of the isogenic AthCV1-free and AthCV1-infected lines assessed at 20 °C and 37 °C and gene expression data collected at 37 °C. AthCV1-free and AthCV1-infected *A. fumigatus* produced only conidia at both temperatures but more than ten-fold reduced compared to the AthCV1-infected line. Conidiation was also significantly reduced in infected lines of *A. nidulans* and *A. niger* at 37 °C. AthCV1-infected lines of *A. thermomutatus* and *A. nidulans* produced large numbers of ascospores at both temperatures, whereas the AthCV1-free line of the former did not produce ascospores. AthCV1-infected lines of all species developed sectoring phenotypes with sclerotia produced in aconidial sectors of *A. niger* at 37 °C. AthCV1 was detected in 18% of sclerotia produced by AthCV1-infected *A. niger* and 31% of ascospores from AthCV1-infected *A. nidulans*. Transcriptome analysis of the naturally AthCV1-infected *A. thermomutatus* and the three AthCV1-transfected *Aspergillus* species showed altered gene expression as a result of AthCV1-infection. The results demonstrate that AthCV1 can infect a range of *Aspergillus* species resulting in reduced sporulation, a potentially useful attribute for a biological control agent.

Keywords: chrysovirus; mycovirus; *Aspergillus*; *A. fumigatus*; *A. nidulans*; *A. niger*; *A. thermomutatus*; biocontrol

1. Introduction

Mycoviruses have been reported in a wide variety of fungi [1]. In nature, mycoviruses may be vertically transmitted through both asexual and sexual spores, although the rate of transmission varies depending upon the fungal species, the virus and the growth conditions [2]. Horizontal transmission is typically via hyphal anastomosis which is often limited by hyphal incompatibility, even within a species [3], so consequently most mycoviruses have a narrow natural host range. However, there is growing evidence that extracellular transmission may occur [4–6] potentially providing a mechanism for inter-species transmission, although to date this has only been demonstrated in a few instances [4–6].

While the vast majority of mycoviruses appear to have little or no obvious impact on their host, there are reports of significant changes in the phenotype and growth characteristics of some fungi due to viral infection. These alterations include reduction in growth and virulence of plant pathogenic fungi [7–11], cytological alterations of cellular organelles [12], changes in pigmentation [13] and enzymatic activities [14], enabling the host fungus to confer heat tolerance to its host plant [15] and production of killer toxins [16,17]. Moreover, dsRNAs of viral origin are known to be interferon inducers [18].

Like many mycoviruses, most chrysoviruses are associated with latent infections of their fungal hosts [19]. However, there are reports of negative impacts on the fungal host attributed to chrysovirus infection, including: attenuated growth, hypovirulence and sector formation caused by *Botryosphaeria dothidea* chrysovirus 1 [20]; weakened growth, altered pigmentation and abnormal hyphal aggregation induced by *Magnaporthe oryzae* chrysovirus 1-A [21]; the absence of aerial hyphae formation and subsequent conidiophore development with impaired mycelial growth and albino hyphae produced due to *Magnaporthe oryzae* chrysovirus 1-B infection [22]; hypovirulence of the host fungal strain caused by *Botryosphaeria dothidea* chrysovirus 1 [23]; aconidial sectoring, decreased pigmentation and reduced biomass induced by *Aspergillus fumigatus* chrysovirus [24]; and aconidial sectoring, low spore production and reduced radial growth caused by *A. niger* chrysovirus [25].

The effects of mycovirus infection on gene expression have been studied for some fungi. In a study of four different *Fusarium graminearum* viruses [26] a hypovirus (FgV1), a chrysovirus (FgV2), a totivirus (FgV3) and a partitivirus (FgV4) were individually used to infect *Fusarium graminearum* strain PH-1 via protoplast fusion. An RNA-Seq-based transcriptome analysis of each of the virus-infected PH-1 cultures showed that all four mycoviruses affected the transcriptome profiles and each mycovirus regulated the expression of a totally different set of host genes. However, transcriptome profiles of the naturally infected *Fusarium graminearum* strains do not appear to have been included in that assessment. Also, Sun et al. [27] showed that levels of Mycoreovirus 1-Cp9B21 genomic dsRNA increased when the virus was co-infecting *Cryphonectria parasitica* together with the Cryphonectria hypovirus 1-EP713.

We previously reported [28] sector formation, with noticeable changes in colony texture and color, plus changes in sporulation rate in cultures of *A. thermomutatus* (isolate Ath-1) infected with *Aspergillus thermomutatus* chrysovirus 1 (AthCV1). This paper presents the results of a subsequent study to investigate the ability of AthCV1 to infect other *Aspergillus* species (*A. fumigatus*, *A. niger* and *A. nidulans*) and the biological and molecular impacts of AthCV1 on these species.

2. Materials and Methods

2.1. Source of Aspergillus Isolates

An isolate of *A. thermomutatus* infected with AthCV1 was used as a source of virus particles for protoplast transfection of *A. fumigatus* (Afu-13, dsRNA-free), *A. niger* (Ang-9, dsRNA-free) and *A. nidulans* (And-1, dsRNA-free). Clinical isolates of Afu-13 and Ang-9 were provided by Dr David Holland, Clinical Director Infection Services, Middlemore Hospital, Auckland, New Zealand and And-1 was provided by Wendy McKinney, Mycology Reference Laboratory, Auckland Hospital. Cultures were maintained on Potato Dextrose Agar (PDA, DifcoTM, BD Diagnostics, Heidelberg, Germany) in the dark at 37 °C and 20 °C.

2.2. Virus Purification and Detection

AthCV1 was purified as follows: Virus-infected *A. thermomutatus* was grown in Yeast Extract Peptone Dextrose broth (YPD Broth, Yeast Extract 5 g/L, Microbiological peptone 3 g/L and Dextrose 10 g/L) and incubated on a shaking incubator at 180 rpm in the dark for 2 days. Approximately 10 g of fungal mycelium was harvested on a filter paper using vacuum filtration, ground to a fine powder in liquid nitrogen and transferred to a 50 mL Falcon tube containing 20 mL of sodium phosphate buffer (SPB) (0.1 M, pH 7.0) and 10 mL chloroform. The mixture was incubated for 30 min on ice on an orbital shaker, at 230 rpm and then centrifuged at 10,000× *g* for 30 min at 4 °C. The upper aqueous phase was centrifuged at 120,000× *g* for 2 h at 4 °C and the resultant pellet was re-suspended in 1 mL SPB (0.02 M pH 7.0) for 4 h at 4 °C. The suspension was clarified by centrifugation at 10,000× *g* for 20 min at 4 °C, the supernatant centrifuged at 120,000× g for 2 h at 4 °C and the pellet was re-suspended in 0.5 mL SPB (0.02 M, pH 7.0) overnight at 4 °C. Following centrifugation at 10,000× g for 20 min a 50-µL drop of the supernatant was negatively stained with 2% uranyl acetate (pH 4.0) and observed for virus particles using a Phillips CM12 TEM. The virus was detected by virus-specific reverse transcription polymerase chain reaction (RT-PCR), using virus specific primers that amplified a 639 bp product from the coat protein (genome segment 2), as described by Ejmal et al. [28].

2.3. Protoplast Preparation and Virus Transfection

Protoplast preparation and virus transfection were undertaken as described by Ejmal et al. [28]. Briefly, 5 g of one-day old mycelium, grown in YPD Broth, was washed once with sterile distilled water and once with protoplast buffer (0.8 M $MgSO_4 \cdot 7H_2O$, 0.2M $C_6H_5Na_3O_7 \cdot 2H_2O$, pH 5.5). The mycelium was then coarsely chopped and transferred to a flask containing 17 mL protoplast buffer. Three mL of filter sterilized Novozyme buffer (1 M Sorbitol, 50 mM Sodium citrate, pH 5.8) containing 200 mg of Lysing Enzymes from *Trichoderma harzianum* (Sigma-Aldrich, St. Louis, MO, USA) was added to the mycelial suspension, which was incubated for 4 h at 28°C in a shaker at 85 rpm. Protoplasts were passed through a 75-µm strainer and collected in a 50-mL tube containing 30 mL KC buffer (0.6 M KCl and 50 mM $CaCl_2$) and centrifuged at 4000× g for 10 min. The protoplast pellet was then washed twice with 10 mL sorbitol-Tris-calcium chloride (STC) buffer (1M Sorbitol, 50 mM Tris, pH 8 and 50 mM $CaCl_2.2H_2O$) and centrifuged at 4000× g for 10 min before being was re-suspended in 0.5 mL STC and kept on ice. For transfection, 130 µL PEG 4000 (60% in sterile water) was mixed with 70 µL potassium chloride-Tris-Calcium Chloride (KTC) buffer (1.8 M KCl, 150 mM Tris pH 8, 150 mM $CaCl_2$) and added to a tube containing 200 µL of purified virus particle suspension and 5 µL 0.05 mM spermidine (Sigma-Aldrich, St. Louis, MO, USA). A 200-µL aliquot of the protoplast suspension was then added to the suspension and mixed by twirling the tube for 10 s before it was incubated on ice for 30 min. Following incubation, a mix of 200 µL of polyethylene glycol (PEG) 4000 and 100 µL KTC was added to the previous suspension and gently twirled again. Following incubation at room temperature for 20 min, 40-µL aliquots of the mixture were added to 5 mL warm agar medium (Stabilized Minimal Medium (SMM) containing 0.7% agar), gently mixed and spread on SMM plates, which were parafilm sealed and incubated at 37 °C. Individual colonies were picked off, grown on fresh PDA plates and sub-cultured three times, at weekly intervals, then checked for the presence of the virus by RT-PCR as described by Ejmal et al. [28]. As negative controls, three plates were spread with a protoplast suspension lacking virus particles, to test protoplast viability. In addition, three plates were spread with no protoplasts in the transfection suspension to test for possible mycelial contamination in the virus particle suspension. As a general contamination check, three plates containing only SMM media were included.

2.4. Quantifying the Biological Impacts of AthCV1 Infection on Three Aspergillus Species

Sporulation rate comparisons were conducted at 20 °C to represent environmental temperature and 37 °C to represent human body temperature. Five virus-free and five virus-infected lines were grown from single spores, inoculated at the edge of PDA plates (9 cm diameter) and incubated in the dark until the mycelium reached the opposite side of the plate. To harvest conidiospores, the plates were washed with 40 mL aqueous 0.05% Tween 80, which was then filtered through cheesecloth and centrifuged at 8000× g for 10 min. The spores were re-suspended in 10 mL distilled water before being counted in a Neubauer chamber, as described by Aneja [29].

For linear growth comparison, five replicates of isogenic virus-free and virus-infected single spore isolates were individually inoculated at the edge of PDA plates and grown at 37 °C and at 20 °C. The growth was measured every 24 h until the mycelium reached the far edge of the plate. At the completion of each experiment cultures were tested for the presence of the virus using RT-PCR as described by Ejmal et al. [28].

To compare biomass production five isogenic virus-free and virus-infected single spore isolates, of each species, were grown at 37 °C and at 20 °C in the dark. Plugs of the resultant mycelium were individually transferred to conical flasks containing 200 mL YPD Broth and incubated with shaking at 180 rpm, in the dark. The resultant mycelium was vacuum filtered and 100 mg of each sample retained for virus screening by one-step RT-PCR by Ejmal et al. [28]. The remainder was dried at 90 °C for 72 h before weighing. Data were analyzed by an independent samples *t*-test, using SPSS version 21 (IBM SPSS statistics, Armonk, NY, USA).

2.5. AthCV1 Transmission Through Ascospores and Sclerotia

AthCV1 transmission through *A. nidulans* ascospores was tested as follows: First ascospores were processed to eliminate contamination from mycelium and conidia, based on the methods of O'Gorman et al. [30] and Girardin et al. [31] and individually germinated on PDA. Once the single ascospore cultures had produced sufficient mycelium they were screened for the presence of AthCV-by RT-PCR using virus specific primers as described by Ejmal et al. [28]. To test for AthCV1 in sclerotia produced by *A. niger*, the sclerotia were isolated from cultures on PDA plates incubated at 37 °C for 2 weeks in the dark, according to the method of Utkhede and Rahe [32], with minor modifications as follows: Sclerotia were picked from the agar plates with forceps and placed in a 100 mL beaker containing 75 mL sterile water, filtered through three layers of cheesecloth, washed with sterile water and surface sterilized with 0.25% sodium hypochlorite for 2 min. The sclerotia were then immediately passed through 3 layers cheesecloth and washed again with 500 mL sterile water. One hundred sclerotia were individually inoculated to PDA plates and incubated at 37 °C for 2 weeks in the dark before the resultant mycelium was screened for AthCV1 infection using one-step RT-PCR as described by Ejmal et al. [28].

2.6. The Effects of AthCV1 on Gene Expression

Both AthCV1-infected and uninfected cultures of *A. thermomutatus*, *A. fumigatus*, *A. niger* and *A. nidulans* were grown on sterile cellophane film overlaid on PDA media. The cultures were incubated at 37 °C for 5 days in the dark, as described by Zhang et al. [33], to provide growth conditions similar to those used for growth and sporulation experiments. To minimize the effects of variability between cultures, three replicates for each treatment were pooled before total RNA was extracted. Thirty milligrams of mycelium from each of three plates was harvested and combined in a 2-mL Eppendorf tube and total RNA extracted using a Spectrum Plant Total RNA Kit (Sigma-Aldrich, St. Louis, MO, USA), as described by the manufacturer. A 0.1 volume of 3 M sodium acetate, pH 5.2 and two volumes of 100% ethanol were added to each extract to precipitate the RNA, which was sent on ice to Macrogen Inc. (Soeul, Korea), for RNA sequencing. An Agilent Technologies 2100 Bioanalyzer 2100 Bioanalyzer (Agilent Technologies, Santa Clara, CA, USA) was used to measure RNA quality and quantity of the original samples and confirm an RNA integrity number (RIN) of 8 or greater. Fragmentation was performed on RNA samples before cDNA synthesis and the cleaved RNA fragments primed with random hexamers. First strand cDNA was transcribed using reverse transcriptase according to the Truseq RNA sample preparation V2 guide (Illumina, San Diego, CA, USA). The second cDNA strand synthesis was performed using DNA polymerase I in the presence of RNase H. Adapters were ligated to the DNA and PCR amplification performed to selectively enrich DNA fragments with adapter molecules on both ends and to amplify the amount of DNA in the library. PCR amplification was conducted using a PCR Primer Cocktail which anneals to the ends of the adapters. An Agilent Technologies 2100 Bioanalyzer (Agilent Technologies, Santa Clara, CA, USA) with a DNA 1000 chip was used to verify the size of PCR-enriched fragments and for library quantification qPCR was used, according to the Illumina qPCR Quantification Protocol Guide (# 11322363) (Illumina, San Diego, CA, USA). Indexed DNA libraries were normalized to 10 nM in the Diluted Cluster Template (DCT) plate and then pooled in equal volumes in the Pooled DCT plate. Sequencing was conducted using the Hiseq2000 platform (Illumina, San Diego, CA, USA) which generated reads of 100 bp × 2 (paired-end) with a total run output of 35–40 Gb. The FastQC quality control tool [34] was used to provide quality control checks on each sequence data file.

For *A. fumigatus*, *A. niger* and *A. nidulans*, where a reference genome was available (fungi.ensembl. org/index.html), the Tuxedo protocol [35] was used. The 100-bp paired-end output reads data files for each treatment (5 GB each) were uploaded individually to the Galaxy-qld platform (galaxy-qld. genome.edu.au) and Tophat2, version 0.6 [36] used to align the sequences (both forward and reverse reads). Each BAM file, containing accepted hits produced by Tophat2, was then assembled using Cufflinks software version 0.0.7 [37]. Cuffmerge software version 0.0.6 [37] was then used to merge all the assembled transcripts together in one output GTF file and Cuffdiff software version 0.0.7 [37] used to identify possible significant changes in transcript expression between AthCV1-negative and positive lines. Following that, identifiers of the differentially expressed genes were used to search for their gene ontology annotations in *Aspergillus* genome databases (www.aspergillusgenome.org), Ensemble Fungi database (fungi.ensembl.org/index.html) and the UniProt Knowledgebase (www.uniprot.org).

For *A. thermomutatus*, where no reference genome was available, the Trinity *de novo* transcriptome assembly software (version 0.0.2) [38,39] was used to create a full-length transcriptome which can be processed by Tophat and cufflinks. The Trinity read normalization tool was used to reduce coverage of highly covered areas and then the Galaxy "concatenate datasets" tool (version 1.0.0) used to combine all normalized forward and reverse reads, for each treatment. The combined reads were then used to produce a fasta file that contained all possible assembled transcripts from both AthCV1-positive and AthCV1-negative samples. The resultant transcriptome was used for the Tuxedo protocol, as described above. From the Cuffdiff output, gene loci identifiers (e.g., comp0_c0_seq1) were used to retrieve their relevant transcripts and the DNA sequences used as queries for nucleotide BLASTn searches in GenBank, to find the closest available gene sequences, which were used to search for gene ontology annotations.

3. Results

3.1. Transfection of Aspergillus Species

AthCV1-specific RT-PCR, conducted on the third serial subcultures of *A. fumigatus* (Afu-13), *A. niger* (Ang-9) and *A. nidulans* (And-1), confirmed that all three species were successfully transfected with AthCV1 purified particles (Figure 1A).

3.2. Impact of AthCV1 on Sporulation of Aspergillus Spp.

The results for the effects of AthCV1 on sporulation of *A. fumigatus* (Afu-13), *A. niger* (Ang-9) and *A. nidulans* (And-1) at 37 °C and 20 °C, together with the results for *A. thermomutatus* previously published [28] are presented in Table 1. In *A. fumigatus*, AthCV1 infection significantly reduced ($p \leq 0.05$) asexual sporulation at both temperatures, with the effects being more extreme at 37 °C than at 20 °C. As would be expected there was no sexual reproduction observed in either the AthCV1-free or AthCV1-infected treatments at either temperature as *A. fumigatus* is a heterothallic species. In the homothallic species *A. nidulans* conidial production was significantly reduced ($p \leq 0.05$) in the AthCV1-infected line at 37 °C but not at 20 °C, while there was a significant increase ($p \leq 0.05$) in the number of ascospores produced by the AthCV1-infected line at both 37 °C and 20 °C. In *A. niger* asexual sporulation was significantly decreased ($p \leq 0.05$) in the AthCV1-infected line at 37 °C while there was no significant difference between AthCV1-free and infected lines at 20 °C. *A. niger* also formed sclerotia in the AthCV1-Infected line at 37 °C but not at 20 °C. There was no sexual reproduction in *A. niger* cultures at either temperature.

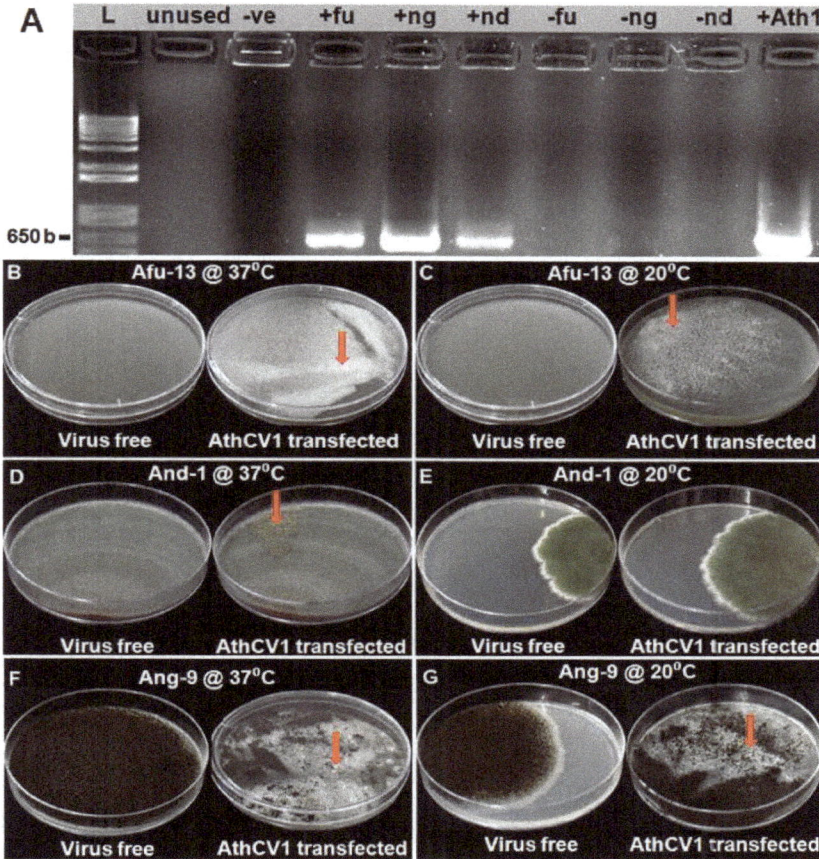

Figure 1. Effects of AthCV1 infection on the colony morphology of *Aspergillus* species. (**A**) polymerase chain reaction (PCR) assay showing successful AthCV1 transfection to different virus-free *Aspergillus* species; reverse transcription PCR, lane L = 1 kb plus DNA ladder; -Ve= PCR negative control (RNA sample was replaced with ultrapure water), +Ath1 = AthCV1 +ve control; +fu = AthCV1 transfected *A. fumigatus* (Afu-13); +ng = AthCV1 transfected *A. niger* (Ang-9); +nd = AthCV1 transfected *A. nidulans* (And-1); –fu, –ng and –nd = original virus-free isolates of *A. fumigatus* (Afu-13), *A. niger* (Ang-9) and *A. nidulans* (And-1), respectively. (**B**) Afu-13 grown at 37 °C; normal growth in the virus-free isolate, sectors formed in isogenic AthCV1-transfected line (arrow). (**C**) Afu-13 grown at 20 °C; normal growth in the virus-free isolate, sectors formed in isogenic AthCV1-transfected line (arrow). (**D**) And-1 grown at 37 °C; normal growth in the virus-free isolate, ascospore-rich sectors formed in isogenic AthCV1-transfected line (arrow). (**E**) And-1 grown at 37 °C; normal growth in both virus-free and AthCV1-transfected lines. (**F**) Ang-9 grown at 37 °C; normal growth in the virus-free isolate, conidial-free sectors with sclerotia formed (arrow) in AthCV1-transfected line. (**G**) Ang-9 grown at 20 °C; normal growth in the virus-free isolate, conidial- and sclerotia-free sectors in AthCV1-transfected line (arrow).

Table 1. Summary of the effects of AthCV1 on four Aspergillus species.

Property	Growth Temp	A. thermomutatus		A. fumigatus		A. niger		A. nidulans	
		Virus-Free	AthCV1 Infected	Virus-Free	AthCV1 Infected	Virus-Free	AthCV1 Infected	Virus-Free	AthCV1 Infected
Conidia production (per plate)	37 °C	$2.1 (\pm 0.36) \times 10^5$	$2.0 (\pm 0.10) \times 10^6$ *	$3.8 (\pm 0.42) \times 10^8$	$1.6 (\pm 0.12) \times 10^7$ *	$1.8 (\pm 0.12) \times 10^8$	$2.0 (\pm 0.14) \times 10^7$ *	$2.2 (\pm 0.62) \times 10^8$	$9.5 (\pm 0.79) \times 10^7$ *
	20 °C	$2.4 (\pm 0.36) \times 10^6$	$2.1 (\pm 0.24) \times 10^5$ *	$1.5 (\pm 0.21) \times 10^8$	$3.6 (\pm 0.31) \times 10^7$ *	$2.2 (\pm 0.26) \times 10^8$	$1.9 (\pm 0.24) \times 10^8$	$1.4 (\pm 0.16) \times 10^8$	$1.4 (\pm 0.12) \times 10^8$
Ascospore production (per plate)	37 °C	0	$7.2 (\pm 0.73) \times 10^5$ *	0	0	0	0	$1.1 (\pm 0.10) \times 10^6$	$2.1 (\pm 0.42) \times 10^6$ *
	20 °C	0	$3.6 (\pm 0.37) \times 10^5$ *	0	0	0	0	$2.9 (\pm 0.22) \times 10^5$	$8.0 (\pm 0.55) \times 10^5$ *
Sclerotia production (per plate)	37 °C	0	0	0	0	0	$54 (\pm 4.2)$ *	0	0
	20 °C	0	0	0	0	0	0	0	0
Growth (mm) [1]	37 °C	$74 (\pm 0.32)$	$75 (\pm 1.00)$	$72 (\pm 0.37)$	$73 (\pm 1.00)$	$77 (\pm 0.20)$	$67 (\pm 0.68)$ *	$76 (\pm 0.49)$	$75 (\pm 0.71)$
	20 °C	$76 (\pm 0.32)$	$75 (\pm 0.32)$	$67 (\pm 0.32)$	$72 (\pm 0.20)$ *	$77 (\pm 0.60)$	$67 (\pm 0.74)$	$33 (\pm 0.45)$	$30 (\pm 0.66)$ *
Fungal Biomass [2] (g dry wt)	37 °C	$1.25 (\pm 0.012)$	$1.28 (\pm 0.009)$	$1.30 (\pm 0.008)$	$1.23 (\pm 0.008)$ *	$1.25 (\pm 0.023)$	$1.30 (\pm 0.007)$	$1.36 (\pm 0.021)$	$1.12 (\pm 0.005)$ *
	20 °C	$1.13 (\pm 0.011)$	$1.13 (\pm 0.009)$	$1.15 (\pm 0.011)$	$1.13 (\pm 0.007)$	$1.09 (\pm 0.004)$	$1.12 (\pm 0.003)$ *	$1.12 (\pm 0.016)$	$1.12 (\pm 0.010)$
Sector formation	37 °C		Creamy, rough ascospore-rich sectors		Clear, elongated sectors in the grey mycelium		Conidia-free sectors with no pigmentation		Ascospore-rich sectors in the green mycelium
	20 °C		Creamy, rough ascospore-rich sectors		Clear, elongated sectors in the grey mycelium		Conidia-free sectors with no pigmentation		No sector formation
Pigment change	37 °C		Creamy sectors		Sectors lack pigmentation		Sectors lack pigmentation		No change
	20 °C		Creamy sectors		Sectors lack pigmentation		Sectors lack pigmentation		No change

Data analyzed by independent samples t-test using SPSS version 21, SEs in brackets. * = significant difference ($p < 0.05$) between virus-free and virus infected treatments. [1] Since growth was measured to the edge of the petri dish the growth period differed for the different species: at 37 °C A. thermomutatus = 6 days, A. fumigatus = 6 days, A. niger = 13 days, A. nidulans = 13 days; at 20 °C A. thermomutatus = 15 days, A. fumigatus = 24 days, A. niger = 29 days, A. nidulans = 24 days. [2] Growth in liquid culture was measured after 4 days at 37 °C and 15 days at 20 °C.

3.3. Effects of AthCV1 on Aspergillus Growth and Morphology

A summary of the effects of AthCV1 on the growth and the cultural characteristics of the *Aspergillus* species at 37 °C and 20 °C, together with the results for *A. thermomutatus* previously published [28], are presented in Table 1. The radial growth of *A. fumigatus* on PDA plates was significantly greater ($p \leq 0.05$) in AthCV1-infected cultures at 20 °C but not at 37 °C. In contrast, there was a significant decrease in mycelial dry weight associated with AthCV1 infection at 37 °C but not at 20 °C. In addition, while the AthCV1-free culture produced a uniform grey-colored mycelium, the mycelium of the AthCV1-infected line formed small white sectors that subsequently merged together. This was more prominent at 37 °C than at 20 °C, eventually covering most of the plate (Figure 1B,C). In *A. nidulans* there was a significant decrease in radial mycelial growth of the AthCV1-infected line at 20 °C while at 37 °C there was no difference in growth between AthCV1-free and AthCV1-infected lines. Mycelial dry weight was lower in AthCV1-infected cultures at 37 °C but there was no change in mycelial dry weight at 20 °C. A sectoring phenotype was observed in AthCV1-infected cultures grown at 37 °C which were very rich in sexual fruiting bodies (Figure 1D) but not in those grown at 20 °C (Figure 1E). In *A. niger* there was a significant reduction of radial mycelial growth on PDA plates in infected cultures at 37°C but no difference in mycelial growth at 20 °C. There was a significant increase in mycelial dry weight in the AthCV1-infected line grown at 20 °C but no difference between virus-infected and virus-free at 37 °C. Sectors with a conidia-free phenotype occurred in the AthCV1-positive culture at both 37 °C and 20 °C (Figure 1F,G). At 37 °C sclerotia were formed in these conidia-free sectors (Figure 1F) but no sclerotia were formed in cultures grown at 20 °C (Figure 1G).

3.4. Transmission of AthCV1 Through Ascospores and Sclerotia

In AthCV1 infected cultures, ascospores were produced by *A. nidulans* but not in *A. fumigatus* or *A. niger*. Of the single ascospore cultures grown at 37 °C, 31 of 100 were AthCV1-infected, while from the culture grown at 20 °C, 34 of 100 ascospores were infected. Sclerotia were produced in AthCV1 infected cultures of *A. niger* but not *A. nidulans* or *A. fumigatus*. AthCV1 was detected in 18 of 100 sclerotia produced by *A. niger* grown at 37 °C.

3.5. Changes in Gene Expression Associated with AthCV1 Infection

The quality and quantity check performed on the original total RNA samples indicated that all of the samples were of acceptable quality (RIN = 6.5–7.6). The average RNA size was *c.* 260–280 bp, of which 126 bp represented the Illumina sequencing adapters, yielding a gene sequence fragment of *c.* 140–160 bp. FastQC indicated an acceptable Phred quality score of ≥ 20 over all samples. FastQC also confirmed that the encoding format of the data in the RNA-seq files was in "Sanger/Illumina 1.9" format as required for the Galaxy platform. De novo transcriptome assembly for *A. thermomutatus*, for which no genome was available by the time of the study, produced a fasta file containing 44430 sequences. FPKM (Fragments Per Kilobase of transcript, per Million mapped reads) results indicated that for AthCV1-infected cultures some genes were highly upregulated while others showed downregulation or were undetectable (i.e., showed a zero FPKM value), compared with the isogenic virus-free line. These included genes related to fungal sporulation (Table 2). The number of differentially expressed genes ($\geq$ 5-fold change or undetectable in one treatment) were 62 each in *A. thermomutatus* and in *A. fumigatus*, 65 in *A. niger* and 34 in *A. nidulans*. The functional analysis of these genes is shown in Figure 2.

Table 2. Regulation of sporulation-related genes in AthCV1-infected *Aspergillus* species compared with their isogenic AthCV1-free line. Genes showing a five-fold or greater difference in expression between infected and non-infected lines are presented in the table.

Gene ID	Verified and Predicted Gene and Function ◇ retrieved from www.aspergillusgenome.org, www.uniprot.org and fungi.ensembl.org/index.html	Fold Change
A. thermomutatus △		
NFIA_070550	*csnB*; orthologue(s) subunit2 of the COP9 signalosome; required for formation of cleistothecia	7 ↑
NFIA_010750	*laeA*; orthologue (*laeA*/AN0807, *A. nidulans*) coordinates asexual development in response to light, involved in the regulation of secondary metabolism and required for the formation Hulle cells.	1913 ↓
NFIA_030070	*fhbA*; flavohemoprotein, regulates sexual sporulation and sterigmatocystin production	100 ↓
NFIA_071100	*MAT1*; positive regulation of mating type specific transcription, DNA-templated	35 ↓
A. fumigatus		
Afu1g14520	Pyridine nucleotide-disulphide oxidoreductase; NADPH dehydrogenase activity, spore germination	25 ↑
Afu2g17560	Hydroxynaphthalene reductase arp2; involved in conidial pigment biosynthesis; conidia-enriched protein	16 ↑
Afu2g17600	Conidial pigment polyketide synthase PksP/Alb1; conidial pigment biosynthesis; conidia wall assembly	13 ↑
Afu2g17550	Heptaketide hydrolase ayg1; conidial pigment; polyketide shortening; conidia-enriched protein	13 ↑
Afu2g17580	Scytalone dehydratase arp1; conidial pigment biosynthesis; conidia formation and sporulation	9 ↑
Afu1g09750	Aldehyde reductase (AKR1), putative; conidia-enriched protein	38 ↓
Afu4g10770	*ppoA*; response to oxidative stress; orthologue (AN1967/*ppoA*) negative regulation of sexual sporulation and positive regulation asexual sporulation.	7 ↓
A. niger		
An18g06650	Orthologue (Afu3g14540, *A. fumigatus*), 30-kilodalton heat shock protein; conidia-enriched protein	157 ↑
An01g12490	Orthologue (*Neurospora crassa*) has NADPH dehydrogenase activity, role in spore germination	18 ↑
An14g02460	*fhbA*; orthologue in *A. nidulans* regulation of sexual sporulation and sterigmatocystin biosynthetic process	12 ↑
An04g07400	Putative C6 zinc finger transcription factor; orthologue (AN1848/*nosA*, *A. nidulans*) positive regulation of sexual development; orthologue (*adv-1*) in *N. crassa* has predicted role in conidium formation and hyphal growth.	326 ↓
An12g00710	Orthologue (AN9121/*esdC*, *A. nidulans*); negative regulation of conidiation and positive regulation of sexual sporulation	200 ↓
An02g05420	*flbC*; putative C2H2 transcription factor; predicted role in conidiation; expressed in germinating conidia	31 ↓
An04g05880	*ppoA*; response to oxidative stress; orthologue (AN1967/*ppoA*) negative regulation of sexual sporulation and positive regulation of asexual sporulation.	20 ↓
An01g04830	*flbD*; Myb-like DNA-binding protein; positive regulation of conidiation; expressed in germinating conidia	12 ↓
An17g01580	Orthologue (AN2290/*steA*, *A. nidulans*) has negative regulation of transcription by RNA polymerase II promoter; regulation of secondary metabolite biosynthetic process; sporocarp development involved in sexual reproduction	5 ↓
An05g00480	*stuA*; positive regulation of conidium formation and conidiophore development	5 ↓

Table 2. *Cont.*

Gene ID	Verified and Predicted Gene and Function ◇ retrieved from www.aspergillusgenome.org, www.uniprot.org and fungi.ensembl.org/index.html	Fold Change
	A. nidulans	
AN7795	gprK; positive regulation of sexual sporulation [40]	39 ↑
AN3387	gprD; Putative G-protein coupled receptor; Deletion of gprD resulted in delayed conidial germination and enhanced sexual development	16 ↑
AN3695	Putative anthranilate synthase with a predicted role in aromatic amino acid biosynthesis and cleistothecium development	13 ↑
AN7169	fhbA; flavohemoprotein; sterigmatocystin biosynthetic process and regulates sexual development	12 ↑
AN5844	Controls conidia germination and adjusts cellular substances which protect conidia against dryness [41]	7 ↑
AN5046	anisin-1; asexual sporulation, response to oxidative stress and defense response	7 ↑
AN5156	pho80; overexpression decreases conidiation and increases formation of cleistothecia	6 ↑
AN3148	PH domain protein; have role in ascospore wall assembly, ascospore-type prospore membrane assembly	44 ↓
AN0387	cryA; negative regulation of cleistothecium development	37 ↓
AN8640	conF; Conidiation protein Con-6, putative; contributes in conidia germination and desiccation resistance	21 ↓
AN1848	nosA; Zinc(II)2Cys6 putative transcription factor involved in the regulation of sexual development	20 ↓
AN2755	MAT1; regulator of sexual development; acts with Mat2 HMG domain protein	16 ↓
AN6046	noxR; P67phox regulatory subunit homolog; required for normal sexual and asexual development	13 ↓
AN6688	aspB; conidiophore development and hyphal growth	12 ↓
AN4163	cpcB; required for sexual development; positive regulation of cleistothecium development	11 ↓
AN4351	palA; pH-response regulator protein palA; orthologue in Saccharomyces cerevisiae (RIM20) role in sporulation resulting in formation of a cellular spore	6 ↓
AN1017	hOGA; Putative mitogen-activated protein kinase; required for sexual development and sporulation	6 ↓
AN7553	dezR; Basic helix-loop-helix transcription factor required for normal conidiophore development	5 ↓
AN10306	camdA-N; role in sexual development, secondary metabolism and light control of asexual development	5 ↓
AN2458	camdA-C; role in sexual development, secondary metabolism and light control of asexual development	5 ↓

Δ = due to the absence of *A. thermomutatus* genome, gene orthologues of *Neosartorya fischeri* (closest BLAST hits) were used. ◇ = function is for the gene if available or for *A. nidulans* gene orthologue. ↑ = up-regulated in the AthCV1-infected line; ↓ = down-regulated in the AthCV1-infected line.

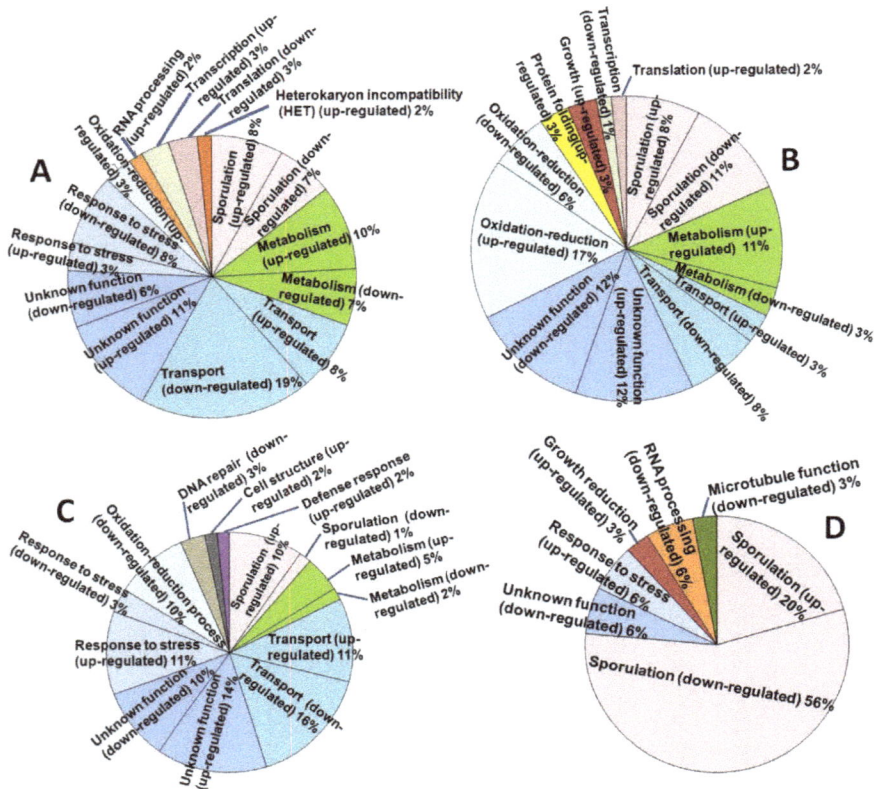

Figure 2. Differentially expressed genes in AthCV1-infected *Aspergillus* species and their functional analysis. (**A**) *A. thermomutatus*; (**B**) *A. niger*; (**C**) *A. fumigatus*; (**D**) *A. nidulans*.

As shown in Table 3, the effect of AthCV1 infection on gene expression varied between the different *Aspergillus* species. Some genes showed expression change in only one species, such as the transcription factor gene *steA* (AN2290), for which changes were observed only in *A. niger*. For other genes, the variation was just in the degree of expression, as for the orthologue of the heat shock protein gene, *hsp30/hsp42* (AN5781). In other instances, the difference was in the direction of regulation, such as the internal alternative NADH dehydrogenase *ndiA* (AN10660), which was completely repressed in *A. nidulans* but upregulated by 25-fold in *A. fumigatus* and by 18-fold in *A. niger*. The orthologue of this gene in *N. crassa* has an NADPH dehydrogenase activity and a role in spore germination with mitochondrial inner membrane localization (Table 2). In yeast (*S. cerevisiae*) overexpression of the orthologue (*ndi-1*) of this gene (*ndiA*) can result in apoptosis-like cell death, which is associated with an increase in the production of reactive oxygen species (ROS) in mitochondria [42] and the transcription factor *nosA*/AN1848, which in the AthCV1-infected lines was reduced 326-fold in *A. niger*, 20-fold in *A. nidulans* and 2-fold in *A. fumigatus* but was unchanged in *A. thermomutatus*.

Table 3. Major changes in the expression of gene orthologues of AthCV1-infected *Aspergillus* species compared to their isogenic AthCV1-free lines.

Gene ID A. nidulans	Fold Change in AthCV1-infected Samples				Gene and Function From www.uniprot.org & www.aspergillusgenome.org
	A. thermomutatus	A. fumigatus	A. nidulans	A. niger	
AN10660	-	25	-	18	nirA; Pyridine nucleotide-disulphide oxidoreductase, putative; NADPH dehydrogenase activity and spore germination.
AN2755	−35		−16	-	MAT1; Alpha-domain mating-type protein; regulator of sexual development; acts with Mat2 HMG domain protein.
AN1652	-	−5	-	-	sebA; C2H2 transcription factor; required for virulence; response to oxidative stress and heat shock.
AN5781	-	31	6	157	Heat shock protein Hsp30, transcript increase during the unfolded-protein response; palA-dependent expression independent of pH
AN2458	3	-	−5	-	camdA-C; orthologue(s) N-terminal subunit of Cand1; sexual development, secondary metabolism and light control of asexual development.
AN10306	5	-	−5	3	camdA-N; role in sexual development and secondary metabolism.
AN10049	-	9	−4	-	mdpB; Probable scytalone dehydratase; involved in conidial pigment biosynthesis, conidium formation and sporulation.
AN5836	-	-	−4	−5	stuA; Positive regulation of asexual sporulation and conidiophore development.
AN1848	-	−2	−20	−326	nosA; Zinc(II)2Cys6 putative transcription factor; positive regulation of sexual development.
AN5156	−2	-	6	-	Pho80-like cyclin; overexpression decreases conidiation and increases cleistothecia.
AN7169	−100	4	12	12	flbA; Flavohemoglobin, negative regulation of sexual sporulation and positive regulation of sterigmatocystin biosynthesis.
AN0807	−1913	−2	-	−4	laeA; Methyltransferase-domain protein; self-methylates; coordinates asexual development in response to light; regulates secondary metabolism and is required for Hulle cell formation.
AN0387	-	2	−37	-	cryA: Negative regulation of cleistothecium development.
AN4783	7	-	−2	-	csnB; COP9 signalosome subunit 2 (CsnB) formation of cleistothecia.
AN2421	4	−3	-	−31	flbC; regulation of conidium formation and spore germination.
AN7553	-	−3	−5	-	derR; Basic helix-loop-helix transcription factor; conidiophore development.
AN5893	−5	2	-	-	flbA; Developmental regulator FlbA, conidiophore development and asexual sporulation.
AN0082	−6	−2	−2	−2	phnA; Phosducin, putative; regulates sporulation
AN3387	-	2	16	−2	gprD: Deletion of gprD resulted in delayed conidial germination and enhanced sexual development.
AN0279	-	−2	2	−12	flbD: Myb-like DNA-binding protein; positive regulation of conidiation.
AN2290	-	-	-	−5	steA; Sexual development transcription factor SteA; Required for cleistothecial development and ascosporogenesis. Not required for conidiation.
AN0170	-	-	-	−24	trxA; Thioredoxin, required for conidiation; expression upregulated after exposure to farnesol.
AN9121	2	-	-	−200	esdC; sexual development protein, involved in early sexual development and regulated by VeA and FlbA.
AN4163	-	2	−11	-	cpcB; G-protein complex beta subunit CpcB; positive regulation of cleistothecium development.
AN1017	2	−2	−6	2	hogA; Putative mitogen-activated protein kinase; required for sexual development and sporulation.
AN7795	-	-	39	-	gprK; positive regulation of sexual sporulation [+].

Up-regulated

Colour	Fold change
	<5
	5 to 20
	21 to 100
	101 to 500
	>500

Down-regulated

Colour	Fold change
	<5
	5 to 20
	21 to 100
	101 to 500
	>500

4. Discussion

Mycoviruses have received much attention in relation to their potential as biological control agents for harmful fungi. Two necessary properties of biocontrol agents are (i) the ability to reduce the growth and virulence of the target fungus and (ii) rapid and efficient transmission of the virus from the inoculum to the target fungal population. Hyder et al. [43] found that the effects of a virus isolate on its fungal host can differ depending upon the strain of the infected fungus and that the extent of those effects can vary according to the environment and ecological conditions. Similarly, Vainio et al. [44] found that conditions (such as temperature) can alter virus impact on its fungal host. Consequently, the development of mycoviruses as effective biological control agents requires an understanding of both the viruses and their interaction with their fungal hosts and environment.

Vegetative incompatibility is a major barrier to studying the effects of viruses on a range of fungal hosts and their use in the field, as it can be a serious barrier to horizontal virus transmission between different fungal species and even different strains of the same fungal species. However, in vitro, this obstacle may be overcome by the use of protoplast fusion [45–48] or transfection of protoplasts with purified virus particles [48,49]. The latter technique has been successfully used for the transfection of virus-free *A. fumigatus* protoplasts with *Aspergillus fumigatus* tetramycovirus (AfuTmV-1) [50] and *Aspergillus fumigatus chrysovirus* (AfuCV) [24], as well as species of other fungal genera, including *Botryosphaeria dothidea* with *Botryosphaeria dothidea chrysovirus* 1 (BdCV1) [20], *Rhizoctonia solani* with *Rhizoctonia solani partitivirus* 2 [51], *Cryphonectria parasitica* with a reovirus from *C. parasitica* [52], *Rosellinia necatrix* with *Rosellinia necatrix partitivirus* (RnPV1) [53] and *Rosellinia necatrix mycoreovirus* 3 (RnMYRV-3) [54]. In the current study, we successfully transfected *A. fumigatus*, *A. niger* and *A. nidulans* with AthCV1 particles from naturally infected *A. thermomutatus*, to produce isogenic virus-free and virus-infected lines, in order to evaluate the effects of the virus on these fungal hosts.

The impact of AthCV1 infection on mycelial growth on PDA plates varied from no change at either 37 °C or 20 °C for *A. thermomutatus*, to significantly reduced growth for *A. niger* at 37 °C and *A. nidulans* at 20 °C and a significant increase in the growth for *A. fumigatus* at 20 °C. An increase in linear growth due to virus infection was previously reported by Nuss [55] where *C. parasitica* (strain Euro7) cultures infected with *Cryphonectria hypovirus* 1 (CHV-1/Euro7) showed faster growth than the isogenic virus-free isolate. However, linear growth on agar plates does not necessarily correlate with total biomass production, as seen in our experiments for *A. fumigatus*, *A. nidulans* and *A. niger* (Table 1), or with growth and/or pathogenicity in vivo [56,57].

In the current investigation, sectoring (areas with different appearance) were observed in the virus-infected cultures of all four *Aspergillus* species. Sectoring has been observed in association with a range of mycoviruses including, BdCV1 [20], *Helminthosporium victoriae* co-infected with the totivirus, *Helminthosporium victoriae virus* 190S (HvV190S) and the chrysovirus, *Helminthosporium victoriae virus* 145s (HvV145S) [56]. In a study by Bhatti et al. [24], *A. fumigatus* formed non-sporulating sectors when infected with *A. fumigatus partitivirus* 1 (AfuPV-1) or with AfuCV. However, the basis of fungal sectoring can have a number of underlying causes other than virus infection.

Although virus infection resulted in changes in sporulation and phenotype of all three species, the different *Aspergillus* species did not respond to AthCV1 infection in the same way. The response was often temperature dependent; *A. fumigatus* reacted consistently at both temperatures with significantly reduced conidial production (Table 1), whereas *A. thermomutatus* showed a significant increase in conidia at 37 °C but produced significantly fewer conidia at 20 °C (Table 1). In the AthCV1-infected *A. nidulans* (homothallic), ascospore production was significantly increased at both 37 °C (90%) and 20 °C (170%) compared to uninfected. *A. thermomutatus* (also homothallic) produced ascospores only when infected with AthCV1 (Table 1), although the number produced at 20 °C was about half that produced at 37 °C (Table 1). Individual isolates of *A. niger* and *A. fumigatus* do not produce ascospores in culture as these species are not homothallic.

Since vertical transmission and dissemination of virus infection is largely dependent on spores, the rate of infection in spores is an important factor. For many fungi, virus transmission through conidia is often high (<100%) [47,58,59], while transmission via ascospores is generally less efficient [60,61]. Coenen et al. [62] concluded that vertical transfer of dsRNA viruses in *A. nidulans* did not occur through ascospores resulting in the exclusion of viruses in the fungal population, presumably because vegetative incompatible groups formed as a result of sexual recombination in the successive generations. However, other studies have shown varying degrees of virus transmission through ascospores, including transmission of *Botrytis Virus* X (BVX) through ascospores in *Botrytis cinerea* [63], dsRNA virus particles in *S. cerevisiae* [64] and two different dsRNA mycoviruses in *Fusarium graminearum* [58]. In the current study AthCV1 was detected in 34% and 31% of ascospores at 20 °C and 37 °C, respectively.

The production of large numbers of ascospores in AthCV-1 infected *A. thermomutatus*, compared to no ascospores in the virus-free line and a significant increase in ascospore production in *A. nidulans*, together with the decrease in conidial production, could be the result of physiological stress due to AthCV1 infection, as stress is often a trigger for a change from asexual to sexual reproduction in fungi [65]. Therefore, AthCV1 could potentially increase sexual diversity of the fungus by stimulating ascospore production, creating the opportunity for increased gene sharing and increased sexual diversity. In addition, sclerotia, which are often produced under nutrient-depleted conditions (e.g., in old cultures) and are thought essential for sexual reproduction in *Aspergillus* sections *Flavi* [40] and *Circumdati* [66] and *A. japonicas* [67], were produced in *A. niger* at 37 °C. Frisvad et al. [66] were the first to report the production of sclerotia by *A. niger* although most of the *A. niger* strains used in their study failed to produce sclerotia despite the multiple attempts to enhance the process. It is possible that mycovirus infection affected their two strains that produced sclerotia, however the strains used in their study were screened for the presence of mycoviruses. In summary, the findings in this study indicate that AthCV1 infection enhances sexual sporulation in *A. nidulans* and *A. thermomutatus* and possibly also in *A. niger* where sclerotia were produced at 37 °C.

In the current study variability in the nature of the sectoring phenotype and differences in sporulation, mycelial growth diameter and mycelial dry weight associated with AthCV1 infection in the four different *Aspergillus* species demonstrate that the effect of AthCV1 is both species dependent and temperature dependent.

Given that AthCV1 has significant effects on asexual and sexual spore reproduction in some *Aspergillus* species, knowledge of which metabolic pathways are affected might possibly be exploited for better control of the fungus. Consequently, preliminary gene expression data, based on pooled samples of three replicate cultures for each treatment, were obtained to identify changes in the expression of known reproduction-related genes as a starting point for future, more extensive studies. In response to AthCV1 infection changes in expression were detected in a total of 223 different genes, across the four *Aspergillus* species examined. Genes specifically related to sporulation and/or reproduction numbered 54, those showing greater than five-fold changes (either + or −) are presented in Table 2.

An important family of genes for sexual reproduction in fungi are the Mating Type (MAT) genes which enable homothallic fungi to switch their mating type. In *A. nidulans*, Paoletti et al. [68] have shown that overexpression of *MAT2* represses MAT1 expression and vice versa. The current findings reflect this with the two homothallic species *A. thermomutatus* and *A. nidulans* where *MAT1*/NFIA_071100 and *MAT1*/AN2755 were downregulated in the AthCV1-infected treatment by 8-fold and 16-fold respectively whereas, *MAT2* was upregulated by 4-fold in the AthCV1-infected *A. nidulans* and expressed in the AthCV1-infected *A. thermomutatus* but not in the virus-free line.

In AthCV1-infected *A. nidulans*, other highly upregulated genes that contribute to the increase in sexual sporulation include: *gprk*/AN7795 [40] (39-fold increase) and AN3695 (www.aspergillusgenome. org) (13-fold), *fhbA*/AN7169 [69] (12-fold). Genes that were undetected and whose repression might have contributed to the high reduction in conidiation include *mst1*/AN5674 (positive regulation of asexual development and negative regulation of sexual development) [70] and *nudF*/AN6197 (positive regulation of conidiation and ascospore production) [71]. In addition, genes that might have a negative impact on sexual reproduction if overexpressed and were somewhat downregulated included *cryA*/AN0387 (37-fold decrease) that codes for a blue light- and UVA-sensing cryptochrome that represses sexual development by regulating other regulators such as VeA, NsdD and RosA [72] and *hogA*/AN1017 (6-fold decrease) which is a putative mitogen-activated protein kinase (MAPK), is highly up-regulated under osmotic stress conditions required for sexual development and sporulation [73]. In AthCV1-infected *A. nidulans* conidiation was hugely reduced and this seems consistent with the increased number of ascospores, as it is known that ascospore production can inhibit conidiation and vice versa [40]. Changes in expression of genes known to be involved in the reduction of conidiation included the downregulation of *conF*/AN8640 (21-fold reduction) which contributes in conidia germination and in the protection of conidia against desiccation [41], *noxR*/AN6046 (13-fold reduction) which is important for conidiophore development and conidia production [74] and the putative septin B, *aspB*/AN6688 (12-fold reduction), which plays a role in growth emergence and conidiation [75,76].

In AthCV1-infected *A. thermomutatus* there was a 7-fold increase in expression of *csnB*/NFIA_070550, which is essential for cleistothecia and ascospore formation [77,78]. The expression of *laeA*/NFIA_010750, an orthologue of the global regulator of secondary metabolism (*laeA*/AN0807), was downregulated by 1913-fold. This gene coordinates asexual development in response to light, is involved in the regulation of secondary metabolism and is required for the formation of Hulle cells [79–83]. The NirA-dependent flavohemoprotein gene (*fhbA*/NFIA_030070), which is involved in the positive regulation of sterigmatocystin production and in the negative regulation of sexual sporulation [69], was downregulated by 100-fold.

Genes of *A. niger* that were affected and possibly contributed to the huge decrease of conidiation in the AthCV1-infected treatment at 37 °C included: the putative C2H2 transcription factor (*flbC*/An02g05420) which showed a 31-fold downregulation and has a predicted role in conidiation and is expressed in germinating conidia [84–86]; An04g07400 (326-fold reduction) an orthologue of *adv-1* in *N. crassa* which is required for normal sexual and asexual development [87]; *ppoA*/An04g05880 (20-fold decrease in expression), the miss expression of the *A. nidulans* orthologue *ppoA*/AN1967 leads to a decreased level of conidiation (http://www.aspergillusgenome.org); and An05g00480 (5-fold reduction) which is expressed in germinating conidia and has a predicted role in positive regulation of conidiophore development and conidium formation (www.aspergillusgenome.org). In addition, An14g02540, the orthologue of the sclerotium regulator in *A. oryzae* (*sclR*/AO090011000215), that encodes a transcription factor with a role in hyphal morphology and the promotion of sclerotial production [88], was upregulated by 11-fold in the AthCV1-infected treatment. This is possibly an indication that the production of sclerotia was also related to stress, as they are survival structures typically produced in response to environmental stress [89]. Also, An18g06650, which has a strong similarity to *A. nidulans* heat shock protein 30 (*hsp30*/AN2530) [86], showed a 157-fold increase in expression. An orthologue of this gene in *A. fumigatus*, Afu3g14540 (*hsp30*), has been found to be expressed in a high abundance in conidia and is known to increase in response to the antifungal amphotericin B and to hydrogen peroxide [90,91].

Transcriptome analysis of the heterothallic *A. fumigatus* revealed the greatest upregulation (6427-fold increase) for Afu1g04410, which has no known function and the greatest downregulation (18094-fold reduction) for the DNA N-glycosylase, putative (Afu7g05320), the orthologue of which in *S. cerevisiae* has a predicted role in oxidized purine nucleobase lesion DNA N-glycosylase activity and molecular function in DNA damage repair (www.uniprot.org). Other *A. fumigatus* genes that showed altered transcription levels due to AthCV1 infection were: *ppoA*/Afu4g10770 (7-fold reduction), an orthologue of *A. nidulans ppoA*/AN1967, which is involved in the response to oxidative stress, negative regulation of sexual sporulation and positive regulation asexual sporulation [92] and Afu1g09750 (38-fold reduction) which is reported to be enriched in conidia compared to mycelium of *A. fumigatus* [90]; and Afu1g14520 (25-fold increase) which is orthologous to *N. crassa* (*ndi-1*) and has a predicted role in oxidation-reduction process and spore germination [93]. In addition, conidial pigmentation biosynthesis genes that were highly expressed in AthCV1-infected *A. fumigatus* included: Scytalone dehydratase *arp1*/Afu2g17580 (9-fold increase), which is involved in conidial pigment biosynthesis and for which mutants display increased C3 complement binding [94,95]; Afu2g17550/*ayg1* which is involved in conidial pigment biosynthesis with a role in polyketide shortening and melanin biosynthesis in *Aspergillus fumigatus* [94,96]; *pksP*/Afu2g17600 (12-fold increase) which is involved in biosynthesis of the conidial pigment and asexual spore wall assembly [97]; *arp2*/Afu2g17560 (16-fold increase) which is involved in conidial pigment biosynthesis and expression of conidia-enriched protein [90,94]. The latter three genes may be important virulence factors in the establishment of infection as mutations in these genes showed increased virulence and failed to inhibit phagolysosome acidification in their insect host *Galleria mellonella* [98].

In response to AthCV1 infection all four of the *Aspergillus* species included in this study exhibited changes in phenotype (Table 1) and gene expression (Table 3). For some factors, similar changes were observed in two or more of the *Aspergillus* species while for others the different species varied in their response. This is similar to the findings of Hyder et al. [43] who studied two viruses of the wood decay fungus *Heterobasidion* and reported that (i) a specific virus strain can cause different effects on different *Heterobasidion* strains and (ii) the impact of a single virus strain on a certain *Heterobasidion* isolate can differ according to the changes in environmental conditions. The diversity in response of the four *Aspergillus* species to AthCV1 supports the proposition that even if a particular mycovirus is found to have a hypovirulent effect on a certain fungal isolate, it does not necessarily follow that it will have the same effect on other isolates of the same or closely related species. Consequently, any evaluation of the potential of mycoviruses as biological control agents should be conducted on a range of fungal isolates in a range of environmental conditions. In the current study, differences in the expression of the internal alternative NADH dehydrogenase NdiA (AN10660) in different *Aspergillus* species is probably a good example of this phenomenon as it was upregulated by 25-fold in *A. fumigatus* and by 18-fold in *A. niger* but entirely repressed in *A. nidulans*.

In-vitro transfection can be used to experimentally extend the host range of a mycovirus in order to investigate the potential effects on new fungal hosts. This ability to overcome the barriers caused by hyphal incompatibility and genetic diversity may, if it results in stable transfected lines, extend the application of mycoviruses, such as AthCV1, for fungal disease control. However, it is important to understand the wider implications of specific phenotypic changes (e.g., spore production), observed under controlled experimental conditions, to the epidemiology of the fungus in the environment. For example, while the low conidiation rate of AthCV1-infected *A. fumigatus* and *A. niger* may reduce the ability of the fungus to spread and compete with uninfected fungal isolates in the field and thereby limit the impact of AthCV1 on *Aspergillus*, the high number of ascospores produced by the AthCV1-positive *A. nidulans* and *A. thermomutatus* and the enhancement of sclerotia production in *A. niger* may improve the survival of these fungi, as these structures typically resist adverse environmental changes better than conidiospores. Moreover, the presence of virus could directly or indirectly, increase diversity of *Aspergillus* by stimulating sexual reproduction and possibly increase the percentage of vegetatively compatible fungal strains allowing virus spread between them.

The use of mycoviruses as effective biological control agents requires consideration of multiple factors (host, virus and environment). The first objective is to find a virus that is capable of inducing serious impact on its fungal host. Once such a virus is identified, intensive research on how and when to use the virus is required, for example, direct application to the site of infection in the case of chestnut blight control [99]. A major constraint on the use of mycoviruses is that natural horizontal virus spread typically requires hyphal anastomosis, which is limited by genetically controlled hyphal incompatibility. Consequently, the direct use of mycoviruses as biological control agents, especially in a clinical context, is very challenging. However, an understanding of the molecular nature of the hypovirulence caused by a particular mycovirus may enable the development of a more direct approach such as targeting the expression of certain genes by pharmaceuticals that mimic the mechanism of hypovirulence.

Author Contributions: Conceptualization, D.J.H. and M.N.P.; Data curation, M.A.E.; Formal analysis, M.A.E., R.M.M. and M.N.P.; Investigation, M.A.E.; Methodology, M.A.E.; Supervision, D.J.H. and M.N.P.; Writing—original draft, M.A.E.; Writing—review & editing, D.J.H., R.M.M. and M.N.P.

Acknowledgments: We thank the School of Biological Sciences at the University of Auckland for funding this study. We also would like to thank Wendy McKinney from LabPlus, Auckland for providing us with the *A. nidulans* And-1 isolate. We appreciate the assistance of Damien Fleetwood and Mostafa Rahnama from the School of Biological Sciences, The University of Auckland, in the initial analysis of the gene expression data.

Conflicts of Interest: The authors declare no conflict of interest.

References

1. Son, M.; Yu, J.; Kim, K.H. Five Questions about Mycoviruses. *PLoS Pathog.* **2015**, *11*, e1005172. [CrossRef] [PubMed]
2. Pearson, M.N.; Beever, R.E.; Boine, B.; Arthur, K. Mycoviruses of filamentous fungi and their relevance to plant pathology. *Mol. Plant. Pathol.* **2009**, *10*, 115–128. [CrossRef] [PubMed]
3. Boddy, L. Interactions between fungi and other microbes. In *The Fungi*; Watkinson, S.C., Money, N., Boddy, L., Eds.; Academic Press: Cambridge, MA, USA, 2016; pp. 337–360.
4. Yu, X.; Li, B.; Fu, Y.; Xie, J.; Cheng, J.; Ghabrial, S.A.; Li, G.; Yi, X.; Jiang, D. Extracellular transmission of a DNA mycovirus and its use as a natural fungicide. *Proc. Natl. Acad. Sci. USA* **2013**, *110*, 1452–1457. [CrossRef] [PubMed]
5. Marzano, S.L.; Nelson, B.D.; Ajayi-Oyetunde, O.; Bradley, C.A.; Hughes, T.J.; Hartman, G.L.; Eastburn, D.M.; Domier, L.L. Identification of Diverse Mycoviruses through Metatranscriptomics Characterization of the Viromes of Five Major Fungal Plant Pathogens. *J. Virol.* **2016**, *90*, 6846–6863. [CrossRef] [PubMed]
6. Liu, S.; Xie, J.; Cheng, J.; Li, B.; Chen, T.; Fu, Y.; Li, G.; Wang, M.; Jin, H.; Wan, H.; et al. Fungal DNA virus infects a mycophagous insect and utilizes it as a transmission vector. *PNAS* **2016**, *113*, 12803–12808. [CrossRef] [PubMed]
7. Boland, G.J.; Mould, M.J.R.; Robb, J. Ultrastructure of a hypovirulent isolate of *Sclerotinia sclerotiorum* containing double stranded RNA. *Physiol. Mol. Plant. Pathol.* **1993**, *43*, 21–32. [CrossRef]
8. Hammar, S. Association of double-stranded RNA with low virulence in an isolate of *Leucostoma persoonii*. *Phytopathology* **1989**, *79*, 568–572. [CrossRef]
9. Boland, G.J. Hypovirulence and double stranded RNA in *Sclerotinia sclerotiorum*. *Can. J. Plant. Pathol.* **1992**, *14*, 10–17. [CrossRef]
10. Bottacin, A.M.; Levesque, C.A.; Punja, Z.K. Characterization of dsRNA in *Chalara elegans* and effects on growth and virulence. *Phytopathology* **1994**, *84*, 303–312. [CrossRef]
11. Punja, Z.K. Influence of double-stranded RNAs on growth, sporulation, pathogenicity and survival of *Chalara elegans*. *Can. J. Bot.* **1995**, *73*, 1001–1009. [CrossRef]
12. Newhouse, J.R.; Hoch, H.C.; Macdonald, W.L. The ultra structure of *Endothia parasitica*. Comparison of a virulent with a hypovirulent isolate. *Can. J. Botany* **1983**, *61*, 389–399. [CrossRef]
13. Chu, M.; Jean, J.; Yea, S.J.; Kim, Y.H.; Lee, Y.W.; Kim, K.H. Double-strand RNA mycoviruses from *Fusarium graminearum*. *Appl. Environ. Microbiol.* **2002**, *68*, 2529–2534. [CrossRef] [PubMed]
14. Rigling, D.; van Alfen, N.K. Extra and intracellular laccase activity of the Chestnut blight fungus, *Cryphonectria paraesitica*. *Appl. Environ. Microbial.* **1993**, *59*, 3634–3639.
15. Márquez, L.M.; Redman, R.S.; Rodriguez, R.J.; Roossinck, M.J. A Virus in a Fungus in a Plant: Three-Way Symbiosis Required for Thermal Tolerance. *Science* **2007**, *315*, 513–515. [CrossRef] [PubMed]
16. Magliani, W.; Conti, S.; Gerloni, M.; Bertolotti, D.; Polonelli, L. Yeast killer systems. *Clin. Microbiol. Rev.* **1997**, *10*, 369–400. [PubMed]
17. Park, C.M.; Banerjee, N.; Koltin, Y.; Bruenn, J.A. The *Ustilago maydis* virally encoded KP1 killer toxin. *Mol. Microbiol.* **1996**, *20*, 957–963. [CrossRef] [PubMed]
18. Banks, G.T.; Buck, K.W.; Chain, E.B.; Darbyshire, J.E.; Himmelweit, F.; Ratti, G.; Sharpe, T.J.; Planterose, D.N. Antiviral activity of double stranded RNA from a virus isolated from *Aspergillus foetidus*. *Nature* **1970**, *227*, 505–507. [CrossRef] [PubMed]

19. Ghabrial, S.A.; Caston, J.R. Family Chrysoviridae, genus Chrysovirus. In *Virus Taxonomy: Classification and Nomenclature of Viruses*; Ninth Report of the International Committee on Taxonomy of Viruses; King, A.M.Q., Adams, M.J., Carstens, E.B., Lefkowitz, E.J., Eds.; Elsevier: Amsterdam, The Netherlands; Academic Press: Cambridge, MA, USA, 2012; pp. 509–513.

20. Wang, L.; Jiang, J.; Wang, Y.; Hong, N.; Zhang, F.; Xu, W.; Wang, G. Hypovirulence of the Phytopathogenic Fungus *Botryosphaeria dothidea*: Association with a Coinfecting *Chrysovirus* and a *Partitivirus*. *J. Virol.* **2014**, *88*, 7517–7527. [CrossRef] [PubMed]

21. Urayama, S.; Kato, S.; Suzuki, Y.; Aoki, N.; Le, M.T.; Arie, T.; Teraoka, T.; Fukuhara, T.; Moriyama, H. Mycoviruses related to *chrysovirus* affect vegetative growth in the rice blast fungus *Magnaporthe oryzae*. *J. Gen. Virol.* **2010**, *91*, 3085–3094. [CrossRef] [PubMed]

22. Urayama, S.; Sakoda, H.; Katoh, Y.; Takai, R.; Le, T.M.; Fukuhara, T.; Arie, T.; Teraoka, T.; Moriyama, H. A dsRNA mycovirus, *Magnaporthe oryzae* chrysovirus1-B, suppresses vegetative growth and development of the rice blast fungus. *Virology* **2014**, *448*, 265–273. [CrossRef] [PubMed]

23. Ding, Z.; Zhou, T.; Guo, L.Y. Characterization of a novel strain of *Botryosphaeria dothidea* chrysovirus 1 from the apple white rot pathogen *Botryosphaeria dothidea*. *Arch. Virol.* **2017**, *162*, 2097–2102. [CrossRef] [PubMed]

24. Bhatti, M.F.; Jamal, A.; Petrou, M.A.; Cairns, T.C.; Bignell, E.M.; Coutts, R.H.A. The effects of dsRNA mycoviruses on growth and murine virulence of *Aspergillus fumigatus*. *Fungal Genet. Biol.* **2011**, *48*, 1071–1075. [CrossRef] [PubMed]

25. Hammond, T.M.; Andrewski, M.D.; Roossinck, M.J.; Keller, N.P. *Aspergillus* mycoviruses are targets and suppressors of RNA silencing. *Eukaryotic Cell* **2008**, *7*, 350–357. [CrossRef] [PubMed]

26. Lee, K.M.; Cho, W.K.; Yu, J.; Son, M.; Choi, H.; Min, K.; Lee, Y.W.; Kim, K.H. A Comparison of Transcriptional Patterns and Mycological Phenotypes following Infection of *Fusarium graminearum* by Four Mycoviruses. *PLoS ONE* **2014**, *9*, e100989. [CrossRef] [PubMed]

27. Sun, L.; Nuss, D.L.; Suzuki, N. Synergism between a mycoreovirus and a hypovirus mediated by the papain-like protease p29 of the prototypic hypovirus CHV1-EP. *J. Gen. Virol.* **2006**, *87*, 3703–3714. [CrossRef] [PubMed]

28. Ejmal, M.A.; Holland, D.J.; MacDiarmid, R.M.; Pearson, M.N. A novel chrysovirus from a clinical isolate of *Aspergillus thermomutatus* affects sporulation. *PLoS ONE*. under review.

29. Aneja, K.R. *Experiments in Microbiology, Plant Pathology and Biotechnology*; New Age International (P) Ltd.: New Delhi, India, 2003.

30. O'Gorman, C.M.; Fuller, H.T.; Dyer, P.S. Discovery of a sexual cycle in the opportunistic fungal pathogen *Aspergillus fumigatus*. *Nature* **2009**, *457*, 471–474. [CrossRef] [PubMed]

31. Girardin, H.; Monod, M.; Latge, J. Molecular Characterization of the Food-Borne Fungus *Neosartorya fischeri* (Malloch and Cain). *Appl. Environ. Microbiol.* **1995**, *61*, 1378–1383. [PubMed]

32. Utkhede, R.S.; Rahe, J.E. Wet-sieving floatation technique for isolation of sclerotia of *Sclerotium cepivorum* from muck soil. *Phytopathology* **1979**, *69*, 295–297. [CrossRef]

33. Zhang, F.; Guo, Z.; Zhong, H.; Wang, S.; Yang, W.; Liu, Y.; Wang, S. RNA-Seq-based transcriptome analysis of aflatoxigenic *Aspergillus flavus* in response to water activity. *Toxins* **2014**, *6*, 3187–3207. [CrossRef] [PubMed]

34. Andrews, S. FastQC: A Quality Control Tool for High Throughput Sequence Data. Available online: http://www.bioinformatics.babraham.ac.uk/projects/fastqc (accessed on 12 December 2016).

35. Trapnell, C.; Roberts, A.; Goff, L.; Pertea, G.; Kim, D.; Kelley, D.R.; Pimentel, H.; Salzberg, S.L.; Rinn, J.L.; Pachter, L. Differential gene and transcript expression analysis of RNA-seq experiments with TopHat and Cufflinks. *Nat. Protoc.* **2012**, *7*, 562–578. [CrossRef] [PubMed]

36. Kim, D.; Pertea, G.; Trapnell, C.; Pimentel, H.; Kelley, R.; Salzberg, S.L. TopHat2: Accurate alignment of transcriptomes in the presence of insertions, deletions and gene fusions. *Genome Biol.* **2013**, *14*, R36. [CrossRef] [PubMed]

37. Trapnell, C.; Williams, B.A.; Pertea, G.; Mortazavi, A.M.; Kwan, G.; van Baren, M.J.; Salzberg, S.L.; Wold, B.; Pachter, L. Transcript assembly and abundance estimation from RNA-Seq reveals thousands of new transcripts and switching among isoforms. *Nat. Biotechnol.* **2010**, *28*, 511–515. [CrossRef] [PubMed]

38. Grabherr, M.G.; Haas, B.J.; Yassour, M.; Levin, J.Z.; Thompson, D.A.; Amit, I.; Adiconis, X.; Fan, L.; Raychowdhury, R.; Zeng, Q.; et al. Full-length transcriptome assembly from RNA-Seq data without a reference genome. *Nat. Biotechnol.* **2011**, *29*, 644–652. [CrossRef] [PubMed]

39. Haas, J.H.; Papanicolaou, A.; Yassour, M.; Grabherr, M.; Blood, P.D.; Bowden, J.; Couger, M.B.; Eccles, D.; Li, B.; Lieber, M.; et al. De novo transcript sequence reconstruction from RNA-seq using the Trinity platform for reference generation and analysis. *Nat. Protoc.* **2013**, *8*, 1494–1512. [CrossRef] [PubMed]

40. Dyer, P.S.; O'Gorman, C.M. Sexual development and cryptic sexuality in fungi: Insights from *Aspergillus* species. *FEMS Microbiol. Rev.* **2012**, *36*, 165–192. [CrossRef] [PubMed]

41. Suzuki, S.; Sarikaya Bayram, Ö.; Bayram, Ö.; Braus, G.H. ConF and conJ contribute to conidia germination and stress response in the filamentous fungus *Aspergillus nidulans*. *Fungal Genet. Biol.* **2013**, *56*, 42–53. [CrossRef] [PubMed]

42. Li, W.; Sun, L.; Liang, Q.; Wang, J.; Mo, W.; Zhou, B. Yeast AMID homologue Ndi1p displays respiration-restricted apoptotic activity and is involved in chronological aging. *Mol. Biol. Cell* **2006**, *17*, 1802–1811. [CrossRef] [PubMed]

43. Hyder, R.; Pennanen, T.; Hamberg, L.; Vainio, E.J.; Piri, T.; Hantula, J. Two viruses of *Heterobasidion* confer beneficial, cryptic or detrimental effects to their hosts in different situations. *Fungal Ecology* **2013**, *6*, 387–396. [CrossRef]

44. Vainio, E.J.; Korhonen, K.; Tuomivirta, T.T.; Hantula, J. A novel putative partitivirus of the saprotrophic fungus *Heterobasidion ecrustosum* infects pathogenic species of the *Heterobasidion annosum* complex. *Fungal Biology* **2010**, *114*, 955–965. [CrossRef] [PubMed]

45. Madhosingh, C. Production of intraspecific hybrids of *Fusarium oxysporum* f. sp. radicis-lycopersici and *Fusarium oxysporum* f. sp. lycopersici by protoplast fusions. *J. Phytopathol.* **1994**, *142*, 301–309. [CrossRef]

46. Van Diepeningen, A.D.; Debets, A.J.; Hoekstra, R.F. Intra- and interspecies virus transfer in *Aspergilli* via protoplast fusion. *Fungal Genet. Biol.* **1998**, *25*, 171–180. [CrossRef] [PubMed]

47. Van Diepeningen, A.D.; Varga, J.; Hoekstra, R.F.; Debets, A.J. Mycoviruses in *Aspergilli*. In *Aspergillus in the genomics era*; Samson, R., Varga, J., Eds.; Wageningen Academic Publishers: Wageningen, The Netherlands, 2008; p. 133. ISBN 978-90-8686-065-4.

48. Kanematsu, S.; Sasaki, A.; Onoue, M.; Oikawa, Y.; Ito, T. Extending the fungal host range of a partitivirus and a mycoreovirus from *Rosellinia necatrix* by Inoculation of protoplasts with virus particles. *Phytopathology.* **2010**, *100*, 922–930. [CrossRef] [PubMed]

49. Lee, K.M.; Yu, J.; Son, M.; Lee, Y.W.; Kim, K.H. Transmission of *Fusarium boothii* mycovirus via protoplast fusion causes hypovirulence in other phytopathogenic fungi. *PLoS ONE* **2011**, *6*, e21629. [CrossRef] [PubMed]

50. Kanhayuwa, L.; Kotta-Loizou, I.; Özkan, S.; Gunning, A.P.; Coutts, R.H. A novel mycovirus from *Aspergillus fumigatus* contains four unique dsRNAs as its genome and is infectious as dsRNA. *Proc. Natl. Acad. Sci. USA* **2015**, *112*, 9100–9105. [CrossRef] [PubMed]

51. Zheng, L.; Zhang, M.; Chen, Q.; Zhu, M.; Zhou, E. A novel mycovirus closely related to viruses in the genus *Alphapartitivirus* confers hypovirulence in the phytopathogenic fungus *Rhizoctonia solani*. *Virology* **2014**, *456–457*, 220–226. [CrossRef] [PubMed]

52. Hillman, B.I.; Supyani, S.; Kondo, H.; Suzuki, N. A Reovirus of the fungus *Cryphonectria parasitica* that is infectious as particles and related to the *Coltivirus* genus of animal pathogens. *J. Virol.* **2004**, *78*, 892–898. [CrossRef] [PubMed]

53. Sasaki, A.; Kanematsu, S.; Onoue, M.; Oyama, Y.; Yoshida, K. Infection of *Rosellinia necatrix* with purified viral particles of a member of *Partitiviridae* (RnPV1-W8). *Arch Virol.* **2006**, *151*, 697–707. [CrossRef] [PubMed]

54. Sasaki, A.; Kanematsu, S.; Onoue, M.; Oikawa, Y.; Nakamura, H.; Yoshida, K. Artificial Infection of *Rosellinia necatrix* with Purified Viral Particles of a Member of the Genus Mycoreovirus Reveals Its Uneven Distribution in Single Colonies. *Phytopathology* **2007**, *97*, 278–286. [CrossRef] [PubMed]

55. Nuss, D.L. Hypovirulence: Mycoviruses at the fungal-plant interface. *Nat. Rev. Microbiol.* **2005**, *3*, 632–642. [CrossRef] [PubMed]

56. Ghabrial, S.A.; Suzuki, N. Viruses of Plant Pathogenic Fungi. *Annu. Rev. Phytopathol.* **2009**, *47*, 353–384. [CrossRef] [PubMed]

57. Romeralo, C.; Bezos, D.; Martínez-Álvarez, P.; Diez, J.J. Vertical Transmission of *Fusarium circinatum* Mitoviruses FcMV1 and FcMV2-2 via Microconidia. *Forests* **2018**, *9*, 356. [CrossRef]

58. Chu, Y.M.; Lim, W.S.; Yea, S.J.; Cho, J.D.; Lee, Y.W.; Kim, K.H. Complexity of dsRNA mycovirus isolated from *Fusarium graminearum*. *Virus Genes* **2004**, *28*, 135–143. [CrossRef] [PubMed]

59. Milgroom, M.G.; Cortesi, P. Biological control of chestnut blight with hypovirulence: A critical analysis. *Annu. Rev. Phytopathol.* **2004**, *42*, 3113–3118. [CrossRef] [PubMed]

60. Van Diepeningen, A.D.; Debets, A.J.; Hoekstra, R.F. Heterokaryon incompatibility blocks virus transfer among natural isolates of black Aspergilli. *Curr. Genet.* **1997**, *32*, 2091–2097. [CrossRef]

61. Refos, J.M.; Vonk, A.G.; Eadie, K.; Lo-Ten-Foe, J.R.; Verbrugh, H.A.; van Diepeningen, A.D.; van de Sande, W.W. Double-Stranded RNA Mycovirus Infection of *Aspergillus fumigatus* is Not Dependent on the Genetic Make-Up of the Host. *PLoS ONE* **2013**, *8*, e77381. [CrossRef] [PubMed]

62. Coenen, A.; Kevei, F.; Hoekstra, R.F. Factors affecting the spread of double-stranced RNA viruses in *Aspergillus nidulans*. *Genet. Res.* **1997**, *69*, 1–10. [CrossRef] [PubMed]

63. Tan, C.M.C.; Pearson, M.N.; Beever, R.E.; Parkes, S.L. Why Fungi Have Sex? In *14th International Botrytis Symposium abstract book*; Cape Town, South Africa, 2007; p. 26.

64. Brewer, B.J.; Fangman, W.L. Preferential inclusion of extrachromosomal genetic elements in yeast meiotic spores. *Proc. Nat. Acad. Sci. USA* **1980**, *77*, 5380–5384. [CrossRef]

65. Bell, G. *The Masterpiece of Nature: The Evolution and Genetics of Sexuality*; University of California Press: Berkeley, CA, USA, 1982.

66. Frisvad, J.C.; Petersen, L.M.; Lyhne, K.; Larsen, T.O. Formation of sclerotia and production of indoloterpenes by *Aspergillus niger* and other species in section *Nigri*. *PLoS ONE* **2014**, *9*, e94857. [CrossRef] [PubMed]

67. Rajendran, C.; Muthappa, B.N. *Saitoa*, a new genus of Plectomycetes. *Proc. Plant Sci.* **1980**, *89*, 185–191.

68. Paoletti, M.; Seymour, F.A.; Alcocer, M.J.C.; Kaur, N.; Calvo, A.M.; Archer, D.B.; Dyer, P.S. Mating type and the genetic basis of self-fertility in the model fungus *Aspergillus nidulans*. *Curr. Biol.* **2007**, *17*, 1384–1389. [CrossRef] [PubMed]

69. Baidya, S.; Cary, J.W.; Grayburn, W.S.; Calvo, A.M. Role of nitric oxide and flavohemoglobin homolog genes in *Aspergillus nidulans* sexual development and mycotoxin production. *Appl. Environ. Microbiol.* **2011**, *77*, 5524–5528. [CrossRef] [PubMed]

70. De Souza, C.P.; Hashmi, S.B.; Osmani, A.H.; Andrews, P.; Ringelberg, C.S.; Dunlap, J.C.; Osmani, S.A. Functional Analysis of the *Aspergillus nidulans* Kinome. *PLoS ONE* **2013**, *8*, e58008. [CrossRef] [PubMed]

71. Xiang, X.; Osmani, A.H.; Osmani, S.A.; Xin, M.; Morris, N.R. NudF, a nuclear migration gene in *Aspergillus nidulans*, is similar to the human LIS-1 gene required for neuronal migration. *Mol. Biol. Cell* **1995**, *6*, 297–310. [CrossRef] [PubMed]

72. Bayram, O.; Biesemann, C.; Krappmann, S.; Galland, P.; Braus, G.H. More than a repair enzyme: *Aspergillus nidulans* photolyase-like CryA is a regulator of sexual development. *Mol. Biol. Cell* **2008**, *19*, 3254–3262. [CrossRef] [PubMed]

73. Kawasaki, L.; Sánchez, O.; Shiozaki, K.; Aguirre, J. SakA MAP kinase is involved in stress signal transduction, sexual development and spore viability in *Aspergillus nidulans*. *Mol. Microbiol.* **2002**, *45*, 1153–1163. [CrossRef] [PubMed]

74. Semighini, C.P.; Harris, S.D. Regulation of apical dominance in *Aspergillus nidulans* hyphae by reactive oxygen species. *Genetics* **2008**, *179*, 1919–1932. [CrossRef] [PubMed]

75. Westfall, P.J.; Momany, M. *Aspergillus nidulans* septin AspB plays pre- and postmitotic roles in septum, branch, and conidiophore development. *Mol. Biol. Cell* **2002**, *13*, 110–118. [CrossRef] [PubMed]

76. Hernández-Rodríguez, Y.; Hastings, S.; Momany, M. The septin AspB in *Aspergillus nidulans* forms bars and filaments and plays roles in growth emergence and conidiation. *Eukaryot Cell* **2012**, *11*, 3112–3113. [CrossRef] [PubMed]

77. Busch, S.; Eckert, S.E.; Krappmann, S.; Braus, G.H. The COP9 signalosome is an essential regulator of development in the filamentous fungus *Aspergillus nidulans*. *Mol. Microbiol.* **2003**, *49*, 717–730. [CrossRef] [PubMed]

78. Busch, S.; Schwier, E.U.; Nahlik, K.; Bayram, O.; Helmstaedt, K.; Draht, O.W.; Krappmann, S.; Valerius, O.; Lipscomb, W.N.; Braus, G.H. An eight-subunit COP9 signalosome with an intact JAMM motif is required for fungal fruit body formation. *Proc. Natl. Acad. Sci. USA* **2007**, *104*, 8089–8094. [CrossRef] [PubMed]

79. Bok, J.W.; Keller, N.P. LaeA, a regulator of secondary metabolism in *Aspergillus* spp. *Eukaryot Cell* **2004**, *3*, 5273–5275. [CrossRef]

80. Bayram, O.; Krappmann, S.; Ni, M.; Bok, J.W.; Helmstaedt, K.; Valerius, O.; Braus-Stromeyer, S.; Kwon, N.J.; Keller, N.P.; Yu, J.H.; et al. VelB/VeA/LaeA complex coordinates light signal with fungal development and secondary metabolism. *Science* **2008**, *320*, 1504–1506. [CrossRef] [PubMed]

81. Sarikaya, B.O.; Bayram, O.; Valerius, O.; Park, H.S.; Irniger, S.; Gerke, J.; Ni, M.; Han, K.H.; Yu, J.H.; Braus, G.H. LaeA control of velvet family regulatory proteins for light-dependent development and fungal cell-type specificity. *PLoS Genet.* **2010**, *6*, e1001226. [CrossRef] [PubMed]

82. Bayram, O.; Braus, G.H. Coordination of secondary metabolism and development in fungi: The velvet family of regulatory proteins. *FEMS Microbiol. Rev.* **2012**, *36*, 12–14. [CrossRef] [PubMed]

83. Patananan, A.N.; Palmer, J.M.; Garvey, G.S.; Keller, N.P.; Clarke, S.G. A novel automethylation reaction in the *Aspergillus nidulans* LaeA protein generates S-methylmethionine. *J. Biol. Chem.* **2013**, *288*, 14032–14045. [CrossRef] [PubMed]

84. Nitsche, B.M.; Jørgensen, T.R.; Akeroyd, M.; Meyer, V.; Ram, A.F. The carbon starvation response of *Aspergillus niger* during submerged cultivation: Insights from the transcriptome and secretome. *BMC Genomics* **2012**, *13*, 380. [CrossRef] [PubMed]

85. Jørgensen, T.R.; Nitsche, B.M.; Lamers, G.E.; Arentshorst, M.; van den Hondel, C.A.; Ram, A.F. Transcriptomic insights into the physiology of *Aspergillus niger* approaching a specific growth rate of zero. *Appl. Environ. Microbiol.* **2010**, *76*, 5344–5355. [CrossRef] [PubMed]

86. Van Leeuwen, M.R.; Krijgsheld, P.; Bleichrodt, R.; Menke, H.; Stam, H.; Stark, J.; Wösten, H.A.; Dijksterhuis, J. Germination of conidia of *Aspergillus niger* is accompanied by major changes in RNA profiles. *Stud. Mycol.* **2013**, *74*, 59–70. [CrossRef] [PubMed]

87. Fu, C.; Iyer, P.; Herkal, A.; Abdullah, J.; Stout, A.; Free, S.J. Identification and characterization of genes required for cell-to-cell fusion in *Neurospora crassa*. *Eukaryot Cell.* **2011**, *10*, 1100–1109. [CrossRef] [PubMed]

88. Wada, R.; Jin, F.J.; Koyama, Y.; Maruyama, J.; Kitamoto, K. Efficient formation of heterokaryotic sclerotia in the filamentous fungus *Aspergillus oryzae*. *Appl. Microbiol. Biotechnol.* **2014**, *98*, 3253–3254. [CrossRef] [PubMed]

89. Carlile, M.J. The Success of the Hypha and Mycelium. In *The Growing Fungus*; Gow, N.A.R., Gadd, G.M., Eds.; Chapman and Hall: London, UK, 1995; pp. 3–19.

90. Teutschbein, J.; Albrecht, D.; Pötsch, M.; Guthke, R.; Aimanianda, V.; Clavaud, C.; Latgé, J.P.; Brakhage, A.A.; Kniemeyer, O. Proteome profiling and functional classification of intracellular proteins from conidia of the human-pathogenic mold *Aspergillus fumigatus*. *J. Proteome Res.* **2010**, *9*, 3427–3442. [CrossRef] [PubMed]

91. Albrecht, D.; Guthke, R.; Brakhage, A.A.; Kniemeyer, O. Integrative analysis of the heat shock response in *Aspergillus fumigatus*. *BMC Genom.* **2010**, *11*, 32. [CrossRef] [PubMed]

92. Tsitsigiannis, D.I.; Zarnowski, R.; Keller, N.P. The lipid body protein, PpoA, coordinates sexual and asexual sporulation in *Aspergillus nidulans*. *J. Biol. Chem.* **2004**, *279*, 11344–11353. [CrossRef] [PubMed]

93. Duarte, M.; Peters, M.; Schulte, U.; Videira, A. The internal alternative NADH dehydrogenase of *Neurospora crassa* mitochondria. *Biochem. J.* **2003**, *371*, 1005–1011. [CrossRef] [PubMed]

94. Tsai, H.F.; Wheeler, M.H.; Chang, Y.C.; Kwon-Chung, K.J. A Developmentally Regulated Gene Cluster Involved in Conidial Pigment Biosynthesis in *Aspergillus fumigatus*. *J. Bacteriol.* **1999**, *181*, 6469–6477. [PubMed]

95. Tsai, H.F.; Washburn, R.G.; Chang, Y.C.; Kwon-Chung, K.J. *Aspergillus fumigatus* arp1 modulates conidial pigmentation and complement deposition. *Mol. Microbiol.* **1997**, *26*, 175–183. [CrossRef] [PubMed]

96. Tsai, H.F.; Fujii, I.; Watanabe, A.; Wheeler, M.H.; Chang, Y.C.; Yasuoka, Y.; Ebizuka, Y.; Kwon-Chung, K.J. Pentaketide melanin biosynthesis in *Aspergillus fumigatus* requires chain-length shortening of a heptaketide precursor. *J. Biol. Chem.* **2001**, *276*, 29292–29298. [CrossRef] [PubMed]

97. Langfelder, K.; Jahn, B.; Gehringer, H.; Schmidt, A.; Wanner, G.; Brakhage, A.A. Identification of a polyketide synthase gene (pksP) of *Aspergillus fumigatus* involved in conidial pigment biosynthesis and virulence. *Med. Microbiol. Immunol.* **1998**, *187*, 798–799. [CrossRef]

98. Jackson, J.C.; Higgins, L.A.; Lin, X. Conidiation color mutants of *Aspergillus fumigatus* are highly pathogenic to the heterologous insect host *Galleria mellonella*. *PLoS ONE* **2009**, *4*, e4224. [CrossRef] [PubMed]

99. Muñoz-Adalia, E.J.; Fernández, M.M.; Diez, J.J. The use of mycoviruses in the control of forest diseases. *Biocontrol. Sci. Technol.* **2016**, *26*, 577–604. [CrossRef]

viruses

MDPI

Article

Identification and Molecular Characterization of a Novel Partitivirus from *Trichoderma atroviride* NFCF394

Jeesun Chun [1], Han-Eul Yang [2] and Dae-Hyuk Kim [1,2,*]

[1] Institute for Molecular Biology and Genetics, Chonbuk National University, 567 Baekje-daero, Jeonju, Chonbuk 54896, Korea; brainyjsc@gmail.com
[2] Department of Bioactive Material Sciences, Chonbuk National University, 567 Baekje-daero, Jeonju, Chonbuk 54896, Korea; yhe0419@naver.com
* Correspondence: dhkim@jbnu.ac.kr; Tel.: +82-63-270-3440; Fax: +82-63-270-4312

Received: 19 September 2018; Accepted: 21 October 2018; Published: 23 October 2018

Abstract: An increasing number of novel mycoviruses have been described in fungi. Here, we report the molecular characteristics of a novel bisegmented double-stranded RNA (dsRNA) virus from the fungus *Trichoderma atroviride* NFCF394. We designated this mycovirus as Trichoderma atroviride partitivirus 1 (TaPV1). Electron micrographs of negatively stained, purified viral particles showed an isometric structure approximately of 30 nm in diameter. The larger segment (dsRNA1) of the TaPV1 genome comprised 2023 bp and contained a single open reading frame (ORF) encoding 614 amino acid (AA) residues of RNA-dependent RNA polymerase (RdRp). The smaller segment (dsRNA2) consisted of 2012 bp with a single ORF encoding 577 AA residues of capsid protein (CP). The phylogenetic analysis, based on deduced amino acid sequences of RdRp and CP, indicated that TaPV1 is a new member of the genus *Alphapartitivirus* in the family *Partitiviridae*. Virus-cured isogenic strains did not show significant changes in colony morphology. In addition, no changes in the enzymatic activities of β-1,3-glucanase and chitinase were observed in virus-cured strains. To the best of our knowledge, this is the first report of an *Alphapartitivirus* in *T. atroviride*.

Keywords: Trichoderma atroviride; Mycovirus; Partitivirus

1. Introduction

The genus *Trichoderma* is one of the core fungal genera, and ubiquitous strains are typically found in soil and root environments. *Trichoderma* spp. are the principal decomposers of the ecosystem, and perform specialized soil mineralizing functions and nutrient cycling functions by decomposing organic matter in various ecological niches. *Trichoderma* and *Aspergillus* spp. are major producers of a number of industrial and pharmaceutical enzymes, such as cellulase and endo-β-1,3-glucanase from *T. reesei*, *T. harzianum*, or *T. longibrachiatum* [1–4]. In addition, various secondary metabolites of *Trichoderma* spp. are applied as food and animal feed additives [5]. Moreover, *Trichoderma* spp., including *T. harzianum* and *T. atroviride*, are known for their potential value as a biocontrol agent through both classical mycoparaticism and other augmentative biocontrol effects [6,7]. However, these fungi are also known as the cause of green mold disease, which results in substantial losses in the yield of cultivated mushrooms.

Mycoviruses, which are fungal viruses, have been detected in all major taxa of filamentous fungi, mushrooms, and yeasts [8,9]. Fungal viruses have various genome types: (1) double-stranded RNA (dsRNA) genomes, which are taxonomically classified into seven families: *Chrysoviridae*, *Endornaviridae*, *Megabirnaviridae*, *Quadriviridae*, *Partitiviridae*, *Reoviridae*, and *Totiviridae*; (2) single-stranded RNA (ssRNA) genomes, which are classified into six families: *Alphaflexiviridae*, *Barnaviridae*, *Gammaflexiviridae*,

Hypoviridae, *Narnaviridae*, and *Mymonaviridae*; and (3) circular single-stranded DNA (ssDNA) genomes, which are assigned into a newly proposed family: *Gemoniviridae*, and as yet unclassified genomes [10]. Although various fungal viruses have recently been described, the number of mycoviruses for which the genome has been characterized is small compared to plant and animal viruses [11]. Despite the extensive research that led to the discovery of diverse fungal viruses, more information is needed to determine the full role of mycoviruses in relation to their hosts.

Recently, the presence of various mycoviruses in *Trichoderma* spp. has been suggested in studies reporting variable dsRNA mycovirus incidence [12]. In addition, studies of the biological function of mycoviruses have been conducted [13,14]. Moreover, only a few mycoviruses identified in this fungus have been characterized at the molecular level. In this study, we report the dsRNA of a novel mycovirus from *Trichoderma atroviride*.

2. Materials and Methods

2.1. Fungal Strains and Culture Conditions

The *T. atroviride* strain NFCF394 infected with mycovirus was isolated from substrates showing green mold symptoms collected from Korean shiitake farms [12]. Fungal isolates were maintained at 25 °C in the dark on potato dextrose agar (PDA). Viral dsRNA was removed through single-sporing followed by the hyphal-tipping technique [15].

2.2. Isolation and Purification of Virus Particles and Transmission Electron Microscopy

To obtain viral particles, 30 g of mycelia was ground and mixed with 100 mM phosphate buffer (pH 7.4). After the removal of the cellular debris, the lysate was subjected to ultracentrifugation at 4 °C for 2 h to obtain the sediment, which was suspended in 100 mM phosphate buffer. The extract was further subjected to ultracentrifugation in sucrose density gradients (100 to 500 mg/mL with intervals of 100 mg/mL) [16]. The fraction containing the virus particles was carefully collected and dialyzed overnight. The dialyzed fractions were collected through ultracentrifugation for 2 h and suspended in 50 µL of 0.05 M phosphate buffer for further analysis. The structure of the virus-like particles was visualized using a transmission electron microscope (TEM) on an H-7650 instrument installed at the Center for University-Wide Research Facilities at Chonbuk National University (Hitachi, Tokyo, Japan) after negative staining with 2% uranyl acetate. The viral dsRNA elements from the crude extract were extracted with phenol, chloroform, and isoamyl alcohol, precipitated with ethanol, and visualized through agarose gel electrophoresis. Viral proteins were detected through 10 % SDS-PAGE analysis.

2.3. Nucleic Acid Extraction and Viral Genome Sequencing

dsRNA extraction and Northern hybridization analysis were performed as previously reported by Park et al. [17]. Purified dsRNA was subjected to cDNA library construction and genome sequencing using the Illumina HiSeq 2000 platform (Macrogen Inc., Seoul, Korea). The Illumina adapter sequence reads were quality checked using FastQC and trimmed using Trimmomatic (ver. 0.32). Qualified reads were assembled to generate contigs with Trinity, and abundance was estimated using RSEM software (v1.2.15) [18], to calculate the fragments per kilobase of transcript per million mapped reads (FPKM)-values. Northern blot analyses using probes corresponding to the sequences of each contig were conducted.

2.4. Rapid Amplification of cDNA Ends (RACE) Analysis

RNA ligase-mediated rapid amplification of cDNA ends (RLM-RACE) was performed to determine the 5'- and 3'-terminal sequences of dsRNA using an RLM-RACE kit (Ambion, Austin, TX, USA). Purified dsRNA was denatured in dimethyl sulfoxide and treated with calf intestine alkaline phosphatase and tobacco acid pyrophosphatase to remove free 5' phosphates and cap structures. The 5' RNA adapter oligonucleotide (5'-GCUGAUGGCGAUGAAUGAACACUGCGUUU

GCUGGCUUUGAUGAAA-3′) was ligated to the decapped RNA using T4 RNA ligase. The ligates were subjected to random-primed reverse transcription and the 5′ end of a specific sequence was amplified. The 3′ terminus of dsRNA was ligated to a 3′ RACE adapter oligonucleotide (5′-GCGAGCACAGAATTAATACGACTCACTATAGGT12VN-3′) and subjected to RT-PCR. The resulting cDNA was amplified by PCR to determine the 3′ end sequence.

2.5. Sequence Analysis

Phylogenetic trees were constructed using the maximum-likelihood methods [19] with the software package MEGA7 [20] after performing multiple sequence alignments using CLUSTAL X (ver. 2.1) [21].

2.6. Assays of Chitinase and β-1,3-Glucanase Activity

Culture supernatants were harvested and subjected to chitinase assays according to the manufacturer's instructions (Sigma-Aldrich, St. Louis, MO, USA). β-1,3-glucanase assays were performed in 0.05 M sodium citrate buffer (pH 4.5) with β-1,3-glucan for 2 h. The reaction was stopped by heating at 100 °C for 5 min, and the amount of reducing sugar liberated was measured using neocuproine.

3. Results and Discussion

3.1. Profile of Virus Particles from T. atroviride NFCF394

Electron microscopy revealed that the isometric virus particles isolated from *T. atroviride* NFCF394 were isometric with a diameter of approximately 30 nm (Figure 1a), which is similar to the 25 to 40–50 nm diameter reported for members of the family *Partitiviridae*. Subsequently, nucleic acids extracted from the viral particles showed a broad ethidium bromide stained band at 2.0 kbp, which contained two dsRNA segments with a similar size of approximately 2.0 kbp. This agarose gel band pattern was identical to that of the dsRNA preparation from whole RNAs of infected fungal cells (Figure 1b). These data indicated that the dsRNAs identified from mycelia were indeed from the virus particles.

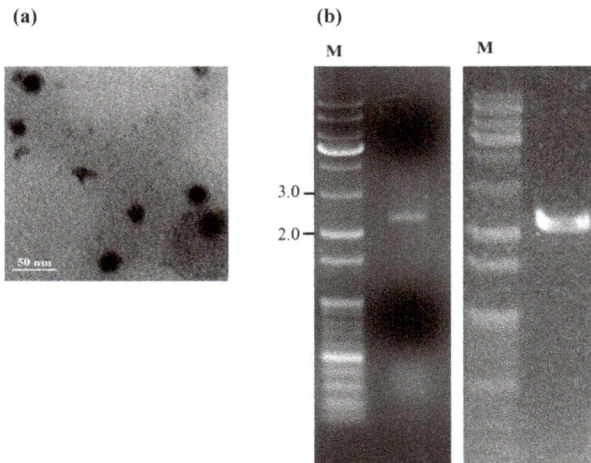

Figure 1. Profiles of isometric viral particles isolated from *T. atroviride* NFCF394. (**a**) Purified virus particles were negative-stained with 2% uranyl acetate and examined using transmission electron microscopy. Scale bar, 50 nm. (**b**) Agarose gel electrophoresis of the dsRNAs extracted from viral particles (left) and mycelia (right) of virus-infected *T. atroviride* NFCF394. Lane M contains the DNA size standard and the numbers indicate the size in kbp.

3.2. Molecular Characterization of Novel Partitivirus from T. atroviride NFCF394

The dsRNA extracted from the mycelia of *T. atroviride* NFCF394 was treated with DNase I and S1 nuclease, and the 2.0 kbp band was resolved with 1% agarose gel electrophoresis (Figure 1b right). The constructed cDNA library from the purified dsRNAs was subjected to next-generation sequencing (NGS) using Illumina HiSeq 2000. A total of 10,494 contigs with an average length of 379 nucleotides was produced. After quality trimming and assembly with Trinity, a total of 21 assembled contigs with significant FPKM values were obtained and further manually assembled. The assembled reads were subjected to a BLASTX search to identify the viral sequence in the NCBI protein database with an E-value cutoff of 0. The results showed two dsRNA segments identified as dsRNA1 and dsRNA2 containing a single open reading frame (ORF) for each segment (Figure 2a). To verify the contig sequences, RT-PCR analyses using corresponding primer pairs based on the two representative contigs (1948 and 1816 bp) were conducted. The resulting specific amplicons were cloned, at least three clones for each amplicon were sequenced, and the near full-length sequences obtained from NGS were verified. Northern hybridization demonstrated that both segments existed in the dsRNA band of the mycoviral genome (Figure 2b). These results confirmed that the dsRNA bands were double bands with similar sizes.

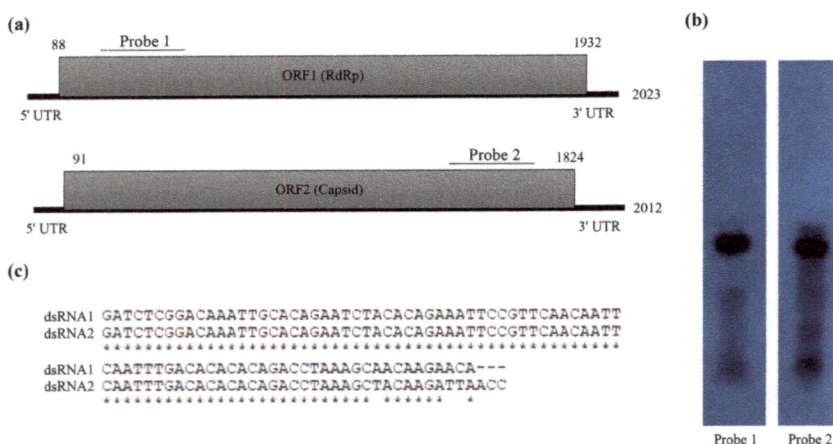

Figure 2. Molecular characteristics of the T. atroviride partitivirus 1 (TaPV1) genomic double-stranded RNAs (dsRNAs). (a) Schematic diagrams of the genomic organization of TaPV1 dsRNA segments. Shaded boxes are open reading frames (ORFs) encoding RNA-dependent RNA polymerase (RdRp) and coat protein (CP). Numbers indicate the total lengths of the TaPV1 genome segments and the positions of the start and stop codons. (b) Northern blot analysis of TaPV1 dsRNA1 and dsRNA2. RNAs were hybridized with probes for dsRNA1 and dsRNA2, and these probes are indicated in panel (b). (c) Alignment of the 5′-terminal untranslated regions of dsRNA1 and dsRNA2 of TaPV1. The conserved sequences are indicated with asterisks.

The sequence analysis of dsRNA1 revealed that it encoded an ORF (ORF1) consisting of 614 amino acids with a predicted molecular mass of 72 kDa and an isoelectric point (pI) of 8.2. Homology searches of the deduced amino acid sequence showed a high similarity to the known sequences of RdRp of *Rosellinia necatrix partitivirus 7* (RnPV7), *Rhizoctonia solani partitivirus 1* (RsPV1), *Rosellinia necatrix partitivirus 5* (RnPV5), and *Sclerotinia sclerotiorum partitivirus S* (SsPV-S) (Table 1). In addition, ORF2 consisted of 577 amino acids with a predicted molecular mass of 65 kDa and pI of 6.7. ORF2 showed similarity to the known sequences of CP of Rosellinia necatrix partitivirus 7 (RnPV7) and *Sclerotinia sclerotiorum partitivirus S* (SsPV-S). Compared to the RdRp amino acid sequences, the CP amino acid sequence showed a lower level of similarity.

The RACE protocol was applied to determine the 5′ and 3′ terminal sequences. The full lengths of the dsRNA sequences for dsRNA1 and dsRNA2 were determined to be 2023 and 2012 bp, respectively. These sequences were deposited in GenBank (accession number MH921573 and MH921574, respectively). The 5′ untranslated region (UTR) of the coding strand of dsRNA1 was 87 nt, and the corresponding 5′ UTR of dsRNA2 was 90 nt. The nucleotide sequences at the 5′-termini of dsRNA1 and dsRNA2 shared a sequence identity of 99% (Figure 2c). A conserved sequence (GAWNW: N, any nt; W, A, or U) at the 5′-terminus of fungal *Alphapartitivirus* was observed at the 5′-termini of dsRNA1 and dsRNA2 (GACAAAUU). These sequences shared a characteristic feature in that the G near the 5′-termini was followed by an A, U, or C but not a G for the next five or six nucleotide positions [22]. In addition, the coding strands of dsRNA1 and dsRNA2 contained 3′ UTRs comprising 91 and 188 nt, respectively. A-rich regions interrupted by non-A residues in the 3′-termini of dsRNA1 (a sequence of 39 A residues in the 3′-terminal 50-nt) and dsRNA2 (a sequence of 26 A residues in the 3′-terminal 50-nt) were found, which is a common characteristic of *Alphapartitivirus* [8].

Table 1. Amino acid sequence identity (%) between T. atroviride alphapartitivirus 1 (TaPV1) ORF1, ORF2 and other viruses from the genus *Alphapartitivirus*.

Virus	Identity (%)	Overlap
ORF1 search		
RnPV7 (LC076694)	60.5	376/622
RsPV1 (AND83003)	52.6	328/623
RnPV5 (BAM36403)	52.8	344/652
SsPV-S (GQ280377)	42.1	263/625
ORF2 search		
RnPV7 (BAT32943)	26.7	178/666
SsPV-S (GQ280378)	21.9	138/630

3.3. Phylogenic Analysis of Amino Acid Sequences Encoded by the ORFs

The phylogenetic analysis of RdRp with top-ranked similar sequences and selected members of *Partitiviridae* was performed using the maximum-likelihood method (Figure 3a). This revealed that RdRp encoded by ORF1 was affiliated with the clade encompassing the genus *Alphapartitivirus* with strong bootstrap support. *Rosellinia necatrix partitivirus 7* and *Sclerotinia sclerotiorum partitivirus S* were the closest phylogenetic neighbors. In addition, the phylogenetic analysis of CP of selected members (Figure 3b) also indicated that this segment belongs to the genus *Alphapartitivirus*.

The genome organization of dsRNAs with bipartite segments, isometric viral particle structure, and virus size indicate that the dsRNAs belong to the family *Paritiviridae*. Together, the genome organization, sequence similarity of RdRp and CP, and the phylogenetic analysis suggest that the dsRNAs are genomic components of the genus *Alphapartitivirus* in the family *Partitiviridae*. Based on the cutoff values of the ICTV criterion for the genus *Alphapartitivirus* in the *Partitiviridae* demarcation (≤ 90% and ≤ 80% amino acid identities in RdRp and CP, respectively), we concluded that the dsRNA in the current study represents a novel species, of which dsRNA1 and dsRNA2 were proposed as genome segments of RdRp and CP in the family *Alphapartitivirus*. Thus, we named our dsRNA Trichoderma atroviride alphapartitivirus 1 (TaPV1). The genome size of TaPV1 is also interesting. The genome segment size of RdRp in TaPV1, which gave a genome length of 2023 bp, fell within the range of genome sizes (1866–2027 bp) of *Alphapartitivirus* members. In addition, the genome size of CP in TaPV1 (2012 bp) exhibited a substantially larger size than previous reports suggested (1708–1866 bp). Thus, the genome size of TaPV1 appears to be the largest of the fungal *Alphapartitivirus* organisms.

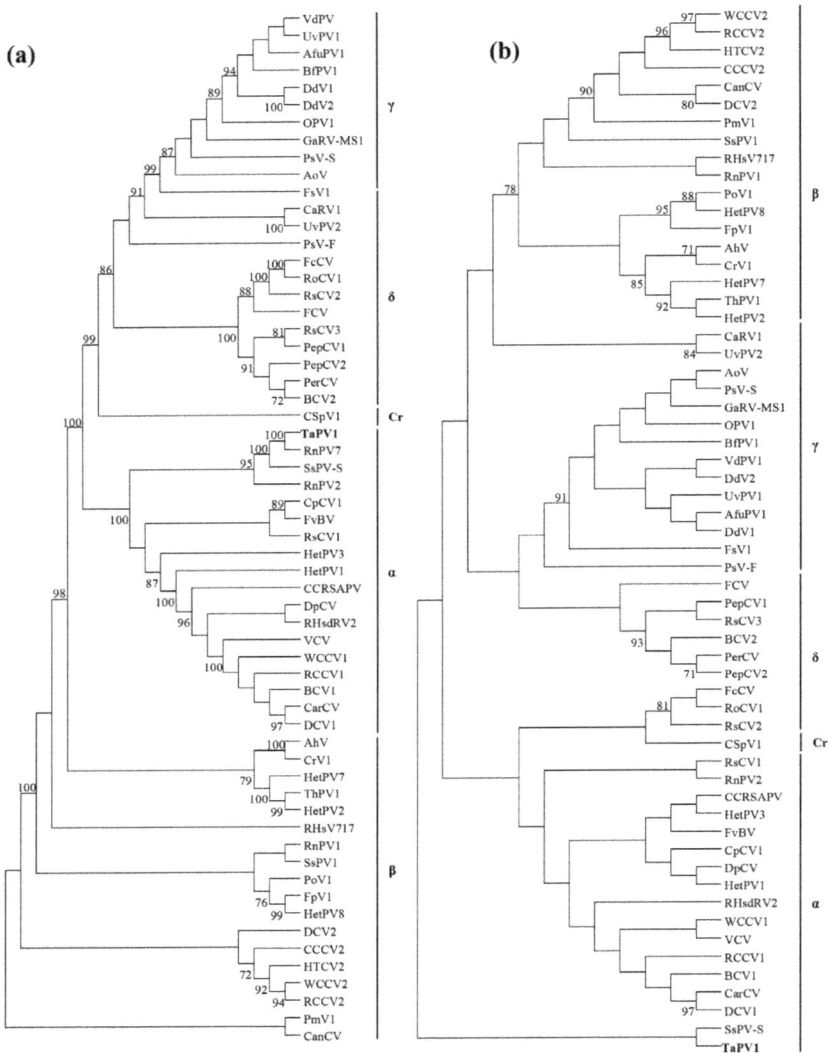

Figure 3. Phylogenetic analysis of TaPV1 and selected dsRNA viruses. (**a**) Maximum-likelihood (ML) tree using LG + F substitution model based on the RdRp amino acid sequences of *Partitiviridae*. α, *Alphapartitivirus*; β, *Betapartitivirus*; γ, *Gammapartitivirus*; δ, *Deltapartitivirus*; Cr, *Cryspovirus*. (**b**) ML tree for CP amino acid sequences of *Partitiviridae*. α, *Alphapartitivirus*; β, *Betapartitivirus*; γ, *Gammapartitivirus*; δ, *Deltapartitivirus*; Cr, *Cryspovirus*. The numbers at the nodes represent bootstrap values out of 1000 replicates; values are shown only if greater than 70%. See Table S1 for a detailed listing of viruses.

3.4. Phenotypic Characteristics of Mycovirus-Cured and -Containing Strains

T. atroviride NFCF394 was observed as white mycelium formed on the whole plate of PDA after incubation for 3 days. To determine whether the mycovirus affects the fungal host phenotype, virus-cured isogenic strains were obtained using single-spore isolation followed by hyphal tipping. Spores were harvested from 7-day-old culture plates containing actively growing *T. atroviride* NFCF394, and 50 expected colony-forming units were spread on a fresh PDA plate. Of the 20 single-spored progenies, TaPV1 dsRNA bands were eliminated in three single-spored progenies, suggesting the vertical transmission of TaPV1 at a rate consistent with our previous data [13]. However, no apparent difference in colony morphology was observed between TaPV1-containing and the three virus-cured strains. The activities of two representative antifungal enzymes, β-1,3-glucanase and chitinase, were also analyzed. No obvious alterations of enzymatic activities were observed between the infected and virus-cured isogenic strains (Figure S1). Comparisons of means were tested using the Student's *t*-test.

Although there are many cases of mycoviruses that induce the viral-specific symptoms in the host, including reduced fungal virulence (hypovirulence) in phytopathogenic fungi [23], infection of mycoviruses is generally thought to be asymptomatic or cryptic in nature [24]. While further studies are required, no changes in fungal growth, colony morphology, or antifungal enzyme activity were observed in this study. However, to clarify the TaPV1-host interaction, studies that include viral transfection with purified viral particles will be necessary.

Supplementary Materials: Supplementary materials can be found at http://www.mdpi.com/1999-4915/10/11/578/s1.

Author Contributions: Supervision; D.-H.K.; Methodology, J.C. and H-E.Y.; Writing and Original Draft Preparation, J.C.; Writing, Review and Editing, D.-H.K.

Funding: This work was supported by the NRF grants from MSIP (2018R1A2A1A05078682).

Acknowledgments: We wish to thank the Institute of Molecular Biology and Genetics at Chonbuk National University for kindly providing the facilities for this research.

Conflicts of Interest: The authors declare that they have no conflict of interest.

References

1. Sivasithamparam, K.; Ghisalberti, E.L. *Trichoderma and Gliocladium: Basic Biology, Taxonomy and Genetics*, 1st ed.; Taylor and Francis: London, UK, 1998; pp. 139–191.

2. Kubicek, C.P.; Penttila, M.E. *Trichoderma and Gliocladium: Enzymes, Biological Control and Commercial Applications*, 1st ed.; Taylor and Francis: London, UK, 1998; pp. 49–71.

3. Gupta, V.K.; Schmoll, M.; Herrera-Estrella, A.; Upadhyay, R.S.; Druzhinina, I.; Tuohy, M.G. *Biotechnology and Biology of Trichoderma*; Elsevier: Oxford, UK, 2014.

4. Schuster, A.; Schmoll, M. Biology and biotechnology of *Trichoderma*. *Appl. Microbiol. Biotechnol.* **2010**, *87*, 787–799. [CrossRef] [PubMed]

5. Contreras-Cornejo, H.A.; Macías-Rodríguez, L.; del-Val, E.; Larsen, J. Ecological functions of *Trichoderma* spp. and their secondary metabolites in the rhizosphere: Interactions with plants. *FEMS Microbiol. Ecol.* **2016**, *92*, fiw036. [CrossRef] [PubMed]

6. Papavizas, G.C. *Trichoderma* and *Gliocladium*: Biology, ecology, and potential for biocontrol. *Annu. Rev. Phytopathol.* **1985**, *23*, 23–54. [CrossRef]

7. Howell, C.R. Mechanisms employed by *Trichoderma* species in the biological control of plant diseases: The history and evolution of current concepts. *Plant Dis.* **2003**, *87*, 4–10. [CrossRef]

8. Ghabrial, S.A.; Castón, J.R.; Jiang, D.; Nibert, M.L.; Suzuki, N. 50-plus years of fungal viruses. *Virology* **2015**, *479–480*, 356–368. [CrossRef] [PubMed]

9. Wickner, R.B. Double-stranded and single-stranded RNA viruses of *Saccharomyces cerevisiae*. *Annu. Rev. Microbiol.* **1992**, *46*, 347–375. [CrossRef] [PubMed]

10. The Online (10th) Report of the International Committee on Taxonomy of Viruses. Available online: https://talk.ictvonline.org/ictv-reports/ictv_online_report (accessed on 12 December 2016).

11. Roossinck, M.J. Metagenomics of plant and fungal viruses reveals an abundance of persistent lifestyles. *Front. Microbiol.* **2015**, *5*, 1–3. [CrossRef] [PubMed]

12. Yun, S.H.; Lee, S.H.; So, K.K.; Kim, J.M.; Kim, D.H. Incidence of diverse dsRNA mycoviruses in *Trichoderma* spp. causing green mold disease of shiitake *Lentinula edodes*. *FEMS Microbiol. Lett.* **2016**. [CrossRef] [PubMed]

13. Lee, S.H.; Yun, S.H.; Chun, J.; Kim, D.H. Characterization of a novel dsRNA mycovirus of *Trichoderma atroviride* NFCF028. *Arch. Virol.* **2017**, *162*, 1073–1077. [CrossRef] [PubMed]

14. Chun, J.; Yang, H.E.; Kim, D.H. Identification of a novel partitivirus of *Trichoderma harzianum* NFCF319 and evidence for the related antifungal activity. *Front. Plant Sci.*. under review.

15. Kim, J.M.; Jung, J.E.; Park, J.A.; Park, S.M.; Cha, B.J.; Kim, D.H. Biological function of a novel chrysovirus, CnV1-Bs122, in the Korean *Cryphonectria nitschkei* BS122 strain. *J. Biosci. Bioeng.* **2013**, *115*, 1–3. [CrossRef] [PubMed]

16. Wang, L.; Jiang, J.; Wang, Y.; Hong, N.; Zhang, F.; Xu, W.; Wang, G. Hypovirulence of the phytopathogenic fungus *Botryosphaeria dothidea*: Association with a coinfecting chrysovirus and a partitivirus. *J. Virol.* **2014**, *88*, 7517–7527. [CrossRef] [PubMed]

17. Park, S.M.; Kim, J.M.; Chung, H.J.; Lim, J.Y.; Kwon, B.R.; Lim, J.G.; Kim, J.A.; Kim, M.J.; Cha, B.J.; Lee, S.H.; et al. Occurrence of diverse dsRNA in a Korean population of the chestnut blight fungus, *Cryphonectria parasitica*. *Mycol. Res.* **2008**, *112*, 1220–1226. [CrossRef] [PubMed]

18. Li, B.; Dewey, C.N. RSEM: Accurate transcript quantification from RNA-Seq data with or without a reference genome. *BMC Bioinform.* **2011**, *12*, 323. [CrossRef] [PubMed]

19. Felsenstein, J. Confidence limits on phylogenies: An approach using the bootstrap. *Evolution* **1985**, *39*, 783–791. [CrossRef] [PubMed]

20. Kumar, S.; Stecher, G.; Tamura, K. MEGA7: Molecular Evolutionary Genetics Analysis version 7.0 for bigger datasets. *Mol. Biol. Evol.* **2016**, *33*, 1870–1874. [CrossRef] [PubMed]

21. Thompson, J.D.; Gibson, T.J.; Plewniak, F.; Jeanmougin, F.; Higgins, D.G. The CLUSTAL_X windows interface: Flexible strategies for multiple sequence alignment aided by quality analysis tools. *Nucleic Acids Res.* **1987**, *25*, 4876–4882. [CrossRef]

22. Nibert, M.L.; Ghabrial, S.A.; Maiss, E.; Lesker, T.; Vainio, E.J.; Jiang, D.; Suzuki, N. Taxonomic reorganization of family *Partitiviridae* and other recent progress in partitivirus research. *Virus Res.* **2014**, *88*, 128–141. [CrossRef] [PubMed]

23. Van Alfen, N.K. Hypovirulence of *Endothia* (*Cryphonectria*) *parasitica* and *Rhizoctonia solani*. In *Fungal Virology*, 1st ed.; Buck, K.W., Ed.; CRC Press: Boca Raton, FL, USA, 1986; pp. 143–162.

24. Pearson, M.N.; Beever, R.E.; Boine, B.; Arthur, K. Mycoviruses of filamentous fungi and their relevance to plant pathology. *Mol. Plant. Pathol.* **2009**, *10*, 115–128. [CrossRef] [PubMed]

Article

Mitovirus and Mitochondrial Coding Sequences from Basal Fungus *Entomophthora muscae*

Max L. Nibert [1],*, Humberto J. Debat [2], Austin R. Manny [1], Igor V. Grigoriev [3,4] and Henrik H. De Fine Licht [5]

[1] Department of Microbiology and Program in Virology, Harvard Medical School, Boston, MA 02115, USA; austinmanny@g.harvard.edu

[2] Instituto de Patología Vegetal, Centro de Investigaciones Agropecuarias, Instituto Nacional de Tecnología Agropecuaria (IPAVE-CIAP-INTA), Córdoba X5020ICA, Argentina; debat.humberto@inta.gob.ar

[3] U.S. Department of Energy Joint Genome Institute, Walnut Creek, CA 94598, USA; ivgrigoriev@lbl.gov

[4] Department of Plant and Microbial Biology, University of California Berkeley, Berkeley, CA 94720, USA

[5] Department of Plant and Environmental Sciences, University of Copenhagen, DK-1871 Frederiksberg, Denmark; hhdefinelicht@plen.ku.dk

* Correspondence: mnibert@hms.harvard.edu; Tel.: +1-617-645-3680

Received: 29 March 2019; Accepted: 15 April 2019; Published: 17 April 2019

Abstract: Fungi constituting the *Entomophthora muscae* species complex (members of subphylum *Entomophthoromycotina*, phylum *Zoopagamycota*) commonly kill their insect hosts and manipulate host behaviors in the process. In this study, we made use of public transcriptome data to identify and characterize eight new species of mitoviruses associated with several different *E. muscae* isolates. Mitoviruses are simple RNA viruses that replicate in host mitochondria and are frequently found in more phylogenetically apical fungi (members of subphylum *Glomeromyoctina*, phylum *Mucoromycota*, phylum *Basidiomycota* and phylum *Ascomycota*) as well as in plants. *E. muscae* is the first fungus from phylum *Zoopagomycota*, and thereby the most phylogenetically basal fungus, found to harbor mitoviruses to date. Multiple UGA (Trp) codons are found not only in each of the new mitovirus sequences from *E. muscae* but also in mitochondrial core-gene coding sequences newly assembled from *E. muscae* transcriptome data, suggesting that UGA (Trp) is not a rarely used codon in the mitochondria of this fungus. The presence of mitoviruses in these basal fungi has possible implications for the evolution of these viruses.

Keywords: database mining; *Entomophthora*; *Entomophthoromycotina*; fungal virus; mitochondrion; mycovirus; virus discovery; *Mitovirus*; *Narnaviridae*

1. Introduction

The classification scheme for kingdom *Fungi* currently applied by the National Center for Biotechnology Information (NCBI; Bethesda, MD, USA) includes eight major phyla: *Ascomycota*, *Basidiomycota*, *Mucoromycota* (subphyla *Glomeromycotina*, *Mortierellomycotina*, and *Mucoromycotina*), *Zoopagomycota* (subphyla *Entomophthoromycotina*, *Kickxellomycotina*, and *Zoopagomycotina*), *Blastocladiomycota*, *Chytridiomycota*, *Cryptomycota*, and *Microsporidia*, in approximate order of increasing time since they emerged as divergent taxa [1,2]. Fungal mitoviruses have been reported to date only from the most recently (apically) diverging phyla, *Ascomycota* and *Basidiomycota*, as well as from subphylum *Glomeromycotina* in phylum *Mucoromycota* [3–5]. We remain interested to discover fungal mitoviruses from other, less recently (basally) diverging phyla or subphyla, as part of an effort to understand the deeper evolutionary history of these viruses.

Mitoviruses are currently classified in genus *Mitovirus*, family *Narnaviridae*. They have small plus-strand RNA genomes and replicate persistently in host cell mitochondria [3,6–8]. Though initially reported only from fungi [3–5], they have recently been found also in plants [9,10]. Several mitoviruses reported also from invertebrates [11,12] might instead have been derived from fungal symbionts of the sampled animals. Mitovirus genomes range between 2.0 and 4.5 kb in length, each encompassing a single long open reading frame that encodes a deduced protein with the conserved motifs of a viral RNA-dependent RNA polymerase (RdRp) [13]. Mitoviruses do not form virions and are thought to persist and replicate instead as ribonucleoprotein complexes inside infected mitochondria, which are transmitted to daughter cells during cell division, as well as horizontally during hyphal anastomosis in the case of fungal mitoviruses and vertically through spores or seeds in the case of fungal or plant mitoviruses, respectively [3,8–10,14]. Mitovirus infections in fungi have been linked to morphological abnormalities in mitochondria, defective in vitro growth, and hypovirulence in some phytopathogenic fungi [8,15–18]. In fact, a recent report based on exhaustive phylogenetic analyses of RNA-dependent polymerases suggests that mitoviruses represent one of the most phylogenetically basal groups of eukaryotic RNA viruses [19].

Entomophthoromycotina is one of three subphyla that currently constitute fungal phylum *Zoopagomycota* [1,2]. A few entomophthoroid fungi that are normally soil saprobes can also cause human infections, including *Conidiobolus coronatus* (order *Entomophthorales*) and *Basidiobolus ranarum* (order *Basidiobolales*) [20,21]. Entomophthoroid fungi are probably best known, however, as insect parasites, which exhibit generally narrow host ranges and can cause epizootic events in which large numbers of susceptible insects are killed within local geographic regions [22]. The *Entomophthora muscae* species complex (order *Entomophthorales*) [23] is one such group of entomopathogenic fungi, which infect and kill their dipteran, commonly muscoid fly, hosts. *E. muscae* is also known to manipulate fly behaviors, in particular inducing the behavior known as "summiting", in which the fly climbs to a high surface where it becomes affixed via fungal outgrowths and then strikes a characteristic pose considered to aid in dispersal of spores from the subsequent carcass [24].

De Fine Licht et al. [25] have reported on the transcriptomics of *E. muscae* isolates, in a broad effort to understand their host specificity and pathogenicity. In that report, the authors mention the presence of several apparent viral transcripts in their transcriptome shotgun assemblies but provide no further descriptions of those viruses. As described in detail below, we subsequently discovered a number of apparent mitovirus sequences from that transcriptome study by searching for mitovirus-like sequences within the Transcriptome Shotgun Assembly (TSA) database at NCBI. Recognizing several interesting features of these apparent new mitoviruses from *E. muscae*, we have gone on to characterize their sequences in detail, as we report here. As a part of this work, we have also identified additional strains of these mitoviruses by assembling sequence reads from two other transcriptome studies of *E. muscae*, as obtained from the Sequence Read Archive (SRA) database at NCBI. Included among our findings is identification of numerous UGA (Trp) codons in the mitovirus sequences from *E. muscae*, as well as in mitochondrial core-gene coding sequences from *E. muscae*, leading us to conclude that UGA (Trp) is not a rarely used codon in *E. muscae*, unlike the case in many other basal fungi [4]. *E. muscae* is hereby the first fungus from phylum *Zoopagomycota*, and also the most phylogenetically basal fungus [1,2], that has been reported to harbor mitoviruses to date, with possible implications for mitovirus evolution.

2. Materials and Methods

2.1. Assembly and Analysis of Mitovirus Sequences from E. muscae

The main steps in this process are detailed in Results. All searches of the TSA, SRA, and NR databases were performed using the BLAST web server at NCBI. The 15 TSA hits from the initial TBLASTN searches were: GEMZ01003603.1, GEMZ01006022.1, GEMZ01006112.1, GEMZ01008256.1, GEMZ01008924.1, GEMZ01011847.1, GEMZ01008924.1, and GEMZ01021189.1; GENA01003603.1, GENA01006022.1, GENA01006112.1, GENA01008256.1, GENA01008924.1, GENA01011847.1, GENA01008924.1, and GENA01021189.1; and GEND01018947.1 [25]. The mitovirus queries for those searches were: BAJ23143.2, BAN85985.1, AVA17449.1–AVA17452.1, and AXY40441.1–AXY40444.1 [5,26–28]. To identify mitovirus-matching reads in SRA accessions prior to contig assembly, searches were performed using Discontiguous MegaBLAST or BLASTN. Sequence reads were then assembled into contigs using CAP3 [29] as implemented at http://biosrv.cab.unina.it/webcap3/ or galaxy.pasteur.fr/. Terminal residues in each assembled contig were trimmed back to those that agree between at least two mitovirus strains of the same species. Following contig assembly, mapped reads were re-identified using MegaBLAST, and coverage depths were determined using the "Map Reads to Reference" tool in CLC Genomics Workbench v7. RPKM and coverage depth values are shown in Table S1.

For generating transcriptome shotgun assemblies from SRA libraries, we used Trinity v2.2.0 [30] with default parameters as implemented through https://usegalaxy.org/ and rnaSPAdes v3.13.0 [31] with default parameters as implemented locally. The contigs from each Trinity assembly were subjected to searches with TBLASTN using mitovirus RdRp sequences as queries. Identified contigs were then extended by iterative mapping of reads from the corresponding BioProject using the "Map to Reference" tool in Geneious v8.1.9 with low-sensitivity parameters. The contigs from each rnaSPAdes assembly were subjected to taxonomic classification using locally implemented DIAMOND v0.9.22.123 [32] in BLASTX mode against the NCBI NR database, with an E-value threshold of 1e−3 and a "top" parameter of 1 to identify the best species-level hits for each contig.

Codon usage was analyzed using the Codon Usage tool in Sequence Manipulation Suite at http://www.bioinformatics.org/sms2/. Pairwise and multiple sequence alignments were performed using MAFFT v7 [33] as implemented at https://mafft.cbrc.jp/alignment/server/. For determining % sequence identity, pairwise alignments were performed using EMBOSS Needleall as implemented at http://www.bioinformatics.nl/cgi-bin/emboss/needleall. Maximum-likelihood phylogenetic analyses were performed using the program IQ-Tree [34], including the "Find best and apply" option (ModelFinder) [35] and the "Bootstrap ultrafast" option (UFboot) [36], as implemented at https://www.hiv.lanl.gov/content/sequence/HIV/HIVTools.html.

2.2. Assembly and Analysis of Mitochondrial Gene Coding Sequences from E. muscae

Programs were used as described above for mitovirus sequences. The following TSA hits arose from the initial TBLASTN searches for mitochondrial gene coding sequences from *E. muscae* and were then used as queries of the SRA transcriptome libraries from *E. muscae* isolate Berkeley: *atp6*, GENA01019820.1; *atp9*, GENA01019993.1; *cob*, GEND01033240.1; *cox1*, GEND01036136.1 and GEND01034538.1; *cox2*, GEND01003800.1; *cox3*, GEND01033140.1; *nad1*, GENA01026764.1 and GENA01028096.1; *nad2*, GENA01003576.1; *nad3*, GENA01020377.1; *nad4*, GEND01006792.1 and GEND01006791.1; *nad5*, GENA01001061.1; and *nad6*, GEND01032932.1 and GENA01021432.1 [25]. RPKM and coverage depth values are shown in Table S2.

2.3. SRA Accessions Used for Assembling Contigs

SRA accessions that we used in this study for assembling mitovirus or mitochondrial transcript contigs from each *E. muscae* isolate are listed in Table S3. For *E. muscae* isolate KVL-14-117, the SRA accessions only from media-grown fungus (not fly-grown fungus) were used for assembling mitovirus contigs, in an effort to ensure that the identified viruses were derived from *E. muscae*, and not the fly hosts or a different fly symbiont. For *E. muscae* isolate Berkeley, the SRA accessions from only (i) flies infected with *E. muscae* (not uninfected flies) and (ii) later times post-infection (72, 96, and 120 h, not 24 and 48 h) were used for assembling mitovirus or mitochondrial transcript contigs, in an effort to increase the proportion of reads that were derived from *E. muscae*, not the fly hosts.

2.4. Newly Reported Sequences

Coding-complete mitovirus sequences from *E. muscae* isolates KVL-14-117, KVL-14-118, HHdFL130914-1, HHdFL050913-1, and Berkeley have been deposited in GenBank as accessions MK682513.1–MK682534.1 and BK010729.1–BK010736.1. Mitochondrial core-gene coding sequences from *E. muscae* isolate Berkeley have been deposited in GenBank as accessions BK010748–BK010759. In addition, the mitovirus and mitochondrial core-gene sequences from all five of these *E. muscae* isolates have been included in the Supplementary Materials for this report as Data S1 and Data S2, respectively.

2.5. Sequencing at the Joint Genome Institute, 1000 Fungal Genomes Project

The sequence reads from *E. muscae* isolate HHdFL130914-1 (accession SRX2782457) have not been reported in a peer-reviewed article to date, and the methods are therefore described here. A stranded cDNA library was generated using an Illumina TruSeq Stranded RNA LT kit (San Diego, CA, USA). mRNA was purified from 1 µg of total RNA using magnetic beads containing poly-T oligos. mRNA was fragmented and reverse-transcribed using random hexamers and Superscript II (Invitrogen, Carlsbad, CA, USA) followed by second strand synthesis. The fragmented cDNA was treated with end repair, A-tailing, adapter ligation, and eight cycles of PCR. The prepared library was quantified using KAPA Biosystem's next-generation sequencing library qPCR kit (Wilmington MA, USA) and run on a Roche LightCycler 480 real-time PCR instrument (Pleasanton CA, USA). The quantified libraries were then prepared for sequencing on the Illumina HiSeq platform using an Illumina TruSeq Rapid paired-end cluster kit. Sequencing of the flow cell was performed on the Illumina HiSeq2500 sequencer using Illumina HiSeq TruSeq SBS sequencing kits, following a 2 × 150-nt indexed run recipe.

3. Results

3.1. Mitovirus-like Sequences in Transcriptome Data from E. muscae

We used TBLASTN to search entries from basal fungi (not phylum *Ascomycota*, phylum *Basidiomycota*, or subphylum *Glomeromycotina*) in the TSA database at NCBI, using RdRp sequences of previously reported fungal mitoviruses as queries. The searches yielded 15 strong hits (E-values, 5e−52 to 2e−10 with different queries; see Materials and Methods for further details), all derived from the same transcriptomics study of muscoid-fly pathogens from the *E. muscae* species complex (BioProject PRJEB10825) [25], including *E. muscae sensu stricto* isolates from house flies (*Musca domestica*) and *E. muscae sensu lato* isolates from cabbage flies (*Delia radicum*). The lengths of these 15 hits range from 2321 to 2813 nt, within the expected interval for mitovirus genomes [3,4].

Before analyzing these apparent new mitovirus sequences, we examined the TSA metadata to learn about the samples and to identify the SRA accessions from which the TSA hits had been assembled. After reviewing the metadata and the sequences themselves, we chose to reassemble the mitovirus-like contigs for each *E. muscae* isolate, using the SRA-derived sequence reads that mapped to each respective TSA hit and including any additional reads that lengthened or shortened each contig. We were thereby able to reassemble seven distinct mitovirus-like sequences from media-grown *E. muscae* house fly isolate KVL-14-117, seven distinct mitovirus-like sequences from fly-grown *E. muscae* house fly isolate KVL-14-118 (one of the seven sequences includes two small gaps in sequencing coverage, as inferred by alignment with related sequences from other isolates), and one mitovirus-like sequence from fly-grown *E. muscae* cabbage fly isolate HHdFL050913-1. A small number of reads (22 total) matching the mitovirus from isolate HHdFL050913-1 were also identified from fly-grown *E. muscae* cabbage fly isolate HHdFL040913-2 [25] but were sufficient in coverage to assemble only a few short contigs.

By reviewing the SRA database for other transcriptome-associated accessions that are annotated as relating to *E. muscae* or other *Entomophthoromycotina* members, we found such accessions from five other BioProjects: PRJNA372837 for *E. muscae* house fly isolate HHdFL130914-1 (also known as KVL-14-115 [37]), PRJNA435715 for *E. muscae* fruit fly isolate Berkeley (originally obtained from *Drosophila hydei*) [24], PRJNA259024 for *Conidiobolus thromboides* isolate FSU 785, PRJNA67455 for *Conidiobolus coronatus* isolate NRRL 28638, and PRJNA501640 for *Zoophthora radicans* isolate ATCC 208865/ARSEF 4784. We therefore used the newly assembled mitovirus-like sequences described above as queries to search the SRA transcriptome libraries from these five other BioProjects and obtained numerous strong hits from two of them, those for *E. muscae* house fly isolate HHdFL130914-1 and *E. muscae* fruit fly isolate Berkeley. Assembling these SRA reads from isolates HHdFL130914-1 and

Berkeley then gave rise to seven distinct mitovirus-like sequences from each of these additional isolates, for a total of 29 mitovirus-like sequences assembled from five different *E. muscae* isolates (Figure 1).

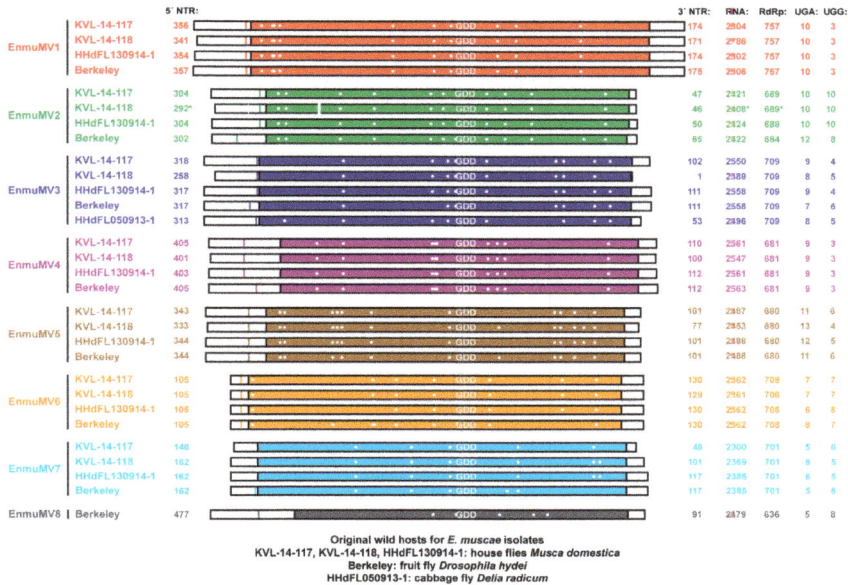

Figure 1. Summary diagram of mitovirus sequences from *E. muscae* isolates. Mitovirus strains from five different *E. muscae* isolates, representing eight apparent new mitovirus species, are labeled at left. The RdRp-encoding open reading frame ORF is shown as a colored box within each assembled RNA sequence (see matching colors in Figure 2), starting with the first in-frame AUG codon by convention. The position of the first in-frame stop codon upstream of the proposed AUG start codon is shown as a colored line. Positions of UGA (Trp) codons in each RdRp ORF are shown as white dots. RNA sequences are aligned with respect to the catalytic motif GDD in each deduced RdRp sequence. Positions of two small gaps in sequencing coverage in the assembled EnmuMV2-KVL-14-118 sequence are shown as white breaks in the boxes; asterisks indicate inferred values consequent to these coverage gaps. For each sequence, lengths of the 5′ NTR, 3′ NTR, and overall RNA sequence are labeled in nt; length of the RdRp sequence in aa; and UGA and UGG codons in raw counts.

Given the large number of mitovirus-like sequences associated with *E. muscae*, we reasoned that there might be sequences of yet other, divergent mitoviruses remaining to be identified in the SRA transcriptome libraries. To address this possibility, we generated our own shotgun assemblies from the libraries analyzed in the preceding paragraph and then searched these new assemblies for other mitoviruses. In this manner, we were able to identify one additional mitovirus-like sequence from *E. muscae* isolate Berkeley (sequence length, 2479 nt) (Figure 1), bringing the overall total to 30 such sequences from the five *E. muscae* isolates.

Especially in plants, fragments of mitovirus genome sequences have been commonly endogenized in host DNA [9,38–40]. To address this possibility for the mitovirus sequences reported here, we noted that there are six SRA libraries at NCBI from PacBio SMRT runs on genomic DNA from *E. muscae* isolate HHdFL130914-1 (BioProject PRJNA346904), an isolate that we showed above to harbor sequences from seven mitovirus strains in its RNA transcriptome data. These six SRA libraries from *E. muscae* DNA were found to register no MegaBLAST hits to the seven mitovirus sequences, providing evidence against these virus sequences being derived from endogenized elements. A small number of mitovirus-matching reads (28 total) found in three SRA libraries arising from Illumina runs on genomic DNA from the same BioProject as the PacBio SMRT runs seem unlikely to be significant, though they may be reminiscent of evidence for non-genomic DNA fragments detected for *Gigaspora margarita* mitoviruses [28]. Contigs for *E. muscae* mitoviruses found as hits from the Whole Genome Shotgun (WGS) database at NCBI are in fact derived from the transcriptome study BioProject PRJEB10825 [25], and thus from RNA not DNA (H.H.D.F.L.).

3.2. Basic Features of the Apparent Mitovirus Sequences from E. muscae

The lengths of our 30 newly assembled mitovirus-like contigs range from 2300 to 2806 nt (Figure 1), within the expected interval for mitovirus genomes [3,4]. Using genetic code 4 in which UGA encodes Trp not "stop" (as expected for translation in the mitochondria of most animals and fungi including members of phylum *Zoopagomycota* [41–43]), a single long open reading frame (ORF) is found in each contig, flanked by one or more stop codon at the plus-strand 5′ end in 27 of the 30 sequences and one or more stop codon at the plus-strand 3′ end in all 30 sequences, suggesting to us that all of the contigs are coding complete. The contigs encode proteins of nine different lengths (assuming that the first in-frame AUG is the start codon in each): four encode a 757-aa protein, five encode a 709-aa protein, four encode a 708-aa protein, four encode a 701-aa protein, three encode a 689-aa protein, one encodes a 684-aa protein, four encode a 681-aa protein, four encode a 680-aa protein, and one encodes a 636-aa protein, suggesting that there may be multiple mitovirus species represented by these sequences. When used in BLASTP searches of the Nonredundant Protein Sequences (NR) database for Viruses at NCBI, each of these deduced protein sequences showed strongest similarities to mitovirus RdRps, and the approximate position of the RdRp catalytic motif GDD is shown and centered for each in Figure 1 (also shown in the multiple sequence alignment of the eight sequences from *E. muscae* isolate Berkeley in Figure S1). For nine contigs, the top hit was to the Erysiphe necator mitovirus 1 (ATS94398; E-values, 1e−91 to 8e−82; identity scores, 37–40%); for nine other contigs, the top hit was to Erysiphe necator mitovirus 2 (ATS94399; E-values, 1e−94 to 5e−70; identity scores, 34–38%); for eight other contigs, the top hit was to Erysiphe necator mitovirus 3 (ATS94400; E-values, 0.0 to 5e−119; identity scores, 36–45%); and for the remaining four contigs, the top hit was to Hubei narna-like virus 25 (APG77157; E-values, 0.0; identity scores, 90%), which is also an apparent mitovirus [10]. These BLASTP findings suggest again that there may be multiple mitovirus species represented by these sequences. The mitovirus-like contigs may be missing some residues at one or both ends relative to the full-length viral genomes, but even so, there is a generally long nontranslated region (NTR) at the plus-strand 5′ end of each contig (105–477 nt, median 317 nt; again assuming that the first in-frame AUG is the start codon in each) and a generally shorter NTR at the plus-strand 3′ end of each contig (1–175 nt, median 106 nt) (Figure 1).

Most fungal mitoviruses contain a number of in-frame UGA codons in the RdRp open reading frame, encoding Trp per genetic code 4; however, some do not, and almost all of those that do not derive from fungal hosts from phylum *Basidiomycota* or subphylum *Glomeromycotina* in which UGA(Trp) is a rarely used mitochondrial codon [4,28]. The 30 mitovirus-like sequences from *E. muscae* described above fit the more typical pattern for fungal mitoviruses, in that a substantial fraction of their Trp codons are UGA (vs. UGG): 38–77%, numbering between 5 and 13 UGA(Trp) codons in the different apparent mitovirus sequences (Figure 1).

3.3. Pairwise and Phylogenetic Comparisons of the Mitovirus Sequences from E. muscae

Features of the mitovirus-like contigs described above led us to suspect that there are multiple new mitovirus species represented by these sequences. To address this possibility, we performed pairwise comparisons of the nt sequences using EMBOSS Needleall. These comparisons grouped the sequences into eight distinct clusters, with pairwise identities >86% within each cluster and <50% between any two clusters (Figure S2). The degree of divergence between these clusters, as well as the low identity scores with previously reported mitoviruses found in the BLASTP searches described above (≤42%), led us to conclude that these eight clusters represent eight new mitovirus species. Phylogenetic analyses involving the *E. muscae* mitoviruses alone corroborated this conclusion by showing the distribution of the 30 viruses into eight well-delimited clades (Figure 2). In addition, by examining which members of each clade derive from which *E. muscae* isolate, we found that one of the *E. muscae* isolates (Berkeley; from fruit fly) harbors strains of all eight of the new mitovirus species, three of the *E. muscae* isolates (KVL-14-117, KVL-14-118, and HHdFL130914-1; from house flies) harbor strains of even of the new mitovirus species, and the remaining one isolate (HHdFL050913-1; from cabbage fly) harbors a strain of only one of the new mitovirus species (Figures 1 and 2). We suggest the names of these new species to be "Entomophthora muscae mitovirus 1" through "Entomophthora muscae mitovirus 8" and further suggest the abbreviations EnmuMV1 through EnmuMV8 for the common names of these viruses. Six of these eight viruses (EnmuMV1, EnmuMV2, EnmuMV4, EnmuMV5, EnmuMV6, and EnmuMV7) are represented by four strains each, one of these viruses (EnmuMV3) is represented by five strains, and the remaining one of these viruses (EnmuMV8) is represented by only one strain in the results presented here. In fact, the few short mitovirus-like contigs assembled from fly-grown *E. muscae* cabbage fly isolate HHdFL040913-2 are sufficient to identify it as a sixth strain of EnmuMV3, most closely related to EnmuMV3-HHdFL050913-1 (95% nt identity), i.e., from another cabbage fly isolate of *E. muscae*.

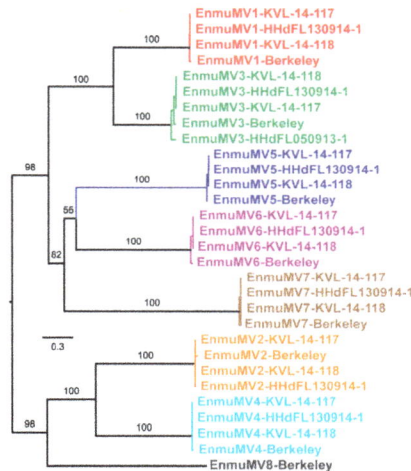

Figure 2. Maximum-likelihood phylogenetic analysis of *E. muscae* mitoviruses. Deduced RdRp sequences were aligned using MAFFT-L-INS-i. The best-fit (per BIC score) aa-substitution model used for the phylogenetic analysis shown here was LG+I+G4. The tree is displayed as a midpoint-rooted rectangular phylogram with branch support values from UFboot (1000 replicates) shown in %; support values for distal branches have been deleted for visual clarity. Scale bar indicates average number of substitutions per alignment position. Labels of different colors highlight the clustering of strains of eight different mitovirus species suggested by these results, as well as by the results in Figure 1 and Figure S2.

Phylogenetic analyses were next performed to ascertain the positions of the eight *E. muscae* mitoviruses relative to others in current genus *Mitovirus*. A representative set of 83 previously reported, full-length mitovirus RdRp sequences were obtained from the Protein database at NCBI and then used in multiple sequence alignments along with the eight mitovirus RdRp sequences from *E. muscae* isolate Berkeley, plus two RdRp sequences from members of current genus *Narnavirus* as an outgroup. Each multiple sequence alignment was analyzed using the ModelFinder module within IQ-TREE to identify the best-fit substitution model, which was then directly applied for maximum-likelihood phylogenetic analyses within IQ-TREE. As shown by the representative phylogram in Figure 3, the results reveal that EnmuMV1–EnmuMV8 (red) all cluster within the same discernible subclade of current genus *Mitovirus*, within a portion of the major clade previously designated Ia. This clustering suggests that all eight of these mitoviruses from *E. muscae* shared a most recent common ancestor, from which they all diverged, near the root of this subclade (red arrow in Figure 3), i.e., fairly early in mitovirus evolution but still well removed from the root.

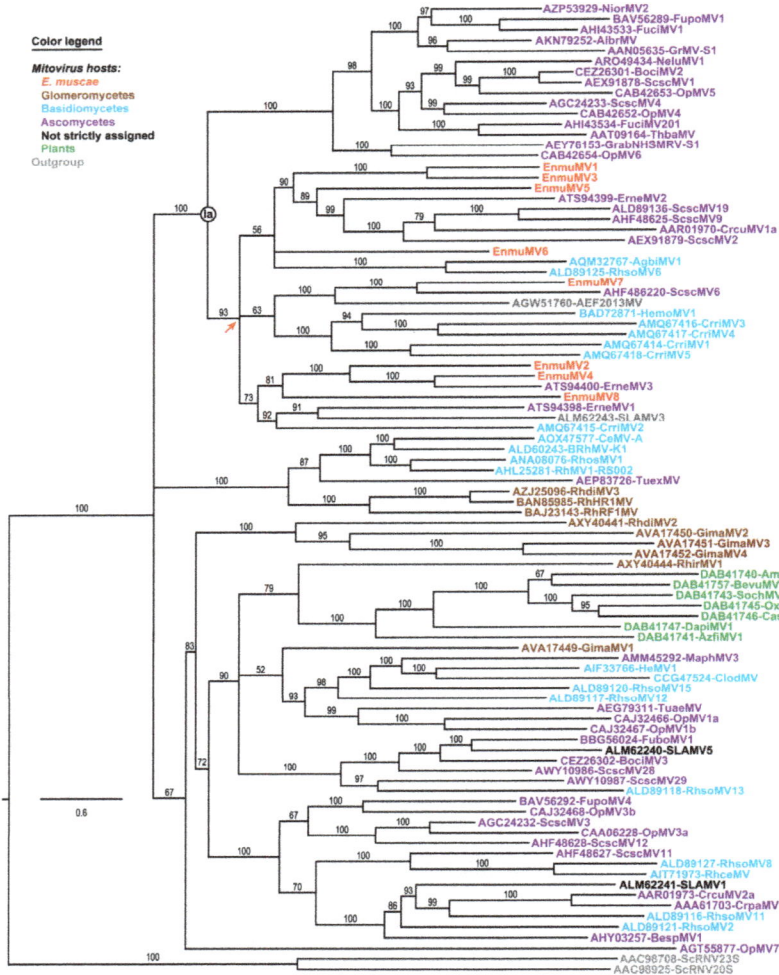

Figure 3. Maximum-likelihood phylogenetic analysis of *E. muscae* mitoviruses and representative other mitoviruses. Deduced RdRp sequences were aligned using MAFFT-L-INS-i. The best-fit substitution model (per BIC score) used for the phylogenetic analysis shown here was LG + F + R8. The site proportions and rates for the FreeRate model were (0.0224, 0.0123) (0.0271, 0.0755), (0.0627, 0.2087), (0.1161, 0.4202), (0.1784, 0.6877), (0.2655, 1.0570), (0.2163, 1.4069), and (0.1115, 2.0462). The tree is displayed as a rectangular phylogram rooted on a set of two members of current genus *Narnavirus* included as outgroup. Branch support values from UFboot (1000 replicates) are shown in %; branches with <50% support have been collapsed to the preceding node. Scale bar indicates average number of substitutions per alignment position. Previously identified clade Ia of current genus *Mitovirus* is labeled. Red arrow, most recent common ancestor shared by all eight *E. muscae* mitoviruses. Virus names matching the sequence accession numbers and abbreviations are provided in Table S6. Color-coding: mitoviruses from *E. muscae* (subphylum *Entomophthoromycotina*, phylum *Zoopagomycota*), red; members of subphylum *Glomeromycotina*, phylum *Mucoromycota*, brown; members of phylum *Basidiomycota*, cyan; members of phylum *Ascomycota*, purple; not assigned to a specific fungal host, black; plants, green; and outgroup, gray.

By aligning the sequences of the different strains within each mitovirus species from *E. muscae*, we observed that the coding sequences for each species are marked by a total absence of indels among the strains, such that RdRp length is conserved within each species (Figure 1). One minor exception is that in EnmuMV2-Berkeley, there is a slightly earlier stop codon (UAA) resulting from a C-to-U substitution relative to the other EnmuMV2 strains, such that the RdRp length of EnmuMV2-Berkeley is reduced by 5 aa. In contrast, within the NTRs, several small indels are seen in the alignments among certain strains from the same species (EnmuMV1, EnmuMV2, EnmuMV3, and EnmuMV7), and especially in their 5′ NTRs (Figure S3). The putative functions of the NTRs in RNA replication, translation, etc., are thus presumably able to accommodate small indels in some positions better than the functions of the coding region and/or the encoded RdRp of each species.

3.4. UGA(Trp) Codons in Mitochondrial Core Genes of E. muscae

A previous report has revealed that UGA(Trp) is a rarely used codon in mitochondrial core genes of many basal fungi [4], including *Zancudomyces* (*Smittium*) *culisetae* from subphylum *Kickxellomycotina*, phylum *Zoopagomycota*, the only member of this basal phylum for which a complete mitochondrial genome sequence had been annotated as such at NCBI at that time (accession NC_006837.1) [41]. Simplistically, one might therefore expect UGA(Trp) to be a rarely used codon also in mitochondrial genes of subphylum *Entomophthoromycotina*, phylum *Zoopagomycota*. The evidence presented above for numerous UGA(Trp) codons in the mitovirus sequences from *E. muscae*, however, run counter to this expectation. To address this possible discrepancy, we sought to determine whether UGA(Trp) codons are found, too, in mitochondrial core genes of *E. muscae*. We first performed TBLASTN searches for transcript contigs representing mitochondrial core genes in TSA database entries for subphylum *Entomophthoromycotina*, using the deduced sequences of 14 mitochondrial core proteins from *Z. culisetae* as queries. Strong hits, all from the *E. muscae* transcriptome study BioProject PRJEB10825 [25], were obtained for 12 of the mitochondrial core genes: *atp6*, *atp9*, *cob*, *cox1*, *cox2*, *cox3*, *nad1*, *nad2*, *nad3*, *nad4*, *nad5*, and *nad6*.

We next performed BLASTP searches of the NR database to discern which of the TSA hits for the 12 mitochondrial core genes appear to have originated from a basal fungus. We then used those TSA accessions (see Materials and Methods for accession numbers) as queries to search the SRA libraries from which the mitovirus sequences from *E. muscae* were assembled as described above, beginning with the SRA transcriptome libraries from *E. muscae* isolate Berkeley (BioProject PRJNA435715) [24]. Using the matching sequence reads that we identified, we were then ultimately able to assemble complete coding sequences for all 12 of these mitochondrial core genes from *E. muscae* isolate Berkeley (Figure 4). Additional BLASTP searches of the NR database strongly supported the conclusion that the mitochondrial protein sequences deduced from these contigs derived from a member of subphylum *Entomophthoromycotina*, phylum *Zoopagomycota*, given that the top-scoring one or two hit(s) for each protein consistently derived from *Conidiobolus* species *C. coronatus* or *C. heterosporus*, the latter for which a complete mitochondrial genome sequence has only recently been deposited and annotated at NCBI (accession MK049352.1) [43] (Table 1). Moreover, phylogenetic analyses using the concatenated sequences of these 12 mitochondrial proteins from *E. muscae*, along with those from other representative fungi, placed the *E. muscae* proteins adjacent to those of *C. heterosporus* within the *Zoopagomycota* clade (Figure S4).

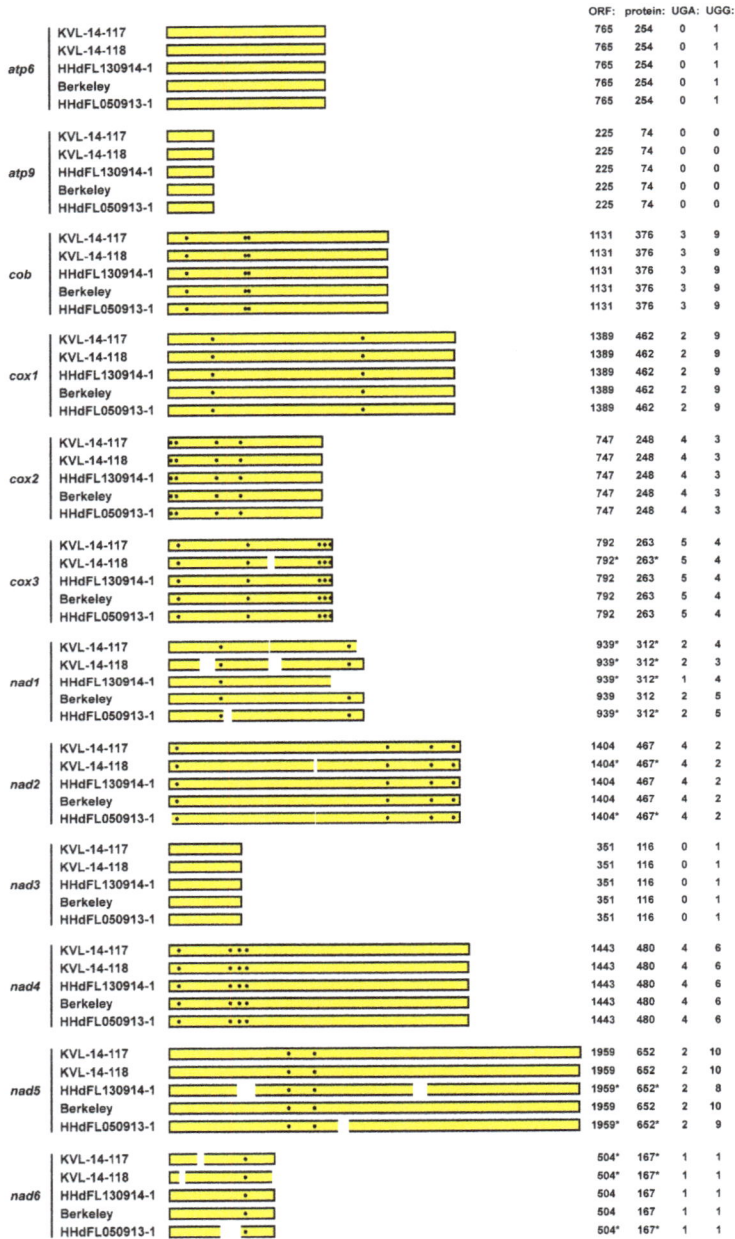

Figure 4. Summary diagram of mitochondrial coding sequences from *E. muscae* isolates. Mitochondrial core genes from each *E. muscae* isolate are labeled at left. The protein-encoding ORF of the respective transcript is shown as a yellow box to represent each assembled RNA sequence, starting with the first in-frame AUG codon by convention. Positions of UGA(Trp) codons in each ORF are shown as white dots. Positions of gaps or truncations in the assembled sequences are indicated by white breaks in the boxes; asterisks indicate inferred values consequent to these breaks. For each sequence, length of the ORF is shown in nt, length of the encoded protein in aa, and UGA and UGG codons in raw counts.

Table 1. Top NR hits for deduced mitochondrial core-protein sequences derived from *E. muscae*.

Gene [a]	Accession no.	Fungal Species	Sub–Phy [b]	E-Value [c]
atp6	AZZ06694.1	*Conidiobolus heterosporus*	Ent–Zoo	2e−115
	KXN65649.1	*Conidiobolus coronatus*	Ent–Zoo	1e−102
atp9	KXN65653.1	*Conidiobolus coronatus*	Ent–Zoo	6e−38
	AZZ06722.1	*Conidiobolus heterosporus*	Ent–Zoo	1e−37
cob	AZZ06693.1	*Conidiobolus heterosporus*	Ent–Zoo	0
cox1	AZZ06707.1	*Conidiobolus heterosporus*	Ent–Zoo	0
cox2	AZZ06713.1	*Conidiobolus heterosporus*	Ent–Zoo	7e−142
cox3	AZZ06710.1	*Conidiobolus heterosporus*	Ent–Zoo	6e−169
nad1	AZZ06725.1	*Conidiobolus heterosporus*	Ent–Zoo	4e−169
nad2	AZZ06721.1	*Conidiobolus heterosporus*	Ent–Zoo	6e−135
nad3	AZZ06726.1	*Conidiobolus heterosporus*	Ent–Zoo	6e−47
nad4	AZZ06717.1	*Conidiobolus heterosporus*	Ent–Zoo	0
nad5	AZZ06724.1	*Conidiobolus heterosporus*	Ent–Zoo	0
nad6	AZZ06724.1	*Conidiobolus heterosporus*	Ent–Zoo	1e−34
	KXN65652.1	*Conidiobolus coronatus*	Ent–Zoo	3e−23

[a] Mitochondrial core genes; [b] Subphylum–Phylum: *Ent–Zoo*, *Entomophthoromycotina–Zoopagomycota*; [c] Top one or two hits in terms of E-value score are shown from searching the NR database with each mitochondrial protein sequence deduced from the apparent mitochondrial core-gene coding sequences from *E. muscae* isolate Berkeley.

To provide additional evidence that these transcript contigs from *E. muscae* isolate Berkeley represent the respective mitochondrial genes of *E. muscae*, we used the contigs to search the SRA transcriptome libraries from the four other *E. muscae* isolates KVL-14-117, KVL-14-118, HHdFL130914-1, and HHdFL050913-1. The results summarized in Figure 4 reflect that nearly identical coding sequences for all 12 mitochondrial genes were assembled from sequence reads from all four other *E. muscae* isolates. Complete coding sequences were assembled for eight to ten of these genes from each isolate and partial coding sequences for the other genes (Figure 4), with single-nt mismatches at only a few nt positions between any two isolates. These findings provide further strong support for the conclusion that these 12 mitochondrial coding sequences indeed derive from *E. muscae*.

We lastly examined the 12 mitochondrial coding sequences from *E. muscae* for UGA (Trp) codons. At least one and as many as five UGA (Trp) codons each are found in nine of these sequences, namely, in all but those for *atp6*, *atp9*, and *nad3*, which also contain no or only one UGG (Trp) codon each (Figure 4). Notably as well, these UGA (Trp) codons are conserved among the sequences assembled from the five different *E. muscae* isolates (Figure 4). Overall in these coding sequences, 35% of the Trp codons are UGA (vs. 65% UGG). We therefore conclude that UGA (Trp) is not a rarely used codon in the mitochondrial core genes of *E. muscae*, consistent with our findings for the mitovirus sequences from this fungus. Our initial generalization that UGA (Trp) might be expected to be a rarely used codon in the mitochondria of all basal fungi thus appears to have been incorrect.

4. Discussion

The eight mitovirus species newly identified in this report are the first to be discovered in a fungal host from phylum *Zoopagomycota* (subphylum *Entomophthoromycotina*). Before this, the most phylogenetically basal fungi [1,2] that had been found to harbor mitoviruses were ones from phylum *Mucoromycota* (subphylum *Glomeromycotina*) [5,26–28]. The mitoviruses in *Entomophthora muscae* might have been acquired by horizontal transmission, perhaps from some more apically diverging fungi given their juxtaposition with ascomycete and basidiomycete mitoviruses in the Figure 3 tree. The explanation that we favor, however, is that an early ancestor of the *E. muscae* mitoviruses first entered the fungal lineage in an ancestral species that predated the divergence of phylum *Zoopagomycota* from the three more apically diverging phyla *Mucoromycota*, *Basidiomycota*, and *Ascomycota*, and was then transmitted largely by vertical means over the subsequent course of fungal evolution, including to *E. muscae*. See Figure S4 for a fungal tree of life based on mitochondrial proteins, which shows a similar pattern of progressive divergence of the fungal phyla to that previously shown for nuclear products [1,2], and which was similarly shown for fungal mitochondrial proteins by Nie et al. [43] in their recent report on the complete mitochondrial genome sequence of entomophthoroid fungus *Conidiobolus heterosporus*. The reciprocal possibility, i.e., that *E. muscae* mitoviruses have been horizontally transmitted to some more apically diverging fungi, must also be considered given their juxtapositions in the Figure 3 tree. For example, note the apparent proximities of Erysiphe necator (ascomycete) mitovirus 1 (ATS94400) to EnmuMV4 and Sclerotinia sclerotiorum (ascomycete) mitovirus 6 (AHF48622) to EnmuMV7.

Four of the *E. muscae* isolates whose mitoviruses were characterized in this study were each found to contain strains of either seven or eight different mitovirus species. Although this large number might seem odd, similar findings have been made in other fungi. For example, strains of at least seven different mitovirus species have been found in the same isolate, Ld, of the Dutch elm disease fungus *Ophiostoma novo-ulmi* (phylum *Ascomycota*) [39,44,45]; strains of six different mitovirus species have been found in the same isolate, AG2-2-IV, of *Rhizoctonia solani* (phylum *Basidiomycota*) [46]; and strains of five different mitovirus species have been found in the same isolate, BC-u3, of the white pine blister rust fungus *Cronartium ribicola* (phylum *Basidiomycota*) [47]. Such multiple co-infections raise interesting questions about possible interactions among the different viruses in each isolate and whether there are selective forces or circumstances that favor multiple co-infections. The fungus *E. muscae* is multinucleate throughout all stages of its life cycle, and both protoplast cells and asexual spores may contain up to 20 nuclei each [25]. Although mitochondrial morphology and membrane potential dynamics are considered independent of nuclear cycle state [48], the consistent presence of multiple nuclei might facilitate a heterogeneous intracellular environment conducive to the presence of multiple mitoviruses. Another, more trivial possibility relates to the fact that the *E. muscae* isolates whose sequence reads were analyzed for this report originated not from individual spores, but from conidial showers from infected flies, meaning that there remains a possibility that these isolates are mixtures of *E. muscae* strains, with differences in mitovirus content between strains.

The findings in this report, including both mitovirus and mitochondrial coding sequences, suggest that the mitochondrial protein synthesis machinery in *E. muscae* can translate codon UGA (Trp) more efficiently than can that in many other basal fungi [4]. In this regard, *E. muscae* mitochondria are more like those of most members of the more apically diverging phyla *Ascomycota* and *Basidiomycota*. As noted in Results, UGA (Trp) has been shown to be a rarely used mitochondrial codon in *Zancudomyces culistae* (subphylum *Kickxellomycotina*, phylum *Zoopagomycota*; no UGA(Trp) codons in the 12 mitochondrial core-gene coding sequences examined here), such that there appears to be some variation in mitochondrial UGA (Trp) codon usage across members of phylum *Zoopagomycota*. In the recently reported mitochondrial genome sequence of *Conidiobolus heterosporus* (subphylum *Entomophthoromycotina*, phylum *Zoopagomycota*) [43], we find that UGA (Trp) is also somewhat uncommonly used (6 UGA (Trp) vs. 62 UGG (Trp) codons in the same 12 mitochondrial coding sequences, 8.8%), even though *C. heterosporus* encodes a mitochondrial tRNA sequence with anticodon UCA, which should allow for more efficient translation of UGA (Trp). A complete mitochondrial genome sequence for *E. muscae*, and for other members of subphylum *Entomophthoromycotina*, may be helpful for understanding why UGA (Trp) codons are more commonly used in the mitochondria of *E. muscae* than in those of related species *C. heterosporus* (same order, *Entomophorales*; different families, *Entomophthoraceae* and *Ancylistaceae*, respectively).

The RNA samples that were analyzed for *E. muscae* isolates KVL-14-118, Berkeley, and HHdFL050913-1 in the original transcriptome studies (BioProjects PRJEB10825 and PRJNA435715) [24, 25] were obtained from infected host flies. One might therefore have some concern that the mitovirus sequences identified from those samples could have originated from the flies, or from a different fly symbiont, rather than from *E. muscae* itself. Importantly, however, the samples from *E. muscae* isolate KVL-14-117 analyzed in the original transcriptome studies (BioProject PRJNA372837) [25] and then further employed here were obtained from media-grown cultures of the fungus, serving to allay such possible concern. Indeed, the general absence of fly-derived sequence reads in the SRA libraries from media-grown *E. muscae* isolate KVL-14-117 [25] was confirmed in this study by use of several fly-derived mitochondrial gene queries. In addition, for *E. muscae* isolate Berkeley, the original transcriptome study (BioProject PRJNA435715) [24] included a number of samples from flies not inoculated with *E. muscae*, and the SRA libraries from those control samples contain no or very few sequence reads matching the *E. muscae* mitoviruses (Table S4). Moreover, the SRA libraries from the samples from inoculated flies in that study contain increasing numbers of sequence reads matching the *E. muscae* mitoviruses at increasing times after inoculation with *E. muscae* isolate Berkeley, consistent with the growth of *E. muscae* in those flies following inoculation (Table S4). Both of these observations provide strong evidence that the mitoviruses are associated with *E. muscae* isolate Berkeley itself, not the fly hosts. Lastly, we have been recently able to confirm the presence of all seven mitovirus sequences in new transcriptome data from conidial samples of *E. muscae* isolate HHdFL130914-1, which will be described in detail in a future article [49] and also provides important evidence for vertical transmission of the *E. muscae* mitoviruses via conidiospores. The consistency of sequence findings that we obtained from the different *E. muscae* isolates, for both mitovirus and mitochondrial coding sequences and despite these isolates having been originally obtained from three different species of host flies, also strongly supports the conclusion that these sequences originated from *E. muscae*.

By convention, each ORF diagrammed in Figures 1 and 4 is considered to start with the first in-frame AUG codon. Because an alternative start codon might be used in some cases, however, it remains possible that some of these ORFs may be longer or shorter at their 5′ ends than suggested here. In this regard, we observed in particular that the *cox1* ORF of *E. muscae* contains a conserved in-frame GUG codon well-upstream of the first in-frame AUG (see Data S2), which would extend that ORF and its encoded protein considerably, to 1587 nt and 528 aa, respectively, bringing those lengths more in line with those of *C. heterosporus* [43] and most other organisms. For the mitoviruses, considering the strains with a longer region of sequence between the first in-frame AUG and the first upstream in-frame stop codon (see Figure 1), use of an alternative start codon upstream of the first in-frame AUG seems most likely for EnmuMV4, EnmuMV5, and EnmuMV8. Reciprocally, considering those viruses with shorter 5′ NTRs preceding the first in-frame AUG (see Figure 1), use of a start codon downstream of the first in-frame AUG seems most probable for EnmuMV6. Apparently notable in these regards, use of an alternative start codon upstream of the first in-frame AUG in EnmuMV4, EnmuMV5, and EnmuMV8 could provide for several additional residues near the RdRp N-terminus that would be conserved among all of the mitovirus strains reported here, and use of the second in-frame AUG codon to initiate translation in EnmuMV6 would retain these conserved residues (Figure S5). We therefore consider it likely that the first in-frame AUG is not the true start codon for RdRp translation in several of these viruses. In particular, for EnmuMV4, EnmuMV5, and EnmuMV8, we propose that an upstream AUU start codon is used instead, and for EnmuMV6, we propose that the second in-frame AUG codon is used instead (see Figure S3, Figure S5, and Data S2).

Coyle et al. [50] recently identified three SRA transcriptome libraries derived from mixtures of wild-caught flies, at least some of which were apparently harboring *E. muscae* when their RNAs were sampled (SRX955881, SRX955902, and SRX1711976; from BioProjects PRJNA277921 and PRJNA318834 [12]). Upon performing MegaBLAST searches of those libraries using the sequences of EnmuMV1–8 as queries, we found that two of the libraries, SRX955902 and SRX1711976, include sequence reads matching the *E. muscae* mitoviruses. Specifically, SRX955902 includes reads matching all eight of these viruses, and SRX1711976 includes reads matching four of these viruses (EnmuMV1, EnmuMV3, EnmuMV6, and EnmuMV7) (Table S5). Only the reads matching EnmuMV7 from SRX1711976 were sufficient in coverage to allow assembly of a complete coding sequence for this virus, which turns out to correspond (99.9% nt-sequence identity) to Hubei narna-like virus 25 (GenBank KX883546; 88.7–91.0% pairwise nt-sequence identity with the four EnmuMV7 strains from *E. muscae* detailed above). Thus, although Hubei narna-like virus 25 has been previously reported as an insect mitovirus [12], it is now shown to derive more likely from the fly-associated entomophthoroid fungus *E. muscae* or perhaps a related fungus. These findings provide further support for the conclusion that these and perhaps other yet-to-be-discovered mitoviruses are widespread in naturally occurring populations of *E. muscae*.

The behavioral effects of *E. muscae* on host flies are fascinating and are indeed the subjects of ongoing genetic and neuroscientific investigations in *Drosophila* [24]. Interestingly, a distinct plus-strand RNA virus from *E. muscae*, an iflavirus (order *Picornavirales*), has been recently reported by Coyle et al. [50] and speculated to contribute to the manipulation of fly-host behaviors by *E. muscae*. The *E. muscae* mitoviruses that we describe here might also be considered as possible contributors in that regard. On a related note, two novel mitoviruses were recently identified in the entomopathogenic fungus *Ophiocordyceps sinensis* (phylum *Ascomycota*) [51], which causes behavioral perturbations and death of its caterpillar hosts. In this case, the dying caterpillar, which had been feeding as normal on plant roots underground, appears to be induced by the infecting fungus to crawl into a position and orientation relative to the soil surface that is ideal for growth of the fungal fruiting body aboveground, promoting efficient dispersal of the fungal spores [52]. These effects are clearly analogous to those of *E. muscae* in flies, and it is fascinating to speculate that again in this case the fungal mitoviruses might in some way contribute to the manipulation of insect host behaviors by the entomopathogenic fungus.

Supplementary Materials: The following are available online at http://www.mdpi.com/1999-4915/11/4/351/s1, Figure S1: Multiple sequence alignment of deduced RdRp sequences of mitoviruses from *E. muscae* isolate Berkeley, Figure S2: Sequence identity scores from pairwise alignments of *E. muscae* mitovirus sequences, Figure S3: Multiple sequence alignments of 5′ and 3′ NTR sequences of *E. muscae* mitoviruses, Figure S4: Fungal tree of life based on mitochondrial core-protein sequences from representative species, Figure S5: Multiple sequence alignment of N-terminal regions of deduced RdRp sequences of *E. muscae* mitoviruses, Table S1: Contig assembly values for coding-complete mitovirus sequences from *E. muscae*, Table S2: Contig assembly values for mitochondrial core-gene coding sequences from *E. muscae*, Table S3: SRA accessions used for contig assemblies in this study, Table S4: Mitovirus-matching reads in samples from fruit flies infected or not with *E. muscae* isolate Berkeley per Elya et al. [24], Table S5: Mitovirus-matching reads in samples from SRA transcriptome libraries derived from mixtures of wild-caught flies apparently infected with *E. muscae* per Coyle et al. [50], Table S6: Mitovirus RdRp sequences used for phylogenetic analyses, Data S1: *E. muscae* mitovirus sequences, Data S2: *E. muscae* mitochondrial coding sequences.

Author Contributions: Conceptualization, M.L.N. and H.J.D.; Investigation, M.L.N., H.J.D., A.R.M., I.V.G., and H.H.D.F.L.; Methodology, M.L.N., H.J.D., A.R.M., I.V.G., and H.H.D.F.L.; Validation, M.L.N., H.J.D., A.R.M., I.V.G., and H.H.D.F.L.; Writing—Original Draft Preparation, M.L.N.; Writing—Review & Editing, M.L.N., H.J.D., A.R.M., I.V.G., and H.H.D.F.L.

Funding: This research was funded in part by NIH Grant T32 AI007245 to the Ph.D. program in Virology at Harvard University (A.R.M.). The work conducted by the U.S. Department of Energy Joint Genome Institute, a DOE Office of Science User Facility, is supported by the Office of Science of the U.S. Department of Energy under Contract No. DE-AC02-05CH11231 (I.V.G.). H.H.D.F.L. was supported by the Independent Research Fund Denmark and the Villum Foundation (Grant No. 10122).

Acknowledgments: We wish to acknowledge and thank those scientists who made this study possible by depositing their sequence reads and shotgun assemblies in the public databases. We also thank Carolyn Elya (Harvard University) for helpful discussions and permission to use her SRA data from NCBI.

Conflicts of Interest: The authors declare no conflict of interest.

References

1. Spatafora, J.W.; Chang, Y.; Benny, G.L.; Lazarus, K.; Smith, M.E.; Berbee, M.L.; Bonito, G.; Corradi, N.; Grigoriev, I.; Gryganskyi, A.; et al. A phylum-level phylogenetic classification of zygomycete fungi based on genome-scale data. *Mycologia* **2016**, *108*, 1028–1046. [CrossRef] [PubMed]

2. Spatafora, J.W.; Aime, M.C.; Grigoriev, I.V.; Martin, F.; Stajich, J.E.; Blackwell, M. The fungal tree of life: From molecular systematics to genome-scale phylogenies. *Microbiol. Spectr.* **2017**, *5*. [CrossRef]

3. Hillman, B.I.; Cai, G. The family *Narnaviridae*: Simplest of RNA viruses. *Adv. Virus Res.* **2013**, *86*, 149–176.

4. Nibert, M.L. Mitovirus UGA(Trp) codon usage parallels that of host mitochondria. *Virology* **2017**, *507*, 96–100. [CrossRef]

5. Ikeda, Y.; Shimura, H.; Kitahara, R.; Masuta, C.; Ezawa, T. A novel virus-like double-stranded RNA in an obligate biotroph arbuscular mycorrhizal fungus: A hidden player in mycorrhizal symbiosis. *Mol. Plant Microbe Interact.* **2012**, *25*, 1005–1012. [CrossRef]

6. Rogers, H.J.; Buck, K.W.; Brasier, C.M. A mitochondrial target for the double-stranded RNAs in diseased isolates of the fungus that causes dutch elm disease. *Nature* **1987**, *329*, 558–560. [CrossRef]

7. Polashock, J.J.; Hillman, B.I. A small mitochondrial double-stranded (ds) RNA element associated with a hypovirulent strain of the chestnut blight fungus and ancestrally related to yeast cytoplasmic T and W dsRNAs. *Proc. Natl. Acad. Sci. USA* **1994**, *91*, 8680–8684. [CrossRef] [PubMed]

8. Polashock, J.J.; Bedker, P.J.; Hillman, B.I. Movement of a small mitochondrial double-stranded RNA element of *Cryphonectria parasitica*: Ascospore inheritance and implications for mitochondrial recombination. *Mol. Gen. Genet.* **1997**, *256*, 566–571. [CrossRef]

9. Nibert, M.L.; Vong, M.; Fugate, K.K.; Debat, H.J. Evidence for contemporary plant mitoviruses. *Virology* **2018**, *518*, 14–24. [CrossRef]

10. Nerva, L.; Vigani, G.; Di Silvestre, D.; Ciuffo, M.; Forgia, M.; Chitarra, W.; Turina, M. Biological and molecular characterization of Chenopodium quinoa mitovirus 1 reveals a distinct sRNA response compared to cytoplasmic RNA viruses. *J. Virol.* **2019**, *93*. [CrossRef] [PubMed]

11. Cook, S.; Chung, B.Y.; Bass, D.; Moureau, G.; Tang, S.; McAlister, E.; Culverwell, C.L.; Glücksman, E.; Wang, H.; Brown, T.D.; et al. Novel virus discovery and genome reconstruction from field RNA samples reveals highly divergent viruses in dipteran hosts. *PLoS ONE* **2013**, *8*, e80720. [CrossRef] [PubMed]

12. Shi, M.; Lin, X.D.; Tian, J.H.; Chen, L.J.; Chen, X.; Li, C.X.; Qin, X.C.; Li, J.; Cao, J.P.; Eden, J.S.; et al. Redefining the invertebrate RNA virosphere. *Nature* **2016**, *540*, 539–543. [CrossRef]

13. Bruenn, J.A. A structural and primary sequence comparison of the viral RNA-dependent RNA polymerases. *Nucleic Acids Res.* **2003**, *31*, 1821–1829. [CrossRef] [PubMed]

14. Vong, M.; Manny, A.R.; Smith, K.L.; Gao, W.; Nibert, M.L. Beta vulgaris mitovirus 1 in diverse cultivars of beet and chard. *Virus Res.* **2019**, *265*, 80–87. [CrossRef] [PubMed]

15. Wu, M.; Zhang, L.; Li, G.; Jiang, D.; Ghabrial, S.A. Genome characterization of a debilitation-associated mitovirus infecting the phytopathogenic fungus *Botrytis cinerea*. *Virology* **2010**, *406*, 117–126. [CrossRef]

16. Wu, M.D.; Zhang, L.; Li, G.Q.; Jiang, D.H.; Hou, M.S.; Huang, H.C. Hypovirulence and double-stranded RNA in *Botrytis cinerea*. *Phytopathology* **2007**, *97*, 1590–1599. [CrossRef] [PubMed]

17. Xie, J.; Ghabrial, S.A. Molecular characterization of two mitoviruses co-infecting a hypovirulent isolate of the plant pathogenic fungus *Sclerotinia sclerotiorum*. *Virology* **2012**, *428*, 77–85. [CrossRef]

18. Xu, Z.; Wu, S.; Liu, L.; Cheng, J.; Fu, Y.; Jiang, D.; Xie, J. A mitovirus related to plant mitochondrial gene confers hypovirulence on the phytopathogenic fungus *Sclerotinia sclerotiorum*. *Virus Res.* **2015**, *197*, 127–136. [CrossRef]

19. Wolf, Y.I.; Kazlauskas, D.; Iranzo, J.; Lucía-Sanz, A.; Kuhn, J.H.; Krupovic, M.; Dolja, V.V.; Koonin, E.V. Origins and evolution of the global RNA virome. *MBio* **2018**, *9*. [CrossRef]

20. Shaikh, N.; Hussain, K.A.; Petraitiene, R.; Schuetz, A.N.; Walsh, T.J. Entomophthoramycosis: A neglected tropical mycosis. *Clin. Microbiol. Infect.* **2016**, *22*, 688–694. [CrossRef]

21. Vilela, R.; Mendoza, L. Human pathogenic Entomophthorales. *Clin. Microbiol. Rev.* **2018**, *31*. [CrossRef]

22. Araújo, J.P.; Hughes, D.P. Diversity of entomopathogenic fungi: Which groups conquered the insect body? *Adv. Genet.* **2016**, *94*, 1–39.

23. Jensen, A.B.; Thomsen, L.; Eilenberg, J. Value of host range, morphological, and genetic characteristics within the *Entomophthora muscae* species complex. *Mycol. Res.* **2006**, *110*, 941–950. [CrossRef] [PubMed]

24. Elya, C.; Lok, T.C.; Spencer, Q.E.; McCausland, H.; Martinez, C.C.; Eisen, M. Robust manipulation of the behavior of *Drosophila melanogaster* by a fungal pathogen in the laboratory. *Elife* **2018**, *7*. [CrossRef] [PubMed]

25. De Fine Licht, H.H.; Jensen, A.B.; Eilenberg, J. Comparative transcriptomics reveal host-specific nucleotide variation in entomophthoralean fungi. *Mol. Ecol.* **2017**, *26*, 2092–2110. [CrossRef]

26. Kitahara, R.; Ikeda, Y.; Shimura, H.; Masuta, C.; Ezawa, T. A unique mitovirus from *Glomeromycota*, the phylum of arbuscular mycorrhizal fungi. *Arch. Virol.* **2014**, *159*, 2157–2160. [CrossRef]

27. Neupane, A.; Feng, C.; Feng, J.; Kafle, A.; Bücking, H.; Lee Marzano, S.Y. Metatranscriptomic analysis and in silico approach identified mycoviruses in the arbuscular mycorrhizal fungus *Rhizophagus* spp. *Viruses* **2018**, *10*, 707. [CrossRef] [PubMed]

28. Turina, M.; Ghignone, S.; Astolfi, N.; Silvestri, A.; Bonfante, P.; Lanfranco, L. The virome of the arbuscular mycorrhizal fungus *Gigaspora margarita* reveals the first report of DNA fragments corresponding to replicating non-retroviral RNA viruses in fungi. *Environ. Microbiol.* **2018**, *20*, 2012–2025. [CrossRef] [PubMed]

29. Huang, X.; Madan, A. CAP3: A DNA sequence assembly program. *Genome Res.* **1999**, *9*, 868–877. [CrossRef]

30. Haas, B.J.; Papanicolaou, A.; Yassour, M.; Grabherr, M.; Blood, P.D.; Bowden, J.; Couger, M.B.; Eccles, D.; Li, B.; Lieber, M.; et al. De novo transcript sequence reconstruction from RNA-seq using the Trinity platform for reference generation and analysis. *Nat. Protoc.* **2013**, *8*, 1494–1512. [CrossRef]

31. Bushmanova, E.; Antipov, D.; Lapidus, A.; Przhibelskiy, A.D. rnaSPAdes: A de novo transcriptome assembler and its application to RNA-Seq data. *bioRxiv* **2018**. [CrossRef]

32. Buchfink, B.; Xie, C.; Huson, D.H. Fast and sensitive protein alignment using DIAMOND. *Nat. Methods* **2015**, *12*, 59–60. [CrossRef] [PubMed]

33. Katoh, K.; Standley, D.M. MAFFT multiple sequence alignment software version 7: Improvements in performance and usability. *Mol. Biol. Evol.* **2013**, *30*, 772–780. [CrossRef] [PubMed]

34. Nguyen, L.T.; Schmidt, H.A.; von Haeseler, A.; Minh, B.Q. IQ-TREE: A fast and effective stochastic algorithm for estimating maximum likelihood phylogenies. *Mol. Biol. Evol.* **2015**, *32*, 268–274. [CrossRef] [PubMed]

35. Kalyaanamoorthy, S.; Minh, B.Q.; Wong, T.K.F.; von Haeseler, A.; Jermiin, L.S. ModelFinder: Fast model selection for accurate phylogenetic estimates. *Nat Methods* **2017**, *14*, 587–589. [CrossRef] [PubMed]

36. Minh, B.Q.; Nguyen, M.A.T.; von Haeseler, A. Ultrafast approximation for phylogenetic bootstrap. *Mol. Biol. Evol.* **2013**, *30*, 1188–1195. [CrossRef] [PubMed]

37. Becher, P.G.; Jensen, R.E.; Natsopoulou, M.E.; Verschut, V.; De Fine Licht, H.H. Infection of *Drosophila suzukii* with the obligate insect-pathogenic fungus *Entomophthora muscae*. *J. Pest Sci.* **2018**, *91*, 781–787. [CrossRef]

38. Marienfeld, J.R.; Unseld, M.; Brandt, P.; Brennicke, A. Viral nucleic acid sequence transfer between fungi and plants. *Trends Genet.* **1997**, *13*, 260–261. [CrossRef]

39. Hong, Y.; Cole, T.E.; Brasier, C.M.; Buck, K.W. Evolutionary relationships among putative RNA-dependent RNA polymerases encoded by a mitochondrial virus-like RNA in the Dutch elm disease fungus, *Ophiostoma novo-ulmi*, by other viruses and virus-like RNAs and by the *Arabidopsis* mitochondrial genome. *Virology* **1998**, *246*, 158–169. [CrossRef]

40. Bruenn, J.A.; Warner, B.E.; Yerramsetty, P. Widespread mitovirus sequences in plant genomes. *PeerJ* **2015**, *3*, e876. [CrossRef]

41. Seif, E.; Leigh, J.; Liu, Y.; Roewer, I.; Forget, L.; Lang, B.F. Comparative mitochondrial genomics in zygomycetes: Bacteria-like RNase P RNAs, mobile elements and a close source of the group I intron invasion in angiosperms. *Nucleic Acids Res.* **2005**, *33*, 734–744. [CrossRef]

42. Schon, E.A. Appendix 4. Mitochondrial genetic codes in various organisms. *Methods Cell Biol.* **2007**, *80*, 831–833.

43. Nie, Y.; Wang, L.; Cai, Y.; Tao, W.; Zhang, Y.J.; Huang, B. Mitochondrial genome of the entomophthoroid fungus *Conidiobolus heterosporus* provides insights into evolution of basal fungi. *Appl. Microbiol. Biotechnol.* **2018**, *103*, 1379–1391. [CrossRef] [PubMed]

44. Hong, Y.; Dover, S.L.; Cole, T.E.; Brasier, C.M.; Buck, K.W. Multiple mitochondrial viruses in an isolate of the Dutch Elm disease fungus *Ophiostoma novo-ulmi*. *Virology* **1999**, *258*, 118–127. [CrossRef] [PubMed]

45. Doherty, M.; Coutts, R.H.A.; Brasier, C.M.; Buck, K.W. Sequence of RNA-dependent RNA polymerase genes provides evidence for three more distinct mitoviruses in *Ophiostoma novo-ulmi* isolate Ld. *Virus Genes* **2006**, *33*, 41–44. [CrossRef] [PubMed]

46. Bartholomäus, A.; Wibberg, D.; Winkler, A.; Pühler, A.; Schlüter, A.; Varrelmann, M. Deep sequencing analysis reveals the mycoviral diversity of the virome of an avirulent isolate of *Rhizoctonia solani* AG-2-2 IV. *PLoS ONE* **2016**, *11*, e0165965. [CrossRef] [PubMed]

47. Liu, J.J.; Chan, D.; Xiang, Y.; Williams, H.; Li, X.R.; Sniezko, R.A.; Sturrock, R.N. Characterization of five novel mitoviruses in the white pine blister rust fungus *Cronartium ribicola*. *PLoS ONE* **2016**, *11*, e0154267. [CrossRef] [PubMed]

48. Gerstenberger, J.P.; Occhipinti, P.; Gladfelter, A.S. Heterogeneity in mitochondrial morphology and membrane potential is independent of the nuclear division cycle in multinucleate fungal cells. *Eukaryot. Cell* **2012**, *11*, 353–367. [CrossRef] [PubMed]

49. Natsopoulou, M.E.; Nunn, A.; Hansen, A.N.; Veselská, T.; Gryganskyi, A.; Groth, M.; Felder, M.; Kolařík, M.; Grigoriev, I.; Voigt, K.; et al. Dual RNA-Seq analysis during infection of house flies with the insect-pathogenic fungus *Entomophthora muscae* reveals unique molecular interactions between host and pathogen. Manuscript in preparation.

50. Coyle, M.C.; Elya, C.N.; Bronski, M.J.; Eisen, M.B. Entomophthovirus: An insect-derived iflavirus that infects a behavior manipulating fungal pathogen of dipterans. *bioRxiv* **2018**. [CrossRef]

51. Gilbert, K.; Holcomb, E.E.; Allscheid, R.L.; Carrington, J. Discovery of new mycoviral genomes within publicly available fungal transcriptomic datasets. *bioRxiv* **2019**. [CrossRef]

52. Zhang, Y.J.; Li, E.W.; Wang, C.S.; Li, Y.L.; Liu, X.Z. *Ophiocordyceps sinensis*, the flagship fungus of China: Terminology, life strategy and ecology. *Mycology* **2012**, *3*, 2–10.

viruses

MDPI

Communication

Mycoviral Population Dynamics in Spanish Isolates of the Entomopathogenic Fungus *Beauveria bassiana*

Charalampos Filippou [1,2,†], Inmaculada Garrido-Jurado [1,3,†], Nicolai V. Meyling [4], Enrique Quesada-Moraga [3], Robert H. A. Coutts [2] and Ioly Kotta-Loizou [1,*]

[1] Department of Life Sciences, Imperial College London, London SW7 2AZ, UK; c_filippou@live.com (C.F.); g72gajui@uco.es (I.G.-J.)
[2] Department of Biological and Environmental Sciences, University of Hertfordshire, Hatfield AL10 9AB, UK; r.coutts@herts.ac.uk
[3] Department of Agricultural and Forestry Sciences, University of Cordoba, 14071 Cordoba, Spain; cr2qumoe@uco.es
[4] Department of Plant and Environmental Sciences, University of Copenhagen, 1871 Frederiksberg C, Denmark; nvm@plen.ku.dk
*   Correspondence: i.kotta-loizou13@imperial.ac.uk
†   These authors contributed equally to this work.

Received: 26 October 2018; Accepted: 21 November 2018; Published: 24 November 2018

Abstract: The use of mycoviruses to manipulate the virulence of entomopathogenic fungi employed as biocontrol agents may lead to the development of novel methods to control attacks by insect pests. Such approaches are urgently required, as existing agrochemicals are being withdrawn from the market due to environmental and health concerns. The aim of this work is to investigate the presence and diversity of mycoviruses in large panels of entomopathogenic fungi, mostly from Spain and Denmark. In total, 151 isolates belonging to the genera *Beauveria*, *Metarhizium*, *Lecanicillium*, *Purpureocillium*, *Isaria*, and *Paecilomyces* were screened for the presence of dsRNA elements and 12 Spanish *B. bassiana* isolates were found to harbor mycoviruses. All identified mycoviruses belong to three previously characterised species, the officially recognised *Beauveria bassiana victorivirus 1* (BbVV-1) and the proposed Beauveria bassiana partitivirus 2 (BbPV-2) and Beauveria bassiana polymycovirus 1 (BbPmV-1); individual *B. bassiana* isolates may harbor up to three of these mycoviruses. Notably, these mycovirus species are under distinct selection pressures, while recombination of viral genomes increases population diversity. Phylogenetic analysis of the RNA-dependent RNA polymerase gene sequences revealed that the current population structure in Spain is potentially a result of both vertical and horizontal mycovirus transmission. Finally, pathogenicity experiments using the Mediterranean fruit fly *Ceratitis capitata* showed no direct correlation between the presence of any particular mycovirus and the virulence of the *B. bassiana* isolates, but illustrated potentially interesting isolates that exhibit relatively high virulence, which will be used in more detailed virulence experimentation in the future.

Keywords: mycovirus; *Beauveria bassiana*; partitivirus; victorivirus; polymycovirus; selection pressure; recombination; transmission

1. Introduction

Recently there has been a resurgence of interest in the use of microbial control agents as alternatives to chemical pesticides or as part of integrated pest management programs [1]. Among the fungal-based biopesticides, entomopathogenic fungi within the Ascomycota and order Hypocreales are represented in over 150 commercially available products [2], mostly based on strains from the genera *Beauveria*, *Metarhizium*, *Isaria*, and *Lecanicillium* [3]. Although entomopathogenic fungi constitute an environmentally friendly alternative to chemical pesticides, it is crucial to optimise their application in order to ensure maximum efficiency and reliability [4].

Mycoviruses are currently classified in seventeen taxa, sixteen families and one genus that does not belong to a family; eight of these taxa accommodate mycoviruses with double-stranded (ds) RNAs as their genome, including the families *Totiviridae* and *Partitiviridae*. Totiviruses have a linear non-segmented genome, while partitiviruses have two genomic segments. The members of these families encode two proteins, an RNA-dependent RNA polymerase (RdRp) required for genome replication, and a capsid protein (CP) that forms icosahedral virions (https://talk.ictvonline.org/ ictv-reports/ictv_online_report). Recently, a novel virus family has been described, provisionally designated as the Polymycoviridae [5,6]. Polymycoviruses have four dsRNA segments as their genome, one encoding the RdRp, and are non-conventionally encapsidated and are as infectious as dsRNA [5]. To date, dsRNA mycoviruses are not known to have an extracellular phase in their replication cycle. They can be transmitted horizontally from one fungal strain to another via hyphal fusion (anastomosis), or vertically from parent to offspring during the formation of sexual or asexual spores. Therefore, dispersion of these mycoviruses depends entirely on their hosts' movements. The majority of mycoviral infections have no discernible effects on the fungal host. Nevertheless, cases of hypovirulence [7–9], and more recently hypervirulence [5,10,11], have been reported, including an increase in growth and virulence of the entomopathogenic fungus *B. bassiana* caused by a polymycovirus [6].

The aim of the present work was to investigate the presence and diversity of mycoviruses in a selection of entomopathogenic fungal isolates belonging to the genera *Beauveria*, *Metarhizium*, *Lecanicillium*, *Purpureocillium*, *Isaria*, and *Paecilomyces*, derived from Spain and Denmark. Partial sequencing of the mycoviruses found and subsequent data analysis provided insights into population structure and dynamics, including selection pressures, recombination and transmission. Finally, a rapid pathogenicity screening against the Mediterranean fruit fly *Ceratitis capitata* highlighted potentially interesting isolates to be used in more detailed virulence experiments in the future.

2. Materials and Methods

2.1. Screening of Fungal Isolates and Molecular Cloning

In total, 151 isolates from two collections of entomopathogenic fungi (Table S1), one from the University of Cordoba, Spain, and one from the University of Copenhagen, Denmark, were used in the present study. The Spanish collection is kept at −80 °C following lyophilisation and the viability of the isolates is checked every two years. The Danish collection is also kept at −80 °C as conidia, suspended in 1:1 of glycerol:10% (w/v) skimmed milk. None of the isolates in this study are being used commercially as biological control agents. All isolates were grown on malt agar extract (MEA) medium at 25 °C and screened for the presence of dsRNA elements using phenol/Sevag treatment and enzymatic digestions with DNase I (Promega) and S1 nuclease (Promega) as described previously [6]. A random reverse transcriptase-chain polymerase reaction process [12] was applied for the initial identification of the dsRNAs, and target-specific primers (Table S2) were subsequently designed with the help of Primer-BLAST [13]. For fungal identification, DNA was extracted using a phenol/Sevag treatment, following disruption of the mycelia in liquid nitrogen [14], and the universal primers ITS1F (5′-CTT GGT CAT TTA GAG GAA GTA A-3′ [15] and ITS4 (5′-TCC TCC GCT TAT TGA TAT GC-3′ [16]) were used to amplify the complete sequence of internal transcribed spacer (ITS) 1, the 5.8S ribosomal RNA gene, and the internal transcribed spacer 2, flanked by the partial sequence of the 18S and 28S ribosomal RNA genes. Amplicons were then cloned using the pGEM-T Easy System (Promega, Madison, WI, USA) and transformed into X10-Gold Ultracompetent Cells (Agilent, Santa Clara, CA, USA). Plasmids were extracted using the QIAprep Spin Miniprep Kit (QIAGEN, Hilden, Germany) and at least three independent clones were sequenced for each amplicon. All sequences were deposited in the European Nucleotide Archive: BbVV-1 strain EABb 01/33-Su accession number LR028007; BbVV-1 strain EABb 01/112-Su accession number LR028008; BbVV-1 strain EABb 00/13-Su accession number LR028009; BbVV-1 strain EABb 07/06-Rf accession number LR028010; BbVV-1 strain EABb 10/57-Fil accession number LR028011; BbVV-1 strain EABb 11/01-Mg accession number LR028012; BbVV-1 strain EABb 00/11-Su accession number LR028013; BbPV-2 strain EABb 10/57-Fil accession number LR028014; BbPV-2 strain EABb 09/07-Fil accession number LR028015; BbPV-2 strain EABb 00/13-Su accession number LR028016; BbPV-2 strain EABb 11/01-Mg accession number LR028017; BbPV-2 strain EABb 07/06-Su accession number LR028018; BbPV-2 strain EABb 00/11-Su accession number LR028019; BbPmV-1 strain EABb 10/57-Fil accession number LR028020; BbPmV-1 strain EABb 10/28-Su accession number LR028021; BbPmV-1 strain EABb 00/13-Su accession number LR028022; BbPmV-1 strain EABb 00/11-Su accession number LR028023; BbPmV-1 strain EABb 07/06-Rf accession number LR028024; BbPmV-1 strain EABb 11/01-Mg accession number LR028025; BbPmV-1 strain EABb 10/01-Fil accession number LR028026; BbPmV-1 strain EABb 10/30-Fil accession number LR028027; ITS strain EABb 10/01-Fil accession number LR028032; ITS strain EABb 09/07-Fil accession number LR028033; ITS strain EABb 00/11-Su accession number LR028034; ITS strain EABb 01/112-Su accession number LR028035; ITS strain EABb 11/01-Mg accession number LR028036; ITS strain EABb 07/06-Rf accession number LR028037; ITS strain EABb 00/13-Su accession number LR028038; ITS strain EABb 10/28-Su accession number LR028039; ITS strain EABb 10/57-Fil accession number LR028040; ITS strain EABb 10/30-Fil accession number LR028041.

2.2. Computational and Phylogenetic Analysis

Sequence similarity searches of the GenBank, Swissprot, and EMBL databases were conducted using the BLASTx program [17]. Phylogenetic analysis of viral nucleotide sequences and p-distance calculations were performed using MEGA 6 [18], following alignment using MUSCLE as implemented by MEGA 6. Maximum likelihood phylogenetic trees were constructed using the K2 + I + G substitution model for victoriviruses, the K2 + G substitution model for partitiviruses ITS sequences and the T92 + I substitution model for polymycoviruses. The most appropriate substitution model for each set of sequences was selected using MEGA 6. The codon-based Z-test of selection on the overall average of viral sequences was also performed using MEGA 6 by computing the number of synonymous (dS) and non-synonymous (dN) substitutions per site. The probability of rejecting the null hypothesis of strict-neutrality (dN = dS) in favour of the alternative hypothesis (dN < dS or dN > dS) were calculated and *p*-values less than 0.05 were considered significant at the 5% level. The test statistic (dS–dN) was calculated and the variance of the difference was computed using the bootstrap method (100 replicates). Analyses were conducted using the Nei-Gojobori method [19]. If the number of synonymous substitutions was significantly higher than the number of non-synonymous substitutions then the population was under purifying/negative selection. Conversely, if the number of synonymous substitutions was significantly lower than the number of non-synonymous substitutions then the population was under positive selection. Recombination events were detected using RDP4 [20].

2.3. Insect Pathogenicity Experiments

The insect pathogenicity experiments were performed following standard protocols [21–26]. Conidial suspensions were prepared from selected isolates by scraping the conidia from Petri plates with malt extract agar (MEA) medium into a sterile aqueous solution of 0.1% Tween 80. Each conidial suspension was filtered through cheesecloth to remove the mycelial mat and adjusted to a concentration of 1.0×10^8 conidia per ml. Cold-anesthetised newly emerged *C. capitata* adults were sprayed with 1 mL of the conidial suspension in a Potter Spray Tower (Burkard Scientific, Uxbridge, UK). Each repetition of 10 insects were treated separately. Control flies were treated with the same volume of a sterile aqueous solution of 0.1% Tween 80. The treated adult *C. capitata* were placed in methacrylate boxes (8 × 8 × 6 cm) containing a circular hole of 2 cm in diameter covered with a net cloth. The bioassay conditions were 26 ± 2 °C, 50–60% RH and a photoperiod of 16:8 (L:D) h. An adult diet (0.1 g of hydrolyzed protein and 0.4 g sucrose with 1.5 mL of distilled water) and water were provided every 24 h. Three replicates of 10 insects were used and mortality was monitored daily for 8 days.

Dead flies were removed daily, immediately surface-sterilised and placed in humidity chambers for observation of mycosis as outlined by Quesada-Moraga et al. (2006) [27]. Data were analyzed using a generalised linear model (distribution = binomial; link = logit) and treatment comparisons were performed applying the χ^2 test ($p < 0.05$).

3. Results and Discussion

3.1. Presence of dsRNA Elements in B. bassiana Isolates from the Iberian Peninsula

Seventy-five *Metarhizium* sp. isolates [28] and two *B. bassiana* isolates [29] from Denmark, together with seventy-four isolates mostly from the Iberian Peninsula [30], including fifty *Beauveria* sp., thirteen *Metarhizium* sp., eight *Purpureocillium lilacinum*, two *Lecanicillium attenuatum*, one *Isaria farinosa*, and one *Paecilomyces marquandii* were screened for the presence of dsRNA elements. Putative mycoviruses were discovered in twelve out of forty (30%) *B. bassiana* isolates of Spanish provenance (Table 1), collected mainly from the south of Spain (Figure 1c). Isolate EABb 01/12-Su is known to be infected by a strain of *Beauveria bassiana victorivirus 1* (BbVV-1; [6], while isolates EABb 01/33-Su and EABb 00/11-Su have previously been reported to harbour uncharacterised dsRNA elements [31]. No dsRNA elements were discovered in any of the Danish *Metarhizium* sp. or *B. bassiana* isolates, although their presence has been documented previously in populations of *Metarhizium* mostly from Brazil [32–39]. The number of isolates from other fungal species screened was very small and no dsRNA elements were found in any of these. A notable presence of dsRNA elements or mycoviruses has been reported previously for *P. lilacinum* [40], *I. farinosa* [41] and other *Isaria* sp. [41], and *Paecilomyces* sp. [37,41]. The reason behind the high prevalence of dsRNA elements in the Spanish population is not clear. In Spain, there are three products registered in the Official Register for Phytosanitary Products and Materials of the Ministry of Agriculture, Forestry and Fisheries, and commercialised based on *B. bassiana*: Naturalis-L (ATCC 74040 strain), Botanigard, and Botanigard 22 WP (GHA strain). These are used to protect fruits and vegetables, such as apple, aubergine, beans, broccoli, cherry, citrus, cauliflower, cotton, cucurbit, grapevine, kaki, lettuce, olive, pear, green pepper, potato, strawberry, and tomato, against aphids, tephritids, thrips, and whiteflies. It should be noted that isolates ATCC 74040 and GHA are both virus-free [6], therefore the high prevalence of dsRNA elements cannot be explained by their potential introduction in the ecosystem via these biological control agents.

Table 1. Mycovirus-infected *B. bassiana* isolates, their habitat and Spanish location.

Species	Isolate	Habitat	Location	Mycovirus *
B. bassiana	EABb 00/11-Su	Soil (scrubland)	Jaen	BbVV-1 + BbPV-2 + BbPmV-1
B. bassiana	EABb 00/13-Su	Soil (woodland)	Jaen	BbVV-1 + BbPV-2 + BbPmV-1
B. bassiana	EABb 01/12-Su	Soil (scrubland)	Seville	BbVV-1
B. bassiana	EABb 01/33-Su	Soil (olive grove)	Cadiz	BbVV-1
B. bassiana	EABb 01/112-Su	Soil (wheat field)	Seville	BbVV-1
B. bassiana	EABb 07/06-Rf	*Rhynchophorus ferrugineus*	Alicante	BbVV-1 + BbPV-2 + BbPmV-1
B. bassiana	EABb 09/07-Fil	Phylloplane (meadow)	Malaga	BbPV-2
B. bassiana	EABb 10/01-Fil	Phylloplane (olive grove)	Malaga	BbPmV-1
B. bassiana	EABb 10/28-Su	Soil (olive grove)	Cordoba	BbPmV-1
B. bassiana	EABb 10/30-Fil	Phylloplane (olive grove)	Cordoba	BbPmV-1
B. bassiana	EABb 10/57-Fil	Phylloplane (meadow)	Cordoba	BbVV-1 + BbPV-2 + BbPmV-1
B. bassiana	EABb 11/01-Mg	*Monochamus galloprovincialis*	Palencia	BbVV-1 + BbPV-2 + BbPmV-1

* BbVV-1: Beauveria bassiana victorivirus 1; BbPV-2: Beauveria bassiana partitivirus 2; BbPmV-1: Beauveria bassiana polymycovirus 1.

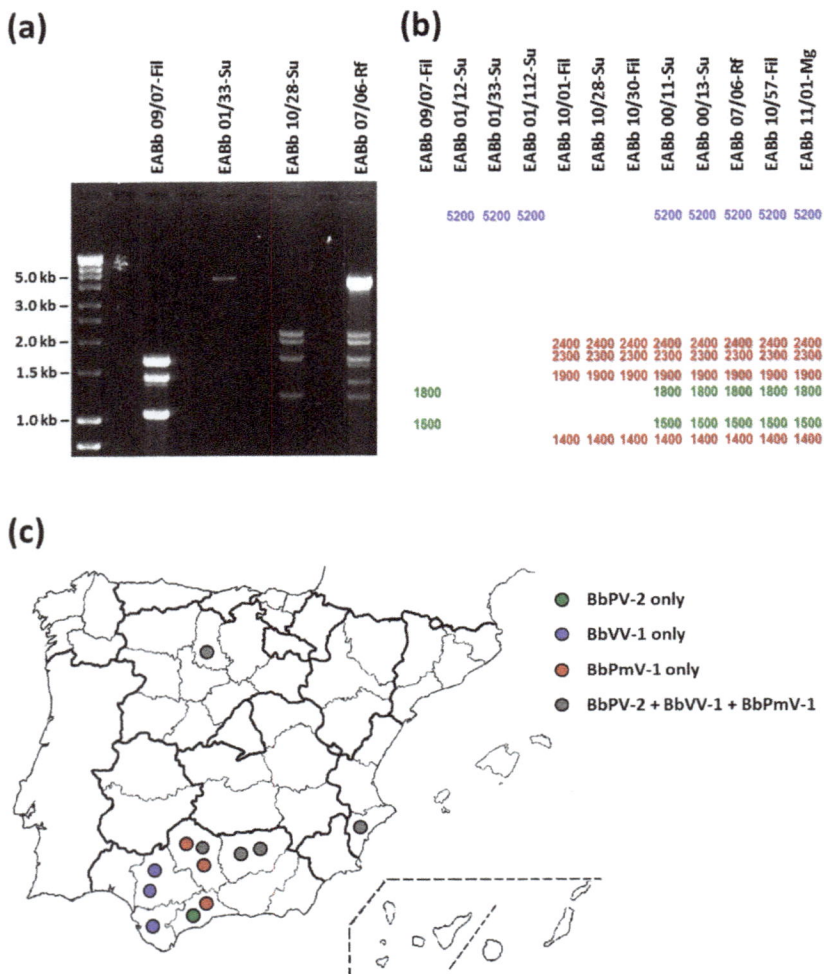

Figure 1. (a) 1% (*w/v*) agarose gel electrophoresis of dsRNA elements from four *Beauveria bassiana* isolates, representative of the four observed electrophoretic profiles. (b) Schematic representation of the observed electrophoretic profiles of dsRNA elements isolated from 12 *B. bassiana* isolates and their relative sizes, green for members of the family *Partitiviridae*, blue for members of the family *Totiviridae* and red for members of the proposed family Polymycoviridae. (c) Geographical distribution of mycoviruses found in Spanish *B. bassiana* isolates.

3.2. Mixed Infections of B. bassiana Isolates with Up to Three Different Mycoviruses

Following agarose electrophoresis of the purified dsRNA elements from all twelve *B. bassiana* isolates, four distinct electrophoretic profiles were noted as depicted in Figure 1a,b. One isolate harbours two dsRNAs 1–2 kbp in size, potentially a member of the family *Partitiviridae*, together with a smaller dsRNA which is probably a satellite RNA. Three isolates harbor a sole large dsRNA approximately 5 kbp in size, potentially a member of the family *Totiviridae*. Three isolates exhibit a banding pattern reminiscent of the proposed family Polymycoviridae. The rest—including isolate EABb 11/01-Mg from the pine sawyer beetle *Monochamus galloprovincialis* that acts as a vector for the parasitic nematode *Bursaphelenchus xylophilus*, causative agent of pine wilt—contain a combination of the above, suggesting the presence of multiple mycovirus infections. Since the profiles were identical to those observed previously [6], and initial molecular characterisation experiments revealed the presence of mycoviruses very similar to already fully sequenced strains, three oligonucleotide primer pairs were used to amplify part of the RdRp genes of BbVV-1 [31], Beauveria bassiana partitivirus 2 (BbPV-2) [6] and Beauveria bassiana polymycovirus 1 (BbPmV-1) [6]. All amplicons were cloned and sequenced and the presence of quasispecies, highly similar but not necessarily identical (Figure S1) strains, of the aforementioned mycoviruses in the fungal isolates was confirmed. Based on this and previous studies, all three viruses appear to be widespread in Spain. BbVV-1-like and BbPmV-1-like strains were discovered in eight out of forty (20%) Spanish *B. bassiana* isolates, while six out of forty (15%) harbor BbPV-2-like strains. To date, BbVV-1-like and BbPmV-1-like strains have been discovered exclusively in Spain, while BbPV-2-like strains have also been found in Asia and South America [6,31]. According to Andino and Domingo (2015), "viral quasispecies are defined as collections of closely related viral genomes subjected to a continuous process of genetic variation, competition among the variants generated, and selection of the most fit distributions in a given environment" [42]. Quasispecies are the result of the high mutation rates of the error-prone RdRps, which characterise all RNA viruses and lead to populations of mutants instead of identical viral genomes [43]. The quasispecies theory has been employed to better understand RNA viral pathogens, e.g., human immunodeficiency virus [44,45], their interactions within the population and with their host, and to design appropriate therapeutic approaches [46]. To our knowledge, this is the first time viral quasispecies have been discussed in the context of mycoviruses. However, since the potential effects of the mycoviruses on their hosts and the molecular mechanisms that underpin those are understudied, it is more difficult to appreciate the significance of this observation, for instance regarding the use of mycoviruses as enhancers of mycoinsecticides.

3.3. Selection Pressures and Recombination Events within the Viral Genome

Interestingly, while BbVV-1-like and BbPV-2-like strains are under positive selection, as indicated by a codon-based Z-test ($p < 0.05$; for BbVV-1-like strains: dN–dS = 6.28; 12 nucleotide sequences analysed, for BbPV-2-like strains: dN–dS = 3.17; 7 nucleotide sequences analysed), BbPmV-1-like strains appear to be under purifying or negative selection ($p < 0.05$; dS–dN = 17.30; 9 nucleotide sequences analysed). Positive selection is inferred by the significantly higher abundance of non-synonymous substitutions as compared to synonymous substitutions in the RdRp gene sequences and indicate an evolving viral population where novel phenotypic traits are evaluated favourably [47]. In contrast, purifying or negative selection does not allow for non-synonymous nucleotide substitutions that alter the protein sequence, which are then eliminated from the population [47], because they are presumably deleterious to the mycovirus. The reason why the polymycovirus RdRps are less tolerant to new variants as compared to those of victoriviruses and partitiviruses is debatable and may not be limited to the functionality of the protein as the enzyme responsible for viral replication. For instance, it has been speculated that the BbPmV-1 RdRp may physically interact with at least some of the other viral proteins (the putative scaffold protein, the methyl transferase, and/or the protein coating the viral genome [6]) and mutations that abolish these interactions would not be tolerated. Moreover, BbPmV-1 causes an increase in growth and virulence of *B. bassiana* [6], the molecular mechanisms of which are as yet unknown. However, hypervirulence may be mediated at least partially via virus-host protein interactions whose disruption is not favoured, since it deprives the fungal host and consequently the virus from this advantage. Finally, another mechanism contributing to the genetic variation of viral quasispecies is recombination [43] and one recombination event was detected in the case of BbPmV-1-like viruses: BbPmV-1 strain EABb 10/30-Fil is a recombinant of BbPmV-1 strain EABb 11/01-Mg (major parent) and BbPmV-1 strain EABb 10/28-Su (minor parent) More specifically, nucleotides 1-625 and 854-876 of the BbPmV-1 strain EABb 10/30-Fil amplified segment are derived from the major parent and nucleotides 626-853 are derived from the minor parent (Figure S2). Notably, this observation suggests that the two BbPmV-1-like viruses must have been simultaneously present in the same fungal isolate for the recombination event to take place. Subsequently, the competition among the three BbPmV-1-like viruses would have led to the loss of the two parents and the establishment of the recombinant. It is feasible that recombination events that transfer large fragments of the protein potentially optimised for their function(s) are more tolerable for polymycoviruses than single amino acid substitutions and therefore an appropriate mechanism for increasing diversity.

3.4. Evidence of Vertical and Horizontal Transmission of Mycoviruses

Since the viral sequences obtained were all very similar but not identical to each other, phylogenetic analysis was conducted to determine their evolutionary relationships and three phylogenetic trees were constructed for BbVV-1-like, BbPV-2-like, and BbPmV-1-like strains, respectively (Figure 2). In all three cases, it is evident that the mycoviruses in the fungal isolates with triple infections are more closely related to each other than to any of the other quasispecies investigated and, at least in the case of BbVV-1-like and BbPmV-1-like strains, they form distinct clusters. This was confirmed with the p-distance matrices (Figure S1), revealing that the co-infecting viral strains are often identical. Since their genomes are highly similar, the implication is that these three mycoviruses are transmitted simultaneously as a complex, either vertically or horizontally, from parent to offspring or from one fungal strain to another. In order to clarify the mode of transmission, the evolutionary relationships among the 12 *B. bassiana* isolates was examined and a phylogenetic tree was constructed using the ITS sequences (Figure S3). There are very few nucleotide substitutions noted among the ITS sequences, therefore the *B. bassiana* isolates are evolutionarily very close to each other. Nevertheless, the isolates harbouring the triple infections no longer cluster together, suggesting the possibility of one horizontal transmission event for the three viruses between more distantly related isolates, while vertical transmission is the most likely explanation between closely related isolates. Vegetative incompatibility studies have been performed in the past on *Beauveria* populations [48,49], revealing a large number of vegetative compatibility groups (VCGs) and a low frequency of genetic exchange. However, specifically regarding mycovirus transmission, different levels of (in)compatibility between fungal strains have been reported [50], and therefore it is feasible that, even if the Spanish *B. bassiana* population is similar to the ones described previously [48,49], this would not prevent spread of the mycoviruses, which is not necessarily constricted by genetic compatibility between fungal strains. As an ascomycete, *B. bassiana* has the capacity to produce both asexual spores, conidia, and sexual spores, and ascospores. However, the sexual stage is rarely observed and *B. bassiana*, being heterothallic, does not possess within a single individual the resources to reproduce sexually, but requires two distinct individuals of opposite mating types [51]. Our difficulties with eliminating *B. bassiana* mycoviruses in the past [6] suggest that they are easily transmitted during conidiation, while studies on other ascomycetes show that transmission via ascospores is variable and often less efficient [52–56].

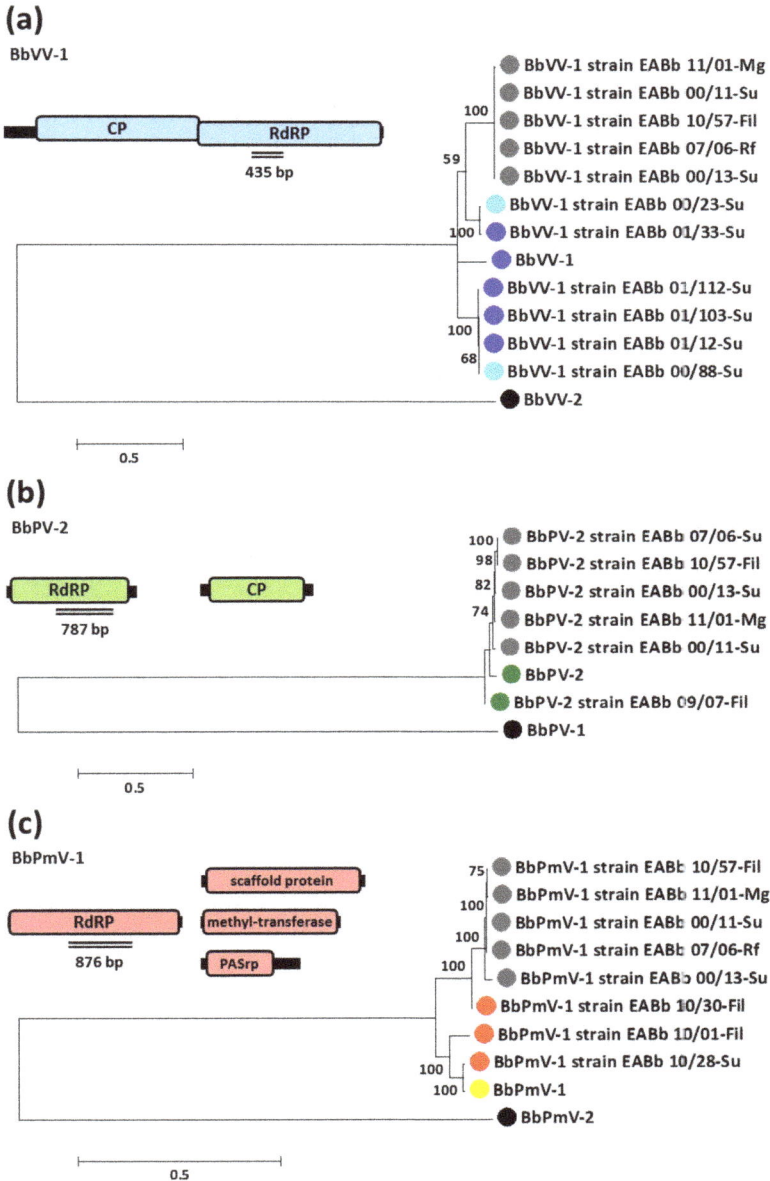

Figure 2. Maximum likelihood phylogenetic trees created based on the alignment of RdRp sequences of (a) members of the family *Partitiviridae*, (b) members of the family *Totiviridae* and (c) members of the proposed family Polymycoviridae infecting *Beauveria bassiana*. At the end of the branches: grey circles indicate that the *B. bassiana* isolate is infected with all three BbPV-2-like, BbVV-1-like, and BbPmV-1-like viruses; single infections of *B. bassiana* isolate is indicated by blue circles for BbVV-1-like virus; green circles for BbPV-2-like virus; red circles for BbPmV-1-like virus; turquoise circles for BbVV-1-like virus and a partitivirus [6]; yellow circles indicate that the *B. bassiana* isolate is infected with a BbPmV-1-like virus and a unirnavirus [57]; black circles indicate the outgroup.

119

3.5. Pathogenicity of Virus-Infected B. Bassiana against the Mediterranean Fruit Fly.

A preliminary bulk assessment of the variability in virulence among mycovirus infected *B. bassiana* isolates was performed using adult *C. capitata* as a model system. Significant differences in total mortality were found between the isolates [χ^2 (13) = 130.7, $p < 0.001$]. The total mortality values ranged between 80.0% and 100.0% for EABb 10/30-Fil and for the isolates EABb 01/112-Su, EABb 07/06-Rf, EABb 09/07-Fil, and EABb 11/01-Mg, respectively (Table 2). Regarding cadavers producing fungal outgrowth, significant differences were also found between the isolates [χ^2 (13) = 92.4, $p < 0.001$]. This mortality ranged between 50.0% and 90.0% for EABb 01/112-Su and EABb 07/06-Rf, respectively. No mortality with fungal outgrowth was observed in the control. The average survival times ranged from 7.5 days for the controls and 4.2 days for EABb 01/112-Su (Table 2). Interestingly, there appeared to be no correlation between the presence of specific mycoviruses and/or mycovirus complexes and the virulence of *B. bassiana* isolates. Additionally, some of the isolates with triple infections caused mortalities higher than 95.0%, suggesting that their heavy viral burdens did not impair pathogenicity. This may be explained by potential hypervirulence caused by the mycoviruses, especially strains of BbPmV-1, as shown previously [6]. Despite this previous demonstration of hypervirulence, it remains unclear if mycovirus infections in *B. bassiana* populations provide any adaptive advantage over wild-types. The absence of mycoviruses in the sampled European populations of other *Beauveria* and *Metarhizium* sp. Indicates, that for these species of entomopathogenic fungi, mycoviruses are not or at least rarely prevalent. Future experiments focusing on curing the *B. bassiana* isolates from the mycovirus infection and comparing the isogenic lines in terms of e.g., growth and germination, UV-tolerance, spore production, and pathogenicity against a range of insects will shed light on any selective advantage or related cost of mycovirus infections in entomopathogenic fungi.

Table 2. Insecticidal activity of mycovirus infected *B. bassiana* isolates to new emerged *C. capitata* adults inoculated with a suspension of 1.0×10^8 conidia mL^{-1}.

Treatment *	Mortality (Mean ± SE)% **		Kaplan-Meier Survival Analysis
	Total Mortality	Fungal Outgrowth	AST *** (d, Mean ± SE)
Control	13.3 ± 3.3 a	0.0 ± 0.0 a	7.5 ± 0.3 a
EABb 10/30-Fil ⬤	80.0 ± 0.0 b	53.3 ± 3.3 b	5.5 ± 0.4 b
EABb 01/33-Su ⬤	86.7 ± 3.3 b	76.7 ± 8.8 c	5.4 ± 0.3 b
EABb 00/11-Su ⬤	90.0 ± 0.0 b	70.0 ± 10.0 b	5.2 ± 0.4 bc
EABb 00/13-Su ⬤	90.0 ± 5.8 b	66.6 ± 12.0 b	5.1 ± 0.3 bc
EABb 01/12-Su ⬤	93.3 ± 6.7 bc	76.7 ± 8.8 c	4.5 ± 0.4 c
EABb 10/01-Fil ⬤	93.3 ± 3.3 bc	80.0 ± 10.0 c	5.3 ± 0.3 bc
EABb 10/28-Su ⬤	96.7 ± 3.3 c	63.3 ± 18.6 b	4.7 ± 0.4 bc
EABb 10/57-Fil ⬤	96.7 ± 3.3 c	70.0 ± 15.2 b	5.0 ± 0.4 bc
EABb 01/112-Su ⬤	100.0 ± 0.0 c	50.0 ± 10.0 b	4.2 ± 0.4 c
EABb 07/06-Rf ⬤	100.0±0.0 c	90.0 ± 0.0 c	4.8 ± 0.3 c
EABb 09/07-Fil ⬤	100.0±0.0 c	83.3 ± 6.7 c	5.2 ± 0.3 bc
EABb 11/01-Mg ⬤	100.0±0.0 c	70.0 ± 5.8 b	5.1 ± 0.1 bc

* Grey circles indicate that the isolate is infected with all three BbPV-2-like, BbVV-1-like, and BbPmV-1-like viruses; single infections of isolates are indicated by blue circles for BbVV-1-like virus, green circles for BbPV-2-like virus and red circles for BbPmV-1-like virus. ** Means within columns with the same letter are not significantly different (χ^2 test, $p \leq 0.05$) according to the generalised linear model. *** AST: Average survival time was limited to 8 days. Means within columns with the same letter are not significantly different ($p \leq 0.05$) according to the Log-Rank test.

Supplementary Materials: The following are available online at http://www.mdpi.com/1999-4915/10/12/665/ s1. Figure S1. Pairwise distance matrix created based on the RdRp sequences of chrysoviruses and related viruses; (A) members of the family *Partitiviridae*, (B) members of the family *Totiviridae* and (C) members of the proposed family Polymycoviridae infecting *Beauveria bassiana*. Figure S2: Schematic representation of potential recombination events between two BbPmV-1 strains: BbPmV-1 strain EABb 10/30-Fil is a recombinant of BbPmV-1 strain EABb 11/01-Mg (major parent) and BbPmV-1 strain EABb 10/28-Su (minor parent). Recombination breakage sites are indicated. Figure S3. Maximum likelihood phylogenetic tree created based on the alignment of

ITS sequences of the 12 mycovirus infected *Beauveria bassiana* isolates. At the end of the branches: grey circles indicate that the *B. bassiana* isolate is infected with all three BbPV-2-like, BbVV-1-like, and BbPmV-1-like viruses; blue circles indicate that the *B. bassiana* isolate is exclusively infected with a BbVV-1-like virus; green circles indicate that the *B. bassiana* isolate is exclusively infected with a BbPV-2-like virus; red circles indicate that the *B. bassiana* isolate is exclusively infected with a BbPmV-1-like virus. Table S1. Fungal isolates screened for the presence of mycoviruses. Table S2. Oligonucleotide primers used for viral sequence amplification.

Author Contributions: Conceptualization, I.K.L.; Methodology, I.G.-J. and I.K.L.; Formal Analysis, I.G.-J. and I.K.L.; Investigation, C.F. and I.G.-J.; Resources, N.V.M. and E.Q.-M.; Writing–Original Draft Preparation, I.K.L.; Writing–Review & Editing, C.F., I.G.-J., N.V.M., E.Q.M. and R.H.A.C; Supervision, I.K.L.; Project Administration, I.K.L.; Funding Acquisition, I.G.-J., R.H.A.C and I.K.L.

Acknowledgments: C.F. is financially supported by a University of Hertfordshire PhD research studentship. I.G.-J. wishes to thank the Ministry of Economy and Competitiveness of the Spanish Government for a Juan de la Cierva postdoctoral grant, Plan Propio de Investigación de la Universidad de Córdoba and Programa Operativo de fondos FEDER Andalucía. I.K-L and R.H.A.C. wish to acknowledge the support of the Steel Charitable Trust, the Elizabeth Creak Charitable Trust, the Morley Trust Foundation and the University of Hertfordshire Diamond Fund.

Conflicts of Interest: The authors declare no conflict of interest.

References

1. Chandler, D.; Bailey, A.S.; Tatchell, G.M.; Davidson, G.; Greaves, J.; Grant, W.P. The development, regulation and use of biopesticides for integrated pest management. *Philos. Trans. R. Soc. Lond. B Biol. Sci.* **2011**, *366*, 1987–1998. [CrossRef] [PubMed]

2. De Faria, M.R.; Wraight, S.P. Mycoinsecticides and Mycoacaricides: A comprehensive list with worldwide coverage and international classification of formulation types. *Biol. Control* **2007**, *43*, 237–256. [CrossRef]

3. Lacey, L.A.; Grzywacz, D.; Shapiro-Ilan, D.I.; Frutos, R.; Brownbridge, M.; Goettel, M.S. Insect pathogens as biological control agents: Back to the future. *J. Invertebr. Pathol.* **2015**, *132*, 1–41. [CrossRef] [PubMed]

4. St. Leger, R.J.; Wang, C. Genetic engineering of fungal biocontrol agents to achieve greater efficacy against insect pests. *Appl. Microbiol. Biotechnol.* **2010**, *85*, 901–907. [CrossRef] [PubMed]

5. Kanhayuwa, L.; Kotta-Loizou, I.; Özkan, S.; Gunning, A.P.; Coutts, R.H.A. A novel mycovirus from *Aspergillus fumigatus* contains four unique dsRNAs as its genome and is infectious as dsRNA. *Proc. Natl. Acad. Sci. USA* **2015**, *112*, 9100–9105. [CrossRef] [PubMed]

6. Kotta-Loizou, I.; Coutts, R.H.A. Studies on the virome of the entomopathogenic fungus *Beauveria bassiana* reveal novel dsRNA elements and mild hypervirulence. *PLoS Pathog.* **2017**, *13*, e1006183. [CrossRef] [PubMed]

7. Nuss, D.L. Biological control of chestnut blight: An example of virus-mediated attenuation of fungal pathogenesis. *Microbiol. Rev.* **1992**, *56*, 561–576. [PubMed]

8. Bhatti, M.F.; Jamal, A.; Petrou, M.A.; Cairns, T.C.; Bignell, E.M.; Coutts, R.H.A. The effects of dsRNA mycoviruses on growth and murine virulence of *Aspergillus fumigatus*. *Fungal Genet. Biol.* **2011**, *48*, 1071–1075. [CrossRef] [PubMed]

9. Liu, L.; Xie, J.; Cheng, J.; Fu, Y.; Li, G.; Yi, X.; Jiang, D. Fungal negative-stranded RNA virus that is related to bornaviruses and nyaviruses. *Proc. Natl. Acad. Sci. USA* **2014**, *111*, 12205–12210. [PubMed]

10. Özkan, S.; Coutts, R.H.A. *Aspergillus fumigatus* mycovirus causes mild hypervirulent effect on pathogenicity when tested on *Galleria mellonella*. *Fungal Genet. Biol.* **2015**, *76*, 20–26. [CrossRef] [PubMed]

11. Okada, R.; Ichinose, S.; Takeshita, K.; Urayama, S.I.; Fukuhara, T.; Komatsu, K.; Arie, T.; Ishihara, A.; Egusa, M.; Kodama, M.; et al. Molecular characterization of a novel mycovirus in *Alternaria alternata* manifesting two-sided effects: Down-regulation of host growth and up-regulation of host plant pathogenicity. *Virology* **2018**, *519*, 23–32. [CrossRef] [PubMed]

12. Froussard, P. A random-PCR method (rPCR) to construct whole cDNA library from low amounts of RNA. *Nucleic Acids Res.* **1992**, *20*, 2900. [CrossRef] [PubMed]

13. Ye, J.; Coulouris, G.; Zaretskaya, I.; Cutcutache, I.; Rozen, S.; Madden, T. Primer-BLAST: A tool to design target-specific primers for polymerase chain reaction. *BMC Bioinform.* **2012**, *13*, 134. [CrossRef] [PubMed]

14. Raeder, U.; Broda, P. Rapid preparation of DNA from filamentous fungi. *Lett. Appl. Microbiol.* **1985**, *1*, 17–20. [CrossRef]

15. Gardes, M.; Bruns, T.D. ITS primers with enhanced specificity for basidiomycetes—Application to the identification of mycorrhizae and rusts. *Mol. Ecol.* **1993**, *2*, 113–118. [CrossRef] [PubMed]

16. White, T.J.; Bruns, T.D.; Lee, S.B.; Taylor, J.W. Amplification and direct sequencing of fungal ribosomal RNA genes for phylogenetics. In *PCR Protocols: A Guide to Methods and Applications*; Innis, M.A., Gelfand, D.H., Sninsky, J.J., White, T.J., Eds.; Academic Press: London, UK, 1990; pp. 315–322.

17. Altschul, S.F.; Madden, T.L.; Schäffer, A.A.; Zhang, J.; Zhang, Z.; Miller, W.; Lipman, D.J. Gapped BLAST and PSI-BLAST: A new generation of protein database search programs. *Nucleic Acids Res.* **1997**, *25*, 3389–3402. [CrossRef] [PubMed]

18. Tamura, K.; Stecher, G.; Peterson, D.; Filipski, A.; Kumar, S. MEGA6: Molecular Evolutionary Genetics Analysis version 6.0. *Mol. Biol. Evol.* **2013**, *30*, 2725–2729. [CrossRef] [PubMed]

19. Nei, M.; Gojobori, T. Simple methods for estimating the numbers of synonymous and nonsynonymous nucleotide substitutions. *Mol. Biol. Evol.* **1986**, *3*, 418–426. [PubMed]

20. Martin, D.P.; Murrell, B.; Golden, M.; Khoosal, A.; Muhire, B. RDP4: Detection and analysis of recombination patterns in virus genomes. *Virus Evol.* **2015**, *1*, vev003. [CrossRef] [PubMed]

21. Garrido-Jurado, I.; Torrent, J.; Barrón, V.; Corpas, A.; Quesada-Moraga, E. Soil properties affect the availability, movement, and virulence of entomopathogenic fungi conidia against puparia of *Ceratitis capitata* (Diptera: Tephritidae). *Biol. Control* **2011**, *58*, 277–285. [CrossRef]

22. Garrido-Jurado, I.; Valverde-García, P.; Quesada-Moraga, E. Use of a multiple logistic regression model to determine the effects of soil moisture and temperature on the virulence of entomopathogenic fungi against pre-imaginal Mediterranean fruit fly *Ceratitis capitata*. *Biol. Control* **2011**, *59*, 366–372. [CrossRef]

23. Quesada-Moraga, E.; Valverde-García, P.; Garrido-Jurado, I. The effect of temperature and soil moisture on the development of the preimaginal Mediterranean fruit fly (Diptera: Tephritidae). *Environ. Entomol.* **2012**, *41*, 966–970. [CrossRef]

24. Yousef, M.; Garrido-Jurado, I.; Quesada-Moraga, E. One *Metarhizium brunneum* strain, two uses to control *Ceratitis capitata* (Diptera: Tephritidae). *J. Econ. Entomol.* **2014**, *107*, 1736–1744. [CrossRef] [PubMed]

25. Yousef, M.; Aranda-Valera, E.; Quesada-Moraga, E. Lure-and-infect and lure-and-kill devices based on *Metarhizium brunneum* for spotted wing *Drosophila* control. *J. Pest Sci.* **2018**, *91*, 227–235. [CrossRef]

26. Yousef, M.; Alba-Ramírez, C.; Garrido Jurado, I.; Mateu, J.; Raya Díaz, S.; Valverde-García, P.; Quesada-Moraga, E. Metarhizium brunneum (Ascomycota; Hypocreales) Treatments Targeting Olive Fly in the Soil for Sustainable Crop Production. *Front. Plant Sci.* **2018**, *9*, 1. [CrossRef] [PubMed]

27. Quesada-Moraga, E.; Ruiz-García, A.; Santiago-Alvarez, C. Laboratory evaluation of entomopathogenic fungi *Beauveria bassiana* and *Metarhizium anisopliae* against puparia and adults of *Ceratitis capitata* (Diptera: Tephritidae). *J. Econ. Entomol.* **2006**, *99*, 1955–1966. [CrossRef] [PubMed]

28. Steinwender, B.M.; Enkerli, J.; Widmer, F.; Eilenberg, J.; Thorup-Kristensen, K.; Meyling, N.V. Molecular diversity of the entomopathogenic fungal Metarhizium community within an agroecosystem. *J. Invertebr. Pathol.* **2014**, *123*, 6–12. [CrossRef] [PubMed]

29. Meyling, N.V.; Lübeck, M.; Buckley, E.P.; Eilenberg, J.; Rehner, S.A. Community composition, host-range and genetic structure of the fungal entomopathogen *Beauveria* in adjoining agricultural and semi-natural habitats. *Mol. Ecol.* **2009**, *18*, 1282–1293. [CrossRef] [PubMed]

30. Garrido-Jurado, I.; Fernández-Bravo, M.; Campos, C.; Quesada-Moraga, E. Diversity of entomopathogenic Hypocreales in soil and phylloplanes of five Mediterranean cropping systems. *J. Invertebr. Pathol.* **2015**, *130*, 97–106. [CrossRef] [PubMed]

31. Herrero, N.; Dueñas, E.; Quesada-Moraga, E.; Zabalgogeazcoa, I. Prevalence and diversity of viruses in the entomopathogenic fungus *Beauveria bassiana*. *Appl. Environ. Microb.* **2012**, *78*, 8523–8530. [CrossRef] [PubMed]

32. Leal, S.C.M.; Bertioli, D.J.; Ball, B.V.; Butt, T.M. Presence of double-stranded RNAs and virus-like particles in the entomopathogenic fungus *Metarhizium anisopliae*. *Biocontrol Sci. Technol.* **1994**, *4*, 89–94. [CrossRef]

33. Bogo, M.R.; Queiroz, M.V.; Silva, D.M.; Giménez, M.P.; Azevedo, J.L.; Schrank, A. Double-stranded RNA and isometric virus-like particles in the entomopathogenic fungus *Metarhizium anisopliae*. *Mycol. Res.* **1996**, *100*, 1468–1472. [CrossRef]

34. Melzer, M.J.; Bidochka, M.J. Diversity of double-stranded RNA viruses within populations of entomopathogenic fungi and potential implications for fungal growth and virulence. *Mycologia* **1998**, *90*, 586–594. [CrossRef]

35. Martins, M.K.; Furlaneto, M.C.; Sosa-Gomes, D.R.; Faria, M.R.; Fungaro, M.H.P. Double-stranded RNA in the entomopathogenic fungus *Metarhizium flavoviride*. *Curr. Genet.* **1999**, *36*, 94–97. [PubMed]

36. De la Paz Giménez-Pecci, M.; Bogo, M.; Santi, L.; De Moraes, C.K.; Corrêa, C.T.; Vainstein, M.H.; Schrank, A. Characterization of mycoviruses and analyses of chitinase secretion in the biocontrol fungus *Metarhizium anisopliae*. *Curr. Microbiol.* **2002**, *45*, 334. [CrossRef] [PubMed]

37. Tiago, P.V.; Fungaro, M.H.; de Faria, M.R.; Furlaneto, M.C. Effects of double-stranded RNA in *Metarhizium anisopliae* var. *acridum* and *Paecilomyces fumosoroseus* on protease activities, conidia production, and virulence. *Can. J. Microbiol.* **2004**, *50*, 335–339. [PubMed]

38. Perinotto, W.M.S.; Golo, P.S.; Coutinho Rodrigues, C.J.B.; Sá, F.A.; Santi, L.; Beys da Silva, W.O.; Junges, A.; Vainstein, M.H.; Schrank, A.; Salles, C.M.C.; et al. Enzymatic activities and effects of mycovirus infection on the virulence of *Metarhizium anisopliae* in *Rhipicephalus microplus*. *Vet. Parasitol.* **2014**, *203*, 189–196. [CrossRef] [PubMed]

39. Santos, V.; Mascarin, G.M.; da Silva Lopes, M.; Alves, M.C.D.F.; Rezende, J.M.; Gatti, M.S.V.; Dunlap, C.A.; Júnior, Í.D. Identification of double-stranded RNA viruses in Brazilian strains of *Metarhizium anisopliae* and their effects on fungal biology and virulence. *Plant Gene* **2017**, *11*, 49–58. [CrossRef]

40. Herrero, N. A novel monopartite dsRNA virus isolated from the entomopathogenic and nematophagous fungus *Purpureocillium lilacinum*. *Arch. Virol.* **2016**, *161*, 3375–3384. [CrossRef] [PubMed]

41. Inglis, P.W.; Valadares-Inglis, M.C. Rapid isolation of double-stranded RNAs from entomopathogenic species of the fungus *Paecilomyces* using a commercial minicolumn system. *J. Virol. Methods* **1997**, *67*, 113–116. [CrossRef]

42. Andino, R.; Domingo, E. Viral quasispecies. *Virology* **2015**, *479–480*, 46–51. [CrossRef] [PubMed]

43. Domingo, E.; Perales, C. Quasispecies and virus. *Eur. Biophys. J.* **2018**, *47*, 443. [CrossRef] [PubMed]

44. Briones, C.; Domingo, E. Minority report: Hidden memory genomes in HIV-1 quasispecies and possible clinical implications. *AIDS Rev.* **2008**, *10*, 93–109. [PubMed]

45. Rios, A. Fundamental challenges to the development of a preventive HIV vaccine. *Curr. Opin. Virol.* **2018**, *29*, 26–32. [CrossRef] [PubMed]

46. Lauring, A.S.; Andino, R. Quasispecies theory and the behavior of RNA viruses. *PLoS Pathog.* **2010**, *6*, e1001005. [CrossRef] [PubMed]

47. Domingo, E.; Sheldon, J.; Perales, C. Viral quasispecies evolution. *Microbiol. Mol. Biol. Rev.* **2012**, *76*, 159–216. [CrossRef] [PubMed]

48. Couteaudier, Y.; Viaud, M. New insights into population structure of *Beauveria bassiana* with regard to vegetative compatibility groups and telomeric restriction fragment length polymorphisms. *FEMS Microbiol. Ecol.* **1997**, *22*, 175–182. [CrossRef]

49. Castrillo, L.A.; Griggs, M.H.; Vandenberg, J.D. Vegetative compatibility groups in indigenous and mass-released strains of the entomopathogenic fungus *Beauveria bassiana*: Likelihood of recombination in the field. *J. Invertebr. Pathol.* **2004**, *86*, 26–37. [CrossRef] [PubMed]

50. Deng, F.; Melzer, M.S.; Boland, G.J. Vegetative compatibility and transmission of hypovirulence-associated dsRNA in *Sclerotinia homoeocarpa*. *Can. J. Plant Pathol.* **2002**, *24*, 481–488. [CrossRef]

51. Xiao, G.; Ying, S.H.; Zheng, P.; Wang, Z.L.; Zhang, S.; Xie, X.Q.; Shang, Y.; St Leger, R.J.; Zhao, G.P.; Wang, C.; et al. Genomic perspectives on the evolution of fungal entomopathogenicity in *Beauveria bassiana*. *Sci. Rep.* **2012**, *2*, 483. [CrossRef] [PubMed]

52. Coenen, A.; Kevei, F.; Hoekstra, R.F. Factors affecting the spread of double-stranded RNA viruses in *Aspergillus nidulans*. *Genet. Res.* **1997**, *69*, 1–10. [CrossRef] [PubMed]

53. Anagnostakis, S.L.; Chen, B.; Geletka, L.M.; Nuss, D.L. Hypovirus transmission to ascospore progeny by field-released transgenic hypovirulent strains of *Cryphonectria parasitica*. *Phytopathology* **1998**, *88*, 598–604. [CrossRef] [PubMed]

54. Varga, J.; Rinyu, E.; Kevei, E.; Tóth, B.; Kozakiewicz, Z. Double-stranded RNA mycoviruses in species of *Aspergillus* sections Circumdati and Fumigati. *Can. J. Microbiol.* **1998**, *44*, 569–574. [CrossRef] [PubMed]

55. Deng, F.; Allen, T.D.; Hillman, B.I.; Nuss, D.L. Comparative analysis of alterations in host phenotype and transcript accumulation following hypovirus and mycoreovirus infections of the chestnut blight fungus *Cryphonectria parasitica*. *Eukaryot. Cell* **2007**, *6*, 1286–1298. [CrossRef] [PubMed]

56. Chun, S.J.; Lee, Y.H. Inheritance of dsRNAs in the rice blast fungus, *Magnaporthe grisea*. *FEMS Microbiol. Lett.* **2009**, *148*, 159–162. [CrossRef]

57. Kotta-Loizou, I.; Sipkova, J.; Coutts, R.H.A. Identification and sequence determination of a novel double-stranded RNA mycovirus from the entomopathogenic fungus *Beauveria bassiana*. *Arch. Virol.* **2015**, *160*, 873–875. [CrossRef] [PubMed]

viruses

MDPI

Article

Metatranscriptomic Analysis and In Silico Approach Identified Mycoviruses in the Arbuscular Mycorrhizal Fungus *Rhizophagus* spp.

Achal Neupane [1], Chenchen Feng [2], Jiuhuan Feng [1,2], Arjun Kafle [1], Heike Bücking [1] and Shin-Yi Lee Marzano [1,2,*]

[1] Department of Biology and Microbiology, South Dakota State University, Brookings, SD 57007, USA; achal.neupane@sdstate.edu (A.N.); jiuhuan.feng@sdstate.edu (J.F.); arjun.kafle@sdstate.edu (A.K.); heike.bucking@sdstate.edu (H.B.)
[2] Department of Agronomy, Horticulture, and Plant Sciences, South Dakota State University, Brookings, SD 57007, USA; chenchen.feng@sdstate.edu
* Correspondence: shinyi.marzano@sdstate.edu or shinyileemarzano@gmail.com; Tel.: +1-605-688-5469

Received: 22 September 2018; Accepted: 7 December 2018; Published: 12 December 2018

Abstract: Arbuscular mycorrhizal fungi (AMF), including *Rhizophagus* spp., can play important roles in nutrient cycling of the rhizosphere. However, the effect of virus infection on AMF's role in nutrient cycling cannot be determined without first knowing the diversity of the mycoviruses in AMF. Therefore, in this study, we sequenced the *R. irregularis* isolate-09 due to its previously demonstrated high efficiency in increasing the N/P uptake of the plant. We identified one novel mitovirus contig of 3685 bp, further confirmed by reverse transcription-PCR. Also, publicly available *Rhizophagus* spp. RNA-Seq data were analyzed to recover five partial virus sequences from family *Narnaviridae*, among which four were from *R. diaphanum* MUCL-43196 and one was from *R. irregularis* strain-C2 that was similar to members of the *Mitovirus* genus. These contigs coded genomes larger than the regular mitoviruses infecting pathogenic fungi and can be translated by either a mitochondrial translation code or a cytoplasmic translation code, which was also reported in previously found mitoviruses infecting mycorrhizae. The five newly identified virus sequences are comprised of functionally conserved RdRp motifs and formed two separate subclades with mitoviruses infecting *Gigaspora margarita* and *Rhizophagus clarus*, further supporting virus-host co-evolution theory. This study expands our understanding of virus diversity. Even though AMF is notably hard to investigate due to its biotrophic nature, this study demonstrates the utility of whole root metatranscriptome.

Keywords: mycorrhizal fungi; mycovirus; mitovirus; *Rhizophagus*

1. Introduction

About eighty percent of land plants form symbiotic relationships with arbuscular mycorrhizal (AM) fungi [1], where obligate mutualistic fungi colonize plant roots for their spores to germinate and form hyphae. Examples of endophytic fungi, including AM fungi, have been shown to help control fungal pathogens [2], resist drought and salinity [3,4], and affect the overall fitness (growth, survival, etc.) of vascular plant families [5,6]. However, it is not well known whether multipartite plant-AM fungi-virus interactions may play a role in the plant's adaptation to biotic and abiotic stresses. Specifically, it remains unclear how AM fungi infections can alter patterns of plant gene expression, or whether superimposed viral infections would have cascading effects on the plant gene expression.

As AM fungi play important roles in carbon/nitrogen/phosphate cycling and compete with pathogens for ecological niches, there is emerging interest in discovering whether they harbor viruses through next generation sequencing [7,8]. It is necessary to recover the virus sequences associated with these fungi before further determining the effect of viral infections on hyphal growth and nutrient uptake of the host plant. Other endophytic fungi forming mutualistic symbiotic relationships with land plants have been shown to harbor viruses and confer heat tolerance when infected by virus(es) [9]. However, the prevalence and effects of virus infection on AM fungi are largely unknown, and the roles they play in the context of carbon/nutrient cycling are still ambiguous. Additionally, the virome of AM fungi is difficult to study partly because of its obligate nature of biotrophic reproduction that requires a large number of hyphae [8] or spores [10].

Likely not mutually exclusive, "virus-host ancient coevolution theory" is one of two hypotheses that have been proposed for mycovirus origin [11], with the other hypothesis suggesting that plant viruses are the origin of mycoviruses [12]. The *Narnaviridae* family of mycovirus is comprised of two genera, namely *Narnavirus* and *Mitovirus*, and include some of the simplest RNA viruses ever identified [13]. Narnaviruses are known to be localized in the cytosol, expressed using standard genetic code [13] and likely evolved from a RNA bacteriophage [14]. Mitoviruses, meanwhile, are known to be found primarily in mitochondria of the fungal host, translated using mold mitochondrial genetic code, and are believed to have evolved as endosymbiont of alphaproteobacteria [13]. Additionally, *Narnaviradae* RdRps are closely related to leviviruses, viruses of bacteria and ourmiaviruses of plants [13–15].

Typical mitoviruses have <3 kb genomes and have been detected in both fungi and plants [16], and either exist endogenously in plant genomes or freely replicate in mitochondria as genuine viruses. Endogenous mitovirus sequences may or may not be transcribed actively [17]. However, mitoviruses detected from mycorrhizal fungi generally have genome sizes greater than 3 kb, and the coding regions can be either translated by a cytosolic/nuclear genetic codon usage table or a mitochondrial table [7,18].

We recently screened soybean leaf-associated viromes and identified 23 nearly full-length mycoviral genomes using RNA-Seq of total RNA even when the plant sequences were present [19]. In order to understand the effects of a tritrophic relationship among plant-AM fungi-virus interactions on soil processes, root-associated viromes should be profiled. Differences in phosphate and nitrogen uptake of AMF were observed even within the same species [20], suggesting that besides genetic variability, there could be microbes, including mycoviruses, hosted by AMF that affect their functions. Notably, Ikeda et al. [21] determined that AM fungi infected by the virus, GRF1V-M, produced two-fold fewer spores compared to the virus-free culture line of *Rhizophagus* spp. strain RF1, and was less efficient in promoting plant growth. Therefore, in this study, we aimed to discover and characterize new mycoviruses infecting AM fungi with combined approaches. We used a culture-independent metatranscriptomics approach to detect viruses infecting *Rhizophagus* spp., and by reanalyzing data from other *Rhizophagus* spp. available as SRR data from the NCBI database (https://www.ncbi.nlm.nih.gov/sra). As *Medicago truncatula* is a host plant for *Rhizophagus* spp., we performed metatranscriptome RNA-Seq on *M. truncatula* roots directly to screen for mycoviruses. This research could provide insight on virus evolution and may help researchers form hypotheses to study the mechanisms of the varying functions from isolates/species of AMF that affect their biofertilizer potential.

2. Materials and Methods

2.1. Plant and Fungal Material

Medicago truncatula (A17) seeds were surface sterilized with concentrated H_2SO_4, rinsed with autoclaved distilled water, and kept at 4 °C overnight. The seeds were then pregerminated on moist filter paper for 7 days until fully grown cotyledons were developed. We transferred the seedlings into pots containing 250 mL of an autoclaved soil substrate mixture of 40% sand, 20% perlite, 20% vermiculite, and 20% soil (v:v:v:v; 4.81 mg/kg P_i after Olsen extraction, 10 mg/kg NH_4^+, 34.40 mg/kg NO_3^-, pH 8.26). At transplanting, each seedling was inoculated with 0.4 g mycorrhizal root material and ~500 spores of *Rhizophagus irregularis* N.C. Schenck & G.S. Sm. (isolate 09 collected from Southwest Spain by Mycovitro S.L. Biotechnología ecológica, Granada, Spain). The roots and the fungal inoculum were produced in axenic Ri T-DNA transformed carrot (*Daucus carota* clone DCI) root organ cultures in Petri dishes filled with mineral medium [22]. After approximately eight weeks, the spores were isolated by blending the medium in 10 mM citrate buffer (pH 6.0)

The plants were grown in a growth chamber with a 25 °C/20 °C day and night cycle, 30% humidity, a photosynthetic active radiation of 225 μmol m^{-2} s^{-1}, and watered when necessary. After seven weeks, the plants were harvested, and mycorrhizal root material was frozen in liquid nitrogen and stored at −80 °C until RNA extraction. To quantify the mycorrhizal colonization, some roots were cleared with 10% KOH solution at 80 °C for 30 min, rinsed with water, and stained with 5% ink at 80 °C for 15 min [23]. We analyzed a minimum of 100 root segments to determine the percentage of AM root colonization by using the gridline intersection method [24].

2.2. High-Throughput Sequencing

Approximately 150 mg of root tissue was ground in liquid nitrogen, and total RNA was extracted using the Qiagen RNeasy Plant Mini Kit (Valencia, CA, USA). RNA samples were treated with DNase I, evaluated for integrity by agarose gel electrophoresis, and rRNAs were removed by the Ribo-Zero Plant Kit (Illumina, San Diego, CA, USA), and used as templates to construct the library with a ScriptSeq RNA sample preparation kit (Illumina, San Diego, CA, USA). The library was submitted to the W. M. Keck Center, University of Illinois for quality check and cleanup and sequenced on an Illumina HiSeq 4000 for 100 bp paired-end reads.

2.3. Sequence Analysis

Sequence reads from the above sequencing run, as well as publicly available data (published by Tisserant et al., 2013 [25]) under SRX312982 (*Rhizophagus diaphanum* MUCL 43196; previously *Glomus diaphanum* [26]), SRX375378 (*Rhizophagus irregularis* DAOM-197198; previously *Glomus intraradices* or *Rhizophagus intraradices* [25,26]) and SRX312214 (*Rhizophagus irregularis* C2) were retrieved from the NCBI database and the paired-end sequence reads (100 nt in length) were trimmed by BBMap tools (https://sourceforge.net/projects/bbmap) and assembled into contigs using the TRINITY de novo transcriptome assembler [27]. Contigs with significant similarity to viral amino acid sequences were identified using USEARCH ublast option [28] with a parameter e-value of 0.0001 and compared to a custom database containing *Rhizophagus irregularis* and viral amino acid sequences from GenBank using BLASTX [29]. The nucleotide sequences of all suspected mycovirus contigs were compared with the NCBI nr database using BLASTX [29] to exclude misidentified sequences. The number of reads aligning to different target sequences was calculated using Bowtie [30]. Predicted amino acid sequences were aligned using ClustalW [31]. Aligned protein sequences were used to reconstruct a maximum likelihood tree with the model WAG + G + I + F using Mega (Molecular Evolutionary Genetics Analysis) version 7.0 software [32]. Statistical support for this analysis was computed based on 100 nonparametric bootstrap replicates. The MEME suite 5.0.1 was used to compare the motifs [33]. The viral sequences were submitted to the GenBank database under the following accession numbers: RdMV1, MH732931; RdMV2, MH732930; RdMV3, MK156099; RdMV4, MK156100; RirMV1 and MH732933.

2.4. Reverse-Transcription PCR (RT-PCR)

To confirm that the RirMV1 sequence detected was not an artifact and indeed derived from the *Medicago* root material, RT-PCR amplified a 3.4 kb amplicon from the RNA extract after DNase treatment by the virus-specific primers, RirMV1-197F (5′-CACCTATGAGCCCGGTTAAA-3′) and RirMV1-3409R (5′-GGAGAATCGTCCTTCCTTCC-3′). For the nested PCR the primers RirMV1-197F and RirMV1-3228R (5′-ACCTTTCCAGGGGAGACCTA-3′) were used. The nested amplicon was submitted for Sanger sequencing to confirm the identity after ExoSap-IT cleanup (Thermofisher, Waltham, MA, USA). Additionally, to confirm that the viral sequence is not from the Medicago host, reverse transcription of cDNA was made by using Maxima H Minus Reverse Transcriptase (Thermo Scientific, Waltham, MA, USA) at 50 °C for 30 min followed by 85 °C for 5 min inactivation. Then PCR was performed using RirMV1-197F and RirMV1-3228R primer set and Phire Plant Direct PCR Kit (Thermo Scientific, Waltham, MA, USA).

2.5. Rapid Amplification of cDNA Ends (RACE)

To complete the genome sequence of RirMV1, the 5′- and 3′-terminal sequences were determined using the FirstChoice RLM-RACE (rapid amplification of cDNA ends) kit (Life Technologies). Primers 336R (5′-AGAGCGGTCGCTTCTGTCTA-3′) and 216R (5′-TTTAACCGGGCTCATAGGTG-3′) were used for 5′-RACE as outer and inner primers, respectively. Primers 3210F (5′-TAGGTCTCCCCTGGAAAGGT-3′) and 3347F (5′-CGACCTCTGGAGGTTGAAAG-3′) were used for 3′-RACE as outer and inner primers, respectively.

3. Results

3.1. Mycoviruses in the Metatranscriptome of Rhizophagus Irregularis Inoculated Roots

After colonization of *R. irregularis* was confirmed by microscopy (Figure S1), sequencing of the mycorrhizal *Medicago truncatula* roots on the Illumina HiSeq4000 platform resulted in a total of 85 million paired-end reads, yielding 12.1 GB of sequence information. The data were submitted to the SRA database at NCBI (accession number SRX4679168). In this data set, we identified one viral contig (RirMV1). To confirm the viral contig assembled from the short reads, RirMV1-3409R primed cDNA was used as a PCR template to amplify most of the viral contig. The primers RirMV1-197F and RirMV1-3409R amplified multiple bands, and among them there was a faint 3 kb band (not shown). The 3 kb band was subsequently excised, and the gel was purified. Nested PCR using RirMV1-197F and RirMV1-3228R resulted in a clear band of 3 kb (Figure 1A). Sanger sequencing using the same primer set confirmed the band as RirMV1 cDNA amplicon. PCR attempts to amplify RirMV1 directly from the DNA of the *R. irregularis* strain 09 infected *Medicago* roots were not successful, indicating that the viral transcript was not derived from virus segments integrated into the host genome that are actively expressed. Instead they are from the genuine virus (Figure 1B). We also attempted to amplify a smaller target using viral-specific primers 197F and 336R for 140 bp amplicon and ran a 1.5% gel to confirm that there was no amplification, leading to the same conclusion that the viral sequence was not from *Medicago*, which confirms that *R. irregularis* is the host of the virus. Additionally, we also attempted RACE amplification of RirMV1 contig, but failed to extend the contig.

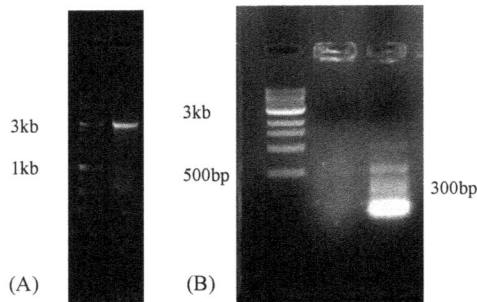

Figure 1. (**A**) Agarose gel electrophoresis of the RT-PCR product showing a ~3 kb nested PCR amplicon that was confirmed by Sanger sequencing as cDNA amplicon of RirMV1. Left lane: 1 kb ladder. Right lane: RirMV1 amplicon of the predicted size of 3 kb and (**B**) Agarose gel electrophoresis of the RT-PCR product showing no amplification, suggesting the viral contig of RNA-Seq was not originated from *Medicago* root without *R. irregularis* strain 09 infection. Left to right lanes: 1 kb ladder, viral primers, plant primers.

3.2. Mycoviruses in the Transcriptomes of Rhizophagus spp.

To identify mycoviruses in *Rhizophagus* spp., we first reanalyzed the publicly available RNA-Seq data sets of *R. irregularis* strain-C2 (SRX312214), *R. irregularis* DAOM-197198 (SRX375378) and *R. diaphanum* MUCL 43196 (SRX312982). No viruses could be identified in the available *Rhizophagus irregularis* DAOM-197198 transcriptome, but we found multiple novel mycoviruses in the transcriptome of *R. diaphanum* MUCL 43196 (Rhizophagus diaphanum mitovirus 1—RdMV1, 3554 nt long; Rhizophagus diaphanum mitovirus 2—RdMV2, 4382 nt long; Rhizophagus diaphanum mitovirus 3—RdMV3, 3652 nt long and Rhizophagus diaphanum mitovirus 4—RdMV4, 3443 nt long) that had similarities to members of the *Mitovirus* genus (Table 1).

Overall, RirMV1 had 41,322 reads and a 0.10% alignment with the sequencing run of the colonized *Medicago* roots. Among the total reads for SRX312982 from *R. diaphanum*, 1475 read-counts aligned to RdMV1 (0.0015%), 3,649 to RdMV2 (0.0036%), 2350 to RdMV3 (0.0023%), and 462 to RdMV4 (0.00045%), see Table 1. The NCBI BLAST results indicate that these contigs are putatively similar in function to previously identified RNA-dependent RNA polymerases of mitoviruses.

Table 1. Identified mycovirus-like sequences, contig lengths, and their putative functions are shown in the table below, including the data source from which the virus sequence was recovered. These new contigs were identified as mitoviruses (MV) and were recovered from two different fungal hosts (Rd, *Rhizophagus diaphanum*; Rir, *Rhizophagus irregularis*).

Contig Name	Data Source	Contig Length (nt)	Read Counts	NCBI Accession	Amino Acid Identity (%)	Putative Function (Most Similar Virus)
RdMV1	SRX312982 (*Rhizophagus diaphanum* MUCL 43196)	3554	1475	MH732931	32	RNA-dependent RNA polymerase [Rhizoctonia solani mitovirus 12]
RdMV2	SRX312982 (*Rhizophagus diaphanum* MUCL 43196)	4382	3649	MH732930	28	RNA-dependent RNA polymerase [Gigaspora margarita mitovirus 2]
RdMV3	SRX312982 (*Rhizophagus diaphanum* MUCL 43196)	3652	2350	MK156099	36	RNA-directed RNA polymerase [Rhizophagus sp. RF1 mitovirus]
RdMV4	SRX312982 (*Rhizophagus diaphanum* MUCL 43196)	3443	462	MK156100	30	RNA-dependent RNA polymerase [Rhizoctonia mitovirus 1]
RirMV1	Sequenced transcriptome (submitted under accession: SRX4679168)	3685	41,322	MH732933	31	RNA-dependent RNA polymerase [Rhizoctonia solani mitovirus 12]

3.3. Phylogenetic Analysis and the Characterization of Conserved RdRp Region of Mitoviruses

To identify the evolutionary lineages among the identified mitoviruses, we analyzed the protein sequences of mitoviruses to reconstruct the phylogenetic tree (Figure 2). While there was no virus found in SRX375378 and SRX312214, there were four partial genome sequences identified from SRX312982 publicly available data similar to viruses from the family *Narnaviridae*. Two of these sequences (RdMV1 and RdMV2) formed a separate clade with RirMV1 and previously identified mitoviruses from *Gigaspora margarita* (GmMV2, GmMV3, and GmMV4). The other two contigs (RdMV3 and RdMV4) were phylogenetically similar to the mitovirus infecting *Rhizophagus clarus* (RcMV1) (Figure 2). We also compared the genome structure of identified mitoviruses to see if the RdRp region is uniformly conserved (Figure 3). To confirm the presence of functionally conserved motifs of RNA-dependent RNA polymerase (RdRp) in identified viruses, we further analyzed and compared six RdRp motifs (A–F) with other mitoviruses of pathogenic fungi in the NCBI database (Figure 4). Three of these motifs (A–C) are among the most conserved motifs of RdRp and include residues involved in catalytic activation and dNTP/rNTP recognition (discussed in detail below) by RdRp [34,35]. Noticeable differences in the amino acid sequence include a histidine in the mitoviruses of *Rhizophagus* spp.

instead of a serine at residue 325 and a glutamic acid instead of an aspartic acid at residue 329 within the RdRp motif F.

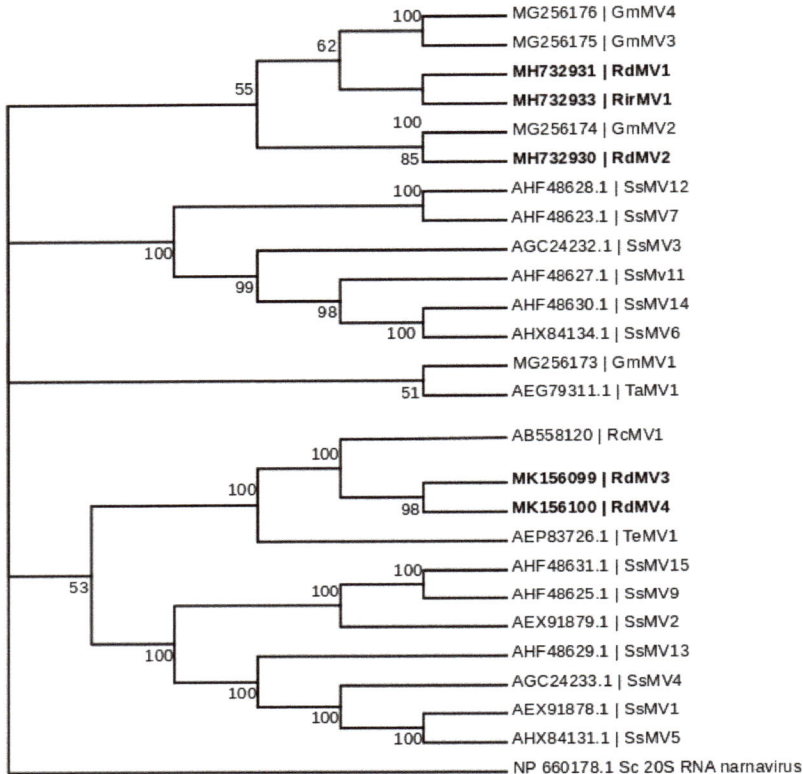

Figure 2. Maximum likelihood tree (with bootstrap consensus) depicting the relationships of the predicted amino acid sequences of RNA-dependent RNA polymerase (RdRp) of the *Rhizophagus* mitoviruses, and other confirmed and proposed members of the *Narnaviridae*. Predicted RdRp amino acid sequences were aligned with ClustalW [31], and the phylogenetic tree was inferred using Mega 7.0 software [32]. Branch lengths are scaled to the expected underlying number of amino acid substitutions per site. The Saccharomyces 20S RNA narnavirus RdRp amino acid sequence was used as an outgroup to root the tree. Five newly identified mitoviruses (in bold) formed two separate monophyletic clusters between the Rhizophagus-associated mitoviruses. The following abbreviations were used for the Mitovirus (MV) sequences: Sc, *Saccharomyces cerevisiae*; Gm, *Gigaspora margarita*; Rd, *Rhizophagus diaphanum*; Rc, *Rhizophagus clarus*; Sc, *Sclerotinia sclerotiorum*; Rir, *Rhizophagus irregularis*; Ta, *Tuber aestivum*; Te, *Tuber excavatum*.

The RdRps can be translated using either a cytosolic or mitochondrial code. The complete coding RdRp was 811 aa long for RirMV1, compared to the average of ~700 aa for the most closely related mitoviruses infecting *Sclerotinia sclerotiorum*. RdMV1 and RirMV1 have nearly identical lengths of RdRp, these being 812 aa and 811 aa, respectively.

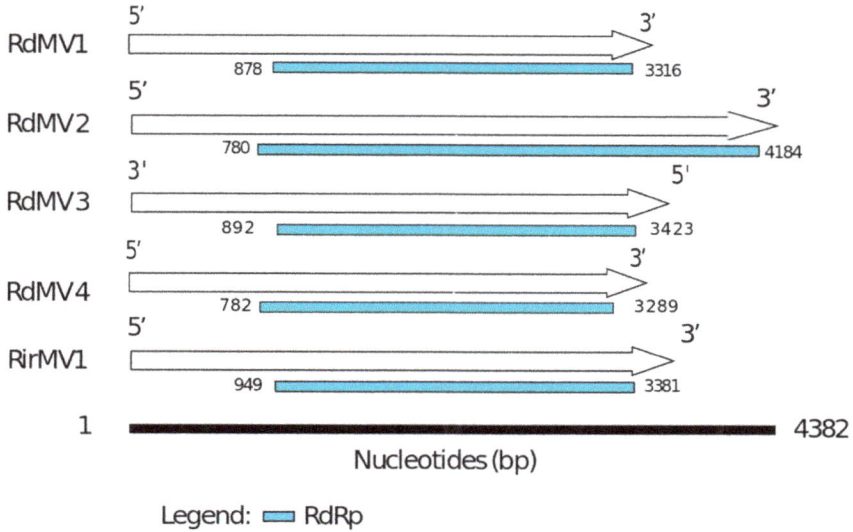

Figure 3. The genome organization of *Rhizophagus* spp. mitoviruses. The comparisons are of the organizations of RdMV1, RdMV2, RdMV3, RdMV4 and RirMV1. RdRp coding regions are labeled in blue (see also Table 1).

F

	310	320	330	340	350	360

```
RdMV1        ...TEWRLAPLANFLNNLSLGRLHVLPEPAGKMRVIAMGTWWVQCMLYPLHKLIYDRLGL
RirMV1       RSCTQWRMAPLSNFVTNLVLGRLHVIPEPAGKMRVVAMGTWWVQCMLYPLHRILYRKLGL
gi_372290782 .........AFHIGTSTNGIGKISVVEDPELKMRPIAMVDYYSQLVLKPIHDGILKKLRT
gi_615814998 .........AFHTGISTNGIGKISIVEDPELKMRPIAMVDYYSQLVLRPIHDGILKKLRN
gi_618765922 .........AFHTGCSTNGIGKISIVEDPELKMRPIAMVDYYSQLVLRPIHDGILQKLKN
gi_440808117 .........IINP.KWTNYTGKLSIVKDPELKRRVIAMVDYHSQFTLKPIHEMLLNKLST
RdMV3        P........SYDSLNDSMIHTGRLHHFEEWGGKVRVVAIVDYWTQILMSPLHNAIFHFLSK
RdMV4        P........GTDVQAGVGLHSGKIHTFEEWGGKIRNVAIVDYWTQILITPLHNTLFSFLKS
gi_568599575 .........VKPCGCKAAPEVRRLSIVKDPECKMVIGMFDWWSQVTLKPLSDSLFKALSK
gi_571798705 .........KFFNNKNCCRRLSCFPDKEGKMVVGVVDHYSQLALKPLHSWLARCLSK
RdMV2        GLRVEPPVDPRDPETPGPILGRLHPIPEAAGKVRVVAIVDYFTQIAMKPVHSHLLSILSK
```

A

	370	380	390	400	410	420

```
RdMV1        IPQDGTWDQEKPIKALACRILEVFKSTGVYPEAFSYDLSAATDRFPVWYQLEVLSFLTN.
RirMV1       IPNDGTWDQSKPLEGMAAKVKEILSS.GGIPQVFSYDLSAATDRFPVWYQVEVLAFLTN.
gi_372290782 LPCDRIFTQD.PFNNWGKTMGHKFWS.........LDLTSATDRFPISLQERVIAHLLGD
gi_615814998 LPQDRIFTQN.PFNNWGKTMGHKFWS.........LDLTSATDRFPISLQEKVIAGLLKD
gi_618765922 LPQDRIFTQD.PFNKWGKTMGHKFWS.........LDLTSATDRFPISLQEKVIHLLSD
gi_440808117 LKCDRIFTQD.PKHSWYQNH.HKFYS.........LDLSAATDRFPLQLQKKLLSYIYEN
RdMV3        IPADGIFDQDAAIARVAAFTRDPNAEVYS......YDLTAATDRLPISVQVEILSFLT.T
RdMV4        IPMDAIFNQDAAAEYIRSLSANENAELYS......VDLSAATDRLPASLQVRILKVLLGN
gi_568599575 IDSDRIYSQD..PAFHFEVDTNLLWS.........IDLTAATDRFPIELQKQILSILTN.
gi_571798705 IRQDCTLEQGKFKKLLFNGDVDIYYS.........VDLSAATDRFPIMLIKQLLKFQLP.
RdMV2        IKTDAIFDQSGRVKEYYESNHSRHWS.........YDLKAATDLIPRALYLEVLAPLLVA
```

B

	430	440	450	460	470

```
RdMV1        ........RRFAEAWRDLLVLPGYSPKPITIIPRGEPLKYGAGQPMGLYSSWAMFSLSHH
RirMV1       ........RRFAETWRDLLIMPRYYTGSITVIPRGDALVYGSGQPMGLYSSWAMFSLAHH
gi_372290782 ........ESKARAWRNILVDRDYKL.....PSGGWTRYSVGQPMGAYSSWTTFTLTHH
gi_615814998 ........PVKAKAWRDILVDRDYKL.....PTGGFTRYSVGQPMGAYSSWTTFTLTHH
gi_618765922 ........DDKARAWRDILIERDYKL.....PNGNLTRYSVGQPMGAYSSWTTFTLTHH
gi_440808117 ........KQFCDAWADLLTSRPYID.....SDGNQHFYNVGQPMGAYSSWAAFISHH
RdMV3        ........RQFANAWANLLVGRSYFHA....PSNKSYEYAVGQPMGSKSSWAMLALTHH
RdMV4        ........DTLAVAWAKIFTERSFKCS....NPPQMVKYAVGLPMGSKSNWAMLALTHH
gi_568599575 ........TAFTRDWADVMVGEEFSYE....GGSISYRVGQPMGAYSSWPMFTLSHH
gi_571798705 ........HEYVEAWGNVMVETPFEYR....GNLYRYAAGNPMGAYSFNSFALTHH
RdMV2        PSETGGEARKRAELWVDVMSDREFLSP.....SKDLWVKYGTGLPMGAYSHWASMALVHH
```

C   D

	480	490	500	510	520

```
RdMV1        MLVQQAASRIG..YKGWYPWYALLGDDLVILGR.DVAMAYKDLCDEIGVKIGLHKSLISS
RirMV1       LLVQQAASRVG..YKGWYPWYALLGDDIVILGE.DVAGAYKDLCDQLQVKIGLAKSLISS
gi_372290782 LVVHYAARLCG..IVDFD.RYILLGDDIVINHD.KVARRYISIMNKLGVDISVAKTHVSK
gi_615814998 MVVHYAAHLCG..IGDFN.KYILLGDDIVINHD.KVARAYIRVMTKLGVDISVAKTHVSK
gi_618765922 LVVHYAAHKCG..IEDFN.RYILLGDDIVINHD.KVAREYIRIMTRLGVDISKQKTHVSK
gi_440808117 LVVAWAAHLCG..EYNFS.QYIILGDDIVIKND.KVANKYITIMTLGVDISINKTHVSK
RdMV3        VIVRQAARSVS..ETPYE.TYALLGDDITLTGS.LIAQAYLKIMSLLDVTINLSKSIVHY
RdMV4        IIIKVAALRAK..LTNYN.TVRVCGDDSSFLGN.TLFVKYTEIMAWYGVPINRSKSIMAV
gi_568599575 ILVRYCGLLNG..IDNFN.RYIMLGDDIVINHN.KVAKTYIRLLHALGVETSQAKTHVSK
gi_571798705 YIIYFCCKELG..MRWKSLPYALLGDDIVIGNA.QVGELYMKVIKSLHVDYSLAKTHKSK
RdMV2        ALVQFAWFRVNGQRLSWYMLYLILGDDLDIAKYPRVAEEYLAICQALGIQIGLHKSLQSN
```

E

	530	540	550	560	570	580

```
RdMV1        NG...SFEFAKRYFVSGI.DCSPISIREYWVAINCLPAFAELIARVKRYIPSLRLADAVR
RirMV1       NG...SFEFAKRYYYKGN.DCSPVSIREYWVSLSSLPAFAEMVMRIKRSVPDIRLSDAVR
gi_372290782 N....TYEFAKRWVRKGI.EISGLPLRGIVSNINVLPVVVKQLVLYLYSNTTLWRGSTT.
gi_615814998 N....TYEFAKRWVRKGV.EISGLPLRGIVSNINSLPVVVKQLASLVLYSNTTRWRGSTT.
gi_618765922 N....TYEFAKRWVRKGI.EITGLPLRGIVSNTNNLPVVVKQLVNYLYSNTTLWRGSTT.
gi_440808117 D....TYEFAKRWIKGGI.EVSGIPLKGILNQWKNPGVVYTTLCSYFDKNP.IQPIPLI.
RdMV3        GDAKPAAEFCKRIFIEGH.ELTAIPVKLIAKTIMN.GRLLPQLQNELFRRGFDTTSTWLL
RdMV4        AGLKSAAEFCKRVFVNGV.ELTTIPVKLIVKTVFN.GRLLSQLQNELHRRRCAFISSQWY
gi_568599575 H....TYEFAKRWIDTRCGEITGLPIKGLLDNIENMSTCFQILFDWQLKNKFWTDRSLV
gi_571798705 D....FFEFAKRIYYKGD.EVTPFPISSLKEVRKSTSALTDLLLEQKSRGWECSSIASSA
RdMV2        LN...AFEFANRRFIPAG.DISPLSLKEELAATSWSQRVEYAKRILERMGTSLKNSALAQ
```

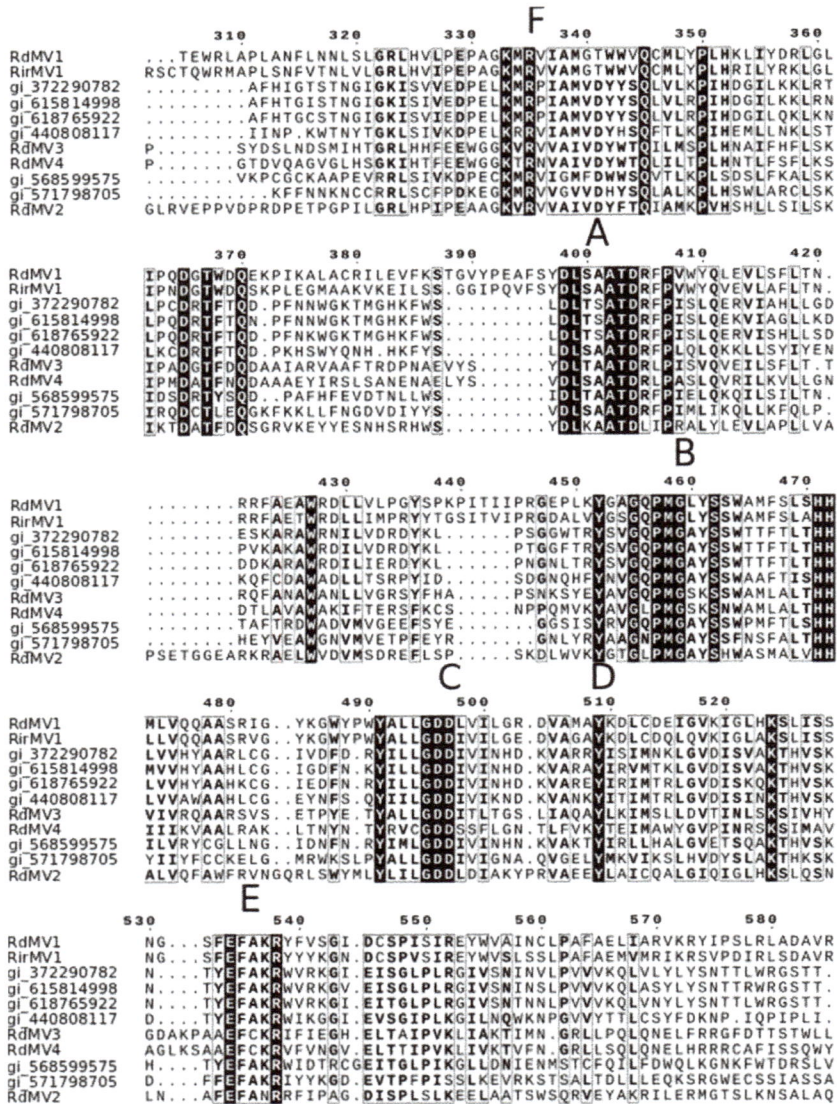

Figure 4. Conserved motifs identified in the RdRp domain of the genus *Mitovirus* based on the multiple sequence alignment of the amino acid sequences. Similar to the other mitoviruses, six conserved motifs were found. These conserved regions were labeled A-F as RdRp associated motifs described previously [36].

4. Discussion

Our own studies revealed a high intraspecfic diversity in the growth and nutrient uptake benefits after colonization with different AM isolates [37]. As we report the identification of mycoviruses in a lab culture and *in silico* from *Rhizophagus* spp., it would be interesting to determine whether mitoviruses play any role in the variability of these responses. In this study, we identified a novel mitovirus from the sequenced transcriptome of *R. irregularis* and confirmed the presence of RT-PCR amplicon of 3kb with gel electrophoresis (Figure 1A). Additionally, agarose gel electrophoresis of the RT-PCR product from the non-mycorrhizal *Medicago* root without *Rhizophagus* infection showed no amplification of viral contig, suggesting the newly identified viral contig (RirMV1) was not from the plant, but from *R. irregularis* (Figure 1B). We also identified four novel mitoviruses: RdMV1, RdMV2, RdMV3 and RdMV4 from *R. diaphanum* from the publicly available SRA database in NCBI (Table 1).

After the viral contigs were assembled, RT-PCR was used to verify the presence of any putative viral sequence. Also, it is necessary to rule out the possibility that the putative mycovirus genomes identified in this study could have been derived from mycoviruses integrated into the AMF genome. Oligonucleotide primers specific to the putative viral sequences need to be used to amplify the sequences using fungal genomic DNA as the template. The infection of these viruses could have resulted in beneficial/neutral effects on the host as they were selected to be sequenced without regards to apparent abnormal growth. These novel viral sequences may be used to establish Koch's postulates in future studies and to provide bases for mechanisms responsible in different nutrient uptake and plant biomass responses.

Using a mitochondrial translation table, the amino acid sequences of the three of five predicted mitovirus-like contigs clustered with mitoviruses of filamentous fungi, and constituted two distinct subclades along with mitoviruses infecting *G. margarita* (Figure 2). Our analysis showed that RdMV3 and RdMV4 are closely related to Rhizophagus clarus mitovirus 1-RF1 (RcMV1; AB558120) which is closely related to another mitovirus that was found in the ectomycorrhizal fungus *Tuber excavatum* (TeMV1; AEP83726.1). Their corresponding fungal hosts are all AM fungi and these mitoviruses all have distinctly longer RdRps than other mitoviruses that use mitochondrial translation code only. Similar to what was reported by Ikeda et al. [21], we found that the largest ORF of the two mitoviruses can also be predicted by applying the universal genetic code. Generally, functional translation of RdRp in mitoviruses involves activation of a mitochondrial genetic code [38], and as a result, tryptophan residues in mitoviruses (such as TeMV, CpMV, and HmMV1-18) are coded either by a UGA (which in universal genetic code means termination) or a UGG codon [7]. Recently identified mitoviruses from AMF, including RcMV1-RF1 [7] and the mitoviruses identified in *G. margarita* [10], all use the UGG codon for tryptophan which is compatible with both cytoplasmic and mitochondrial translation. All five novel mitoviruses identified from our study were found to use UGG for tryptophan. Interestingly, Nibert (2017) provided sequence-based explanation of this subgroup of mitoviruses using UGG instead of UGA for tryptophan in mitochondria, which is likely due to the mitochondrial codon of UGA for tryptophan in the respective fungal hosts that is correspondingly rare. Therefore, Nibert (2017) speculated that this unique group of mitoviruses do not actually replicate in cytosol [39].

Our results support the virus-host coevolution theory for the origin of these mitoviruses infecting nonpathogenic AMF fungi because the viruses they harbor do not cluster with mitoviruses from pathogenic fungi. First, amino acid sizes for the RdRps are very similar between the three viruses, RirMV1, RdMV1 and RdMV3, and the RdRp motifs are highly conserved between them (Figure 3). Besides the conserved motifs I to IV identified in Kitahara et al. [7] and Gorbalenya et al. [34], we also identified the motifs D and E (Figure 4) as shown in Bartholomaus et al. (2016) [36]. Out of these six motifs, A, B and C are among the most conserved structural motifs of the palm subdomain of RdRp with active catalytic sites [34]. Motif A (DX_5D) contains two aspartic acid residues separated by any five amino acids, while motif C contains two aspartic acid residues, consecutively. These residues are known to form divalent bonds between Mg^{2+} ions and/or Mn^{2+} ions for catalytic activation of the domain [34]. Similarly, motif B is known to form a long and conserved alpha-helix sequence with an asparagine residue which is indispensable for discriminating dNTPs and rNTPs that determine whether DNA or RNA is produced [35]. Although these motifs are known to be required for polymerase activity, the other three motifs (D, E and F) are not well studied in terms of their function.

Experiments to assess the impacts of viral infection on fungal colonization and sporulation and on the ability of virus-infected fungal isolates to affect nutrient cycling in host crops can be done by using virus-induced gene silencing (VIGS) approaches to knockdown the expression of viral transcripts. VIGS systems have been successfully applied in *R. clarus* to study fungal gene functions [40], which could be modified to instead silence RirMV1 using *Nicotiana benthamiana* as the plant host to deliver the silencing construct through the Cucumber mosaic virus Y strain based VIGS system [41]. The silencing effect will be effective if RirMV1 replicates in the cytoplasm as well, but will be ineffective if it replicates only in mitochondria since the RNA silencing machinery is not present in mitochondria and the double-layered membrane is a barrier. This may also resolve the long-standing question of whether mitoviruses in AMF replicate in the cytoplasm, which can shed light on the evolution of capsidless positive-strand RNA viruses.

Supplementary Materials: The following are available online at http://www.mdpi.com/1999-4915/10/12/707/s1, Figure S1: Stain of cross-section of the roots to confirm the AM fungal infection showing the (A) density of arbuscules as the small oval-shaped objects and (B) close-up view of hyphae and connected arbuscules.

Author Contributions: A.N. and S.-Y.L.M. conceived and designed the experiments; A.N., C.F., A.K. and J.F. performed the experiments; A.N. and S.-Y.L.M. analyzed the RNA-Seq data; H.B., A.N. and S.-Y.L.M. wrote the paper.

Funding: This study was supported in part by National Sclerotinia Initiative Grant SA1800330 (to Shin-Yi Lee Marzano) and SDSU startup from USDA Hatch fundSD00H606-16 project number with Accession Number 1009451 (to Shin-Yi Lee Marzano). The authors also acknowledge support from the South Dakota Research and Promotion Council (to H.B.) for these studies.

Conflicts of Interest: The authors declare no conflict of interest.

References

1. Smith, S.; Read, D. *Mycorrhizal Symbiosis*; Academic: San Diego, CA, USA, 1997.
2. Azcón-Aguilar, C.; Barea, J. Arbuscular mycorrhizas and biological control of soil-borne plant pathogens—An overview of the mechanisms involved. *Mycorrhiza* **1997**, *6*, 457–464. [CrossRef]
3. Augé, R.M. Water relations, drought and vesicular-arbuscular mycorrhizal symbiosis. *Mycorrhiza* **2001**, *11*, 3–42. [CrossRef]
4. Porcel, R.; Aroca, R.; Ruiz-Lozano, J.M. Salinity stress alleviation using arbuscular mycorrhizal fungi. A review. *Agron. Sustain. Dev.* **2012**, *32*, 181–200. [CrossRef]
5. SCHÜßLER, A.; Daniel, S.; Christopher, W. A new fungal phylum, the Glomeromycota: Phylogeny and evolution. *Mycol. Res.* **2001**, *105*, 1413–1421. [CrossRef]
6. Hildebrandt, U.; Kaldorf, M.; Bothe, H. The Zinc Violet and its Colonization by Arbuscular Mycorrhizal Fungi. *J. Plant Physiol.* **1999**, *154*, 709–717. [CrossRef]
7. Kitahara, R.; Ikeda, Y.; Shimura, H.; Masuta, C.; Ezawa, T. A unique mitovirus from Glomeromycota, the phylum of arbuscular mycorrhizal fungi. *Arch. Virol.* **2014**, *159*, 2157–2160. [CrossRef] [PubMed]
8. Ezawa, T.; Ikeda, Y.; Shimura, H.; Masuta, C. Detection and characterization of mycoviruses in arbuscular mycorrhizal fungi by deep-sequencing. In *Plant Virology Protocols*; Humana Press: New York, NY, USA, 2015; pp. 171–180.
9. Márquez, L.M.; Redman, R.S.; Rodriguez, R.J.; Roossinck, M.J. A virus in a fungus in a plant: Three-way symbiosis required for thermal tolerance. *Science* **2007**, *315*, 513–515. [CrossRef] [PubMed]
10. Turina, M.; Ghignone, S.; Astolfi, N.; Silvestri, A.; Bonfante, P.; Lanfranco, L. The virome of the arbuscular mycorrhizal fungus Gigaspora margarita reveals the first report of DNA fragments corresponding to replicating non-retroviral RNA viruses in fungi. *Environ. Microbiol.* **2018**. [CrossRef]
11. Pearson, M.N.; Beever, R.E.; Boine, B.; Arthur, K. Mycoviruses of filamentous fungi and their relevance to plant pathology. *Mol. Plant Pathol.* **2009**, *10*, 115–128. [CrossRef] [PubMed]
12. Roossinck, M.J. Evolutionary and ecological links between plant and fungal viruses. *New Phytol.* **2018**, *221*, 86–92. [CrossRef] [PubMed]
13. Hrabakova, L.; Koloniuk, I.; Petrzik, K. Phomopsis longicolla RNA virus 1—Novel virus at the edge of myco- and plant viruses. *Virology* **2017**, *506*, 14–18. [CrossRef] [PubMed]
14. Koonin, E.V.; Dolja, V.V.; Krupovic, M. Origins and evolution of viruses of eukaryotes: The ultimate modularity. *Virology* **2015**, *479–480*, 2–25. [CrossRef] [PubMed]

15. Rastgou, M.; Habibi, M.K.; Izadpanah, K.; Masenga, V.; Milne, R.G.; Wolf, Y.I.; Koonin, E.V.; Turina, M. Molecular characterization of the plant virus genus Ourmiavirus and evidence of inter-kingdom reassortment of viral genome segments as its possible route of origin. *J. Gen. Virol.* **2009**, *90*, 2525–2535. [CrossRef] [PubMed]

16. Nibert, M.L.; Vong, M.; Fugate, K.K.; Debat, H.J. Evidence for contemporary plant mitoviruses. *Virology* **2018**, *518*, 14–24. [CrossRef] [PubMed]

17. Bruenn, J.A.; Warner, B.E.; Yerramsetty, P. Widespread mitovirus sequences in plant genomes. *PeerJ* **2015**, *3*, e876. [CrossRef] [PubMed]

18. Lakshman, D.K.; Jian, J.H.; Tavantzis, S.M. A double-stranded RNA element from a hypovirulent strain of Rhizoctonia solani occurs in DNA form and is genetically related to the pentafunctional AROM protein of the shikimate pathway. *Proc. Natl. Acad. Sci. USA* **1998**, *95*, 6425–6429. [CrossRef] [PubMed]

19. Marzano, S.-Y.L.; Domier, L.L. Reprint of "Novel mycoviruses discovered from metatranscriptomics survey of soybean phyllosphere phytobiomes". *Virus Res.* **2016**, *219*, 11–21. [CrossRef] [PubMed]

20. Mensah, J.A.; Koch, A.M.; Antunes, P.M.; Kiers, E.T.; Hart, M.; Bucking, H. High functional diversity within species of arbuscular mycorrhizal fungi is associated with differences in phosphate and nitrogen uptake and fungal phosphate metabolism. *Mycorrhiza* **2015**, *25*, 533–546. [CrossRef] [PubMed]

21. Ikeda, Y.; Shimura, H.; Kitahara, R.; Masuta, C.; Ezawa, T. A novel virus-like double-stranded RNA in an obligate biotroph arbuscular mycorrhizal fungus: A hidden player in mycorrhizal symbiosis. *Mol. Plant Microbe Interact.* **2012**, *25*, 1005–1012. [CrossRef] [PubMed]

22. St-Arnaud, M.; Hamel, C.; Vimard, B.; Caron, M.; Fortin, J. Enhanced hyphal growth and spore production of the arbuscular mycorrhizal fungus Glomus intraradices in an in vitro system in the absence of host roots. *Mycol. Res.* **1996**, *100*, 328–332. [CrossRef]

23. Vierheilig, H.; Coughlan, A.P.; Wyss, U.; Piché, Y. Ink and vinegar, a simple staining technique for arbuscular-mycorrhizal fungi. *Appl. Environ. Microbiol.* **1998**, *64*, 5004–5007. [PubMed]

24. McGonigle, T.P.; Miller, M.H.; Evans, D.G.; Fairchild, G.L.; Swan, J.A. A New Method which Gives an Objective Measure of Colonization of Roots by Vesicular-Arbuscular Mycorrhizal Fungi. *New Phytol.* **1990**, *115*, 495–501. [CrossRef]

25. Tisserant, E.; Malbreil, M.; Kuo, A.; Kohler, A.; Symeonidi, A.; Balestrini, R.; Charron, P.; Duensing, N.; Frey, N.F.D.; Gianinazzi-Pearson, V.; et al. Genome of an arbuscular mycorrhizal fungus provides insight into the oldest plant symbiosis. *Proc. Natl. Acad. Sci. USA* **2013**, *110*, 20117–20122. [CrossRef] [PubMed]

26. Schüßler, A.; Walker, C. *The Glomeromycota: A Species List with New Families and New Genera*; Libraries at The Royal Botanic Garden Edinburgh, The Royal Botanic Garden Kew, Botanische Staatssammlung Munich, and Oregon State University: Kew, UK, 2010; pp. 1–56.

27. Grabherr, M.G.; Haas, B.J.; Yassour, M.; Levin, J.Z.; Thompson, D.A.; Amit, I.; Adiconis, X.; Fan, L.; Raychowdhury, R.; Zeng, Q.; et al. Full-length transcriptome assembly from RNA-Seq data without a reference genome. *Nat. Biotechnol.* **2011**, *29*, 644–652. [CrossRef] [PubMed]

28. Edgar, R.C. Search and clustering orders of magnitude faster than BLAST. *Bioinformatics* **2010**, *26*, 2460–2461. [CrossRef]

29. Altschul, S.F.; Gish, W.; Miller, W.; Myers, E.W.; Lipman, D.J. Basic local alignment search tool. *J. Mol. Biol.* **1990**, *215*, 403–410. [CrossRef]

30. Langmead, B.; Trapnell, C.; Pop, M.; Salzberg, S.L. Ultrafast and memory-efficient alignment of short DNA sequences to the human genome. *Genome Biol.* **2009**, *10*, R25. [CrossRef]

31. Larkin, M.A.; Blackshields, G.; Brown, N.P.; Chenna, R.; McGettigan, P.A.; McWilliam, H.; Valentin, F.; Wallace, I.M.; Wilm, A.; Lopez, R.; et al. Clustal W and clustal X version 2.0. *Bioinformatics* **2007**, *23*, 2947–2948. [CrossRef]

32. Kumar, S.; Stecher, G.; Tamura, K. MEGA7: Molecular Evolutionary Genetics Analysis Version 7.0 for Bigger Datasets. *Mol. Biol. Evol.* **2016**, *33*, 1870–1874. [CrossRef]

33. Bailey, T.L.; Boden, M.; Buske, F.A.; Frith, M.; Grant, C.E.; Clementi, L.; Ren, J.; Li, W.W.; Noble, W.S. MEME SUITE: Tools for motif discovery and searching. *Nucleic Acids Res.* **2009**, *37*, W202–W208. [CrossRef]

34. Gorbalenya, A.E.; Pringle, F.M.; Zeddam, J.L.; Luke, B.T.; Cameron, C.E.; Kalmakoff, J.; Hanzlik, T.N.; Gordon, K.H.J.; Ward, V.K. The palm subdomain-based active site is internally permuted in viral RNA-dependent RNA polymerases of an ancient lineage. *J. Mol. Biol.* **2002**, *324*, 47–62. [CrossRef]

35. Gohara, D.W.; Crotty, S.; Arnold, J.J.; Yoder, J.D.; Andino, R.; Cameron, C.E. Poliovirus RNA-dependent RNA polymerase (3Dpol): Structural, biochemical, and biological analysis of conserved structural motifs A and B. *J. Biol. Chem.* **2000**, *275*, 25523–25532. [CrossRef] [PubMed]

36. Bartholomaus, A.; Wibberg, D.; Winkler, A.; Puhler, A.; Schluter, A.; Varrelmann, M. Deep Sequencing Analysis Reveals the Mycoviral Diversity of the Virome of an Avirulent Isolate of Rhizoctonia solani AG-2-2 IV. *PLoS ONE* **2016**, *11*, e0165965. [CrossRef]

37. Fellbaum, C.R.; Mensah, J.A.; Cloos, A.J.; Strahan, G.E.; Pfeffer, P.E.; Kiers, E.T.; Bücking, H. Fungal nutrient allocation in common mycorrhizal networks is regulated by the carbon source strength of individual host plants. *New Phytol.* **2014**, *203*, 646–656. [CrossRef]

38. Shackelton, L.A.; Holmes, E.C. The role of alternative genetic codes in viral evolution and emergence. *J. Theor. Biol.* **2008**, *254*, 128–134. [CrossRef] [PubMed]

39. Nibert, M.L. Mitovirus UGA (Trp) codon usage parallels that of host mitochondria. *Virology* **2017**, *507*, 96–100. [CrossRef]

40. Kikuchi, Y.; Hijikata, N.; Ohtomo, R.; Handa, Y.; Kawaguchi, M.; Saito, K.; Masuta, C.; Ezawa, T. Aquaporin-mediated long-distance polyphosphate translocation directed towards the host in arbuscular mycorrhizal symbiosis: Application of virus-induced gene silencing. *New Phytol.* **2016**, *211*, 1202–1208. [CrossRef]

41. Otagaki, S.; Arai, M.; Takahashi, A.; Goto, K.; Hong, J.-S.; Masuta, C.; Kanazawa, A. Rapid induction of transcriptional and post-transcriptional gene silencing using a novel Cucumber mosaic virus vector. *Plant Biotechnol.* **2006**, *23*, 259–265. [CrossRef]

viruses

MDPI

Review

Chrysoviruses in *Magnaporthe oryzae*

Hiromitsu Moriyama [1,*], Syun-ichi Urayama [1], Tomoya Higashiura [1], Tuong Minh Le [2,3] and Ken Komatsu [2]

[1] Laboratory of Molecular and Cellular Biology, Department of Applied Biological Sciences, Tokyo University of Agriculture and Technology, 3-5-8, Saiwaicho, Fuchu, Tokyo 184-8509, Japan; urayama.shunichi.gn@u.tsukuba.ac.jp (S.-I.U.); tomoya.higashiura@gmail.com (T.H.)

[2] Laboratory of Plant Pathology, Department of Applied Biological Sciences, Tokyo University of Agriculture and Technology, 3-5-8, Saiwaicho, Fuchu, Tokyo 184-8509, Japan; lmtuong@ctu.edu.vn (T.M.L.); akomatsu@cc.tuat.ac.jp (K.K.)

[3] Department of Plant Protection, College of Agriculture & Applied Biology, Can tho University, Can tho city 900000, Vietnam

* Correspondence: hmori714@cc.tuat.ac.jp; Tel.: +81-42-367-5622; Fax: +81-42-360-8830

Received: 1 October 2018; Accepted: 6 December 2018; Published: 8 December 2018

Abstract: *Magnaporthe oryzae*, the fungus that causes rice blast, is the most destructive pathogen of rice worldwide. A number of *M. oryzae* mycoviruses have been identified. These include *Magnaporthe oryzae* viruses 1, 2, and 3 (MoV1, MoV2, and MoV3) belonging to the genus, *Victorivirus*, in the family, *Totiviridae*; *Magnaporthe oryzae* partitivirus 1 (MoPV1) in the family, *Partitiviridae*; *Magnaporthe oryzae* chrysovirus 1 strains A and B (MoCV1-A and MoCV1-B) belonging to cluster II of the family, *Chrysoviridae*; a mycovirus related to plant viruses of the family, *Tombusviridae* (*Magnaporthe oryzae* virus A); and a (+)ssRNA mycovirus closely related to the ourmia-like viruses (*Magnaporthe oryzae* ourmia-like virus 1). Among these, MoCV1-A and MoCV1-B were the first reported mycoviruses that cause hypovirulence traits in their host fungus, such as impaired growth, altered colony morphology, and reduced pigmentation. Recently we reported that, although MoCV1-A infection generally confers hypovirulence to fungi, it is also a driving force behind the development of physiological diversity, including pathogenic races. Another example of modulated pathogenicity caused by mycovirus infection is that of Alternaria alternata chrysovirus 1 (AaCV1), which is closely related to MoCV1-A. AaCV1 exhibits two contrasting effects: Impaired growth of the host fungus while rendering the host hypervirulent to the plant, through increased production of the host-specific AK-toxin. It is inferred that these mycoviruses might be epigenetic factors that cause changes in the pathogenicity of phytopathogenic fungi.

Keywords: Mycovirus; rice blast fungus; *Magnaporthe oryzae* chrysovirus 1; double-stranded RNA virus; hypovirulence

1. Introduction

Similar to animals and plants, fungi are often infected by viruses. In general, viral infections in higher eukaryotes result in easily detectable alterations, such as disease, whereas when a simpler eukaryote, such as a fungus acts as a host, it is often hard to recognize the effects of infection. However, numerous viruses of yeasts and filamentous fungi have been reported to cause epigenetic phenomena [1]. Prions, or "infectious proteins", are also known to affect fungi [2]. Mycoviruses (viruses that infect fungi) were originally described in diseased *Agaricus bisporus* mushrooms [3], in *Penicillium chrysogenum* [4], and in the brewing yeast, *Saccharomyces cerevisiae* [5]. In fact, yeast killer viruses have been used in fermentation production as bio-controllers. Mycoviruses that infect plant pathogenic fungi were initially discovered in the rice blast fungus, *Magnaporthe oryzae* [6], and have since been found in many other plant pathogenic fungi. The hypovirus CHV1-EP713 that infects the chestnut blight fungus, *Cryphonectria parasitica*, is a famous example of a biological control agent used to attenuate host fungal infection [7]. Mycoviruses have also been found in the human pathogenic fungus, *Aspergillus fumigatus*, since 1990 [8].

In many cases, double-stranded RNA (dsRNA) molecules are either the actual genome of the mycovirus or its replication intermediate. Similar to animals and plants, many fungi have antiviral capabilities, which act through RNA interference mechanisms targeting dsRNA degradation [7]. Therefore, it is of interest to investigate the relationship between mycovirus propagation and host defence mechanisms. In recent years, there have been many reports on the interactions between fungal viruses and RNA interference [9–11].

It is plausible that there are more than 1.5 million species of fungi and oomycetes on Earth. Many of these organisms, including phytopathogenic fungi, cause enormous damage to agricultural crops [12]. To date, about 8000 phytopathogenic fungi have been reported in Japan alone and they are the most common type of plant pathogen, over 10 times more frequent than viruses and bacteria [13]. In contrast, many more viruses and bacteria infect humans than fungi. Considering that each phytopathogenic fungus is infected with multiple mycoviruses, the number of newly identified mycoviruses will certainly increase in the future.

To date, three distinct dsRNA viruses have been reported to infect *M. oryzae*, including from the genus, *Victorivirus*, *Magnaporthe oryzae* virus 1 [14] and *Magnaporthe oryzae* virus 2 [15], *Magnaporthe oryzae* virus 3, a mycovirus that belongs to the family, *Partitiviridae* (*Magnaporthe oryzae* partitivirus 1) [16], and a mycovirus that belongs to cluster II of the family, *Chrysoviridae* *Magnaporthe oryzae* chrysovirus 1 (MoCV1) [17,18]. Recently, a mycovirus related to plant viruses of the family, *Tombusviridae* (*Magnaporthe oryzae* virus A) [19], and a mycovirus closely related to ourmia-like viruses (*Magnaporthe oryzae* ourmia-like virus 1) [20] has also been reported.

In this review, we discuss MoCV1, a mycovirus that causes growth inhibition in the rice blast fungus, *M. oryzae*. We focus on its molecular genetic characteristics, the influence of viral proteins on host cells, and our methodology of investigating physiological activity using a yeast heterologous expression system.

2. Effects of Magnaporthe Chrysovirus on the Rice Blast Fungus

The rice blast fungus, *M. oryzae*, is a plant pathogen that causes significant damage to rice production annually worldwide. In Japan, the annual rice yield is almost 9 million tons, and almost 1% is typically lost to infection with rice blast [21]. We first investigated the prevalence of mycoviruses in 58 isolates of *M. oryzae* collected in the Mekong Delta area of Vietnam. We screened for dsRNA in cell extracts using a simple purification method [22] and found dsRNA elements in 11 isolates (Figure 1) [17]. Among these, a virus with a dsRNA genome of 2.6 kbp to 3.6 kbp was nominated as *Magnaporthe oryzae* chrysovirus 1 (MoCV1). The *M. oryzae* isolates were also infected with a partitivirus whose three dsRNA genomic segments ranged in size from 1.8 kbp to 2.4 kbp. Some isolates were infected with only one of these two mycoviruses while others were infected with both (Figure 2, top left). Among the MoCV1 strains detected in Vietnam, the one that was most stably maintained in liquid medium or on agar plates was nominated MoCV1-A. In another strain nominated MoCV1-B, the content of dsRNA obtained from cells in liquid culture was sometimes lower than that obtained with MoCV1-A (Figure 2) [18]. Since the hyphal morphology of rice blast fungus infected with MoCV1-B showed remarkable white pigmentation on agar medium (Figure 2, right), melanin biosynthesis appears to be suppressed by MoCV1-B infection. Melanin biosynthesis mutants of *M. oryzae* are unable to invade rice leaves due to deficiencies in appressorium formation, and thus have greatly reduced infectivity. Since the fungal melanin biosynthesis pathway is different from the human one, specific compounds, such as sytalone or 1,3,6,8-tetrahydroxynaphthalene, can be used as pesticides for controlling the rice blast fungus [23,24]. Also, no conidial formation was observed in strains infected with MoCV1-B on PDA (Potato dextrose agar) medium [18]. Since *M. oryzae* propagates by spreading conidia in the air, we would expect that the suppression of conidial formation would limit the spread of the rice blast fungus. Furthermore, the cell wall of a MoCV1-B infected strain, as observed under the microscope, was loose and enlarged [18], and staining with calcofluor-white showed that the cell wall was damaged (Figure 3).

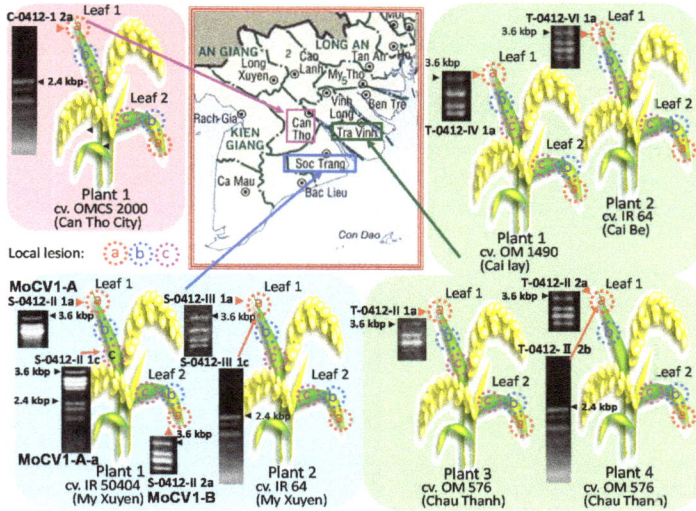

Figure 1. Distribution map of rice blast fungus infected with mycoviruses. Mycoviruses with dsRNA genomes were found in 11 *M. oryzae* isolates from three provinces in the Mekong Delta region of Vietnam. The plant diagrams show the sites where disease lesions were sampled. The agarose gel shows viral dsRNA segments present in the infected *M. oryzae* isolates. Two mycoviruses were identified: MoCV1 had five dsRNA genomic segments of 2.6 kbp to 3.6 kbp, and a partitivirus had three segments of 1.8 kbp to 2.4 kbp. The *M. oryzae* isolates, S-0412-II 1a, S-0412-II 1c, and S-0412-II 2a, were infected with the MoCV1 strains, MOCV1-A, MOCV1-A-a, and MOCV1-B, respectively. Isolate S-0412-II 1c was also infected with the partitivirus.

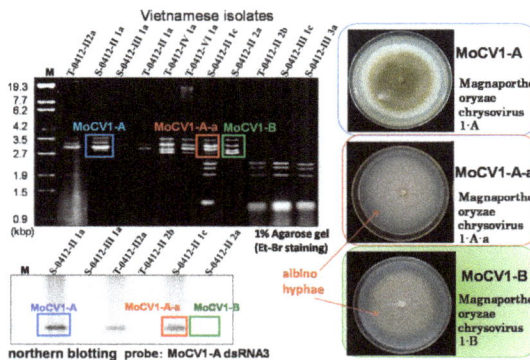

Figure 2. Analysis of mycoviruses isolated from *M. oryzae* in Vietnam. Top left, agarose gel electrophoresis of dsRNAs derived from mycoviruses purified from rice blast isolates. Lower left, northern hybridisation with a cDNA probe derived from MoCV1-A dsRNA3, showing a weak signal for MoCV1-B dsRNA3. Right, fungal flora of the *M. oryzae* isolates infected with MoCV1-A (top), MoCV1-A-a (middle), and MoCV1-B (bottom). These isolates were cultured on PDA medium at 26 °C for 10 days.

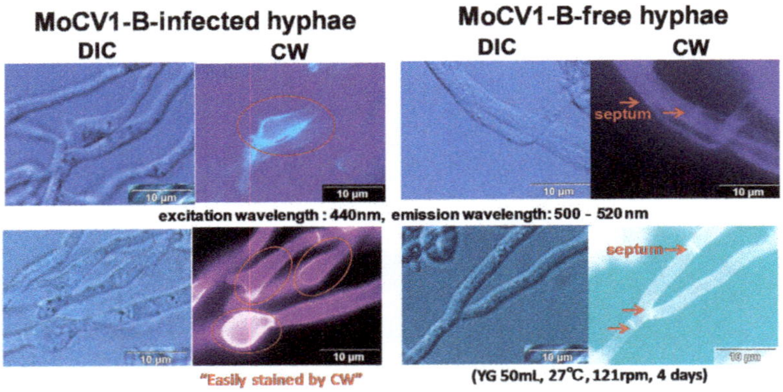

Figure 3. Influence of MoCV1-B on cell wall formation in *M. oryzae* hyphae. Infected and non-infected hyphae were stained with calcofluor-white (Sigma Chemical, St. Louis, MO, USA) and examined at 1000× magnification under a light microscope (Olympus IX71, Tokyo, Japan) with differential interference contrast (DIC) optics. Calcofluor-white (CW) binds strongly to structures containing cellulose and chitin.

3. Molecular Properties of MoCV1-A and MoCV1-B

The genome of MoCV1-A has five dsRNA segments (dsRNAs 1–5), and dsRNA1 encodes an RNA-dependent RNA polymerase (RdRp) and dsRNA3 and dsRNA4 each encode a separate coat protein (CP1 and CP2) (Figure 4) [25]. This RdRp has ca. 30% similarity to the RdRps of *Helminthosporium victoriae* virus 145S (HvV145S) and *Penicillium chrysogenum* virus (PcV), which belong to the family, *Chrysoviridae*. Therefore, we classified MoCV1-A in the same family. MoCV1-B dsRNAs 1–4 showed ca. 75% identity with their MoCV1-A counterparts, whereas dsRNA5 had 96% identity between MoCV1-A and MoCV1-B. This suggested that dsRNA5 might have migrated as a satellite RNA between strains of MoCV1 located in the Soc Trang province of Vietnam. Whilst subculturing *M. oryzae* isolate, S-0412-II 2a, which was infected with MoCV1-B, we discovered a MoCV1-B derivative strain lacking a dsRNA5 segment. This derivative strain also formed viral particles that could be released from the fungal cells, suggesting that dsRNA5 is not essential for viral maintenance [18]. The *M. oryzae* isolate, S-0412-II 1c, was co-infected with both MoCV1 and the partitivirus whose genomic segments were 2.4, 2.2, and 1.8 kbp in size. This MoCV1 strain was named MoCV1-A-a because its five dsRNA genomic segments showed more than 99% identity with the counterpart dsRNAs of MoCV1-A.

Initially, the MoCV1-related viruses, MoCV1-A, MoCV1-A-a, and MoCV1-B, were discovered in rice blast isolates from Vietnam. We then used reverse transcription-PCR and a loop-mediated isothermal amplification method that was developed for the specific detection of these mycoviruses, and discovered a MoCV1-related virus (MoCV1-D) in rice blast isolates from Japan [26]. MoCV1-D has five dsRNAs as its genome and dsRNAs 1–4 showed 75%–81% identity with the corresponding dsRNAs from MoCV1-A and MoCV1-B. Conversely MoCV1-D dsRNA5 possessed relatively low identity (63%) and was dispensable for virus propagation, as was the case with MoCV1-B dsRNA5. Whereas the Vietnamese MoCV1 strains were sometimes found together with a partitivirus in the same *M. oryzae* isolates, MoCV1-D was found in combination with a victorivirus, MoV2, and some different partitiviruses. Recently, a rice blast fungus isolated in Hubei, China was found to be co-infected with a victorivirus (MoV3) and a chrysovirus that was designated MoCV1-C; however, MoCV1-C has not yet been sequenced [27].

The family, *Chrysoviridae*, includes a single genus with two large distinct clusters. Cluster I contains members of the genus, *Chrysovirus*, and related unclassified viruses with three genomic segments, while cluster II comprises related, unclassified viruses with four or five genomic segments [28].

Phylogenetic analysis revealed that MoCV1-A and MoCV1-B are members of cluster II, together with the following viruses: Botryosphaeria dothidea chrysovirus 1 (BdCV1) [29], Penicillium janczewski chrysoviruses 1 and 2 (PjCV1, PjCV2) [30], Aspergillus mycovirus 1816 (AmV1816) [31], Agaricus bisporus virus 1 (AbV1) [32], Fusarium oxysporum f. sp. mycovirus 1 (FodV1) [33–35], Fusarium graminearum mycovirus-China 9 (FgV-Ch-9) [36], Tolypocladium cylindrosporum viruses 1 and 2 (TcV1, TcV2) [37], and Alternaria alternata chrysovirus 1 (AaCV1) [38]. MoCV1-A and AaCV1 are reported to affect fungal pathogenicity [37,39]. FgV-ch9 and Aspergillus thermomutatus chrysovirus 1 (AthCV1) are also reported to attenuate the growth of their host fungi [40,41].

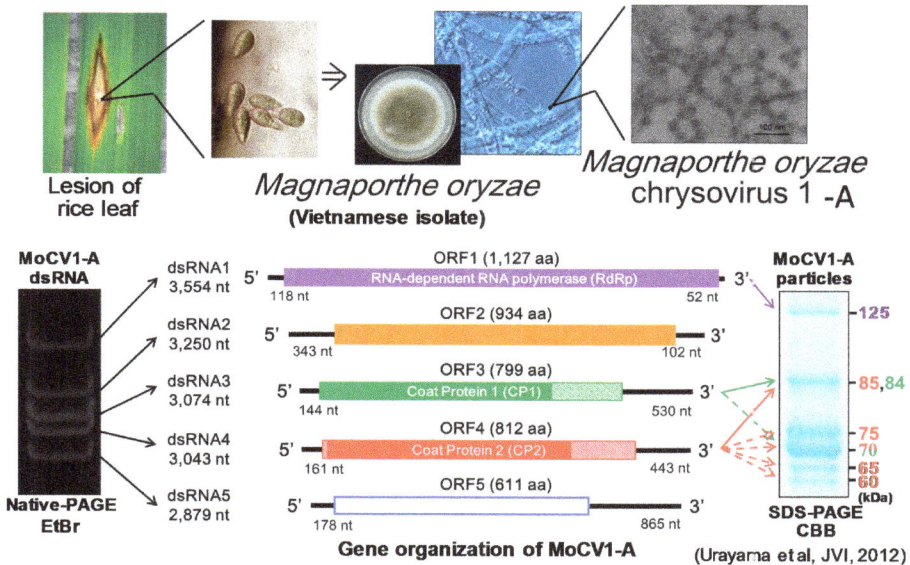

Figure 4. Isolation and analysis of mycovirus MoCV1-A from the rice blast fungus, *M. oryzae*. Upper panels, process of isolation and purification of the mycovirus. Lower left, dsRNA genomic segments extracted from purified MoCV1-A virus particles were subjected to 5% (*w*/*v*) native PAGE. Lower middle, open reading frames (ORFs) within each of the five genomic segments. Lower right, MoCV1-A viral proteins were separated by denaturing PAGE.

4. Virus Particles Containing dsRNAs and Multiform Structural Proteins

Purification of the MoCV1-A and MoCV1-B virus particles was performed using standard methods without solvents to avoid denaturation of viral structural proteins. We found that the buoyant densities of the particles depended on the sizes of the packaged dsRNA segments within them, suggesting that the five dsRNA segments were packaged separately in individual virus particles [17]. Conserved sequences present in the 5'-(GCAAAAAAGAGAAUAAAGC–UUC UCCUUUUUGCA) and 3'-(AAGUACC) terminal regions of each dsRNA may include common packaging signals, replication sites, or ribosomal entry sites.

We extracted and purified MoCV1-A particles from a 14-day-old liquid culture of mycelia. Coomassie Blue staining of SDS-PAGE gels revealed that the particles contained four major proteins 125 kDa, 70 kDa, 65 kDa, and 60 kDa in size [25]. Edman degradation analysis showed that the N-terminus of the 125 kDa protein was blocked, since no phenylthiohydantoin (PTH)-amino acid derivatives were observed. The protein was purified, subjected to trypsin digestion, and analyzed by HPLC. A tryptic peptide in the chromatographic peaks was subjected to Edman degradation, and its sequence matched the amino acid sequence of the protein encoded by ORF1 (the open reading frame of dsRNA1). The 70 kDa protein was also subjected to Edman degradation and was proven to be encoded by ORF4, although it lacked 14 amino acids in the N-terminal region. This suggests that these amino acids are cleaved by post translational modification. The 65 kDa protein was confirmed to be a derivative of the 70 kDa protein following immunoblotting using an antiserum to the ORF4 protein [25].

The 60 kDa protein was also subjected to Edman degradation and was identified as a partially degraded form of the ORF3 protein. Its N-terminal sequence was consistent with the N-terminus encoded by dsRNA3, however its apparent molecular mass (*Mr*) was smaller than the deduced *Mr* of the ORF3 protein (799 aa, 84 kDa). Mass spectrometry of tryptic peptides revealed that the C-terminus of the deduced ORF3 protein (M565 to L799) was absent from the 60 kDa protein (Figure 5). Then, MALDI-TOF MS (Matrix Assisted Laser Desorption/Ionization-Time of Flight Mass Spectrometry) was performed to determine the exact *Mr* of the 60 kDa protein, and a resultant ion signal was observed at *m/z* 62,559. The *Mr* of the deduced amino acid sequence is 62,530 for residues G1 to V590, and 62,658 for residues G1 to Q591. Together, the results suggest that the C-terminal end of the 60 kDa protein might be located around residues M565 to Q591 of the ORF3 protein [25].

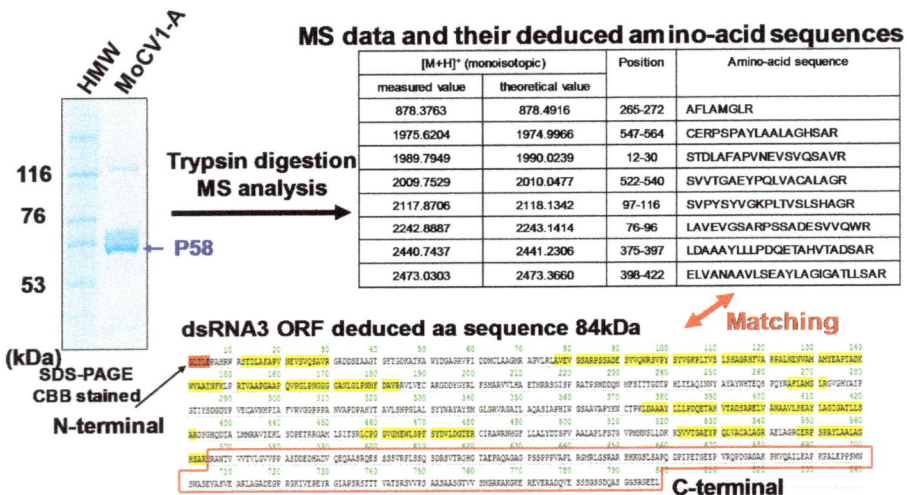

Figure 5. Determination of the C-terminal residue of the MoCV1-A 60 kDa protein (labeled P58 in the figure). The C-terminus of P58 is a truncated version of the ORF3 protein missing 200 amino acids. Mass spectrometry of tryptic peptides derived from P58 identified peptides specific to the deduced amino acid sequence of ORF3, but no peptide sequence mapped to the C terminus of the deduced amino acid sequence of ORF3 (M565 to L799).

To investigate whether the full-size proteins are components of MoCV1-A particles, we purified virus particles from fresh mycelia that had been grown in a fermenter for two days. The purified isometric virus particles had buoyant densities of 1.38 g cm^{-3} to 1.40 g cm^{-3} in CsCl and diameters of about 35 nm. Anti-ORF3 antiserum detected an 84 kDa protein corresponding to the full-size protein encoded by ORF3, in addition to 75, 70, 66, and 60 kDa proteins. Anti-ORF4 antiserum detected an 85 kDa protein corresponding to the full-size ORF4 protein as well as a 70 kDa protein. These results indicated that full-size ORF3 and ORF4 proteins might be components of MoCV1 particles as coat protein 1 and coat protein 2, respectively. The smaller protein bands that were detected by the antisera represent degraded forms of the full-size proteins (Figure 6) [25].

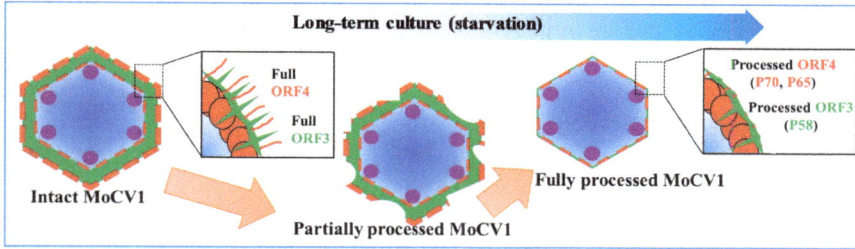

Figure 6. Model for partial degradation of the ORF3 and ORF4 proteins. In the early stage of culture, the MoCV1-A viral particles contained full-length ORF3 and ORF4 proteins. After long-term culture and nutrient depletion, the viral particles were composed of partially degraded ORF3 and ORF4 proteins.

We did not observe any degradation products of the ORF1 (125 kDa) protein. We performed RdRp (RNA-dependent RNA polymerase) assays using [α-^{32}P] UTP on MoCV1-A particles purified from 2-day-old and 14-day-old mycelia. Autoradiography of the native PAGE gel revealed radioactive signals from both types of MoCV1-A particles.

5. Release of Mycoviral dsRNA Genomes from the Mycelium of Mycovirus-Infected *M. oryzae* into the Culture Supernatant

In general, fungal viruses have no extracellular phase, and indeed, we have never detected virions other than the MoCV1 strains. It is believed that the extracellular phase is not required for the spread of mycovirus infection because the virus is propagated via hyphal fusion (anastomosis) between compatible individuals [7]. Hyphal fusion is limited to cases where the hyphae are compatible. In the filamentous fungus, *Podospora anserine*, the [Het-s*] phonotype, which determines compatibility for hyphal fusion, can spontaneously transition to the [Het-s] phenotype due to the prion-like action of the *het-s* gene product. As a result, mutually compatible strains suddenly become incompatible, with the development of a hyphal boundary line and programmed cell death [42]. Hyphal anastomosis seems to have the disadvantage that viruses present in one colony are readily transmitted throughout the other. To limit this problem, the process is completed only by closely related colonies that are likely to already carry the same viruses [43,44]. This can be thought of as a clever self-defence mechanism that prevents the spread of virus infection caused by hyphal fusion.

We discovered that MoCV1-A is an exceptional case because the dsRNAs can exist and survive in the cell-free fraction of the liquid culture medium (Figure 7) [17]. The release of mycoviral dsRNA into the culture supernatant was also found for MoCV1-B [45]. The amount of virus-derived dsRNA detected in the culture supernatant peaked at 1 μg/mL four to five weeks after inoculation of the liquid medium. MoCV1-A and MoCV1-B viral proteins were also detected in the cell-free culture supernatant [17,18]. We investigated this phenomenon in the following additional mycoviruses: *Magnaporthe oryzae* virus 2 (MoV2) [46], Alternaria alternata victorivirus 1 (AaVV1) [47], Alternaria alternata virus 1 (AaV1) [48], AaCV1, and L-A virus of *Saccharomyces cerevisiae* [49]. No virus-derived dsRNA was detected in the culture supernatant with any of these viruses.

The release of viruses into the liquid culture medium used for their propagation is a well-known phenomenon for animal viruses. Generally, viruses that infect animal cells reproduce rapidly and then lyse the cells so that the virions are released rapidly over a few days into the extracellular environment [50]. Conversely, MoCV1 virions are released gradually into liquid cultures in a process caused by nitrogen and carbon starvation, which takes ca. 14 days for mini-jar cultures and four to eight weeks for larger flask cultures. Examination of the protein composition of extracellular MoCV1-A virions revealed that ca. 200 amino acids at the C terminus of the ORF 3 protein had been degraded.

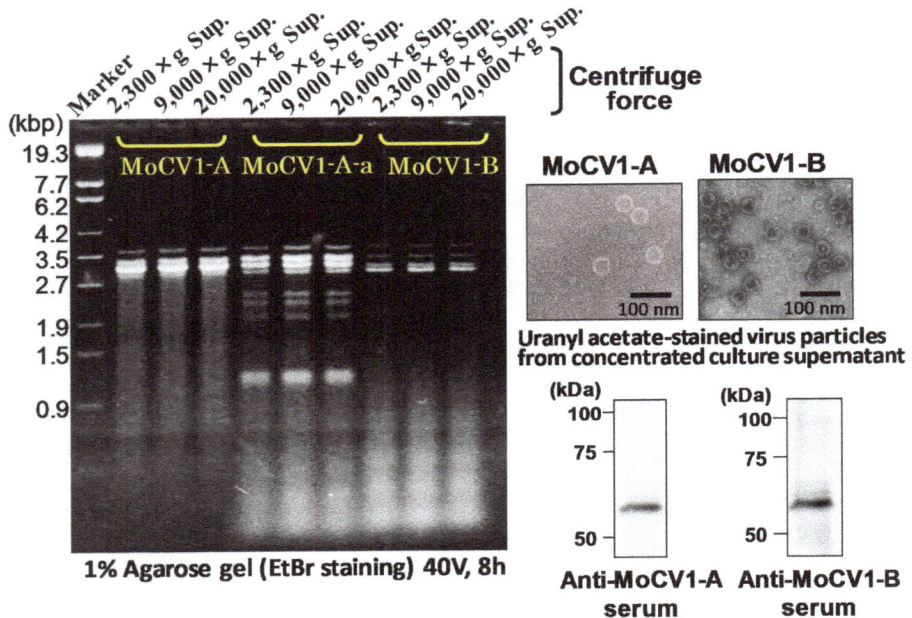

Figure 7. Release of mycoviruses from the mycelium of mycovirus-infected *M. oryzae* into the culture supernatant. The mycoviruses gradually appeared and accumulated in the liquid medium during the long-term suspension culture. After five weeks of liquid culture, 250 µL samples of the culture supernatant were subjected to stepwise centrifugation treatments. Total nucleic acids were extracted from each supernatant and then subjected to agarose gel electrophoresis.

6. Heterologous Expression of Mycoviral Proteins Induced Cytological Damage in the Yeast, *Saccharomyces cerevisiae*

We attempted to use yeast (*S. cerevisiae*) to investigate the functions of the MoCV1-A encoded proteins. Of the five putative MoCV1-A proteins, ORF1 was found to have eight conserved motifs characteristic of RdRps and is assumed to function in this role. However, based on BLAST searches, we were unable to predict the functions of the other four putative proteins. Therefore, shuttle vectors were constructed by ligating each of the four MoCV1-A ORF sequences of unknown function downstream of a low expression promoter (*ADH1*) or a high expression promoter (*TDH3*) for expression in yeast cells [25]. The influence of each protein on yeast cell growth was investigated following mini jar culture in 1 L semi-synthetic media. Sampling was carried out every 2 h from 8–46 h post inoculation. Optical density, viable cell number, glucose concentration, and pH were measured together with observations of cell morphology and immunological analyses. Abnormalities in cell morphology, such as the appearance of enlarged vacuoles and vesicles, were observed when the ORF4 protein was over expressed in yeast cells (Figure 8). A series of cultivation tests revealed that ORF4 expression also caused a decrease in the rate of cell proliferation and a decrease in cell life span [38,51].

We also examined the effects of MoCV1-A ORF4 expression in the human pathogenic budding yeast, *Cryptococcus neoformans*. As with *S. cerevisiae*, over expression of the protein in *C. neoformans* caused a decrease in the growth rate, increase in emergence, and enlargement of vacuoles. Additionally, the formation of capsules, which are involved in the pathogenicity of *C. neoformans*, was also reduced in the ORF4-expressing cells, suggesting a reduction in pathogenic potential [45]. Expression of an ORF4-GFP fusion protein in *S. cerevisiae* resulted in the formation of abnormal yeast cell aggregates. In MoCV1-A, ORF4 expression also resulted in significant inhibition of growth at high temperatures

(35 °C and 37 °C) as compared to cells grown at the optimal temperature (30 °C) [51] plus reduced expression of stress-response genes and increased expression of translation-related genes. We also observed the generation of reactive oxygen species in these cells (Figure 8). Further investigations on the mechanisms of cytotoxicity and growth inhibition caused by expression of the ORF4 protein in yeast and rice blast fungus cells are planned.

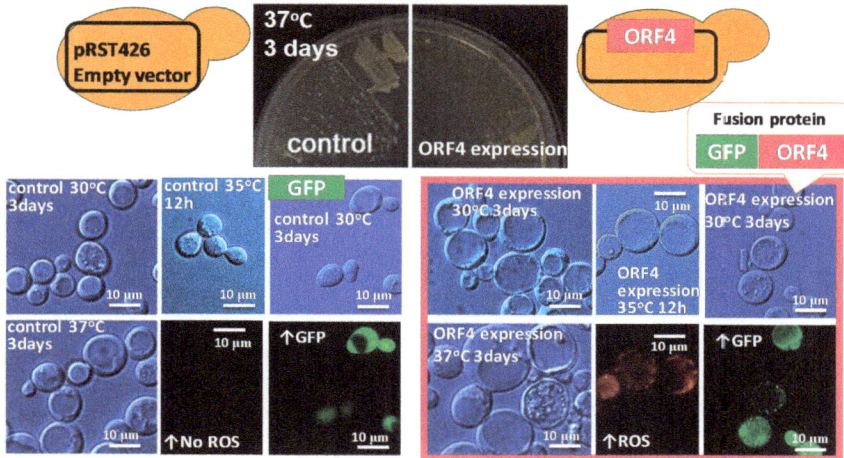

Figure 8. Expression of the MoCV1-A ORF4 protein induced cytological damage in yeast (*S. cerevisiae*) cells. The MoCV1-A ORF4 gene was inserted in a high-expression vector (2 μ ori, *TDH3* promoter) and the vector was used to transform the yeast strain, W303-1A. The morphology and growth of the yeast cells were observed. The MoCV1-A ORF4-GFP fusion protein caused aggregation in the yeast cells.

The MoCV1-A ORF4 protein showed significant similarity to related proteins from other viruses in cluster II of the genus, *Chrysovirus*. Multiple alignments of the ORF4-related protein sequences showed that their central regions (aa 210–591 in MoCV1-A ORF4) are relatively conserved. Indeed, yeast transformants expressing the conserved central region of the MoCV1-A ORF4 protein (325 aa–575 aa) showed similar impaired growth phenotypes to those observed in yeast expressing the full-length MoCV1-A ORF4 protein [51].

When the MoCV1-A ORF4 protein was over expressed in *E. coli*, the recombinant protein (rORF4p) was present in the insoluble fraction and when the culture reached an OD_{600} of 1.0, the yield of rORF4p was ca. 1 mg/50 mL of LB medium, which is ca. >50% of the total amount of protein in the culture. In the secretory production system utilizing *Pichia pastoris*, it is possible to produce ca. 1 mg recombinant protein per 100 mL of liquid medium [51].

7. Influence of MoCV1-A on the Pathogenicity of *Magnaporthe oryzae*

There are several reports showing that mycovirus infection can reduce the pathogenicity of the host fungus in plants, and this phenomenon is called hypovirulence. Hypovirulence has also been reported in the human pathogenic fungus, *A. fumigatus* [52]. In contrast, mycovirus infections sometimes confer hypervirulence to host fungi, characterized by enhanced growth and pathogenicity. Examples of hypervirulence are found in pathogens of both plants and humans [53,54]. However, few studies have examined the effects of mycoviral infection in cases of gene-for-gene interactions between plants and their fungal pathogens.

Pathogenic races of *M. oryzae* are determined by a gene-for-gene system, where an *avirulence* gene in the pathogen induces disease resistance in a rice variety with a corresponding resistance (*R*) gene [55]. To examine whether MoCV1-A infection affects pathogenic races of *M. oryzae*, we inoculated different rice varieties with a virus-free and a MoCV1-A-infected *M. oryzae* strain. Here, inoculation of the *R* gene-free rice variety, Lijiangxintuanheigu (LTH), showed that MoCV1-A infection resulted in reduced fungal virulence and this result was supported by an analysis of invasive hyphal development on onion epidermal cells. However, when spray or leaf-sheath inoculation methods were used to inoculate monogenic rice lines carrying different *R* genes, the MoCV1-A-infected and MoCV1-A-free *M. oryzae* strains caused different lesion types (resistance to susceptible or vice versa) on individual rice varieties. These data suggest that MoCV1-A infection can alter the pathogenicity of the host *M. oryzae* from avirulence to virulence, or from virulence to avirulence, depending on the rice variety (Figure 9) [39]. These results are consistent with the frequent emergence of new pathogenic races of rice blast fungus [56]. However, we did not find any gain or loss of the fungal avirulence genes, which determine pathogenic races of the fungus.

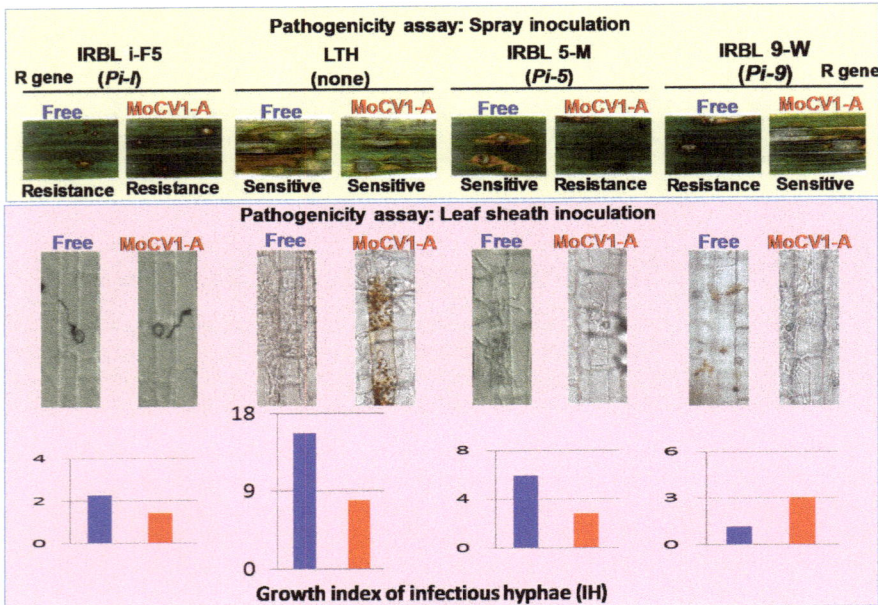

Figure 9. Influence of MoCV1-A on pathogenicity of *M. oryzae*. Upper panel: Spray inoculation assays revealed that MoCV1-A infection altered *M. oryzae* pathogenicity from virulence to avirulence (sensitive to resistance) in IRBL 5-M rice, and from avirulence to virulence (resistance to sensitive) in IRBL 9-W rice. Lower panel: The changes in pathogenicity were confirmed by changes in the growth indices of infectious hyphae in leaf sheath inoculation assays (lower panel). Details about the pathogenicity assays and statistical analyses are given in Aihara et al. [39].

In a recent investigation, we discovered that infection of the Japanese pear pathotype fungus, *Alternaria alternate*, with Alternaria alternata chrysovirus 1 (AaCV1) simultaneously impaired growth of the host fungus and increased levels of the host-specific AK-toxin [38,57]. This is another example of a mycovirus infection causing changes in pathogenic races of the host fungus, because *A. alternata* can infect some specific varieties of Japanese pear, but not others [58]. It is likely that the enhancement of host fungal pathogenicity in some varieties, without any mutation in the avirulence genes, is one of the strategies used by mycoviruses to survive in the agricultural ecosystem, where humans cultivate many plant varieties with different *R* genes to reduce damage by plant disease. Mycoviruses may increase survival rates by retaining the diversity of avirulence genes in their host fungi, which are important for fungal adaptation to plants [59].

8. Conclusions

The presence of mycoviruses infecting fungi is often detected by the appearance of morphological changes in the fungi, such as changes in their growth on agar medium. Mycoviruses that infect plant pathogenic fungi can also alter the pathogenicity of the host fungus sometimes, by changing the mode of plant infection by the fungus. We found that MoCV1-A changes the pathogenicity of its host, the rice blast fungus, *M. oryzae*. We observed growth inhibition in yeast cells due to heterologous expression of the MoCV1-A ORF4 protein. The results suggest that over expression of the ORF4 protein induces a modulation in the transcription of pathogenicity-associated genes in the host fungus. Likewise, totiviruses, partitiviruses, and chrysoviruses similar to MoCV1 are widely distributed in nature. It is possible that these mycoviruses can function as epigenetic factors that elicit changes in the pathogenicity of phytopathogenic fungi. These changes can be accompanied by alterations in the fungal host's susceptibility to fungicides [60].

Funding: This research was supported by a grant from the New Energy and Industrial Technology Development Organization (No. 08C46503c), a grant from the Adaptable and Seamless Technology Transfer Program (No. AS242Z01400N), and by a Grant-in-Aid for Scientific Research (C) from the Japan Society for the Promotion of Science (No. 15K07838).

Conflicts of Interest: The authors declare no conflict of interest.

References

1. Ghabrial, S.A.; Castón, J.R.; Jiang, D.; Nibert, M.L.; Suzuki, N. 50-plus years of fungal viruses. *Virology* **2015**, *479–480*, 356–368. [CrossRef] [PubMed]

2. Wickner, R.B. [URE3] as an altered URE2 protein: Evidence for a prion analog in *Saccharomyces cerevisiae*. *Science* **1994**, *264*, 566–569. [CrossRef] [PubMed]

3. Hollings, M. Viruses associated with a die-back disease of cultivated mushroom. *Nature* **1962**, *196*, 962–965. [CrossRef]

4. Wood, H.A.; Bozarth, R.F. Properties of virus-like particles of *Penicillium chrysogenum*: One double-stranded RNA molecule per particle. *Virology* **1972**, *47*, 604–609. [CrossRef]

5. Herring, A.J.; Bevan, E.A. Virus-like particles associated with the double-stranded RNA species found in killer and sensitive strains of the yeast *Saccharomyces cerevisiae*. *J. Gen. Virol.* **1974**, *22*, 387–394. [CrossRef] [PubMed]

6. Yamashita, S.; Doi, Y.; Yora, K. A polyhedral virus found in rice blast fungus, *Pyricularia oryzae* cavara. *Ann. Phytopathol. Soc. Jpn.* **1971**, *37*, 356–359. [CrossRef]

7. Nuss, D.L. Hypovirulence: Mycoviruses at the fungal–plant interface. *Nat. Rev. Microbiol.* **2005**, *3*, 632–642. [CrossRef]

8. Kotta-Loizou, I.; Coutts, R.H.A. Mycoviruses in *Aspergilli*: A Comprehensive Review. *Front. Microbiol.* **2017**, *8*, 1699. [CrossRef]

9. Dawe, A.L.; Nuss, D.L. Hypovirus molecular biology: From Koch's postulates to host self-recognition genes that restrict virus transmission. *Adv. Virus Res.* **2013**, *86*, 109–147. [CrossRef]

10. Ghabrial, S.A.; Dunn, S.E.; Li, H.; Xie, J.; Baker, T.S. Viruses of *Helminthosporium* (Cochlioblus) *victoriae*. *Adv. Virus Res.* **2013**, *86*, 289–325. [CrossRef]

11. Drinnenberg, I.A.; Weinberg, D.E.; Xie, K.T.; Mower, J.P.; Wolfe, K.H.; Fink, G.R.; Bartel, D.P. RNAi in budding yeast. *Science* **2009**, *326*, 544–550. [CrossRef] [PubMed]

12. Hawksworth, D.L. The fungal dimension of biodiversity: Magnitude, significance, and conservation. *Mycol. Res.* **1991**, *95*, 641–655. [CrossRef]

13. Sato, T. Plant diseases and their pathogenic microbes in Japan. *Microbiol. Cult. Coll.* **2015**, *29*, 79–90. (In Japanese)

14. Yokoi, T.; Yamashita, T.; Hibi, T. The nucleotide sequence and genome organization of *Magnaporthe oryzae* virus 1. *Arch. Virol.* **2007**, *152*, 2265–2269. [CrossRef]

15. Maejima, K.; Himeno, M.; Komatsu, S.; Kakizawa, Y.; Yamaji, H.; Hamamoto, H.; Namba, S. Complete nucleotide sequence of a new double-stranded RNA virus from the rice blast fungus. *Magnaporthe oryzae*. *Arch. Virol.* **2007**, *153*, 389–391. [CrossRef] [PubMed]

16. Du, Y.; He, X.; Zhou, X.; Fang, S.; Deng, Q. Complete nucleotide sequence of *Magnaporthe oryzae* partitivirus 1. *Arch. Virol.* **2016**, *161*, 3295–3298. [CrossRef] [PubMed]

17. Urayama, S.; Kato, S.; Suzuki, Y.; Aoki, N.; Le, M.T.; Arie, T.; Teraoka, T.; Fukuhara, T.; Moriyama, H. Mycoviruses related to chrysovirus affect vegetative growth in the rice blast fungus *Magnaporthe oryzae*. *J. Gen. Virol.* **2010**, *91*, 3085–3094. [CrossRef] [PubMed]

18. Urayama, S.; Sakoda, H.; Takai, R.; Katoh, Y.; Minh Le, T.; Fukuhara, T.; Arie, T.; Teraoka, T.; Moriyama, H. A dsRNA mycovirus, *Magnaporthe oryzae* chrysovirus 1-B, suppresses vegetative growth and development of the rice blast fungus. *Virology* **2014**, *448*, 265–273. [CrossRef]

19. Ai, Y.P.; Zhong, J.; Chen, C.Y.; Zhu, H.J.; Gao, B.D. A novel single-stranded RNA virus isolated from the rice-pathogenic fungus *Magnaporthe oryzae* with similarity to members of the family *Tombusviridae*. *Arch. Virol.* **2016**, *161*, 725–729. [CrossRef]

20. Illana, A.; Marconi, M.; Rodríguez-Romero, J.; Xu, P.; Dalmay, T.; Wilkinson, M.D.; Ayllón, M.; Sesma, A. Molecular characterization of a novel ssRNA ourmia-like virus from the rice blast fungus *Magnaporthe oryzae*. *Arch. Virol.* **2017**, *162*, 891–895. [CrossRef]

21. Fujikawa, T.; Nishimura, M. Pathogenic fungi fleeing the plant's immune system "Stealth strategy". *Shokubutu Boeki.* **2010**, *64*, 740–744. (In Japanese)

22. Okada, R.; Kiyota, E.; Moriyama, H.; Fukuhara, T.; Natsuaki, T. A simple and rapid method to purify viral dsRNA from plant and fungal tissue. *J. Gen. Plant. Pathol.* **2015**, *81*, 103–107. [CrossRef]

23. Wheeler, M.H. Melanin biosynthesis in *Verticillium dahliae*: Dehydration and reduction reactions in cell-free homogenates. *Exp. Mycol.* **1982**, *6*, 171–179. [CrossRef]

24. Hamada, T.; Asanagi, M.; Satozawa, T.; Araki, N.; Banba, S.; Higashimura, N.; Akase, T.; Hirase, K. Action mechanism of the novel rice blast fungicide tolprocarb distinct from that of conventional melamin biosynthesis inhibitors. *J. Pestic. Sci.* **2014**, *39*, 152–158. [CrossRef]

25. Urayama, S.; Ohta, T.; Onozuka, N.; Sakoda, H.; Fukuhara, T.; Arie, T.; Teraoka, T.; Moriyama, H. Characterization of *Magnaporthe oryzae* chrysovirus 1 structural proteins and their expression in *Saccharomyces cerevisiae*. *J. Virol.* **2012**, *86*, 8287–8295. [CrossRef] [PubMed]

26. Komatsu, K.; Urayama, S.; Katoh, Y.; Fuji, S.; Hase, S.; Fukuhara, T.; Arie, T.; Teraoka, T.; Moriyama, H. Detection of *Magnaporthe oryzae* chrysovirus 1 in Japan and establishment of a rapid, sensitive and direct diagnostic method based on reverse transcription loop-mediated isothermal amplification. *Arch. Virol.* **2016**, *161*, 317–326. [CrossRef] [PubMed]

27. Tang, L.; Hu, Y.; Liu, L.; Wu, S.; Xie, J.; Cheng, J.; Fu, Y.; Zhang, G.; Ma, J.; Wang, Y.; et al. Genomic organization of a novel victorivirus from the rice blast fungus *Magnaporthe oryzae*. *Arch. Virol.* **2015**, *160*, 2907–2910. [CrossRef]

28. Ghabrial, S.A.; Castón, J.R.; Coutts, R.H.A.; Hillman, B.I.; Jiang, D.; Kim, D.H.; Moriyama, H. ICTV Virus Taxonomy Profile: *Chrysoviridae*. *J. Gen. Virol.* **2018**, *99*, 19–20. [CrossRef]

29. Wang, L.; Jiang, J.; Wang, Y.; Hong, N.; Zhang, F.; Xu, W.; Wang, G. Hypovirulence of the phytopathogenic fungus *Botryosphaeria dothidea*: Association with a coinfecting chrysovirus and a partitivirus. *J. Virol.* **2014**, *88*, 7517–7527. [CrossRef]

30. Nerva, L.; Ciuffo, M.; Vallino, M.; Margaria, P.; Varese, G.C.; Gnavi, G.; Turina, M. Multiple approaches for the detection and characterization of viral and plasmid symbionts from a collection of marine fungi. *Virus Res.* **2016**, *219*, 22–38. [CrossRef]

31. Hammond, T.M.; Andrewski, M.D.; Roossinck, M.J.; Keller, N.P. Aspergillus mycoviruses are targets and suppressors of RNA silencing. *Eukaryot. Cell.* **2008**, *7*, 350–357. [CrossRef] [PubMed]

32. Van der Lende, T.R.; Duitman, E.H.; Gunnewijk, M.G.; Yu, L.; Wessels, J.G. Functional analysis of dsRNAs (L1, L3, L5, and M2) associated with isometric 34-nm virions of *Agaricus bisporus* (white button mushroom). *Virology* **1996**, *217*, 88–96. [CrossRef] [PubMed]

33. Lemus-Minor, C.G.; Cañizares, M.C.; García-Pedrajas, M.D.; Pérez-Artés, E. Complete genome sequence of a novel dsRNA mycovirus isolated from the phytopathogenic fungus *Fusarium oxysporum* f. sp. *dianthi*. *Arch. Virol.* **2015**, *160*, 2375–2379. [CrossRef] [PubMed]

34. Lemus-Minor, C.G.; Canizares, M.C.; García-Pedrajas, M.D.; Pérez-Artés, E. *Fusarium oxysporum* f. sp. dianthi virus 1 Accumulation Is Correlated with Changes in Virulence and Other Phenotypic Traits of Its Fungal Host. *Phytopathology* **2018**, *108*, 957–963. [CrossRef]

35. Lemus-Minor, C.G.; Cañizares, M.C.; García-Pedrajas, M.D.; Pérez-Artés, E. Horizontal and vertical transmission of the hypovirulence-associated mycovirus *Fusarium oxysporum* f. sp. dianthi virus 1. *Eur. J. Plant Pathol.* **2018**. [CrossRef]

36. Darissa, O.; Willingmann, P.; Schafer, W.; Adam, G. A novel double-stranded RNA mycovirus from *Fusarium graminearum*: Nucleic acid sequence and genomic structure. *Arch. Virol.* **2011**, *156*, 647–658. [CrossRef]

37. Herrero, N.; Zabalgogeazcoa, I. Mycoviruses infecting the endophytic and entomopathogenic fungus *Tolypocladium cylindrosporum*. *Virus Res.* **2011**, *160*, 409–413. [CrossRef]

38. Okada, R.; Ichinose, S.; Takeshita, K.; Syun-ichi Urayama, S.; Fukuhara, T.; Komatsu, K.; Arie, T.; Ishihara, A.; Egusa, M.; Kodama, M.; et al. Molecular characterization of a novel mycovirus in *Alternaria alternata* manifesting two-sided effects: Down-regulation of host growth and up-regulation of host plant pathogenicity. *Virology* **2018**, *519*, 23–32. [CrossRef]

39. Aihara, M.; Urayama, S.I.; Le, M.T.; Katoh, Y.; Higashiura, T.; Fukuhara, T.; Arie, T.; Teraoka, T.; Komatsu, K.; Moriyama, H. Infection by *Magnaporthe oryzae* chrysovirus 1 strain A triggers reduced virulence and pathogenic race conversion of it host fungus, *Magnaporthe oryzae*. *J. Gen. Plant Pathol.* **2018**, *84*, 92–103. [CrossRef]

40. Bormann, J.; Heinze, C.; Blum, C.; Mentges, M.; Brockmann, A.; Alder, A.; Landt, S.K.; Josephson, B.; Indenbirken, D.; Spohn, M.; et al. Expression of a structural protein of the mycovirus FgV-ch9 negatively affects the transcript level of a novel symptom alleviation factor and causes virus-infection like symptoms in *Fusarium graminearum*. *J. Virol.* **2018**. [CrossRef]

41. Ejmal, M.A.; Holland, D.J.; MacDiarmid, R.M.; Pearson, M.N. The Effect of *Aspergillus thermomutatus* Chrysovirus 1 on the biology of three Aspergillus species. *Viruses* **2018**, *10*, 539. [CrossRef] [PubMed]

42. Coustou, V.; Deleu, C.; Saupe, S.; Begueret, J. The protein product of the het-s heterokaryon incompatibility gene of the fungus *Podospora anserina* behaves as a prion analog. *Proc. Natl. Acad. Sci. USA* **1997**, *94*, 9773–9778. [CrossRef] [PubMed]

43. Wickner, R.B. A new prion controls fungal cell fusion incompatibility. *Proc. Natl. Acad. Sci. USA* **1997**, *94*, 10012–10014. [CrossRef] [PubMed]

44. Wickner, R.B.; Edskes, H.K.; Maddelein, M.L.; Taylor, K.; Moriyama, H. Prion of yeast and fungi: Proteins as genetic material. *J. Biol. Chem.* **1999**, *274*, 555–558. [CrossRef] [PubMed]

45. Urayama, S.; Fukuhara, T.; Moriyama, H.; Toh, E.A.; Kawamoto, S. Heterologous expression of a gene of *Magnaporthe oryzae* chrysovirus 1 strain A disrupts growth of the human pathogenic fungus *Cryptococcus neoformans*. *Microbiol. Immunol.* **2014**, *58*, 294–302. [CrossRef] [PubMed]

46. Himeno, M.; Maejima, K.; Komatsu, K.; Ozeki, J.; Hashimoto, M.; Kagiwada, S.; Yamaji, Y.; Namba, S. Significantly low level of small RNA accumulation derived from an encapsidated mycovirus with dsRNA genome. *Virology* **2010**, *396*, 69–75. [CrossRef] [PubMed]

47. Komatsu, K.; Katayama, Y.; Omatsu, T.; Mizutani, T.; Fukuhara, T.; Kodama, M.; Arie, T.; Teraoka, T.; Moriyama, H. Genome sequence of a novel victorivirus identified in the phytopathogenic fungus *Alternaria arborescens*. *Arch. Virol.* **2016**, *161*, 1701–1704. [CrossRef]

48. Aoki, N.; Moriyama, H.; Kodama, M.; Arie, T.; Teraoka, T.; Fukuhara, T. A novel mycovirus associated with four double-stranded RNAs affects host fungal growth in *Alternaria alternata*. *Virus Res.* **2009**, *140*, 179–187. [CrossRef]

49. Icho, T.; Wickner, R.B. The double-stranded RNA genome of yeast virus L-A encodes its own putative RNA polymerase by fusing two open reading frames. *J. Biol. Chem.* **1989**, *264*, 6716–6723.

50. Nagai, M.; Shimada, S.; Fujii, Y.; Moriyama, H.; Oba, M.; Katayama, Y.; Tsuchiaka, S.; Okazaki, S.; Omatsu, T.; Furuya, T.; et al. H2 genotypes of G4P[6], G5P[7], and G9[23] porcine rotaviruses show super-short RNA electropherotypes. *Vet. Microbiol.* **2015**, *176*, 250–256. [CrossRef]

51. Urayama, S.; Kimura, Y.; Katoh, Y.; Ohta, T.; Onozuka, N.; Fukuhara, T.; Arie, T.; Teraoka, T.; Komatsu, K.; Moriyama, H. Suppressive effects of mycoviral proteins encoded by *Magnaporthe oryzae* chrysovirus 1 strain A on conidial germination of the rice blast fungus. *Virus Res.* **2016**, *223*, 10–19. [CrossRef] [PubMed]

52. Bhatti, M.F.; Jamal, A.; Petrou, M.A.; Cairns, T.C.; Bignell, E.M.; Coutts, R.H.A. The effects of dsRNA mycoviruses on growth and murine virulence of *Aspergillus fumigatus*. *Fungal Genet. Biol.* **2011**, *48*, 1071–1075. [CrossRef] [PubMed]

53. Ahn, I.P.; Lee, Y.H. A viral double-stranded RNA up regulates the fungal virulence of *Nectria radicicola*. *Mol. Plant-Microbe Interact* **2001**, *14*, 496–507. [CrossRef] [PubMed]

54. Kanhayuwa, L.; Kotta-Loizou, I.; Özkan, S.; Gunning, A.P.; Coutts, R.H.A. A novel mycovirus from *Aspergillus fumigatus* contains four unique dsRNAs as its genome and is infectious as dsRNA. *Proc. Natl. Acad. Sci. USA* **2015**, *112*, 9100–9105. [CrossRef] [PubMed]

55. Yamada, M.; Kiyosawa, S.; Yamaguchi, T.; Hirano, T.; Kobayashi, T.; Kushibuchi, K.; Watanabe, S. Proposal of a new method for differentiating races of *Pyricularia oryzae* Cavara in Japan. *Ann. Phytopathol. Soc. Jpn.* **1976**, *42*, 216–219. [CrossRef]

56. Kiyosawa, S. Pathogenic variations of *Pyricularia oryzae* and their use in genetic and breeding studies. *SABRAO J.* **1976**, *8*, 53–67.

57. Fuke, K.; Takeshita, K.; Aoki, N.; Fukuhara, T.; Egusa, M.; Kodama, M.; Moriyama, H. The presence of double-stranded RNAs in *Alternaria alternata* Japanese pear pathotype is associated with morphological changes. *J. Gen. Plant Pathol.* **2011**, *77*, 248–252. [CrossRef]

58. Akimitsu, K.; Tsuge, T.; Kodama, M.; Yamamoto, M.; Otani, H. *Alternaria* host-selective toxins: Determinant factors of plant disease. *J. Gen. Plant Pathol.* **2014**, *80*, 109–122. [CrossRef]

59. Asai, S.; Shirasu, K. Plant cells under siege: Plant immune system versus pathogen effectors. *Curr. Opin. Plant Biol.* **2015**, *28*, 1–8. [CrossRef]

60. Niu, Y.; Yuan, Y.; Mao, J.; Yang, Z.; Cao, Q.; Zhang, T.; Wang, S.; Liu, D. Characterization of two novel mycoviruses from *Penicillium digitatum* and the related fungicide resistance analysis. *Sci. Rep.* **2018**, *8*, 5513. [CrossRef]

Communication

Complete Nucleotide Sequence of a Partitivirus from *Rhizoctonia solani* AG-1 IA Strain C24

Chen Liu, Miaolin Zeng, Meiling Zhang, Canwei Shu * and Erxun Zhou *

Guangdong Province Key Laboratory of Microbial Signals and Disease Control, College of Agriculture,
South China Agricultural University, Guangzhou 510642, China; chenliu56@163.com (C.L.);
miulum@163.com (M.Z.); meilingzhangsy@163.com (M.Z.)
* Correspondence: shucanwei@scau.edu.cn (C.S.); exzhou@scau.edu.cn (E.Z.)

Received: 29 September 2018; Accepted: 7 December 2018; Published: 11 December 2018

Abstract: The complete genome of a novel double-stranded (ds) RNA mycovirus, named as Rhizoctonia solani partitivirus 5 (RsPV5), isolated from rice sheath blight fungus *R. solani* AG-1 IA strain C24, was sequenced and analysed. RsPV5 consists of two segments, dsRNA-1 (1899 nucleotides) and dsRNA-2 (1787 nucleotides). DsRNA-1 has an open reading frame (ORF) 1 that potentially codes for a protein of 584 amino acid (aa) containing the conserved motifs of a RNA-dependent RNA polymerase (RdRp), and dsRNA-2 also contains a ORF 2, encoding a putative capsid protein (CP) of 513 aa. Phylogenetic analysis revealed that RsPV5 clustered together with six other viruses in an independent clade of the genus *Alphapartitivirus*, indicating that RsPV5 was a new member of the genus *Alphapartitivirus*, within the family *Partitiviridae*.

Keywords: *Rhizoctonia solani* AG-1 IA; mycovirus; dsRNA; *Alphapartitivirus*; genomic structure analysis

1. Introduction

Mycoviruses (fungal viruses or viruses of fungi) are widely distributed in fungi, of which, only a few affect their fungal hosts resulting in alterations of growth rate or enhanced virulence or hypovirulence [1,2]. Mycoviruses with hypovirulent traits are anticipated to be important biological control agents against plant fungal diseases in the future [3]. Since the successful application of a mycovirus for biological control of chestnut blight [4], mycovirus research from major groups of fungi has attracted attention. A wide range of mycoviruses, such as Rosellinia necatrix partitivirus 1-W8 [5], Botrytis cinerea mitovirus 1 [6], and Rhizoctonia solani partitivirus 2 [7], have been discovered in different fungi. Currently, mycoviruses are mainly classified into 14 families, including 7 families of double-stranded (dsRNA) viruses (*Partitiviridae, Totiviridae, Reoviridae, Chrysoviridae, Quadriviridae, Megabirnaviridae, Endornaviridae*), 5 families of positive-strand RNA (+ssRNA) viruses (*Barnaviridae, Alphaflexiviridae, Hypoviridae, Narnaviridae, Gammaflexiviridae*), 1 family of negative-strand RNA (-ssRNA) viruses (*Mycomononegaviridae*) and family *Amalgaviridae* [8,9]. However many mycoviruses remain unclassified [8], and the first circular ssDNA mycovirus from the phytopathogenic fungus *Sclerotinia sclerotiorum* was discovered in 2010 [10].

Rhizoctonia solani Kühn [teleomorph: *Thanatephorus cucumeris* (Frank) Donk] is an economically important soil-borne fungal pathogen that causes severe plant diseases and disastrous economic losses in a wide variety of commercial crops including rice, maize and wheat [11,12]. *R. solani* is a common mycovirus host [7,13]. Investigations concerning the association of dsRNA with *Rhizoctonia* decline revitalised research on *R. solani* mycoviruses [14] which in turn revealed that mycoviruses are ubiquitous in natural *R. solani* isolates [7,13,15–19]. Subsequently complete genome sequences of several *R. solani* mycoviruses have been documented and their sequence properties and phylogene have been analysed. Thus far the *R. solani* mycoviruses described mainly belong to the genera *Partitivirus* [13,18–20], *Mitovirus* [12,21] and

Endornavirus [22,23], along with some unclassified mycoviruses [7,24,25]. To date, the complete genomes of six mycoviruses, *i.e.* Rhizoctonia solani dsRNA virus 1 (RsRV1) [7], Rhizoctonia solani partitivirus 2 (RsPV2) [13], Rhizoctonia solani RNA virus HN008 (RsRV-HN008) [24], Rhizoctonia solani dsRNA virus 3 (RsRV3) [18], Rhizoctonia solani partitivirus 3 (RsPV3) [19] and Rhizoctonia solani partitivirus 4 (RsPV4) [19], from *R. solani* AG-1 IA have been reported, of which, RsRV1 and RsRV-HN008 are unclassified, while RsPV2, RsRV3, RsPV3 and RsPV4 belong to the genus *Alphapartitivirus*.

Investigations on the AG-1 IA isolate of *R. solani*, the causal agent of rice sheath blight revealed the presence of three novel mycoviruses in the authors' laboratory [7,13,18]. In this study we describe the complete nucleotide sequence of another partitivirus nominated Rhizoctonia solani partitivirus 5 (RsPV5), isolated from *R. solani* AG-1 IA strain C24. The sequences of the two genomic components of RsPV5 were analysed and a phylogenetic tree was constructed based on the derived amino acid sequence of the putative RNA-dependent RNA polymerase (RdRp) to clarify the phylogenetic status of RsPV5. The phylogenetic analysis indicated that RsPV5 has the closest relationship with members of the genus *Alphapartitivirus*.

2. Materials and Methods

2.1. Fungal Strain

The C24 strain of *R. solani* AG-1 IA was used in this study, which was isolated from rice leaves with typical symptoms of rice sheath blight collected from Zhangzhou city, Fujian province, China, in 1999 and stored at $-20\,^{\circ}$C.

2.2. Isolation and Sequencing of Mycovirus dsRNA

Mycelia of the strain C24 were cultured on cellophane covered on potato dextrose agar (PDA) plates at $28\,^{\circ}$C. After cultivation for 5 days, the mycelia were harvested and stored at $-80\,^{\circ}$C for further use. The lyophilized mycelia were ground into a fine powder with a mortar and pestle in the presence of liquid nitrogen. Viral dsRNAs were extracted using a slightly modified version of a CF-11 cellulose chromatography method as described by Morris and Dodds [26]. To remove contaminating DNA and single stranded RNA, the extracts were treated with DNase I and SI nuclease, and viral dsRNAs were separated and analysed by gel electrophoresis and visualization with ethidium bromide staining. The cDNA library was constructed using random primer (5′-CCTGAATTCGGATCCTCCNNNNNN-3′) along with reverse transcriptase, and amplified with specific primer (5′-CCTGAATTCGGATCCTCC-3′). To sequence the 5′ and 3′-termini of the dsRNA, a RACE procedure modified from that described by Potgieter et al. [27] was used. All PCR amplicons were cloned into the pMD18-T vector and transformed into *Escherichia coli* strain JM109. Plasmid DNA from recombinant clones was isolated and at least three clones for each fragment of sequence were sequenced in both directions. The complete nucleotide sequences of the two genomic components of RsPV5 were assembled and deposited in GenBank database with the accession numbers of MH715946 and MH715947.

2.3. Data Analysis

Sequence analysis and multiple alignments were actualized by DNAMAN and ClustalX. A phylogenetic tree was constructed on the basis of neighbor-joining (NJ) method using MEGA 6 with 1000 replicates.

3. Results

3.1. Genomic Structure Analysis

Sequence analysis revealed that *R. solani* strain C24 was infected by a novel virus, RsPV5, belonging the family *Partitiviridae*. The complete genome of RsPV5 is composed of two segments, designated dsRNA-1 and dsRNA-2, respectively (Figure 1a,b). A comparison of both dsRNA segments demonstrated that both 5′- and 3′-termini are conserved (Figure 1c). Additionally the 3′-ends of both dsRNAs were interrupted by poly(A) tails, a feature similar to some other members in the family *Partitiviridae* [13,28].

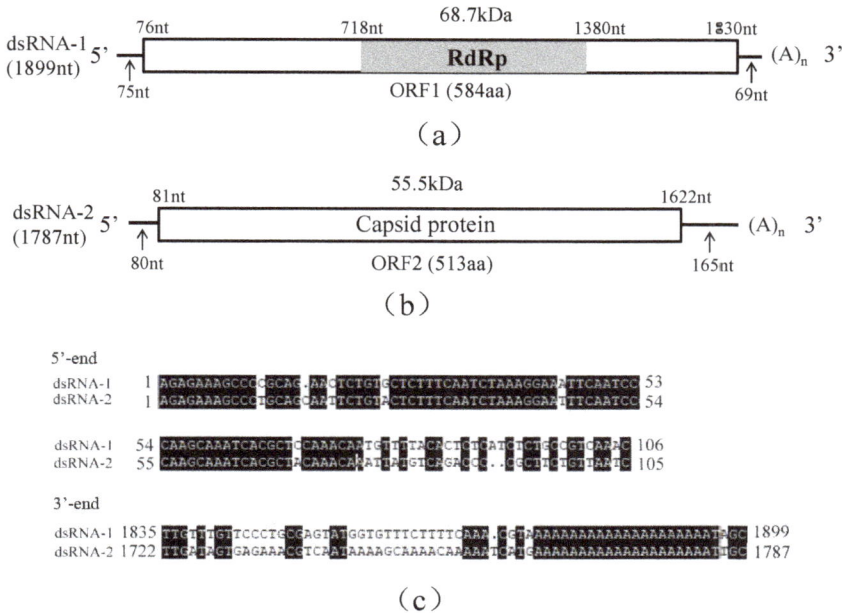

(a)

(b)

(c)

Figure 1. Schematic representation of the genomic organization of RsPV5 isolated from R. solani AG-1 IA strain C24, the causal agent of rice sheath blight. (**a**) Schematic representation of the genomic organization of dsRNA-1. The rectangle represents open reading frame (ORF 1) and the nucleotide positions of the start and end codons are listed above the box. The gray bar represents the conserved RNA-dependent RNA polymerase (RdRp), the predicted molecular masse and the nucleotide positions of the start and termination codons are listed above the bar. The arrows under the single lines represent the length of the non-coding sequence. (**b**) Schematic representation of the genomic organization of dsRNA-2. The rectangle represents the open reading frame (ORF 2) and its encoded protein, capsid protein (CP), the nucleotide positions of the start and end codons are listed above the box. The arrows under the single lines represent the length of the non-coding sequences. (**c**) Alignments of 5′- and 3′-untranslated regions (UTRs) of RsPV5 dsRNA-1 and dsRNA-2. The letters with black shading represent conserved sequences at both ends.

Analysis of the full-length cDNA sequence of dsRNA-1 indicated that it comprises 1899 nucleotides (nt), with a GC content of 45.98%, and contains an open reading frame 1 (ORF1), starting at nt 76 and terminating at nt 1830. ORF1 potentially encodes a 68.7 kDa protein of 584 amino acids (aa) containing sequence-conserved motifs characteristic for RNA-dependent RNA polymerase (RdRp; Figure 1a). The 5′-untranslated region (UTR) and 3′-UTR of dsRNA-1 consists of 75 nt and 69 nt, respectively. Homology searches with BLASTp confirmed that the protein was closely related to the RdRps of partitiviruses including Rhizoctonia solani dsRNA virus 3 (RsRV3, GenBank accession number: YP_009329886.2) with an aa identity of 82%, Heterobasidion partitivirus 12 (HetPV12, GenBank accession number: YP_009508051.1) with an aa identity of 61%, Heterobasidion partitivirus 13 (HetPV13, GenBank accession number: AHL25155.1) with an aa identity of 59%.

Analysis of the full-length cDNA sequence of dsRNA-2 indicated that it is 1787 bp in length with a GC content of 52.20% containing a single open reading frame 2 (ORF2) starting at nt 81 and terminating at nt 1622. DsRNA-2 potentially encodes a putative capsid protein (CP) of 513 aa that has an estimated molecular mass of 55.5 kDa (Figure 1b). The 5'-UTR and 3'-UTR of dsRNA2 are respectively 80 nt and 165 nt in length. BLASTp search revealed that this protein has 65% identity to RsRV3 CP gene (GenBank accession number: YP_009329885.2) and 30% identity to the HetPV13 CP gene (GenBank accession number: AHL25156.1).

3.2. Phylogenetic Analysis

To confirm the taxonomic status of RsPV5, a phylogenetic tree was constructed based on the aa sequences of RdRp regions of RsPV5 and 24 other selected viruses in the families *Partitiviridae* and *Totiviridae* as well as the unclassified viruses (Figure 2). The result of phylogenetic analysis showed that RsPV5, Rhizoctonia solani dsRNA virus 3, Rhizoctonia solani dsRNA virus 2, Rhizoctonia solani partitivirus 3, Rhizoctonia solani partitivirus 4, White clover cryptic virus 1 and Carrot cryptic virus were clustered together in a distinct group belonging to the genus *Alphapartitivirus*. The phylogenetic tree illustrated that RsPV5 is a new member of the genus *Alphapartitivirus* in the family *Partitiviridae*. In addition, RsPV5 was placed in the same clade with RsRV3, a mycovirus of *R. solani* AG-1 IA previously identified in our laboratory [18], indicating that these two viruses have a close relationship. Furthermore, RsRV1 [7] and RsRV-HN008 [24] belong to the subclade of unclassified family. *R. solani* AG-1 IA appears to have an extensive virome which might expand further in the future with the advent of next generation sequencing.

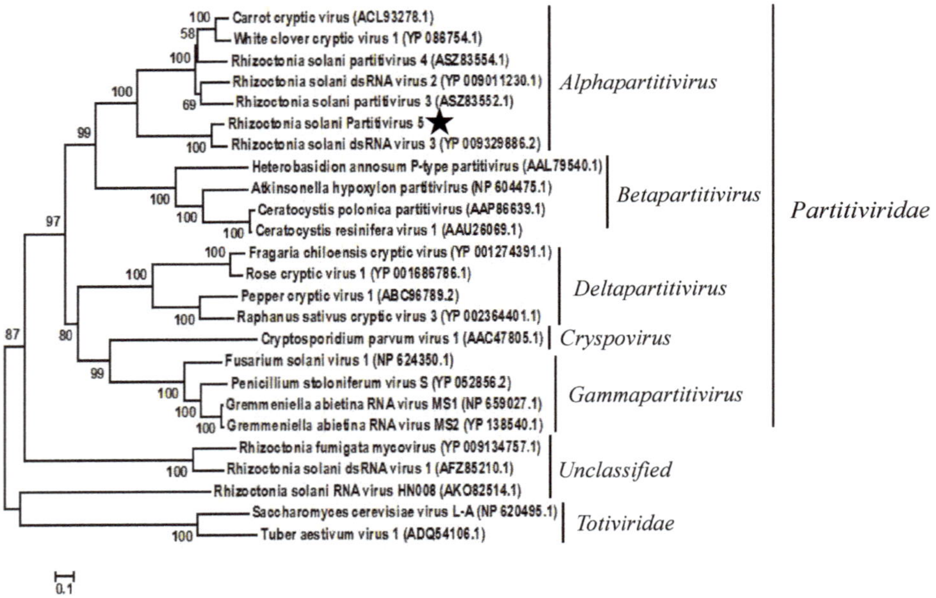

Figure 2. Phylogenetic analysis of RsPV5. A phylogenetic tree was generated for the putative amino acid sequences of the deduced RdRp proteins using the neighbor-joining method with the program MEGA 6.0 and Bootstrap 1000 replicates. The RdRp sequences were obtained from GenBank and the accession numbers of viruses are given in the brackets behind the virus names. The scale means a genetic distance of 0.1 amino acid substitutions per site. Viral lineages are marked based on their taxonomic status.

4. Discussion

It can be seen from the above study, RsPV5, which infects *R. solani* AG-1 IA, is a novel dsRNA mycovirus of the genus *Alphapartitivirus* in the family *Partitiviridae*. So far, seven mycoviruses four of them were found in the authors' laboratory [7,13,18] infecting *Rhizoctonia solani* AG1-IA have been reported and have proved to belong to different viral family [7,13,18,19,24], indicating a rich diversity of mycoviruses in *R. solani* AG-1 IA.

Author Contributions: C.L. performed the experiments, analysed the data, sequenced and analysed the virus, wrote the draft paper; M.Z. (Miaolin Zeng) and M.Z. (Meiling Zhang) provided advices for experimental operation, assisted in data analysis and paper revision; C.S. and E.Z. conceived, designed and supervised the experiments, reviewed and revised the paper.

Funding: This study is supported by the National Natural Science Foundation of China, "Genome structure and function analysis of dsRNA mycoviruses of *Rhizoctonia solani* AG-1 IA, the causal agent of rice sheath blight" (No. 31470247).

Conflicts of Interest: The authors declare no conflict of interest.

References

1. Pearson, M.N.; Beever, R.E.; Boine, B.; Arthur, K. Mycoviruses of filamentous fungi and their relevance to plant pathology. *Mol. Plant Path.* **2009**, *10*, 115–128. [CrossRef] [PubMed]
2. Herrero, N.; Márquez, S.S.; Zabalgogeazcoa, I. Mycoviruses are common among different species of endophytic fungi of grasses. *Arch. Virol.* **2009**, *154*, 327–330. [CrossRef] [PubMed]
3. Nuss, D.L. Hypovirulence: mycoviruses at the fungal-plant interface. *Nat. Rev. Microbiol.* **2005**, *3*, 632–642. [CrossRef] [PubMed]
4. Grente, J.; Berthelay-Sauret, S. Biological control of chestnut blight in France. In Proceedings of the American Chestnut Symposium, Morgantown, WV, USA, 4–5 January 1978; pp. 30–34.
5. Sasaki, A.; Miyanishi, M.; Ozaki, K.; Onoue, M.; Yoshida, K. Molecular characterization of a partitivirus from the plant pathogenic ascomycete *Rosellinia necatrix*. *Arch. Virol.* **2005**, *150*, 1069–1083. [CrossRef] [PubMed]
6. Wu, M.; Zhang, L.; Li, G.; Jiang, D.; Ghabrial, S.A. Genome characterization of a debilitation-associated mitovirus infecting the phytopathogenic fungus *Botrytis cinerea*. *Virology* **2010**, *406*, 117–126. [CrossRef] [PubMed]
7. Zheng, L.; Liu, H.; Zhang, M.; Cao, X.; Zhou, E. The complete genomic sequence of a novel mycovirus from *Rhizoctonia solani* AG-1 IA strain B275. *Arch. Virol.* **2013**, *158*, 1609–1612. [CrossRef] [PubMed]
8. Ghabrial, S.A.; Castón, J.R.; Jiang, D.; Nibert, M.L.; Suzuki, N. 50-plus years of fungal viruses. *Virology* **2015**, *479*, 356–368. [CrossRef] [PubMed]
9. Puchades, A.V.; Carpino, C.; Alfaro-Fernandez, A.; Font-San-Ambrosio, M.I.; Davin, S.; Guerri, J.; Rubio, L.; Galipienso, L. Detection of Southern tomato virus by molecular hybridisation. *Ann. Appl. Biol.* **2017**, *171*, 172–178. [CrossRef]
10. Yu, X.; Li, B.; Fu, Y.; Jiang, D.; Ghabrial, S.A.; Li, G.; Peng, Y.; Xie, J.; Cheng, J.; Huang, J.; et al. A geminivirus-related DNA mycovirus that confers hypovirulence to a plant pathogenic fungus. *Proc. Natl. Acad. Sci. USA* **2010**, *107*, 8387–8392. [CrossRef] [PubMed]
11. Cubeta, M.A.; Vilgalys, R. Population biology of the *Rhizoctonia solani* complex. *Phytopathology* **1997**, *87*, 480–484. [CrossRef]
12. Abdoulaye, A.H.; Cheng, J.; Jiang, D.; Xie, J. Complete genome sequence of a novel mitovirus from the phytopathogenic fungus *Rhizoctonia oryzae-sativae*. *Arch. Virol.* **2017**, *162*, 1409–1412. [CrossRef] [PubMed]
13. Zheng, L.; Zhang, M.; Chen, Q.; Zhu, M.; Zhou, E. A novel mycovirus closely related to viruses in the genus *Alphapartitivirus* confers hypovirulence in the phytopathogenic fungus *Rhizoctonia solani*. *Virology* **2014**, *456*, 220–226. [CrossRef] [PubMed]
14. Castanho, B.; Butler, E.E.; Shepherd, R.J. Association of double-stranded-RNA with *Rhizoctonia* decline. *Phytopathology* **1978**, *68*, 1515–1519. [CrossRef]
15. Bharathan, N.; Tavantzis, S.M. Genetic diversity of double-stranded RNA from *Rhizoctonia solani*. *Phytopathology* **1990**, *80*, 631–635. [CrossRef]

16. Robinson, H.L.; Deacon, J.W. Double-stranded RNA elements in *Rhizoctonia solani* AG 3. *Mycol. Res.* **2002**, *106*, 12–22. [CrossRef]

17. Bharathan, N.; Saso, H.; Gudipati, L.; Bharathan, S.; Whited, K.; Anthony, K. Double-stranded RNA: distribution and analysis among isolates of *Rhizoctonia solani* AG-2 to-13. *Plant Pathol.* **2005**, *54*, 196–203. [CrossRef]

18. Zhang, M.; Zheng, L.; Liu, C.; Shu, C.W.; Zhou, E.X. Characterization of a novel dsRNA mycovirus isolated from strain A105 of *Rhizoctonia solani* AG-1 IA. *Arch. Virol.* **2018**, *163*, 427–430. [CrossRef] [PubMed]

19. Lyu, R.L.; Zhang, Y.; Tang, Q.; Li, Y.; Cheng, J.; Fu, Y.; Chen, T.; Jiang, D.; Xie, J. Two alphapartitiviruses co-infecting a single isolate of the plant pathogenic fungus *Rhizoctonia solani*. *Arch. Virol.* **2018**, *163*, 515–520. [CrossRef]

20. Strauss, E.E.; Lakshman, D.K.; Tavantzis, S.M. Molecular characterization of the genome of a partitivirus from the basidiomycete *Rhizoctonia solani*. *J. General Virol.* **2000**, *81*, 549–555. [CrossRef]

21. Lakshman, D.K.; Jian, J.; Tavantzis, S.M. A double-stranded RNA element from a hypovirulent strain of *Rhizoctonia solani* occurs in DNA form and is genetically related to the pentafunctional AROM protein of the shikimate pathway. *Proc. Natl. Acad. Sci. USA* **1998**, *95*, 6425–6429. [CrossRef]

22. Li, W.; Zhang, T.; Sun, H.; Deng, Y.; Zhang, A.; Chen, H.; Wang, K. Complete genome sequence of a novel endornavirus in the wheat sharp eyespot pathogen *Rhizoctonia cerealis*. *Arch. Virol.* **2014**, *159*, 1213–1216. [CrossRef] [PubMed]

23. Das, S.; Falloon, R.E.; Stewart, A.; Pitman, A.R. Molecular characterisation of an endornavirus from *Rhizoctonia solani* AG-3PT infecting potato. *Fungal Biol.* **2014**, *118*, 924–934. [CrossRef] [PubMed]

24. Zhong, J.; Chen, C.Y.; Gao, B.D. Genome sequence of a novel mycovirus of *Rhizoctonia solani*, a plant pathogenic fungus. *Virus Genes* **2015**, *51*, 167–170. [CrossRef] [PubMed]

25. Li, Y.; Xu, P.; Zhang, L.; Xia, Z.; Qin, X.; Yang, G.; Mo, X. Molecular characterization of a novel mycovirus from *Rhizoctonia fumigata* AG-Ba isolate C-314 Baishi. *Arch. Virol.* **2015**, *160*, 2371–2374. [CrossRef] [PubMed]

26. Morris, T.; Dodds, J. Isolation and analysis of double-stranded RNA from virus-infected plant and fungal tissue. *Phytopathology* **1979**, *69*, 854–858. [CrossRef]

27. Potgieter, A.C.; Page, N.A.; Liebenberg, J.; Wright, I.M.; Landt, O.; van Dijk, A.A. Improved strategies for sequence-independent amplification and sequencing of viral dsRNA genomes. *J. General Virol.* **2009**, *90*, 1423–1432. [CrossRef] [PubMed]

28. Lim, W.S.; Jeong, J.H.; Jeong, R.D.; Yoo, Y.B.; Yie, S.W.; Kim, K.H. Complete nucleotide sequence and genome organization of a dsRNA partitivirus infecting *Pleurotus ostreatus*. *Virus Res.* **2005**, *108*, 111–119. [CrossRef]

viruses

MDPI

Communication

Identification, Molecular Characterization, and Biology of a Novel Quadrivirus Infecting the Phytopathogenic Fungus *Leptosphaeria biglobosa*

Unnati A. Shah [1], Ioly Kotta-Loizou [1,2,*], Bruce D. L. Fitt [1] and Robert H. A. Coutts [1]

[1] Department of Biological and Environmental Sciences, University of Hertfordshire, Hatfield AL10 9AB, UK; unnatishah009@gmail.com (U.A.S.); b.fitt@herts.ac.uk (B.D.L.F.); r.coutts@herts.ac.uk (R.H.A.C.)
[2] Department of Life Sciences, Imperial College London, London SW7 2AZ, UK
* Correspondence: i.kotta-loizou@imperial.ac.uk

Received: 2 December 2018; Accepted: 22 December 2018; Published: 25 December 2018

Abstract: Here we report the molecular characterisation of a novel dsRNA virus isolated from the filamentous, plant pathogenic fungus *Leptosphaeria biglobosa* and known to cause significant alterations to fungal pigmentation and growth and to result in hypervirulence, as illustrated by comparisons between virus-infected and -cured isogenic fungal strains. The virus forms isometric particles approximately 40–45 nm in diameter and has a quadripartite dsRNA genome structure with size ranges of 4.9 to 4 kbp, each possessing a single ORF. Sequence analysis of the putative proteins encoded by dsRNAs 1–4, termed P1–P4, respectively, revealed modest similarities to the amino acid sequences of equivalent proteins predicted from the nucleotide sequences of known and suspected members of the family *Quadriviridae* and for that reason the virus was nominated Leptosphaeria biglobosa quadrivirus-1 (LbQV-1). Sequence and phylogenetic analysis using the P3 sequence, which encodes an RdRP, revealed that LbQV-1 was most closely related to known and suspected quadriviruses and monopartite totiviruses rather than other quadripartite mycoviruses including chrysoviruses and alternaviruses. Of the remaining encoded proteins, LbQV-1 P2 and P4 are structural proteins but the function of P1 is unknown. We propose that LbQV-1 is a novel member of the family *Quadriviridae*.

Keywords: fungal viruses; dsRNA mycoviruses; hypervirulence; *Leptosphaeria biglobosa* quadrivirus

1. Introduction

Fungal viruses or mycoviruses are ubiquitous and have been detected in all major groups of fungi including members of the divisions Ascomycota, Basidiomycota and Glomeromycota [1]. There are at least five established mycovirus families, and one established genus, whose members have multi-segmented dsRNA genomes: families *Reoviridae*, *Partitiviridae*, *Chrysoviridae*, *Quadriviridae*, and *Megabirnaviridae*, and the genus *Botybirnavirus* [1], which possess genome segment numbers of 11-12, 2, 4, 4, 2, and 2, respectively. In addition, there are two proposed families Alternaviridae and Polymycoviridae [2–4], which possess 4 and 4–8 genomic components, respectively. All of these viruses are encapsidated in rigid, spherical virus particles apart from polymycoviruses, which are not conventionally encapsidated [4,5]. Although many mycoviruses have no or few obvious effects on their host fungi, some induce phenotypic alterations including hypovirulence (attenuated virulence) or hypervirulence (enhanced virulence).

Phoma stem canker (blackleg) is an internationally important disease of oilseed rape (*Brassica napus*, canola, rapeseed), causing serious losses in Europe, Australia, and North America. For instance UK losses of ca. £100 million per season have been estimated using national disease survey data and a yield loss formula [6]. Phoma stem canker pathogen populations comprise two main species, *Leptosphaeria maculans*, associated with damaging stem base cankers, and *Leptosphaeria biglobosa*, often associated with less damaging upper stem lesions [6].

A collection of over 70 field isolates of *Leptosphaeria* spp. from *B. napus* were first classified as being either *L. maculans* or *L. biglobosa* and were then screened for the presence of dsRNA using a small scale isolation procedure. Several *L. biglobosa* isolates were found to contain dsRNA species whose electrophoretic pattern and sizes were reminiscent of those described previously for members of the *Quadriviridae* [7,8] and here we describe the complete molecular characterisation of a new quadrivirus isolated from a Chinese field strain of *L. biglobosa*. Moreover we carried out a complete analysis of the genome organisation and phylogeny of the virus, while its effects on the host phenotype and pathogenicity are described in detail elsewhere [9].

2. Materials and Methods

2.1. Fungal Strains, Culture Conditions, and Dsrna Extraction

The Chinese *L. biglobosa* isolate W10 was grown on V8 agar plates and incubated for five days at 20 °C in darkness to produce confluent cultures. The isolate was confirmed as *L. biglobosa* following sub-culturing on potato dextrose agar (PDA) plates on the basis of morphological phenotype and by PCR amplification of the ribosomal RNA region incorporating the internal transcribed spacer (ITS) regions and the 5.8S rRNA gene [10]. The dsRNA elements known to be present in W10 [9] were purified by LiCl fractionation [11] and gel electrophoretic analysis of dsRNA was done according to standard protocols [4]. A virus-cured isogenic line was obtained using a combination of treatment with cycloheximide and hyphal tipping [4,9].

2.2. Virus Purification and Transmission Electron Microscopy (TEM)

Mycelia grown in PD broth with shaking at 25 °C were harvested five days after inoculation using sterile Miracloth (Merck Millipore, Danvers, MA, USA) and rapidly frozen in liquid N_2 and kept at −80 °C until processing. Virus purification was performed as described before [5]. Purified virus was negatively stained with 1% uranyl acetate on carbon-coated 400-mesh copper grids and examined using a FEI Tecnai 20 transmission electron microscope. Isolation of dsRNA from purified virus was performed using phenol/chloroform treatment. DNase I (Promega, Madison, WI, USA) and S1 nuclease (Promega, Madison, WI, USA) treatments of purified dsRNAs were performed according to the manufacturer's instructions.

2.3. cDNA Cloning and RNA Ligase-Mediated Rapid Amplification of cDNA Ends (RLM-RACE)

After electrophoretic separation on agarose gels dsRNAs were used, either collectively or individually, as templates for cDNA synthesis and PCR amplification of products using random priming, sequence-specific priming and RLM-RACE, which were subsequently cloned and sequenced [12,13]. At least three different clones were sequenced, covering the same part of each segment of the viral genome.

2.4. Sequence and Structure Bioinformatic Analysis

Sequence similarity searches of the GenBank, Swissprot and EMBL databases were conducted using the BLAST programs [14]. Searches for protein motifs were conducted using the Pfam database [15]. For phylogenetic analysis the RdRP protein sequences were aligned with MUSCLE as implemented by MEGA 6 [16], the alignment was improved manually and all positions with less than 30% site coverage were eliminated. Maximum likelihood phylogenetic trees were constructed in MEGA 6 using the LG + G + I substitution model. Structural models of the putative capsid proteins encoded by dsRNA 2 and dsRNA 4 were constructed using Phyre2 [17] and molecular graphics images were produced using the UCSF Chimera package from the Computer Graphics Laboratory, University of California, San Francisco (supported by NIH P41 RR-01081) [18].

2.5. SDS-Polyacrylamide Analysis of Purified Virus and Peptide Mass Fingerprinting (PMF)

Proteins obtained by the virus purification procedure from virus-infected strain W10 were analysed by gradient 4–15% SDS-PAGE on Mini-PROTEAN precast gels (BIORAD, Inc., Hercules, CA, USA) stained with the highly sensitive fluorescent SYPRO® Ruby protein gel stain (Thermo-Fisher Scientific, Waltham, MA, USA) and visualised using the Fujifilm FLA-5000 Fluorescent Image Analyser. The PageRuler™ Prestained Protein Ladder (Thermo-Fisher Scientific, Waltham, MA, USA) was digitalised using EPSON Scan. Proteins were digested with trypsin and subjected to PMF broadly as described previously [7].

3. Results

3.1. Isolation of a dsRNA Mycovirus and Biological Comparison Of Virus-Infected And Virus-Free Isogenic Strains

Following extraction of dsRNA from the Chinese *L. biglobosa* isolate W10, it was discovered that it contained four dsRNA elements 4.9–4.0 kbp in size designated as dsRNA 1 to dsRNA 4, based on their gel mobility. Similar electrophoretic patterns have been reported for 10 more *L. biglobosa* isolates originating from China and the UK [9]. A representative agarose electrophoresis of the dsRNA elements isolated from *L. biglobosa* isolate W10 is shown in Figure 1A. As described previously, isolate W10 was successfully freed of infection with mycovirus dsRNAs using a combination of cycloheximide treatment and hyphal tipping to generate strain W10-VF-1 which was confirmed by electrophoretic isolation of dsRNA, Northern blotting and RT-PCR [9]. Comparison of the colony morphologies of the isogenic *L. biglobosa* W10 and W10-VF-1 lines revealed significantly different phenotypes associated with virus infection (Figure S1). A comparison of the mycelial growth rates of the two isogenic lines by examining radial expansion growth curves and biomass production emphatically demonstrated that infection of *L. biglobosa* with the mycovirus results in increased growth rate, as illustrated in Figure S1, and concomitant hypervirulence of the host fungus, phenomena which have been investigated in detail elsewhere [9].

3.2. Virus Particles, Genomic dsRNAs, and Their Sequences

Purified virus particles had an icosahedral structure and particle diameter was estimated to be ca. 40–45 nm (Figure 1B) and most particles were peripherally penetrated by the stain giving them a doughnut-like appearance. The chemical nature of purified preparations of dsRNA, either prepared directly from mycelia or isolated from virions, was confirmed by its insusceptibility to DNase 1 and S1 nuclease. The agarose gel electrophoretic patterns of the four dsRNA species with size estimates of 4.9, 4.4, 4.4, and 4.0 kbp were reminiscent of those seen for two members of the family *Quadriviridae*, *Rosellinia necatrix* quadrivirus strains W1118 and W1075, where the accumulation of dsRNA 1 and dsRNA 4 were also consistently less than those of dsRNA 2 and dsRNA 3 (Figure 1A). Based on these findings, the virus under investigation was nominated Leptosphaeria biglobosa quadrivirus-1 (LbQV-1).

A random cDNA library prepared from a mixture of LbQV-1 dsRNAs 1 to 4 was constructed. Approximately 50 clones with cDNA inserts of 500–1000 bp were selected, sequenced and assembled into contigs. The contigs covered almost the entire length of each dsRNA segment and gaps were filled by RT-PCR using oligonucleotides based on known sequences to generate amplicons, which were subsequently cloned and sequenced [4]. Terminal sequences of the four dsRNAs were determined by 3′-RLM-RACE and fragments for each end of both strands of the dsRNAs were cloned and sequenced. Because of heterogeneity at the extreme termini at least 12 clones were analysed, together with cloned internal amplicons and amplicons linking RACE clones with the assembled contigs.

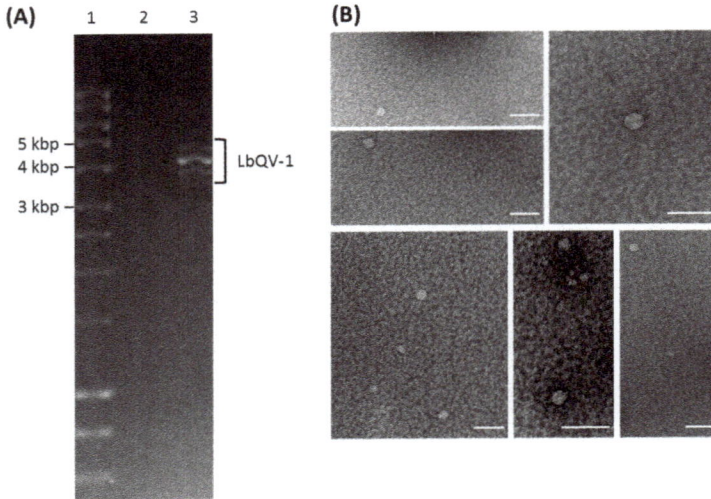

Figure 1. LbQV-1 particle morphology and dsRNA genome. (**A**) Agarose gel electrophoretic fractionation of the LbQV-1 genomic dsRNA segments. RNA was isolated from purified LbQV-1 preparations and the dsRNA profile of the virus-infected Chinese isolate W10 is shown in lane 3 while a dsRNA preparation from the same *L. biglobosa* isolate freed from virus infection and nominated strain W10-VF-1 is shown in lane 2. Lane 1 contains the DNA marker Hyperladder (Bioline), the sizes of which are shown to the left of the gel. (**B**) Electron micrograph of LbQV-1. Virus particles purified from *Leptosphaeria biglobosa* strain W10 were examined in the electron microscope after staining with uranyl acetate. Size bars indicate 100 nm.

The complete nucleotide sequences of the dsRNA 1-4 segments are 4917, 4543, 4490, and 4048 bp in size and each respectively contains an ORF on the plus-strand, potentially encoding proteins 1559 amino acids (aa; 172 kDa), 1383 aa (152 kDa), 1367 aa (153 kDa), and 1111 aa (120 kDa) in size, flanked by 5′- and 3′-untranslated regions (UTRs). Following the nomenclature adopted for other quadriviruses [7,8], these four proteins were nominated P1–P4. The 5′-UTRs of the four segments are respectively 48, 69, 42 and 58 bp long while the 3′-UTRs are 189, 322, 344 and 654 bp long (Figure 3; Table 1). Irrespective of which LbQV-1 dsRNA was used as template all 3′-RLM-RACE clones corresponding to the 5′-terminus of the plus-sense strand had sequences 5′-N/ACGA- (Figure 4) and are identical to the 5′-terminal sequences of the prototype *Rosellinia necatrix* quadrivirus strains W1118 and W1075 (RnQV1-W118 and W1075; [7,8]) and four large (L) dsRNAs isolated from plants infected with Amasya cherry disease (ACD-L), which probably represent the incomplete genomes of two closely-related quadriviruses of cherry or fungal origin [19]. However, the 3′-terminal sequences of the plus-sense strand of the LbQV-1 dsRNAs were highly variable and were extremely heterogeneous (Figure 4). Interestingly, however, there was significant homology between regions upstream of the 3′-termini of the LbQV-1 dsRNAs and these regions also contained characteristic sequence motifs including some small stem loop structures that were also identified in the 3′-UTRs of two isolates of RnQV1 and ACD-L dsRNAs 1–4. Heterogeneity at the extreme 3′termini is also a feature of the genomes of both RnQV1 and ACD-L dsRNAs 1–4.

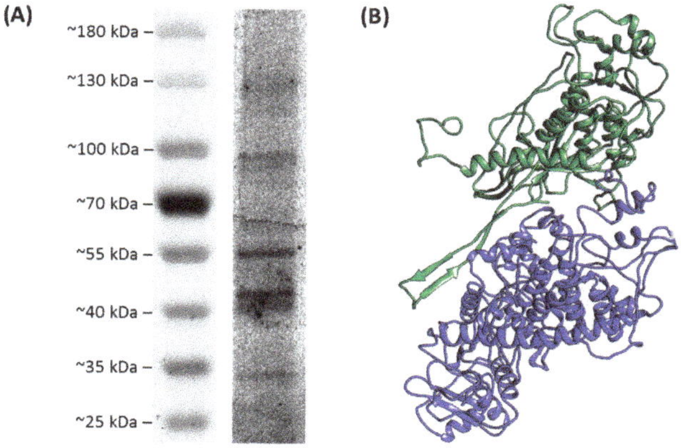

Figure 2. (**A**) SDS-PAGE pattern and modelling of LbQV-1 structural proteins. Proteins in a purified preparation of LbQV-1 were separated using a 4–25% gradient SDS-PAGE, stained with SYPRO® Ruby and visualized by using the Fujifilm FLA-5000 fluorescent image analyser. The prestained protein ladder molecular weight sizes are shown to the left of the gel, following digitalization with EPSON Scan. (**B**) The partial structures and interaction between the two LbQV-1 structural proteins were predicted and visualized; the green and the blue protein domains are encoded by dsRNA 2 and dsRNA 4, respectively.

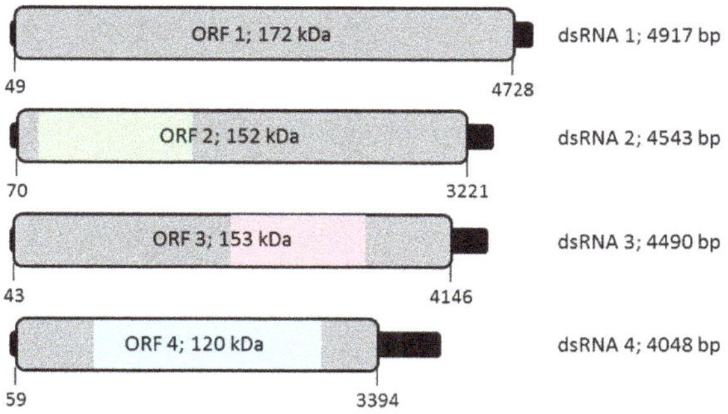

Figure 3. Schematic representation of the genomic organisation of LbQV-1. The LbQV-1 genome consists of four dsRNA segments, 4917, 4543, 4490, and 4048 bp in size, each containing one ORF shown as an open box flanked by 5′- and 3′-UTRs. A pink box on dsRNA 3 represents the RdRP domain. A blue and a green box on dsRNA 2 and dsRNA 4, respectively, illustrate the domains of the putative capsid proteins that were structurally modelled, as shown in Figure 2.

5′-terminal sequences
```
dsRNA1 5'-UACGAAAAACACAAUAUCACCUACAGUACCACGCCGACCAGAGACAGCAUGATCATCGGCAATATCGACTCC
dsRNA2 5'-CACGAAUUACCAAGAUAGACCACUUGUACUUCAAUAACUUCAACACACAACGCAGACGUUAACGCAAACAUG
dsRNA3 5'-CACGAUUUACUGACAUAUACCAAUAGUACAUCACACAUUUGAAUGGCACACCACGAATTCATAAGCAAGGCG
dsRNA4 5'-AACGAAUUACCAGAUACUACAAACUGUAUUUCAACCAAACGAACAAAACAUCUGCAAUAUGGACAGCATAAT
               ****...**          **...   ***. .*.    *     .*
```
3′-terminal sequences
```
dsRNA1 UGA...3nt.UCCCCCAUGGCUCCCAAGUGGUGGAACCCG--AAAGGGUGAGAAC-----------------3' 4917
dsRNA3 UAG.270nt.UCCCCCAUGGCGCCCAAGUGGUGGAGCUUU--AAUUAGCGAGAACGAGACUGU-AUAUGCG-3' 4543
dSRNA2 UAA.286nt.GCCCCCACGGCUCCCAAGUGGUGGAACCAGGCAACUGGUGAGAACGAGCCAA----------3' 4490
dSRNA4 UAG.593nt.GCCCUCAUGGCUCCCAAGUGAUGGAACCAGGAAACUGGUGAGAACGAAACAGGAAAACGCA-3' 4048
                     ***.**.*** ********.****.*.      **   .*.******
```

Figure 4. Terminal sequence domains of the LbQV-1 genome. Sequences of the 5′- (**A**) and 3′-termini (**B**). Identical sequences are denoted by asterisks and sequences present in $\frac{3}{4}$ dsRNAs are shown as dots. CAA repeats in the 5′terminal region are underlined and the possible initiation AUG codons and termination codons UGA, UAA and UAG are shown in italics.

Table 1. Features of the LbQV-1 dsRNA segments 1–4 and comparisons with other known and suspected quadriviruses.

Segment	Virus	Nucleotide Sequence		UTR (nt)		ORF Properties		Accession Number
		Size (nt)	Identity (%)	5′-end	3′-end	Size (nt)	Size (aa)	
dsRNA1	RnQV1-W1075	4942	22	39	94	4809	1602	AB620061
	RnQV1-W1118	4971	20	39	126	4806	1601	AB744677
	ACD-L-1	5121	27	35	199	4867	1628	AM085136
	ACD-L-2	5047	26	35	199	4862	1620	AM085137
	GaTV-1	4218	28	91	-	-	850	GU108585
	LbQV-1	4917	100	48	189	4725	1559	LR028028
dsRNA2	RnQV1-W1075	4352	23	106	175	4071	1356	AB620062
	RnQV1-W1118	4307	22	41	195	4071	1357	AB744678
	LbQV-1	4543	100	69	322	4149	1383	LR028029
dsRNA3	RnQV1-W1075	4099	35	25	141	3933	1310	AB620063
	RnQV1-W1118	4093	35	23	140	3930	1309	AB744679
	ACD-L-3	4458	45	59	307	3992	1363	AM085134
	ACD-L-4	4303	44	73	345	3885	1294	AM085135
	LbQV-1	4490	100	42	344	4149	1367	LR028030
dsRNA4	RnQV1-W1075	3685	24	74	425	3186	1061	AB620064
	RnQV1-W1118	3468	24	45	246	3177	1058	AB744680
	LbQV-1	4048	100	58	654	3333	1111	LR028031

3.3. Assignment of Structural Protein Genes

SDS-PAGE analysis of purified LbQV-1 showed the presence of two major proteins corresponding to ca. 130 kDa and 100 kDa as well as minor proteins of ca. 55 kDa, 40 kDa, and 35 kDa (Figure 2A). It is unclear whether all the observed proteins are viral since some of them may be derived from the host. Following PMF, four and six tryptic peptides were found to be derived from the N-terminal amino acid sequence encoded by dsRNAs 2 and 4, respectively (Table S1), verifying the presence of these proteins in the purified LbQV-1 preparation. Further PMF analysis of the proteins displayed on SDS-PAGE was not pursued but, by analogy with the full molecular characterization of the prototype quadrivirus RnQV1 [7,8], it can be assumed that the smaller proteins represent proteolytic degradation products of both proteins encoded by LbQV-1 dsRNA 2 and dsRNA 4 plus some host proteins. The known cryo-EM structure at 3.7 Å resolution of RnQV1 (PDBID 5ND1) was used as template for predicting the structure of the LbQV-1 putative capsid proteins and their interactions (Figure 2B).

3.4. Amino Acid Similarities and Phylogenetic Analysis

PSI-BLAST searches of the amino acid sequences of LbQV-1 P1–P4 generally showed modest similarities to the amino acid sequences of equivalent proteins predicted from the nucleotide sequences of known and suspected quadrivirus dsRNAs (Table 1). LbQV-1 P1 was distantly related to RnQV1-W1075 and RnQV1-W1118 (22% and 20% identity), ACD-L dsRNA 1 and ACD-L dsRNA 2 (26% and 27% identity) P1 equivalents and the sequence of one dsRNA component of a grapevine associated totivirus 1 (GaTV-1; 28% identity; [20]). Similarly a PSI-BLAST search with P2 showed only 23% and 22% amino acid sequence identity to equivalent proteins of RnQV1-W1118 and RnQV1-W1075. The results of PSI-BLAST searches with LbQV-1 P3 showed that it shared significant sequence similarity to known and suspected quadriviruses and was more closely related to the putative RdRPs encoded by ACD-L dsRNA 3 and -L dsRNA 4 (44% and 45% identity) and cherry chlorotic rusty spot (CCRS) L dsRNA 3 and -L dsRNA 4 (43% and 44% identity) as compared to RnQV1-W1118 and RnQV1-W1075 (35% and 36% identity). The PSI-BLAST search also showed that LbQV-1 P3 shares modest amino acid sequence identities of 20% to 30% to RdRPs from other mycoviruses within the families *Totiviridae*, *Chrysoviridae*, and *Megabirnaviridae* and an alignment of representative RdRP domains was generated (Figure S2; [21]). A Pfam search showed that LbQV-1 P3 belongs to the viral RdRP family (RdRp_4).

Sequences of representative RdRPs from fungal dsRNA viruses were used to construct maximum likelihood phylogenetic trees. As shown in Figure 5, the tree placed LbQV-1 P3 together with RdRPs encoded by ACD-L dsRNA 3, -L ds RNA 4, RnQV1-W1118, RnQV1-W1075, and GaTV-1 in a separate clade from other fungal dsRNAs. A PSI-BLAST search with LbQV-1 P4 showed only 24% amino acid sequence identity to equivalent proteins of both RnQV1-W1118 and RnQV1-W1075.

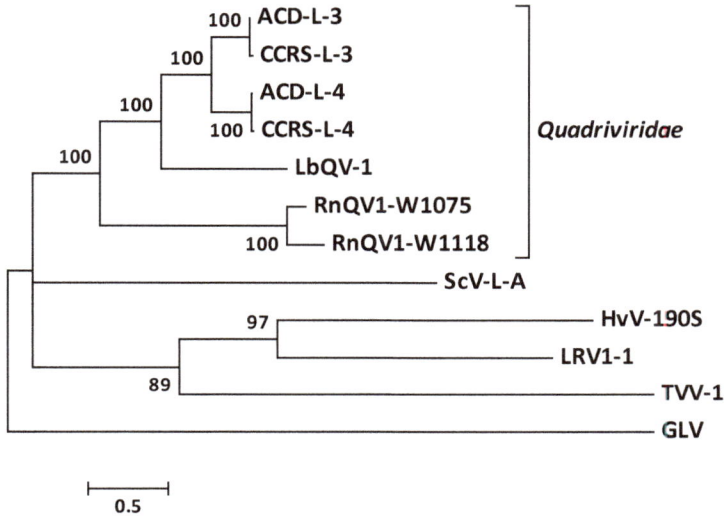

Figure 5. Phylogenetic analysis of Leptosphaeria biglobosa quadrivirus-1 (LbQV-1) RdRP and other known and suspected quadriviruses, together with selected members of the family *Totiviridae* based on the amino acid sequences of their RdRPs. The phylogenetic tree was constructed as described in the text. Only bootstrap percentages >50 are shown. Abbreviations and GenBank accession numbers for the sequences of the viral RdRPs used in the phylogenetic analysis are as follows: Amasya cherry disease-associated large dsRNA 3 (ACD-L-3; AM085134), Amasya cherry disease-associated large dsRNA 4 (ACD-L-4; AM085135), cherry chlorotic rusty spot-associated large dsRNA 3 (CCRS-L-3; AM181141), cherry chlorotic rusty spot-associated large dsRNA 4 (CCRS-L-4; CAJ57274), Rosellinia necatrix quadrivirus-1 strain W1118 (RnQV1-W1118; AB620063), Rosellinia necatrix quadrivirus-1 strain W1075 (RnQV1-W1075; AB744679), Saccharomyces cerevisiae virus L-A (ScV-L-A; J04692), Helminthosporium victoriae virus 190S (HvV-190S; U41345), Leishmania RNA virus 1-1 (LRV1-1; M92355), Trichomonas vaginalis virus-1 (TVV-1; U57898), and Giardia lamblia virus, GLV (L13218).

4. Discussion

The *Quadriviridae* is a family of non-enveloped spherical viruses with quadripartite dsRNA genomes of 3.5–5.0 kbp, comprising 16.8–17.1 kbp in total [22], which includes one genus with two strains of a single prototype species RnQV1 [7,8]. Here we have characterised a novel quadrivirus LbQV-1 from the filamentous fungus *L. biglobosa*, which shares some, but not all, biological and molecular features of the *Quadriviridae*. For instance, unlike RnQV1, LbQV-1 can be transmitted intracellularly via anastomosis and vertically through conidia, and it elicits alterations to the growth and pathogenicity of *L. biglobosa* resulting in the uncommon occurrence of hypervirulence (Figure S1; [9]). As with RnQV1, it is not known whether LbQV-1 has an extracellular phase. Transcription and replication are presumably carried out in a spherical particle by virion-associated RdRP.

Similar to RnQV1, LbQV-1 particles consist of rigid spherical virions ca. 40–45 nm in diameter (Figure 1B) which are encapsidated by two major structural proteins P2 and P4 encoded by dsRNA 2 and dsRNA 4, respectively (Figure 3). In both RnQV1 strains W1075 and W1118, the P2 and P4 proteins, which are more than 80% identical, form an asymmetric hetero-dimer subunit, and 60 of these build a T = 1 capsid [23]. We assume that LbQV-1 possesses a similar virion structure, having demonstrated moderate sequence identity between the equivalent proteins involved (Figure 3; Table 1). It is not known whether LbQV-1 P2 and P4 have acquired new functions through the insertion of complex domains at preferential insertion sites on the capsid outer surface as has been demonstrated for RnQV1 [23–25]. Both RnQV1 strains and LbQV-1 possess four linear dsRNA genome segments, termed dsRNA 1 to dsRNA 4 in a decreasing order of length from 5.0–3.5 kbp and each contains a single large ORF on the positive-sense strand of each dsRNA genomic segment (Figure 3). It is anticipated that each genomic dsRNA of LbQV-1 and both strains of RnQV1 are encapsidated separately [7,8]. LbQV-1, RnQV1-W1118, and RnQV1-W1075 genomic dsRNAs all had the same 5′-terminal motif sequence of 5′-N/ACGA- (Figure 4). The same motif is also present in the ACD-L dsRNAs, which probably represent the dsRNA 1 and 3 segments of two closely-related quadriviruses of cherry or fungal origin [19]. The lengths of all of the 5′-UTRs of fully sequenced quadrivirus dsRNAs are relatively small ranging in size from 35 to 106 nucleotides and those in both isolates of RnQV1 and LbCV-1 dsRNA 2 and 4 segments contain $(CAA)_n$ repeats which are presumably translational enhancers, as reported for chrysoviruses [26]. Examination of the 3′-UTRs of the four dsRNA segments of two strains of RNQV1, LbQV-1 and ACD-L L reveal heterogeneity in the very terminal end sequences (Figure 4). However, as compared to RNQV1 and ACD-L the sequence heterogeneity found in all four LbQV-1 dsRNAs is extensive and regions of identity between them were only discovered some distance upstream of the very terminal end sequences (Figure 4). We believe that these observations for LbQV-1 and the discovery of an extraordinarily 645 nt long 3′-UTR for dsRNA 4 are not artefactual as we obtained the same results with a large number of representative clones in several different experiments. The occurrence of a large number of repeat sequences with the motif $CA(A)_{n = 2-8}$ towards the 5′-terminus of the LbQV-1 dsRNA 4 3′-UTR is another novel feature for a quadrivirus the significance of which is unknown.

Sequence and phylogenetic analysis using the P3 sequence, which encoded the RdRP, revealed that LbQV-1 was most closely related to known and suspected quadriviruses and members of the genus *Totivirus* in the family *Totiviridae* [7,8] rather than other quadripartite mycoviruses, including chrysoviruses and alternaviruses (Figure 5; Figure S2; [7,8]). Of the remaining LbQV-1 encoded proteins, whereas LbQV-1 P2 and P4 are structural proteins the function of P1 is unknown. Based on the findings presented in this report, we propose that LbQV-1 is a novel member of the genus *Quadrivirus* in the family *Quadriviridae*.

Supplementary Materials: The following are available online at http://www.mdpi.com/1999-4915/11/1/9/s1, Figure S1: Colony morphology of virus-infected and virus-free *L. biglobosa* isolate W10 and W10-VF-1, respectively, on PDA plates following incubation at 20 °C for 26 days. Figure S2: Alignment of the region containing conserved motifs in the LbMV-1 dsRNA 3 RdRP with the RdRPs of some related dsRNA viruses. Table S1: Tryptic peptides derived from LbQV-1 P2 and P4 following PMF of the purified virus.

Author Contributions: I.K.-L., B.D.L.F., and R.H.A.C. conceived and designed the experiments; U.A.S. and I.K.-L. performed the experiments; U.A.S., B.D.L.F., and R.H.A.C. analysed the data; and I.K.-L. and R.H.A.C. wrote the paper.

Funding: Additional financial support for B.D.L.F. was provided by the Biotechnology and Biological Sciences Research Council (BBSRC)/ERA-CAPS, grant ref: BB/N005112/1, BB/M028348/1, BBP00489X/1, and BB/I017585/2.

Acknowledgments: We would like to thank Yongju Huang for supplying UK *Leptosphaeria* isolates, Guoqing Li for supplying Chinese *L. biglobosa* isolates, Georgia Mitrousia for advice on their identification using PCR amplification procedures, and Hadrien Peyret for TEM.

Conflicts of Interest: The authors declare no conflict of interest.

References

1. Sato, Y.; Caston, J.R.; Suzuki, N. The biological attributes, genome architecture and packaging of diverse multi-component fungal viruses. *Curr. Opin. Virol.* **2018**, *33*, 55–65. [CrossRef] [PubMed]

2. Aoki, N.; Moriyama, H.; Kodama, M.; Arie, T.; Teraoka, T.; Fukuhara, T. A novel mycovirus associated with four double-stranded RNAs affects host fungal growth in *Alternaria alternata*. *Virus Res.* **2009**, *140*, 179–187. [CrossRef] [PubMed]

3. Kozlakidis, Z.; Herrero, N.; Ozkan, S.; Kanhayuwa, L.; Jamal, A.; Bhatti, M.F.; Coutts, R.H.A. Sequence determination of a quadripartite dsRNA virus isolated from *Aspergillus foetidus*. *Arch. Virol.* **2013**, *158*, 267–272. [CrossRef]

4. Kotta-Loizou, I.; Coutts, R.H.A. Studies on the virome of the entomopathogenic fungus *Beauveria bassiana* reveal novel dsRNA elements and mild hypervirulence. *PLoS Pathog.* **2017**, *13*. [CrossRef] [PubMed]

5. Kanhayuwa, L.; Kotta-Loizou, I.; Özkan, S.; Gunning, A.P.; Coutts, R.H.A. A novel mycovirus from *Aspergillus fumigatus* contains four unique dsRNAs as its genome and is infectious as dsRNA. *Proc. Natl. Acad. Sci. USA* **2015**, *112*, 9100–9105. [CrossRef] [PubMed]

6. Fitt, B.D.L.; Brun, H.; Barbetti, M.J.; Rimmer, S.R. World-wide importance of phoma stem canker (*Leptosphaeria maculans* and *L. biglobosa*) on oilseed rape (*Brassica napus*). *Europ. J. Plant Path.* **2006**, *114*, 3–15. [CrossRef]

7. Lin, Y.-H.; Chiba, S.; Tani, A.; Kondo, H.; Sasaki, A.; Kanematsu, S.; Suzuki, N. A novel quadripartite dsRNA virus isolated from a phytopathogenic filamentous fungus, *Rosellinia necatrix*. *Virology* **2012**, *426*, 42–50. [CrossRef] [PubMed]

8. Lin, Y.-H.; Hisano, S.; Yaegashi, H.; Kanematsu, S.; Suzuki, N. A second quadrivirus strain from the phytopathogenic filamentous fungus *Rosellinia necatrix*. *Arch. Virol.* **2013**, *158*, 1093–1098. [CrossRef]

9. Shah, U.A.; Kotta-Loizou, I.; Fitt, B.D.L.; Coutts, R.H.A. Mycovirus induced hypervirulence of *Leptosphaeria biglobosa* enhances systemic acquired resistance to *Leptosphaeria maculans* in *Brassica napus*. *Mol. Plant Microbe Int.* under revision.

10. Liu, S.Y.; Liu, Z.; Fitt, B.D.L.; Evans, N.; Foster, S.J.; Huang, Y.J.; Latunde-Dada, A.O.; Lucas, J.A. Resistance of *Leptosphaeria maculans* (phoma stem canker) in *Brassica napus* (oilseed rape) induced by *L. biglobosa* and chemical defence activators in field and controlled environments. *Plant Pathol.* **2006**, *55*, 402–412. [CrossRef]

11. Diaz-Ruiz, J.R.; Kaper, J.M. Isolation of viral double-stranded RNAs using a LiCl fractionation procedure. *Prep. Biochem.* **1978**, *8*, 1–17. [CrossRef] [PubMed]

12. Froussard, P. A random-PCR method (rPCR) to construct whole cDNA library from low amounts of RNA. *Nucleic Acids Res.* **1992**, *20*, 2900. [CrossRef] [PubMed]

13. Coutts, R.H.A.; Livieratos, I.C. A rapid method for sequencing the 5′- and 3′-termini of dsRNA viral templates using RLM-RACE. *J. Phytopathol.* **2003**, *151*, 525–527. [CrossRef]

14. Altschul, S.F.; Madden, T.L.; Schäffer, A.A.; Zhang, J.; Zhang, Z.; Miller, W.; Lipman, D.J. Gapped BLAST and PSI-BLAST: A new generation of protein database search programs. *Nuc. Acids Res.* **1997**, *25*, 3389–3402. [CrossRef]

15. Finn, R.D.; Bateman, A.; Clements, J.; Coggill, P.; Eberhardt, R.Y.; Eddy, S.R.; Heger, A.; Hetherington, K.; Holm, L.; Mistry, J.; et al. Pfam: The protein families database. *Nucleic Acids Res.* **2014**, *42*, D222–D2230. [CrossRef] [PubMed]

16. Tamura, K.; Stecher, G.; Peterson, D.; Filipski, A.; Kumar, S. MEGA6: Molecular Evolutionary Genetics Analysis version 6.0. *Mol. Biol. Evol.* **2013**, *30*, 2725–2729. [CrossRef] [PubMed]

17. Kelley, L.A.; Mezulis, S.; Yates, C.M.; Wass, M.N.; Sternberg, M.J. The Phyre2 web portal for protein modeling, prediction and analysis. *Nat. Protoc.* **2015**, *10*, 845–858. [CrossRef] [PubMed]

18. Pettersen, E.F.; Goddard, T.D.; Huang, C.C.; Couch, G.S.; Greenblatt, D.M.; Meng, E.C.; Ferrin, T.E. UCSF Chimera—A Visualization System for Exploratory Research and Analysis. *J. Comput. Chem.* **2004**, *25*, 1605–1612. [CrossRef] [PubMed]

19. Kozlakidis, Z.; Covelli, L.; Di Serio, F.; Citir, A.; Açikgöz, S.; Hernández, C.; Ragozzino, A.; Flores, R.; Coutts, R.H.A. Molecular characterization of the largest mycoviral-like double-stranded RNAs associated with Amasya cherry disease, a disease of presumed fungal aetiology. *J. Gen. Virol.* **2006**, *87*, 3113–3117. [CrossRef] [PubMed]

20. Al Rwahnih, M.; Daubert, S.; Urbez-Torres, J.R.; Cordero, F.; Rowhani, A. Deep sequencing evidence from single grapevine plants reveals a virome dominated by mycoviruses. *Arch. Virol.* **2011**, *156*, 397–403. [CrossRef] [PubMed]

21. Bruenn, J.A. A closely related group of RNA-dependent RNA polymerases from double-stranded RNA viruses. *Nucleic Acids Res.* **1993**, *21*, 5667–5669. [CrossRef] [PubMed]

22. Chiba, S.; Castón, J.R.; Ghabrial, SA.; Suzuki, N. ICTV Report Consortium. ICTV virus taxonomy profile: *Quadriviridae*. *J. Gen. Virol.* **2018**, *9*. [CrossRef]

23. Luque, D.; Mata, C.P.; González-Camacho, F.; González, J.M.; Gómez-Blanco, J.; Alfonso, C.; Rivas, G.; Havens, W.M.; Kanematsu, S.; Suzuki, N.; et al. Heterodimers as the structural unit of the T = 1 capsid of the fungal double-stranded RNA *Rosellinia necatrix* Quadrivirus 1. *J Virol.* **2016**, *90*, 11220–11230. [CrossRef] [PubMed]

24. Mata, C.P.; Luque, D.; Gómez-Blanco, J.; Rodríguez, J.M.; González, J.M.; Suzuki, N.; Ghabrial, S.A.; Carrascosa, J.L.; Trus, B.L.; Castón, J.R. Acquisition of functions on the outer capsid surface during evolution of double-stranded RNA fungal viruses. *PLoS Pathog.* **2017**, *13*, e1006755. [CrossRef] [PubMed]

25. Luque, D.; Mata, C.P.; Suzuki, N.; Ghabrial, S.A.; Castón, J.R. Capsid structure of dsRNA fungal viruses. *Viruses* **2018**, *10*, 481. [CrossRef] [PubMed]

26. Ghabrial, S.A.; Castón, J.R.; Coutts, R.H.A.; Hillman, B.I.; Jiang, D.; Kim, D.H.; Moriyama, H. ICTV Report Consortium. ICTV virus taxonomy profile: *Chrysoviridae*. *J. Gen. Virol.* **2018**, *1*, 19–20. [CrossRef] [PubMed]

viruses

MDPI

Article

Novel Mitoviruses and a Unique Tymo-Like Virus in Hypovirulent and Virulent Strains of the Fusarium Head Blight Fungus, *Fusarium boothii*

Yukiyoshi Mizutani [1], **Adane Abraham** [2,†], **Kazuma Uesaka** [3], **Hideki Kondo** [2], **Haruhisa Suga** [4], **Nobuhiro Suzuki** [2] and **Sotaro Chiba** [1,5,*]

[1] Graduate School of Bioagricultural Sciences, Nagoya University, Nagoya 464-8601, Japan; m.yukiyosi@gmail.com

[2] Institute of Plant Science and Resources, Okayama University, Kurashiki 710-0046, Japan; adaneab2016@gmail.com (A.A.); hkondo@okayama-u.ac.jp (H.K.); nsuzuki@okayama-u.ac.jp (N.S.)

[3] Center for Gene Research, Nagoya University, Nagoya 464-8601, Japan; uesaka@gene.nagoya-u.ac.jp

[4] Life Science Research Center, Gifu University, Gifu 501-1193, Japan; suga@gifu-u.ac.jp

[5] Asian Satellite Campuses Institute, Nagoya University, Nagoya 464-8601, Japan

* Correspondence: chiba@agr.nagoya-u.ac.jp; Tel.: +81-(52)-789-5525

† Present address: Department of Biotechnology, Addis Ababa Science and Technology University, P.O. Box 16417, Addis Ababa, Ethiopia.

Received: 2 October 2018; Accepted: 23 October 2018; Published: 26 October 2018

Abstract: Hypovirulence of phytopathogenic fungi are often conferred by mycovirus(es) infections and for this reason many mycoviruses have been characterized, contributing to a better understanding of virus diversity. In this study, three strains of Fusarium head blight fungus (*Fusarium boothii*) were isolated from Ethiopian wheats as dsRNA-carrying strains: hypovirulent Ep-BL13 (>10, 3 and 2.5 kbp dsRNAs), and virulent Ep-BL14 and Ep-N28 (3 kbp dsRNA each) strains. The 3 kbp-dsRNAs shared 98% nucleotide identity and have single ORFs encoding a replicase when applied to mitochondrial codon usage. Phylogenetic analysis revealed these were strains of a new species termed Fusarium boothii mitovirus 1 in the genus *Mitovirus*. The largest and smallest dsRNAs in Ep-BL13 appeared to possess single ORFs and the smaller was originated from the larger by removal of its most middle part. The large dsRNA encoded a replicase sharing the highest amino acid identity (35%) with that of Botrytis virus F, the sole member of the family *Gammaflexiviridae*. Given that the phylogenetic placement, large genome size, simple genomic and unusual 3′-terminal RNA structures were far different from members in the order *Tymovirales*, the virus termed Fusarium boothii large flexivirus 1 may form a novel genus and family under the order.

Keywords: Fusarium head blight; mycovirus; RNA genome; mitovirus; *Tymovirales*; Ethiopia

1. Introduction

The number of extant fungal species is currently estimated at up to 1.5 million. Viruses that infect fungi are known as mycoviruses or fungal viruses and are ubiquitous across the kingdom Fungi [1]. Partitiviruses, and mitoviruses in particular, are omnipresent. Interestingly, mycoviruses were often found to have unique features such as genomic structures, virion structures, anti-viral defence strategies, etc. that differ from major plant and animal viruses [2–4]. Most mycoviruses infect fungal hosts asymptomatically and the remainder rarely exert a positive or negative impact on fungal host traits such as virulence, asexual/sexual spore production, pigmentation and growth [5–7]. Virocontrol, a method of biological control using mycoviruses that attenuate the virulence of phytopathogenic fungi (conferring hypovirulence), has attracted the attention of researchers for the last few decades [8–11], and information on mycovirus diversity has expanded and deepened dramatically through intensive mycovirus hunting aiming at screening for useful virocontrol agents [1]. However, while mycovirus hunting has been extensive in Asian, European, American and Oceanian countries, its exploration in the African continent has been limited: Diaporthe ambigua RNA virus (an unclassified RNA virus), Diplodia scrobiculata RNA virus 1 (a fusagravirus) and Sphaeropsis sapinea RNA virus 1 and 2 (victoriviruses) are reported in South Africa [12–16] with the remainder being mycovirus-like sequences obtained by metagenomic virome approaches [17].

Fusarium head blight (FHB) is an economically important disease of wheat caused by members of the *Fusarium graminearum* species complex (FGSC) that consists of at least 16 species including *F. boothii*. The disease symptoms are reduced yields, shrivelled kernels, reduced seed quality, and diminished seed weight. In addition, the pathogen leaves mycotoxins such as deoxynivalenol (DON) and nivalenol, type B trichothecenes, on grain surface which can cause diarrhoea and vomiting in animals ingesting them. An increasing number of mycoviruses have been reported to infect various *Fusarium* species, including FGSC from different regions of the world. These viruses include members of *Chrysoviridae, Hypoviridae, Partitiviridae, Narnaviridae, Totiviridae*, and *Mymonaviridae* families; proposed Alternaviridae and Fusariviridae families; and other unclassified virus groups [9,18–27]. Those with the greatest potential as virocontrol agents are: Fusarium graminearum virus 1 strain DK21 (FgV1-DK21), Fusarium graminearum virus strain China 9, Fusarium graminearum hypovirus 2 strain JS16, and Fusarium virguliforme dsRNA mycovirus 1 and 2 [18,21,23,28].

As abovementioned, mitoviruses in the genus *Mitovirus* in the family *Narnaviridae* are omnipresent in filamentous fungi and the members have approximately 3 kb positive sense single stranded RNA ((+)ssRNA) genomes that possess the single open reading frame (ORF) encoding an RNA-dependent RNA polymerase (RdRp) [29]. The ORF accommodates "UGA" triplets that is an interrupting stop codon when applied nuclear gene code but translated as tryptophan in the mitochondrial gene code. Cryphonectria parasitica mitovirus 1 (CpMV1) is the first mitovirus described, and co-presence of virus-derived double stranded RNA (dsRNA; considered as a replicative form) with subcellularly fractionated mitochondria suggests the localization of CpMV1 in mitochondria [30]. Khalifa and Pearson [31] observed the accumulation of filamentous structures in mitochondria of *Sclerotinia sclerotiorum* under electron microscope, specifically when a mitovirus infected to the host. Therefore, mitoviruses are expected to uniquely replicate in the host mitochondria. Botrytis cinerea mitovirus 1, Thanatephorus cucumeris mitovirus (TcMV, known also as Rhizoctonia M2 virus), Sclerotinia sclerotiorum mitovirus 1/KL-1, Sclerotinia sclerotiorum mitovirus 2/KL-1, Sclerotinia sclerotiorum mitovirus 3/NZ1, and Sclerotinia sclerotiorum mitovirus 4/NZ1 were shown to have a hypovirulence-conferring ability but effects of most mitoviruses on host fungal phenotypes have not been elucidated because mitoviruses are generally hard to cure and introduce [7,31–35]. Several mitoviruses have also been reported from *Fusarium* species including FGSC but biological properties of those are largely unknown [25,36,37].

The order *Tymovirales* contains viruses that have a non-segmented (+)ssRNA genome with single or multiple ORFs and its members mostly infect plants. The order consists of five families—*Alphaflexiviridae, Betaflexiviridae, Gammaflexiviridae, Deltaflexiviridae* and *Tymoviridae* in International Committee on Taxonomy of Viruses (ICTV)—based on the genome organisation, phylogenetic relationships and virion morphology. As of now, *Alpha-, Gamma-* and *Deltaflexiviridae* include mycoviruses and, to our knowledge, at least 10 such mycoviruses have been characterised including the ones associated with hosts that exhibit hypovirulent phenotypes [20,27,38–45].

Botrytis virus F (BVF, *Gammaflexiviridae*) was isolated from a hypovirulent strain of *Botrytis cinerea* as the first mycovirus taxonomically classified in the order *Tymovirales*, and was followed by Botrytis virus X (BVX, *Alphaflexiviridae*) [39,40]. *S. sclerotiorum* also hosts mycoviruses of this order, namely Sclerotinia sclerotiorum debilitation-associated RNA virus (SsDRV, *Alphaflexiviridae*) and Sclerotinia sclerotiorum deltaflexivirus 1 (SsDFV1, *Deltaflexiviridae*). The former is associated with hypovirulence of its host and the latter is recognised as the first mycovirus in the order *Tymovirales* that has no capsid protein (CP) [41,42]. Recently, Rhizoctonia solani flexivirus 1 (RsFV-1, unclassified deltaflexivirus-like) was reported to have the largest genome in the order and its unique genomic organisation broadened our perception of genome plasticity within the order [38]. In addition, two mycoviruses, Fusarium graminearum mycotymovirus 1 (FgMTV1, unclassified tymovirus-like) and Fusarium graminearum deltaflexivirus 1 (FgDFV1, *Deltaflexiviridae*), have been isolated from the FHB pathogen, *F. graminearum* [20,27]. Biological assays demonstrated that FgMTV1 infection does not alter the host virulence, but decreases the growth rate and DON production. Finally, although oyster mushroom spherical virus (OMSV) was earlier reported to be a tymo-like mycovirus [44], its taxonomic placement is inconclusive till date.

This study characterised a novel member of *Tymovirales*, potentially forming a new genus or family, and members of a new species in the genus *Mitovirus* in the family *Narnaviridae*. These virus classes were identified for the first time in *Fusarium boothii* strains from the African continent. It is noteworthy that the tymo-like virus possesses the largest genome size (over 12.5 kb) among the currently reported *Tymovirales* members. This study contributes to our understanding of *Tymovirales* virus and mitovirus diversity and evolution.

2. Materials and Methods

2.1. Fungal Strains and Culture

F. boothii strains (Ep-BL13, Ep-BL14 and Ep-N28) were isolated from wheats with the FHB symptoms in Hadiya region of southern Ethiopia. A Japanese isolate of *F. graminearum* sensu stricto (s.str.) strain was also used as a reference. All strains were maintained at 28 °C on potato dextrose agar (PDA, BD Difco Laboratories, Detroit, MI, USA) and synthetic low nutrient agar (SNA) and stored at 4 °C in the dark. Mycelia were cultured in potato dextrose broth (PDB, BD Difco Laboratories, Detroit, MI, USA) liquid medium at room temperature for 4–7 days for nucleic acid extractions.

2.2. Biological Assessment of FGSC Strains

Growth rate and colony morphology. Mycelial plugs were transplanted on PDA and SNA plates (three replicates) and incubated at 20 °C for 1 week. The longest and shortest diameters of each colony were measured 3–7 days post-transplanting and the mean diameter was used for the calculation of colony area.

Pathogenicity test. Conidial spores (1×10^5 spores/mL) suspended in a 1% (v/v) Triton solution were used for inoculation, and the FHB-sensitive wheat cultivar Apogee was grown at 27 °C in a chamber (KG-50HLA; Koito Electric Industries, Yokohama, Japan) to avoid possible release of inocula into the environment [46,47]. Briefly, 10 µL of spore suspension was injected into wheat spikelets, and the number of diseased spikelets was counted at 15 days post-inoculation. Alternatively, the spore suspension was sprayed three times (500 µL/push) to a wheat head, and disease severity was evaluated according to Ban and Suenaga's criteria (index, mild to severe; 0, 5, 10, 20, 30, 50, 60, 80, 100) at 15 days post-inoculation [48]. Inoculated plants were placed in a transparent humid box for one day (injection method) or two days (spray method) to promote fungal germination, prior to the maintenance in the chamber.

2.3. Genomic Polymerase Chain Reaction

Fungal genomic DNA was purified by conventional CTAB (cetyl trimethylammonium bromide) extraction. The partial internal transcribed spacer (ITS) of rDNA and transcription elongation factor 1-α gene (*TEF1-α*) sequences were PCR-amplified with the oligonucleotide primers reported by O'Donnel et al. [49,50]. Primers used in this study were: primer 1 (5′-TCAAAATGGGTAAGGA(A/G)GACAAGAC-3′) and primer 2 (5′-GCCTGGGA(G/A)GTAC CAGT(G/C)ATCATGTT-3′) for ITS; and primer 3 (5′-GTGGGGCATTTACCCCGCC-3′) and primer 4 (5′-GAGTGGCGGGGTAAATGCC-3′) for *TEF1-α*. Fungal species were identified by homology search with and by phylogenetic analysis with sequences from public databases.

2.4. dsRNA Extraction and cDNA Library Construction

dsRNA fraction was extracted from fungal mycelia by conventional cellulose column chromatography. Filtrated mycelia were homogenised in the presence of liquid nitrogen, and total RNA fractions were obtained by treatment with one round each of phenol, phenol–chloroform–isoamylalcohol (25:24:1), and chloroform–isoamylalcohol (24:1) extraction. dsRNAs were further isolated from the total RNA fractions by using Cellulose Powder C (Advantec, Tokyo, Japan). To eliminate fungal chromosomal DNA and ssRNA species, these fractions were treated with RNase-free RQ1 DNase I (Promega, Madison, WI, USA) and S1 Nuclease (Promega). Conventional cDNA library construction was performed for viral sequence determination. The dsRNAs extracted from *F. boothii* Ep-BL13, Ep-BL14 and EP-N28 and the 3 kbp- and 2.5-kbp dsRNA bands were gel-purified using a Zymoclean™ Gel RNA Recovery kit (Zymo Research, Irvine, CA, USA). cDNA libraries of each dsRNA fragment were constructed using a non-PCR and a PCR-based method. After denaturation of dsRNA templates at 99 °C for 5 min, cDNA was synthesised using ReverTra Ace-α-(Toyobo, Osaka, Japan) with an adapter-tagged primer (5'-CCTGAATTCGGATCCTCCNNNNNNN-3'). The resulting cDNA fragments were purified from an agarose gel with a size exclusion (1.0–2.0 kbp) and amplified by PCR with an adapter primer (5'-CCTGAATTCGGATCCTCC-3'). The PCR conditions are as follows: 94 °C (2 min), then 1 cycle at 94 °C (2 min)/65 °C (1 min)/72 °C (1 min), followed by 35 cycles at 94 °C (40 s)/52 °C (30 s)/72 °C (3 min), and a final extension at 72 °C for 7 min. DNA fragments were cloned into the pGEM-Teasy (Promega) or pCR-Blunt (Thermo Fisher, Waltham, MA, USA) cloning vectors and used for the transformation of *Escherichia coli* strain DH5α for Sanger sequencing analyses.

2.5. RNA Seq, Reverse Transcription PCR and Rapid Amplification of cDNA Ends

RNA seq. The *F. boothii* Ep-BL13 virome was established on an Illumina-Hiseq 2500 system (Illumina, San Diego, CA, USA) by a pair-end sequencing run (2 × 100 bp) using 10 ng of dsRNA as a template. cDNA and library construction were outsourced to BGI Japan. Fastqc version 0.11.5 software (https://www.bioinformatics.babraham.ac.uk/projects/fastqc/) was used to check the quality of the read data [51]. Reads were filtered using Sickle version 1.33 software (https://github.com/najoshi/sickle) by requiring an average Phred quality (Q score) of at least 30 and to only retain reads of 20 nucleotides (nt) or greater in length [52]. De novo assembly was carried out using SPAdes version 3.11.1 software (http://cab.spbu.ru/software/spades/) at the default settings [53].

Gap-filling RT-PCR. Target cDNAs were produced using ReverTra Ace and a gene-specific primer set from given dsRNA fragments and amplified with DNA polymerases such as PrimeSTAR (Takara, Shiga, Japan), KOD FX Neo (Toyobo) or GO-Taq (Promega).

3'-RLM-RACE (RNA ligase mediated RACE). The terminal sequences of dsRNAs were determined by this method. Pre-denatured dsRNAs in dimethyl sulphoxide (DMSO) (90%, 65 °C) were ligated at their 3'-ends with a 5'-phosphorylated oligodeoxynucleotide, 3'-RACE-adaptor (5'-CAATACCTTCTGACCATGCAGTGACAGTCAGCATG-3'), using T4 RNA ligase (Takara) at 16 °C for 16 h. Ligated DNA–RNA strands were DMSO-denatured in the presence of the oligonucleotide 3'-RACE-1st (5'-CATGCTGACTGTCACTGCAT-3') and used as templates for cDNA synthesis. The resulting cDNA was amplified by PCR with 3'-RNACE-2nd (5'-TGCATGGTCAGAAGGTATTG-3') and gene-specific primers. Plasmid clones and genomic- or RT-PCR amplicons were purified with spin columns (GeneElute, Sigma, St. Louis, MO, USA; Nucleospin, Takara) and used for BigDye (ABI, ThermoFisher) sequencing by following the manufacturer's instructions. Sanger sequencing was carried out on a 3100-Avant sequencer (ABI/HITACHI). All virus-specific oligonucleotide primers used in this study were designed based on sequences obtained from cDNA library sequencing and Illumina sequencing.

2.6. Bioinformatics Analyses

Sequence assemblies were manipulated using GeneStudio (http://genestudio.com/), SPAdes and Genetyx (Genetyx, Tokyo, Japan) software. Sequence homology searches were performed with the Basic Local Alignment Search Tool (BLAST) programme provided by the National Center for Biotechnology Information (NCBI, http://www.ncbi.nlm.nih.gov/) and the Genetyx. Motif searching was conducted using the online tool, InterPro: protein sequence analysis & classification (https://www.ebi.ac.uk/interpro/) [54]. For secondary RNA structure prediction, the 3′- and 5′-terminal viral sequences were submitted to mfold (http://unafold.rna.albany.edu/?q=mfold/rna-folding-form) and potential stem-loops were predicted [55].

2.7. Phylogenetic Analysis

Multiple amino acid or nucleotide sequence alignments were constructed using MUSCLE [56] and with the MEGA X [57]. Phylogenetic analysis was conducted by using PhyML ver. 3.0 (http://www.atgc-ontpellier.fr/phyml/) with the best-fit models suggested by Smart Model Selection in PhyML (http://www.atgc-montpellier.fr/sms/) [58], and the results were visualised with FigTree and further enhanced in Adobe Illustrator.

3. Results

3.1. Detection of dsRNAs in F. boothii Strains

The fungal strains Ep-BL13, Ep-BL14 and Ep-28 were isolated from wheats exhibiting FHB in Ethiopia by using *Fusarium*-selecting medium, which were found to be dsRNA-positive. Agarose gel electrophoresis of dsRNA fractions extracted from these strains showed a 3 kbp common to all three strains and additional >10 and 2.5 kbp bands only in the Ep-BL13 strain (Figure 1A), thereby suggesting infections by unknown RNA mycoviruses. These fungal strains were identified as *F. boothii* based on comparison of partial *TEF1-α* and ITS sequences (data not shown) and by phylogenetic analysis of partial *TEF1-α* sequences (Figure S1).

3.2. Biological Characteristics of F. boothii Strains

Colony morphology of the *F. boothii* strains on nutrient-rich (PDA) and -poor (SNA) solid media varied as shown in Figure 1B,C. The Ep-BL13 strain grew most slowly on PDA, followed by Ep-BL14 and then Ep-28 (Figure 1B,D). The strains also exhibited different phenotypes on SNA, with Ep-BL13 and Ep-N28 growing faster and slower than those on PDA, respectively (Figure 1C). The pathogenicity of the three *F. boothii* isolates and of a reference *F. graminearum* s.str. strain on wheat head was investigated with injection and spray methods. Plants inoculated with *F. boothii* Ep-BL14 and Ep-28 exhibited typical FHB symptoms similar to that of the plant infected by *F. graminearum* s.str., whereas the Ep-BL13 showed apparently weak virulence (Figure 1E,F). The effects of the dsRNA elements on host fungal growth and pathogenicity to wheats are yet to be determined.

3.3. Analysis of Viral Sequences: A Tymo-Like Virus

The genome organisation of the coding strand of the largest dsRNA segment harboured in Ep-BL13 is shown in Figure 2A. The complete genomic sequence of the virus was 12,579 nt in length and the G + C content of the dsRNA is 64.93% (accession No. LC425115). A BLASTX search of the whole nucleotide sequence showed that the highest amino acid (aa) identity (35%) was to the replicase of BVF, which is the sole member of the family *Gammaflexiviridae* (Table 1). Based on this result, the virus was tentatively named Fusarium boothii large flexivirus 1 (FbLFV1). The whole genome of FbLFV1 encompasses a single large ORF putatively encoding a polypeptide of 4070 aa (162–12,374 nt) with a calculated molecular mass of 443.6 kDa. However, BVF, the closest virus to FbLFV1 in a BLAST search, possesses two ORFs encoding a putative replicase (a potential readthrough product) and a putative

CP. Methyltransferase (Met), helicase (Hel) and RNA-dependent RNA polymerase (RdRp) domains of the FbLFV1 polypeptide showed low aa identities (22–36%, 20–38% and 25–37%, respectively) to those of representative viruses in the order *Tymovirales* and related mycoviruses, as shown in Table 1. Three replicase domains, Met, Hel and RdRp, are located in this order, from N to C terminus, and this is typical of the "alphavirus-like" superfamily that includes many other plant-infecting viruses (Figure 2A) [59]. There were two regions in the polypeptide with similarity to a PHA03247 domain (N-terminal part, E-value: $3.94 \times e^{-6}$; C-terminal part, E-value: $5.01 \times e^{-3}$), which was found in the large tegument protein UL36 of herpesviruses (a large dsDNA virus), although the function of PHA03247-like domains in FbLFV1 is unknown.

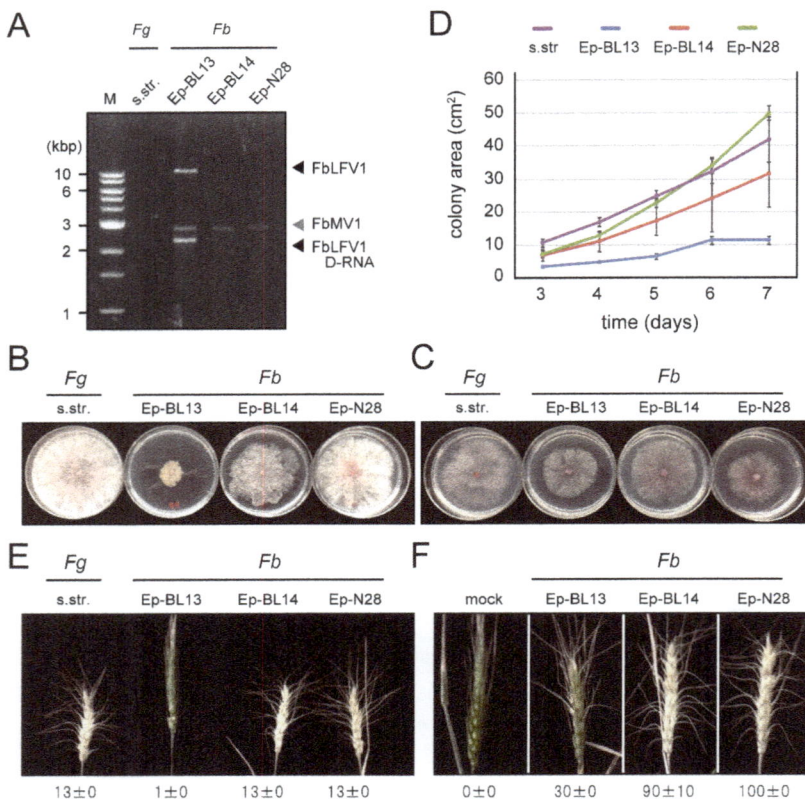

Figure 1. Biological properties of *F. boothii* strains and dsRNA-banding profiles. (**A**) The dsRNA-banding profiles of each *F. boothii* strain (Ep-BL13, Ep-BL14 or Ep-28) and a reference pathogenic strain. *F. graminearum* s.str. Three *F. boothii* strains all carried dsRNA fragments of approximately 3 kbp. Aside from the 3-kbp dsRNA band, Ep-BL13 harboured two additional dsRNA bands of over 10 kbp and about 2.5 kbp. (**B,C**) The colony morphologies of *F. boothii* strains and *F. graminearum* s.str. on PDA nutrition-rich media (**B**) and SNA nutrition-poor media (**C**) that were photographed at four days after transplanting. The three *F. boothii* strains exhibited various colony morphologies in terms of colony size, the amount of aerial mycelium and pigmentation. (**D**) Growth rate of *Fusarium* strains on PDA media. Colony sizes were measured at 3–7 days post-transplantation. (**E**) Pathogenicity test I. A wheat spikelet was inoculated with *Fusarium* strains by injection and photographed at 15 days post-inoculation. The number of spikelets exhibiting symptoms is presented. (**F**) Pathogenicity test II. Wheat photographs taken as in (**E**), but after spray inoculation. Disease indices (0–100) as per Ban and Suenaga (2000) [48] are shown.

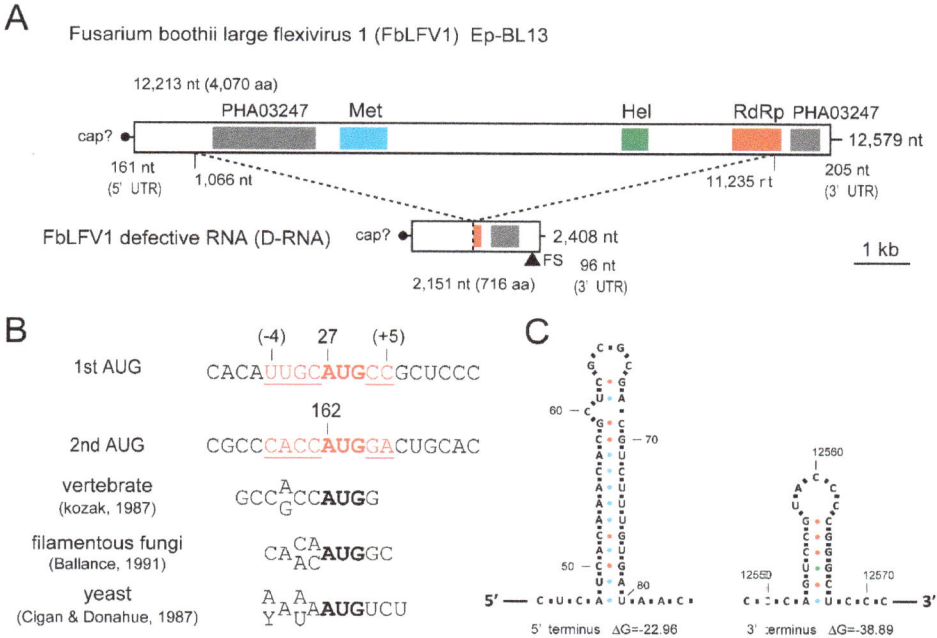

Figure 2. Schematic representation of the genome structure of the large *F. boothii* mycovirus. (**A**) Genome maps of FbLFV1. Thick bars are non-coding sequences and open boxes represent ORFs. Colour-highlighted boxes in ORFs represent the predicted protein domains as indicated. The length of the whole genome, ORF and UTR are given. The deleted region of FbLFV1 in defective RNA (D-RNA) is denoted by the dashed lines and nucleotide positions. Met, methyltransferase domain; Hel, helicase domain; RdRp, RNA-dependent RNA polymerase domain, FS, frame-shift. (**B**) Comparison of nucleotide contexts surrounding potential initiation codons (red bold letters, potential initiation codon; red underlined letters, surrounding nucleotides) of FbLFV1. Favourable contexts of initiation codon in vertebrate, filamentous fungi and yeast are shown as references. (**C**) Structure of terminal, non-coding RNA regions of FbLFV1. 5′- and 3′-terminal sequences of the genomes of the FbLFV1 were analysed with secondary structure prediction tools. Representatives of putative stem-loop structures were depicted. Dots in different colors indicate hydrogen bonds between different base pairs: blue, A-U pairs; red, G-C pairs; green, G-U pair.

The initiation codon at 162–164 nt of the large ORF was in a favourable context, CACCAUGGa, for filamentous fungi, although another potential initiation codon at 27–29 nt appeared in frame to the ORF with a weak context uugCAUGcG (capital and small letters represent good and poor residues in the context, respectively; underlined letters are the putative initiation codon; Figure 2B) [60–62]. FbLFV1 had short 5′ and 3′ untranslated regions (UTRs) that are 161 and 205 nt, respectively (Figure 2A). Although experimental evidence has not been obtained, the presence of a Met domain suggested the capping of 5′-end of FbLFV1 genomic RNA. The Met domain of the members of the order *Tymovirales* is generally located at the extreme N-terminal region of the replicase but that of FbLFV1 was distant from N-terminus of the polypeptide (aa positions 1261–1528) (Figure 2A).

Table 1. List of viruses in the order Tymovirales and related unassigned viruses with similarity to FbLFV1.

Taxon	Virus Name (Abbreviation)	Accession	Met *	Hel *	RdRp *
Alphaflexiviridae					
Allexivirus	Shallot virus X (SVX)	NC_003795	58/198 (29%)	37/154 (24%)	74/221 (33%)
Botrexvirus	Botrytis virus X (BVX)	NC_005132	60/196 (30%)	45/178 (25%)	76/221 (34%)
Lolavirus	Lolium latent virus (LLV)	NC_010434	64/193 (33%)	50/241 (20%)	80/238 (33%)
Mandarivirus	Indian citrus ringspot virus (ICRV)	NC_003093	59/162 (36%)	63/244 (25%)	84/229 (36%)
Platypuvirus	Donkey orchid symptomless virus (DOSV)	NC_022894	60/201 (29%)	49/189 (25%)	69/225 (30%)
Potexvirus	Potato virus X (PVX)	NC_011620	60/198 (30%)	58/245 (23%)	79/227 (34%)
Sclerodarnavirus	Sclerotinia sclerotiorum debilitation-associatedRNA virus (SsDARV)	NC_007415	63/198 (31%)	49/191 (25%)	71/227 (31%)
Betaflexiviridae					
Quinvirinae					
Carlavirus	Potato virus M (PVM)	NC_001361	55/217 (25%)	35/127 (27%)	62/224 (27%)
Foveavirus	Apple stem pitting virus (ASPV)	NC_001749	60/222 (27%)	17/45 (37%)	63/222 (28%)
Robigovirus	Cherry necrotic rusty mottle virus (CNRMV)	NC_002468	52/224 (23%)	44/159 (27%)	61/219 (27%)
Tritrvirinae					
Capillovirus	Apple stem grooving virus (ASGV)	NC_003462	64/214 (29%)	33/97 (34%)	72/231 (31%)
Chordovirus	Carrot Ch virus 1 (CChV)	NC_025649	54/214 (25%)	33/118 (27%)	66/225 (29%)
Citrivirus	Citrus leaf blotch virus (CLBV)	NC_003877	46/205 (22%)	46/160 (28%)	68/223 (30%)
Divavirus	Diuris virus A (DVA)	NC_019029	53/217 (24%)	26/90 (28%)	66/224 (29%)
Prunevirus	Apricot vein clearing associated virus (AVCSV)	NC_023295	49/183 (26%)	24/62 (38%)	66/224 (29%)
Tepovirus	Potato virus T (PVT)	NC_011062	56/186 (30%)	34/115 (29%)	67/219 (30%)
Trichovirus	Apple chlorotic leaf spot virus (ACLSV)	NC_001409	62/195 (31%)	43/172 (25%)	64/218 (29%)
Vitivirus	Grapevine virus A (GVA)	NC_003604	50/201 (24%)	22/57 (38%)	65/218 (29%)
Gammaflexiviridae					
Mycoflexivirus	Botrytis virus F (BVF)	NC_002604	63/193 (32%)	71/252 (28%)	82/224 (36%)
Deltaflexiviridae					
Deltaflexivirus	Sclerotinia sclerotiorum deltaflexivirus 1 (SsDFV1)	NC_038977	54/202 (26%)	61/253 (24%)	45/174 (25%)
	Sclerotinia sclerotiorum deltaflexivirus 2 (SsDFV2)	MH299810	63/222 (28%)	44/184 (23%)	59/207 (28%)
	Fusarium graminearum deltaflexivirus 1 (FgDFV1)	NC_030654	67/204 (32%)	37/157 (23%)	57/211 (27%)
	Soybean leaf-associated deltaflexivirus 1 (SlaDFV1)	KT598226	54/203 (26%)	63/260 (24%)	46/174 (26%)
Tymoviridae					
Maculavirus	Grapevine fleck virus (GFV)	NC_003347	57/203 (28%)	43/144 (29%)	54/143 (37%)
Marafivirus	Maize rayado fino virus (MRFV)	NC_002786	58/197 (29%)	57/201 (28%)	45/140 (32%)
Tymovirus	Turnip yellow mosaic virus (TUMV)	NC_004063	56/196 (28%)	58/214 (27%)	61/204 (29%)
Unassigned					
Mycotymovirus	Fusarium graminearum mycotymovirus 1 (FgMTV1)	KT360947	60/196 (30%)	57/197 (26%)	73/219 (33%)
Unassigned	Rhizoctonia solani flexivirus 1 (RnFV1)	NC_030655	57/198 (28%)	53/228 (23%)	65/224 (29%)
Unassigned	Rhizoctonia solani flexivirus 2 (RnFV2)	KX349069	N/A **	N/A **	50/177 (28%)
Unassigned	Oyster muschroom spherical virus (OMSV)	AY182001	48/189 (25%)	61/237 (25%)	73/219 (33%)
Unassigned	Fusarium boothii large flexivirus 1 (FbLFV1)	LC425115			

* Results of homology search against FbLFV1 for amino acid cover region and identity are shown. ** N/A indicates "not applicable" due to the unavailability of homologous sequences.

3.4. Presence of a Defective Segment of FbFV1 in the Ep-BL13

The cDNA-library-based sequencing of the approximately 2.5 kbp dsRNA revealed that the segment was of a defective RNA (D-RNA) of FbLFV1. The total length of the dsRNA was 2408 bp, and the positive-sense RNA retained an ORF encoding polypeptide with a potential molecular mass of 76.8 kDa (716 aa; Figure 2A) (accession No. LC425116). This occurred because of the loss of aa 303–3693 of the FbLFV1 large polypeptide (removal of FbLFV1 genome positions 1066–11,235 in-frame) and the addition of 36 aa because of a frame-shift from a single nucleotide deletion at −5 nt position relative to the original termination codon, leading to a new downstream stop codon. Since the RNA sequence was almost identical to the corresponding region of FbLFV1 and the junction of the upstream/downstream parts was uniform, this D-RNA of FbLFV1 might have arisen very recently, but at least before the isolation of Ep-BL13. The segment has the potential to facilitate viral replication because the coded protein retained partial RdRp motifs (122 aa residues of C-terminal end; Figure 2A); however, it is also possible that it may interfere with replication. As BVF is reported to carry similar D-RNA species with unknown function [39], it is interesting to investigate the significance of D-RNAs in life cycles of mycoviruses in the order *Tymovirales*.

3.5. Predicted RNA Structure of the Terminal Regions of FbLFV1 Genome

The UTRs of RNA viruses generally carry important *cis*-elements that form secondary/tertiary structures and/or distant base-pairing. In many cases, these are essential for encapsidation, transcription, translation and replication. To investigate whether FbLFV1 contains such structures, UTR sequences were subjected to secondary structure prediction using mfold software. The 5′ UTR of FbLFV1 did not consistently exhibit stem loops in several outputs, except for the one shown in Figure 2C. Similarly, those of the 3′ UTR were mostly inconsistent but a stem-loop predicted near the 3′-end was the most probable (Figure 2C). It is known that some members of *Tymoviridae*, such as turnip yellow mosaic virus (TYMV), have tRNA-like structures (TLS) in their 3′-UTR that play an important role in viral replication [63,64]. Given that TLSs are difficult to predict (e.g., the TYMV-TRL (6234–6318 nt, NC_004063) was not detected by tRNAscan-SE software [65]), more extensive base-pairing analysis of the FbLFV1 3′-end should be conducted before drawing any definitive conclusions.

3.6. Molecular Phylogenetic Analysis of FbLFV1

Amino acid sequences of whole FbLFV1 replicase and representative mycoviruses in the order *Tymovirales*, including taxonomically unassigned members, were aligned and the Met, Hel and RdRp domains were extracted (Table 1 and Figure 3). These three domains were found to be well conserved in the deduced protein sequence of FbLFV1 replicase. The evolutionarily conserved relationship of FbLFV1 to the *Tymovirales* members was characterised by maximum likelihood (ML) phylogenetic analyses based on the alignments of each of the three domains (Figure 4). As expected from BLAST analyses, the FbLFV1 Met Hel and RdRp domains were distantly associated with those of viruses in the order *Tymovirales*, with relatively close relationship to a gammaflexivirus (Figure 4A–C). The concatenate sequence of these domains finally revealed that FbLFV1 was placed independently from the clades of established families in the order *Tymovirales*, although the branching support value was not sufficiently high in the ML tree (Figure 4D). Based on these phylogenetic analyses, FbLFV1 is expected to belong to a novel virus species "Fusarium boothii large flexivirus 1" that may form an independent genus, or even a new virus family, in the order *Tymovirales*.

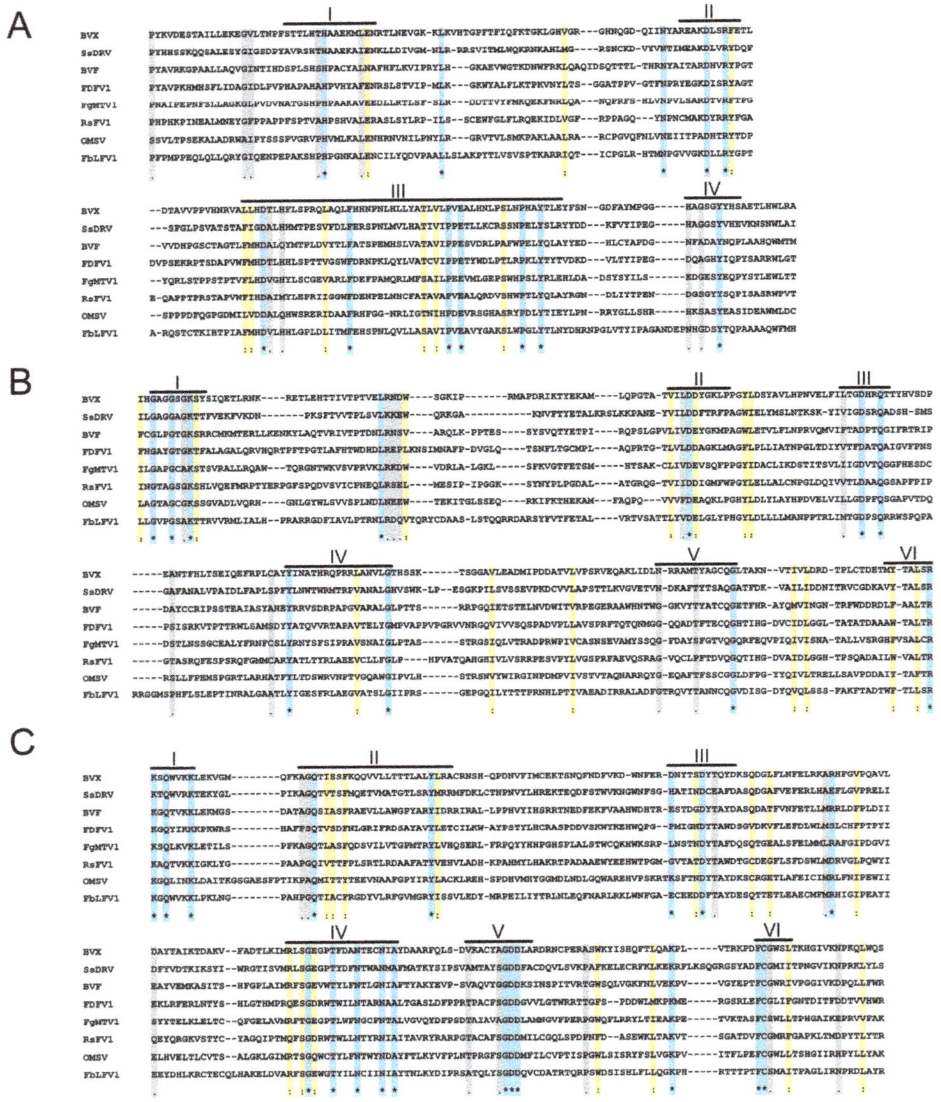

Figure 3. Amino acid alignment of Met, Het and RdRp domains of FbLFV1. The positions of the motifs are indicated by lines above the sequence alignments with the motif numbers from I to VI. Identical residues are colour-highlighted with blue and indicated by asterisks; conserved and semi-conserved amino acid residues are colour-highlighted with yellow and gray, and by colons and dots, respectively. (**A–C**) Deduced aa sequences corresponding to Met (**A**), Hel (**B**) and RdRp (**C**) domains were extracted from the FbLFV1 large polypeptide and aligned with those from mycoviruses of the order Tymovirales. Abbreviations of viral names are the same as in Table 1.

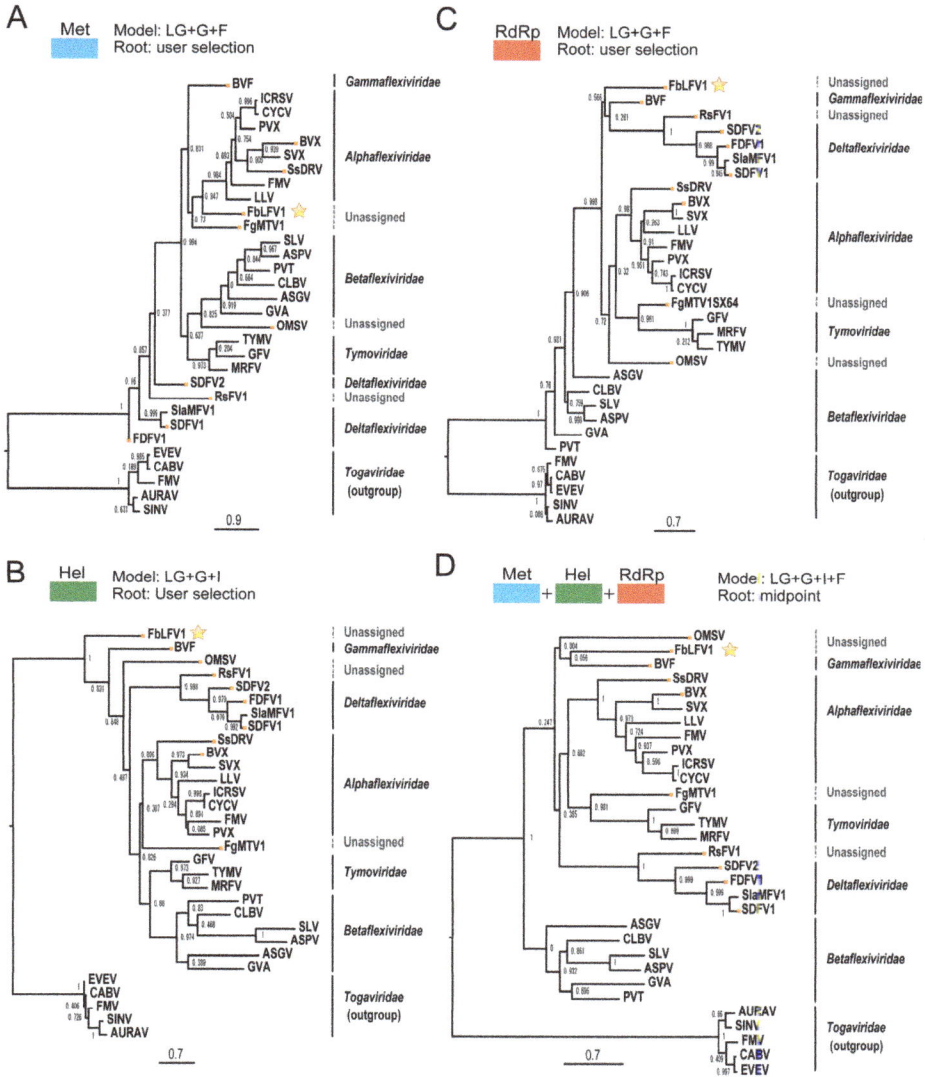

Figure 4. Phylogenetic analyses of FbLFV1. (**A–C**) Phylogenetic trees were constructed based on the alignment of aa sequences of Met (**A**), Hel (**B**) and RdRp (**C**) domains. The ML trees with the best model and rooting methods are shown. Numbers at the nodes represent branch supporting values with an aLRT SH-like method (>0.9 is significant). (**D**) Phylogenetic tree of Met + Hel + RdRp flanking sequences. Yellow dots indicate mycoviruses and stars indicate FbLFV1. The sequences used for (**A–C**) were joined and subjected to alignment and phylogenetic analyses as in (**A–C**). Abbreviations of viral names and accession numbers are listed in Table 1.

3.7. Analysis of Viral Sequences: Three Mitoviruses

The full-length sequences of the 3 kbp-dsRNAs in *F. boothii* strains Ep-BL13, Ep-BL14 and Ep-N28 were obtained by cDNA library sequencing and 3′ RLM-RACE analyses (accession Nos. LC425112, LC425113 and LC425114, respectively). The viral dsRNAs accumulated in strains Ep-BL13 and Ep-N28 were 2802 bp long and that in strain Ep-BL14 was 2801 bp long. Each positive sense strand sequence possessed a single ORF potentially encoding RdRp when using mitochondrial codon usage (Figure 5A). The length of the ORF and 3′-UTRs was the same among three sequences, being 2472 nt (823 aa) and 136 nt, respectively (Figure S2). A few nucleotide deletions at different positions in the 5′-UTRs were found in the three viral sequences; nevertheless, the 5′-UTR of Ep-BL13 and Ep-N28 were of the same length (194 nt), whereas that of strain Ep-BL14 was 1 nt shorter (193 nt; Figure S2). The nucleotide similarities between these three viruses were 98% in any combinations. A BLASTP search found that the highest aa sequence identity of the RdRp coded by three viral RNAs of Ep-BL13, Ep-BL14 and Ep-N28 were 41% (E-value = $1 \times e^{-158}$), 40% (E-values = $8 \times e^{-158}$) and 41% (E-value = $2 \times e^{-159}$), respectively, to the RdRp sequence of a tentative mitovirus of uncertain origin, soybean leaf-associated mitovirus 5. The highest aa identity with a definitive mitovirus species was 30% to Ophiostoma novo-ulmi mitovirus 3a (OnuMV3a). Thus, these 3 kbp-dsRNAs are apparently replication intermediates of new mitovirus strains that are closely related to each other, and these viruses were tentatively termed Fusarium boothii mitovirus 1 (FbMV1).

Some mitoviruses form a long pan-handle structure between the 5′ and 3′ termini [66]. In the terminal regions of FbMV1 genome, tetra nucleotides at extreme termini 5′-GGGG and CCCC-3′ may have potential to form base pairs. Another common characteristic of mitovirus genomes is the presence of stem-loop structures at near the terminal regions. Both the 5′ and 3′ UTRs of the FbMV1 genome clearly folded into two potential stem-loop structures, as shown in Figure 5B. None of the sequences was predicted to form pseudoknot.

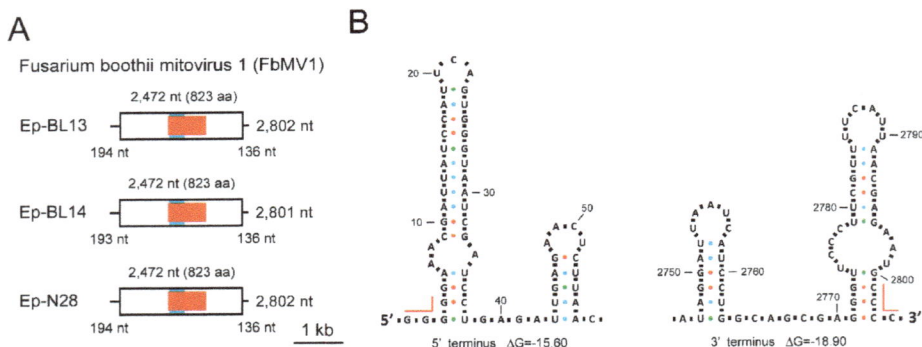

Figure 5. Schematic representation of the genome structure of *F. boothii* mitoviruses. (**A**) Genome maps of FbMV1. Thick bars are non-coding sequences and open boxes represent ORFs. Highlighted boxes in ORFs represent the predicted RdRp (red) and AroE (blue) domains. The length of the whole genome, ORF and UTR are given. (**B**) Structure of terminal, non-coding RNA regions of FbMV1. 5′-and 3′-terminal sequences of the genomes of FbMV1 Ep-BL13 were analysed with secondary structure prediction tools. Representatives of putative stem-loop structures were depicted. Red lines indicate nucleotide residues that potentially form base-pairing between 5′- and 3′-ends. Dots in different colours indicate hydrogen bonds between different base pairs: blue, A-U pairs; red, G-C pairs; green; G-U pairs.

3.8. Molecular Phylogenetic Analysis of Mitoviral RdRps

Phylogenetic analysis of the RdRps of the three new mitoviruses and some representative mitoviruses and narnaviruses (members of the family *Narnaviridae*) was conducted using PhyML (Figure 6 and Table S1). According to Xu et al. (2015) [34], mitoviruses are divided into three major clades I, II and III, and the new mitoviruses were included in clade II accommodating OnuMV3a. Mitoviruses found from *Fusarium* species have been mostly grouped into clade III, however, FbMV1 together with recently reported Fusarium poae mitovirus 3 and 4 (FpMV3 and FpMV4 in Figure 6) [25] exemplified a broad diversity of *Fusarium* mitoviruses. Taken together, these data suggest that the three viruses characterised in this study are strains of a new virus species Fusarium boothii mitovirus 1 in the genus *Mitovirus* in the family *Narnaviridae*. This is consistent with the species demarcation criteria within the genus *Mitovirus* (more than 90% sequence identity with RdRp refers strains) in the 9th report of ICTV [29].

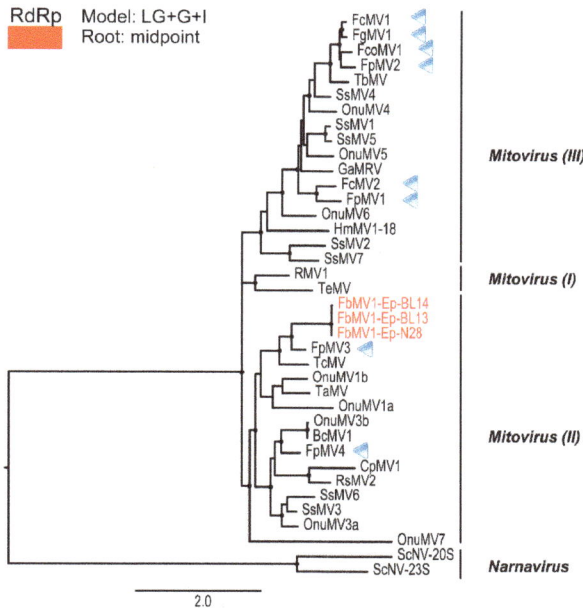

Figure 6. Phylogenetic analyses of FbMV1. A phylogenetic tree was constructed based on the alignment of RdRp domain sequences of narnaviruses, mitoviruses and FbMV1 strains. The midpoint-rooted ML tree is shown. Dot at the nodes represents reliable aLRT values over 0.9 for branch supporting. The best model of the ML tree constructs is indicated at the upper left. Clade clustering is referred to a previous report by Xu et al. (2015) [34]. Blue arrowheads indicate mitoviruses isolated from the *Fusarium* species. Abbreviations of viral names and accession numbers are listed in Table S1. Red letters represent mitoviruses characterized in this study.

4. Discussion

Virus hunting from fungal hosts has been widely conducted, and recent advances in sequencing technologies has accelerated the discovery of novel types of mycoviruses [1,17,25,43,67]. The virus-like sequences mining from NGS-assisted metadata are highly useful in understanding the diversity and evolution of viruses although there is an argument as to what kind of information can validate the presence of "actual viruses" and their classification, even within ICTV [68]. Indeed, recent examples of virus characterization in *Rosellinia necatrix* and *S. sclerotiorum* revealed identification of novel mycoviruses including taxonomically unique individuals, and demonstrated effectiveness of NGS [69,70]. Hence, the accumulation of more virome data from African biological materials will facilitate our understanding of mycovirus diversity and may lead to discovery of unexpectedly unique populations of mycoviruses, as is the case for FbLFV1. There are many barriers in the establishment of effective virocontrol of herbaceous annual plant diseases, including FHB, as discussed by Pearson et al. [71]. Thus, it will be important to screen for virocontrol agents which have much better properties such as higher transmissibility and broader cross-species specificity.

Characterization of mycoviruses has significantly contributed to the taxonomy of viruses including the order *Tymovirales*. Those include creation of the genus *Mycoflexivirus* and the family *Gammaflexiviridae* (BVF); creation of the genus *Deltaflexivirus* and the family *Deltaflexiviridae* (SsDFV1, SsDFV2, FgDFV1 and soybean leaf-associated mycoflexivirus 1); creation of the genera *Sclerodarnavirus* and *Botrexvirus* under the family *Alphaflexiviridae* (SsDARV and BVX, respectively); and proposal of new genus Mycotymovirus under the family *Tymoviridae* (FgMTV1) (Table 1; Figure 7) [20,27,39–42,67]. This study additionally reports a novel member of the order *Tymovirales* that has a potential to further expand taxonomic diversity in the order.

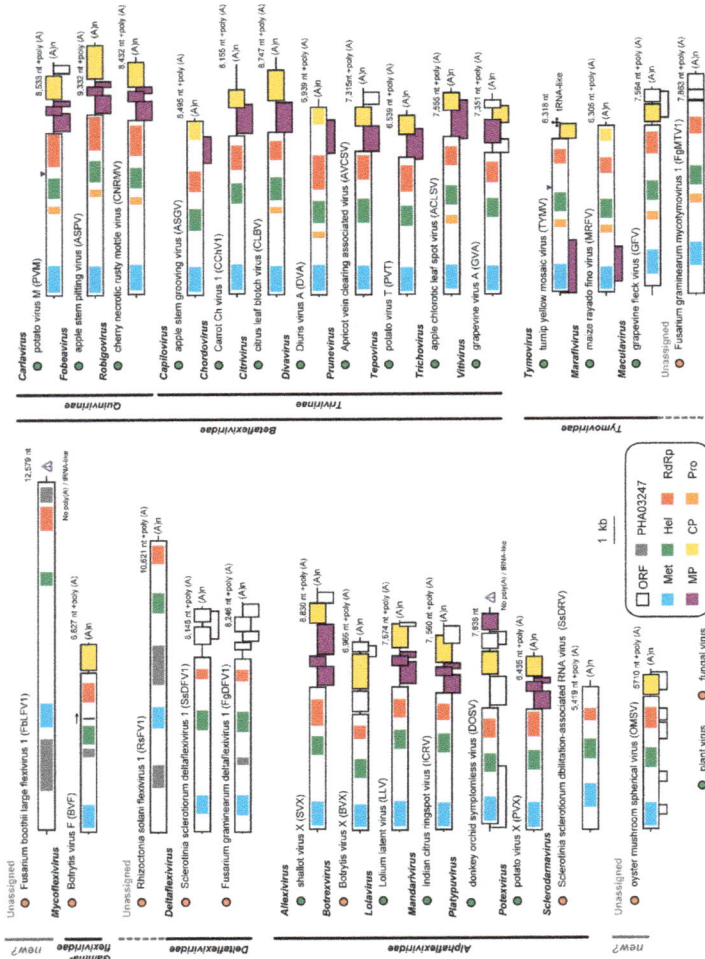

Figure 7. Comparison of genomic structures of the *Tymovirales* members and related unclassified viruses. Representative members of *Tymovirales* were selected as shown in Table 1 and their genomic structures are illustrated. Viral families, subfamilies and genera are shown in bold letters, and 'new?' represents a potential new taxon. Colour differentiated symbols are as indicated. The arrow on BVF replicase indicates potential readthrough translation. The blue triangle points to a known proteolytic cleavage site. Gray triangles at 3′-end of FbLFV1 and DOSV genomes indicate uncommon form that is not polyadenylated or tRNA-like structured.

FbLFV1 showed similar characteristic to and distinct features from present five families in the order as follows. The careful assessment of phylogenetic relationship between FbLFV1 and other *Tymovirales* members revealed its close-yet-distinct correlations to the gammaflexivirus BVF and deltaflexiviruses (Figures 3 and 4). The presence of PHA03247 domain in these viruses further supported their evolutionarily close association. However, the genome length and organization of FbLFV1 and BVF differed from each other (FbLFV1, 12,579 nt with single ORF vs. BVF, 6827 nt + poly-A tail with two ORFs) [39]. The absence of CP or CP domain of trichovirus/tymoviruses (see those of apple stem grooving virus, Diuris virus A and maize rayado fino marafivirus in Figure 7) in the FbLFV1 large polypeptide is highly possible and this feature is similar to deltaflexiviruses, an exceptional alphaflexivirus SsDRV and a tymo-like virus FgMTV1, highlighting the uniqueness of capsid-less nature of some mycoviruses, which is often observed (Figure 7) [20,27,38,41,42,45,67]. All *Tymovirales* viruses possess poly-A tail at the 3′ end with exceptions of those of the genus *Tymovirus* in the family *Tymoviridae*, which have TLSs, and of the genus *Platypuvirus* in the family *Alphaflexiviridae*, which have no obvious RNA structures (Figure 7) [72]. The 3′ end of FbLFV1 genome does has neither a poly-A tail nor TLS, further discriminating FbLFV1 from established genera and families in the order *Tymovirales* (Figures 2 and 7). In contrast, 5′ ends of *Tymovirales* virus genomes poorly conserve characteristics, except for the cap structure, across the families. In the genus *Tymovirus*, 5′ UTRs generally carry one to four protonatable hairpins nearby initiation codons that are probably required for dicistronic expression of overlapping ORFs (Figure 7, *Tymovirus*) [73]. The stem-loop structure predicted in 5′ UTR of FbLFV1 was placed at 81 nt upstream of start codon (Figure 2C) and there is no potential overlapping ORF, suggesting a distinct role of the structure in multiplication process of FbLFV1.

The genome length of the members in the order *Tymovirales* ranges from about 5.9 to 9.0 kb [59] whereas the recently reported unassigned member, a *Rhizoctonia* virus RsFV1, has the largest genome (10,621 nt + poly-A tail) [38]. The discovery of FbFLV1 increases the maximum genome size in the order by about 2 kb. RsFV1 and FbFLV1 share a similar genomic structure: the large RNA genome of over 10 kb; the single large ORF encoding replicase; the presence of a functionally unknown region at N-terminus in the replicase; and a potential of capsid-less nature (Figure 7). The N-terminal regions contain a PHA03247 protein motif that is similar to a portion of the UL36 large tegument protein found in herpesviruses that has also been found from other mycoviruses BVF and SsDRV (Figure 7) [74]. Despite the above-mentioned similarity, RsFV1 is evolutionarily closer to *deltaflexivirus*, whereas FbLFV1 is phylogenetically independent. In addition, a taxonomically floating virus, CMSV, appeared to be monophyletic, despite its apparently tymovirus (*Tymoviridae*)-like characteristics with respect to genome organisation and spherical virion structure (Figures 4 and 7) [44]. Thus, FbLFV1 and OMSV will become key viruses when taxonomic reconsideration within the order *Tymovirales* is required.

The strains of FbMV1 were found in three independent *F. boothii* isolates, and TcMV exhibited a high similarity to FbMV1 (35%). It has been reported that the TcMV RdRp has an AroE domain, which is one of the domains on AROM protein encoded by *Saccharomyces cerevisiae* [75]. AROM is a pentafunctional polypeptide that catalyses a sequence of five reactions in prechorismate polyaromatic amino acid biosynthesis, which is the former half of shikimate pathway. Although the function of the AroE-like domain in TcMV RdRp is unclear, similar sequences have been found in phylogenetically close relatives such as CpMV1 and OnuMV3a, as well as FbMV1 (Figure 5 and Figure S3), further supporting the involvement of FbMV1 in the mitovirus group of clade II. TcMV was identified in a basidiomycete fungus, *Thanatephorus cucumeris* (anamorph *Rhizoctonia solani*), which is phylogenetically very distant from ascomycetous *Fusarium* species. These points support an earlier hypothesis by Deng et al. [76] that mitoviruses have evolved not only through co-evolution with the viruses strictly confined within their host fungi and close relatives, but also through host-switching and adaptation to phylogenetically distant hosts via unknown routes of transmissions.

Overall, this study sequenced three isolates of a new mitovirus FbMV1 harboured in three independent *F. boothii* strains, as well as a novel tymo-like virus, FbLFV1, with the largest genome size thus far described in the order *Tymovirales*. Detailed analyses of gene expression, potential polyprotein processing, particle formation and pathogenicity of FbLFV1 are the next challenges to understand the properties of this new taxon in the order *Tymovirales*. Biological property of FbLFV1 may be able to be assessed by protoplast-fusion method as reported for FgV1-DK21 [77].

Supplementary Materials: The following are available online at http://www.mdpi.com/1999-4915/10/11/584/s1, Table S1. List of viruses in the family *Narnaviridae* and related unassigned members used in the phylogenetic study shown in Figure 6. Figure S1. Phylogenetic analysis of partial *TEF1-α* gene sequences derived from *Fusarium* samples. Products from the genomic PCR targeting *TEF1-α* were subjected to direct sequencing and the resultant nucleotide sequences were aligned in MAFFT ver. 7 with publicly available sequences of various *F. graminearum* species complex members. The alignment was used for maximum likelihood phylogenetic analysis with the recommended model shown at upper left. Figure S2. Nucleotide sequence alignment of three FbMV1 strains Ep-BL13, Ep-BL14 and Ep-N28. The entire nucleotide sequences of FbMV1 strains aligned with MAFFT ver.7. Small letters, sequence of non-coding region; capital letters, replicase coding region. Figure S3. The putative AROM domains in mitovirus RdRps. The amino acid sequence of AroE domain of AROM protein encoded by *S. cerevisiae*, and corresponding sequences of FbMV1, TcMV and close relatives were aligned with MAFFT ver.7.

Author Contributions: Conceptualization, A.A., N.S. and S.C.; Methodology, K.U., H.K. and S.C.; Software, Y.M., K.U. and H.K.; Validation, Y.M., H.S. and S.C.; Formal Analysis, S.C.; Investigation, Y.M., A.A., K.U., H.K. and H.S.; Resources, A.A. and H.S.; Data Curation, Y.M. and S.C.; Writing—Original Draft Preparation, Y.M.; Writing—Review and Editing, Y.M., H.K., H.S., N.S. and S.C.; Visualization, Y.M. and S.C.; and Supervision/Project Administration/Funding Acquisition, N.S. and S.C.

Acknowledgments: This research was supported in part by the Ministry of Education, Culture, Sports, Science and Technology (MEXT) as part of Joint Research Program implemented at the Institute of Plant Science and Resources, Okayama University in Japan (to N.S. and S.C.); Japan Society for the Promotion of Science (JSPS) and National Research Council of Thailand (NRCT) under the NRCT-JSPS joint research program (to S.C.); JSPS Grants-In-Aid for Scientific Research (B) (KAKENHI 17H03950 to S.C.); and Grants-in-Aid for Scientific Research on Innovative Areas from the Japanese MEXT (16H06436, 16H06429, and 16K21723 to N.S. and H.K.). A.A. is a recipient of the JSPS fellowship (L13555).

Conflicts of Interest: The authors declare no conflict of interest.

References

1. Ghabrial, S.A.; Castón, J.R.; Jiang, D.; Nibert, M.L.; Suzuki, N. 50-plus years of fungal viruses. *Virology* **2015**, *479–480*, 356–368. [CrossRef] [PubMed]

2. Pan, J.; Dong, L.; Lin, L.; Ochoa, W.F.; Sinkovits, R.S.; Havens, W.M.; Nibert, M.L.; Baker, T.S.; Ghabrial, S.A.; Tao, Y.J. Atomic structure reveals the unique capsid organization of a dsRNA virus. *Proc. Natl. Acad. Sci. USA* **2009**, *106*, 4225–4230. [CrossRef] [PubMed]

3. Hillman, B.I.; Cai, G. The Family Narnaviridae: Simplest of RNA Viruses. *Adv. Virus Res.* **2013**, *86*, 149–176. [PubMed]

4. Hillman, B.I.; Suzuki, N. Viruses of the chestnut blight fungi, *Cryphonectria parasitica*. *Adv. Virus Res.* **2004**, *63*, 423–472. [PubMed]

5. Suzaki, K.; Ikeda, K.; Sasaki, A.; Kanematsu, S.; Matsumoto, N.; Yoshida, K. Horizontal transmission and host-virulence attenuation of totivirus in violet root rot fungus *Helicobasidium mompa*. *J. Gen. Plant Pathol.* **2005**, *71*, 161–168. [CrossRef]

6. Moleleki, N.; van Heerden, S.W.; Wingfield, M.J.; Wingfield, B.D.; Preisig, O. Transfection of *Diaporthe perjuncta* with Diaporthe RNA virus. *Appl. Environ. Microbiol.* **2003**, *69*, 3952–3956. [CrossRef] [PubMed]

7. Wu, M.D.; Zhang, L.; Li, G.Q.; Jiang, D.H.; Hou, M.S.; Huang, H.C. Hypovirulence and Double-Stranded RNA in *Botrytis cinerea*. *Phytopathology* **2007**, *97*, 1590–1599. [CrossRef] [PubMed]

8. Castro, M.; Kramer, K.; Valdivia, L.; Ortiz, S.; Castillo, A. A double-stranded RNA mycovirus confers hypovirulence-associated traits to *Botrytis cinerea*. *FEMS Microbiol. Lett.* **2003**, *228*, 87–91. [CrossRef]

9. Chu, Y.M.; Jeon, J.J.; Yea, S.J.; Kim, Y.H.; Yun, S.H.; Lee, Y.W.; Kim, K.H. Double-stranded RNA mycovirus from *Fusarium graminearum*. *Appl. Environ. Microbiol.* **2002**, *68*, 2529–2534. [CrossRef] [PubMed]

10. Hong, Y.; Dover, S.L.; Cole, T.E.; Brasier, C.M.; Buck, K.W. Multiple Mitochondrial Viruses in an Isolate of the Dutch Elm Disease Fungus *Ophiostoma novo-ulmi*. *Virology* **1999**, *258*, 118–127. [CrossRef] [PubMed]

11. Yu, X.; Li, B.; Fu, Y.; Jiang, D.; Ghabrial, S.A.; Li, G.; Peng, Y.; Xie, J.; Cheng, J.; Huang, J.; et al. A geminivirus-related DNA mycovirus that confers hypovirulence to a plant pathogenic fungus. *Proc. Natl. Acad. Sci. USA* **2010**, *107*, 8387–8392. [CrossRef] [PubMed]

12. De Wet, J.; Bihon, W.; Preisig, O.; Wingfield, B.D.; Wingfield, M.J. Characterization of a novel dsRNA element in the pine endophytic fungus *Diplodia scrobiculata*. *Arch. Virol.* **2011**, *156*, 1199–1208. [CrossRef] [PubMed]

13. Preisig, O.; Wingfield, B.D.; Wingfield, M.J. Coinfection of a Fungal Pathogen by Two Distinct Double-Stranded RNA Viruses. *Virology* **1998**, *252*, 399–406. [CrossRef] [PubMed]

14. Vainio, E.J.; Martínez-Álvarez, P.; Bezos, D.; Hantula, J.; Diez, J.J. *Fusarium circinatum* isolates from northern Spain are commonly infected by three distinct mitoviruses. *Arch. Virol.* **2015**, *160*, 2093–2098. [CrossRef] [PubMed]

15. Smit, W.A.; Wingfield, B.D.; Wingfield, M.J. Reduction of laccase activity and other hypovirulence-associated traits in dsRNA-containig strains of *Diaporthe ambigua*. *Phytopathology* **1996**, *86*, 1311–1316. [CrossRef]

16. Preisig, O.; Moleleki, N.; Smit, W.A.; Wingfield, B.D.; Wingfield, M.J. A novel RNA mycovirus in a hypovirulent isolate of the plant pathogen *Diaporthe ambigua*. *J. Gen. Virol.* **2000**, *81*, 3107–3114. [CrossRef] [PubMed]

17. Coetzee, B.; Freeborough, M.J.; Maree, H.J.; Celton, J.M.; Rees, D.J.G.; Burger, J.T. Deep sequencing analysis of viruses infecting grapevines: Virome of a vineyard. *Virology* **2010**, *400*, 157–163. [CrossRef] [PubMed]

18. Darissa, O.; Adam, G.; Schäfer, W. A dsRNA mycovirus causes hypovirulence of *Fusarium graminearum* to wheat and maize. *Eur. J. Plant Pathol.* **2012**, *134*, 181–189. [CrossRef]

19. Wang, L.; He, H.; Wang, S.; Chen, X.; Qiu, D.; Kondo, H.; Guo, L. Evidence for a novel negative-stranded RNA mycovirus isolated from the plant pathogenic fungus *Fusarium graminearum*. *Virology* **2018**, *518*, 232–240. [CrossRef] [PubMed]

20. Li, P.; Lin, Y.; Zhang, H.; Wang, S.; Qiu, D.; Guo, L. Molecular characterization of a novel mycovirus of the family Tymoviridae isolated from the plant pathogenic fungus *Fusarium graminearum*. *Virology* **2016**, *489*, 86–94. [CrossRef] [PubMed]

21. Li, P.; Zhang, H.; Chen, X.; Qiu, D.; Guo, L. Molecular characterization of a novel hypovirus from the plant pathogenic fungus *Fusarium graminearum*. *Virology* **2015**, *481*, 151–160. [CrossRef] [PubMed]

22. Yu, J.S.; Lee, K.M.; Son, M.I.; Kim, K.H. Molecular Characterization of Fusarium Graminearum Virus 2 Isolated from *Fusarium graminearum* Strain 98-8-60. *Plant Pathol. J.* **2011**, *27*, 285–290. [CrossRef]

23. Kwon, S.J.; Lim, W.S.; Park, S.H.; Park, M.R.; Kim, K.H. Molecular Characterization of a dsRNA Mycovirus, *Fusarium graminearum* Virus-DK21, which Is Phylogenetically Related to Hypoviruses but Has a Genome Organization and Gene Expression Strategy Resembling Those of Plant Potex-like Viruses. *Mol. Cells* **2007**, *23*, 304–315. [CrossRef] [PubMed]

24. Yu, J.S.; Kwon, S.J.; Lee, K.M.; Son, M.; Kim, K.H. Complete nucleotide sequence of double-stranded RNA viruses from *Fusarium graminearum* strain DK3. *Arch. Virol.* **2009**, *154*, 1855–1858. [CrossRef] [PubMed]

25. Osaki, H.; Sasaki, A.; Nomiyama, K.; Tomioka, K. Multiple virus infection in a single strain of *Fusarium poae* shown by deep sequencing. *Virus Genes* **2016**, *52*, 835–847. [CrossRef] [PubMed]

26. Wang, S.; Kondo, H.; Liu, L.; Guo, L.; Qiu, D. A novel virus in the family Hypoviridae from the plant pathogenic fungus *Fusarium graminearum*. *Virus Res.* **2013**, *174*, 69–77. [CrossRef] [PubMed]

27. Chen, X.; He, H.; Yang, X.; Zeng, H.; Qiu, D.; Guo, L. The complete genome sequence of a novel Fusarium graminearum RNA virus in a new proposed family within the order Tymovirales. *Arch. Virol.* **2016**, *161*, 2899–2903. [CrossRef] [PubMed]

28. Marvelli, R.A.; Hobbs, H.A.; Li, S.; McCoppin, N.K.; Domier, L.L.; Hartman, G.L.; Eastburn, D.M. Identification of novel double-stranded RNA mycoviruses of *Fusarium virguliforme* and evidence of their effects on virulence. *Arch. Virol.* **2014**, *159*, 349–352. [CrossRef] [PubMed]

29. King, A.M.Q.; Adams, M.J.; Carstens, E.B.; Lefkowitz, E.J. Narnaviridae. In *Virus Taxonomy: Ninth Report of the International Committee on Taxonomy of Viruses*, 1st ed.; Elsevier: Amsterdam, The Netherlands, 2011; pp. 1055–1060.

30. Polashock, J.J.; Hillman, B.I. A small mitochondrial double-stranded (ds) RNA element associated with a hypovirulent strain of the chestnut blight fungus and ancestrally related to yeast cytoplasmic T and W dsRNAs. *Proc. Natl. Acad. Sci. USA* **1994**, *91*, 8680–8684. [CrossRef] [PubMed]

31. Khalifa, M.E.; Pearson, M.N. Molecular characterization of three mitoviruses co-infecting a hypovirulent isolate of *Sclerotinia sclerotiorum* fungus. *Virology* **2013**, *441*, 22–30. [CrossRef] [PubMed]

32. Wu, M.; Zhang, L.; Li, G.; Jiang, D.; Ghabrial, S.A. Genome characterization of a debilitation-associated mitovirus infecting the phytopathogenic fungus *Botrytis cinerea*. *Virology* **2010**, *406*, 117–126. [CrossRef] [PubMed]

33. Xie, J.; Ghabrial, S.A. Molecular characterizations of two mitoviruses co-infecting a hyovirulent isolate of the plant pathogenic fungus *Sclerotinia sclerotiorum*. *Virology* **2012**, *428*, 77–85. [CrossRef] [PubMed]

34. Xu, Z.; Wu, S.; Liu, L.; Cheng, J.; Fu, Y.; Jiang, D.; Xie, J. A mitovirus related to plant mitochondrial gene confers hypovirulence on the phytopathogenic fungus *Sclerotinia sclerotiorum*. *Virus Res.* **2015**, *197*, 127–136. [CrossRef] [PubMed]

35. Jian, J.; Lakshman, D.K.; Tavantzis, S.M. Association of Distinct Double-Stranded RNAs with Enhanced or Diminished Virulence in *Rhizoctonia solani* Infecting Potato. *Mol. Plant-Microbe Interact.* **1997**, *10*, 1002–1009. [CrossRef]

36. Martınetz, P.; Vainio, E.J.; Botella, L.; Hantula, J.; Diez, J.J. Three mitovirus strains infecting a single isolate of *Fusarium circinatum* are the first putative members of the family *Narnaviridae* detected in a fungus of the genus *Fusarium*. *Arch. Virol.* **2014**, *159*, 2153–2155. [CrossRef] [PubMed]

37. Osaki, H.; Sasaki, A.; Nomiyama, K.; Sekiguchi, H.; Tomioka, K.; Takehara, T. Isolation and characterization of two mitoviruses and a putative alphapartitivirus from *Fusarium* spp. *Virus Genes* **2015**, *50*, 466–473. [CrossRef] [PubMed]

38. Bartholomäus, A.; Wibberg, D.; Winkler, A.; Pühler, A.; Schlüter, A.; Varrelmann, M. Identification of a novel mycovirus isolated from *Rhizoctonia solani* (AG 2-2 IV) provides further information about genome plasticity within the order *Tymovirales*. *Arch. Virol.* **2017**, *162*, 555–559. [CrossRef] [PubMed]

39. Howitt, R.L.J.; Beever, R.E.; Forster, R.L.S.; Pearson, M.N. Genome characterization of Botrytis virus F, a flexuous rod-shaped mycovirus resembling plant 'potex-like' viruses. *J. Gen. Virol.* **2001**, *82*, 67–78. [CrossRef] [PubMed]

40. Howitt, R.L.J.; Beever, R.E.; Pearson, M.N.; Forster, R.L.S. Genome characterization of a flexuous rod-shaped mycovirus, Botrytis virus X, reveals high amino acid identity to genes from plant 'potex-like' viruses. *Arch. Virol.* **2006**, *151*, 563–579. [CrossRef] [PubMed]

41. Xie, J.; Wei, D.; Jiang, D.; Fu, Y.; Li, G.; Ghabrial, S.; Peng, Y. Characterization of debilitation-associated mycovirus infecting the plant-pathogenic fungus *Sclerotinia sclerotiorum*. *J. Gen. Virol.* **2006**, *87*, 241–249. [CrossRef] [PubMed]

42. Li, K.; Zheng, D.; Cheng, J.; Chen, T.; Fu, Y.; Jiang, D.; Xie, J. Characterization of a novel *Sclerotinia sclerotiorum* RNA virus as the prototype of a new proposed family within the order *Tymovirales*. *Virus Res.* **2015**, *219*, 92–99. [CrossRef] [PubMed]

43. Bartholomäus, A.; Wibberg, D.; Winkler, A.; Pühler, A.; Schlüter, A.; Varrelmann, M. Deep sequencing analysis reveals the mycoviral diversity of the virome of an avirulent isolate of *Rhizoctonia solani* AG-2-2 IV. *PLoS ONE* **2016**, *11*, e0165965. [CrossRef] [PubMed]

44. Yu, H.J.; Lim, D.; Lee, H.S. Characterization of a novel single-stranded RNA mycovirus in *pleurotus ostreatus*. *Virology* **2003**, *314*, 9–15. [CrossRef]

45. Hamid, M.; Xie, J.; Wu, S.; Maria, S.; Zheng, D.; Assane Hamidou, A.; Wang, Q.; Cheng, J.; Fu, Y.; Jiang, D.; et al. A Novel Deltaflexivirus that Infects the Plant Fungal Pathogen, *Sclerotinia sclerotiorum*, Can Be Transmitted Among Host Vegetative Incompatible Strains. *Viruses* **2018**, *10*, 295. [CrossRef] [PubMed]

46. Mackintosh, C.A.; Garvin, D.F.; Radmer, L.E.; Heinen, S.J.; Muehlbauer, G.J. A model wheat cultivar for transformation to improve resistance to Fusarium Head Blight. *Plant Cell Rep.* **2006**, *25*, 313–319. [CrossRef] [PubMed]

47. Suga, H.; Kageyama, K.; Shimizu, M.; Hyakumachi, M. A Natural Mutation Involving both Pathogenicity and Perithecium Formation in the *Fusarium graminearum* Species Complex. *G3 (Bethesda)*. **2016**, *6*, 3883–3892. [CrossRef] [PubMed]

48. Ban, T.; Suenaga, K. Genetic analysis of resistance to Fusarium head blight caused by *Fusarium graminearum* in Chinese wheat cultivar Sumai 3 and the Japanese cultivar Saikai 165. *Euphytica* **2000**, *113*, 87–99. [CrossRef]

49. O'Donnell, K.; Kistler, H.C.; Cigelnik, E.; Ploetz, R.C. Multiple evolutionary origins of the fungus causing Panama disease of banana: Concordant evidence from nuclear and mitochondrial gene genealogies. *Proc. Natl. Acad. Sci. USA* **1998**, *95*, 2044–2049. [CrossRef] [PubMed]

50. O'Donnell, K.; Cigelnik, E.; Nirenberg, H.I. Molecular Systematics and Phylogeography of the *Gibberella fujikuroi* Species Complex. *Mycologia* **1998**, *90*, 465–493. [CrossRef]

51. Andrews, S. FastQC: A Quality Control Tool for High Throughput Sequence Data. Available online: https://www.bioinformatics.babraham.ac.uk/projects/fastqc/ (accessed on 25 October 2018).

52. Joshi, N.; Fass, J. Sickle: A Sliding-Window, Adaptive, Quality-Based Trimming Tool for FastQ Files (Version 1.33). Available online: https://github.com/najoshi/sickle (accessed on 25 October 2018).

53. Bankevich, A.; Nurk, S.; Antipov, D.; Gurevich, A.A.; Dvorkin, M.; Kulikov, A.S.; Lesin, V.M.; Nikolenko, S.I.; Pham, S.; Prjibelski, A.D.; et al. SPAdes: A New Genome Assembly Algorithm and Its Applications to Single-Cell Sequencing. *J. Comput. Biol.* **2012**, *19*, 455–477. [CrossRef] [PubMed]

54. Finn, R.D.; Attwood, T.K.; Babbitt, P.C.; Bateman, A.; Bork, P.; Bridge, A.J.; Chang, H.Y.; Dosztányi, Z.; El-Gebali, S.; Fraser, M.; et al. InterPro in 2017—Beyond protein family and domain annotations. *Nucleic Acids Res.* **2017**, *45*, D190–D199. [CrossRef] [PubMed]

55. Zuker, M. Mfold web server for nucleic acid folding and hybridization prediction. *Nucleic Acids Res.* **2003**, *31*, 3406–3415. [CrossRef] [PubMed]

56. Edgar, R.C. MUSCLE: Multiple sequence alignment with high accuracy and high throughput. *Nucleic Acids Res.* **2004**, *32*, 1792–1797. [CrossRef] [PubMed]

57. Kumar, S.; Stecher, G.; Li, M.; Knyaz, C.; Tamura, K.; Battistuzzi, F.U. MEGA X: Molecular Evolutionary Genetics Analysis across Computing Platforms. *Mol. Biol. Evol.* **2018**, *35*, 1547–1549. [CrossRef] [PubMed]

58. Guindon, S.; Dufayard, J.F.; Lefort, V.; Anisimova, M.; Hordijk, W.; Gascuel, O. New Algorithms and Methods to Estimate Maximum-Likelihood Phylogenies: Assessing the Performance of PhyML 3.0. *Syst. Biol.* **2010**, *59*, 307–321. [CrossRef] [PubMed]

59. King, A.M.Q.; Adams, M.J.; Carstens, E.B.; Lefkowitz, E.J. Tymovirales. In *Virus Taxonomy: Ninth Report of the International Committee on Taxonomy of Viruses*, 1st ed.; Elsevier: Amsterdam, The Netherlands, 2011; pp. 920–941.

60. Ballance, D.J. Sequences important for gene expression in filamentous fungi. *Yeast* **1986**, *2*, 229–236. [CrossRef] [PubMed]

61. Kozak, M. At least six nucleotides preceding the AUG initiator codon enhance translation in mammalian cells. *J. Mol. Biol.* **1987**, *196*, 947–950. [CrossRef]

62. Mark Cigan, A.; Donahue, T.F. Sequence and structural features associated with translational initiator regions in yeast—A review. *Gene* **1987**, *59*, 1–18. [CrossRef]

63. Dreher, T.W. Role of tRNA-like structures in controlling plant virus replication. *Virus Res.* **2009**, *139*, 217–229. [CrossRef] [PubMed]

64. Matsuda, D.; Dreher, T.W. The tRNA-like structure of Turnip yellow mosaic virus RNA is a 3′-translational enhancer. *Virology* **2004**, *321*, 36–46. [CrossRef] [PubMed]

65. Lowe, T.M.; Chan, P.P. tRNAscan-SE On-line: Integrating search and context for analysis of transfer RNA genes. *Nucleic Acids Res.* **2016**, *44*, W54–W57. [CrossRef] [PubMed]

66. Osaki, H.; Nakamura, H.; Nomura, K.; Matsumoto, N.; Yoshida, K. Nucleotide sequence of a mitochondrial RNA virus from the plant pathogenic fungus, *Helicobasidium mompa* Tanaka. *Virus Res.* **2005**, *107*, 39–46. [CrossRef] [PubMed]

67. Marzano, S.Y.L.; Domier, L.L. Novel mycoviruses discovered from metatranscriptomics survey of soybean phyllosphere phytobiomes. *Virus Res.* **2016**, *213*, 332–342. [CrossRef] [PubMed]

68. Khalifa, M.E.; Varsani, A.; Ganley, A.R.D.; Pearson, M.N. Comparison of Illumina de novo assembled and Sanger sequenced viral genomes: A case study for RNA viruses recovered from the plant pathogenic fungus *Sclerotinia sclerotiorum*. *Virus Res.* **2016**, *219*, 51–57. [CrossRef] [PubMed]

69. Mu, F.; Xie, J.; Cheng, S.; You, M.P.; Barbetti, M.J.; Jia, J.; Wang, Q.; Cheng, J.; Fu, Y.P.; Chen, T.; et al. Virome Characterization of a Collection of *S. sclerotiorum* from Australia. *Front. Microbiol.* **2018**, *8*, 2540. [CrossRef] [PubMed]

70. Arjona-Lopez, J.M.; Telengech, P.; Jamal, A.; Hisano, S.; Kondo, H.; Yelin, M.D.; Arjona-Girona, I.; Kanematsu, S.; Lopez-Herrera, C.J.; Suzuki, N. Novel, diverse RNA viruses from Mediterranean isolates of the phytopathogenic fungus, *Rosellinia necatrix*: Insights into evolutionary biology of fungal viruses. *Environ. Microbiol.* **2018**, *20*, 1464–1483. [CrossRef] [PubMed]

71. Pearson, M.N.; Beever, R.E.; Boine, B.; Arthur, K. Mycoviruses of filamentous fungi and their relevance to plant pathology. *Mol. Plant Pathol.* **2009**, *10*, 115–128. [CrossRef] [PubMed]

72. Wylie, S.J.; Li, H.; Jones, M.G.K. Donkey Orchid Symptomless Virus: A Viral 'Platypus' from Australian Terrestrial Orchids. *PLoS ONE* **2013**, *8*, e79587. [CrossRef] [PubMed]

73. Tzanetakis, I.E.; Tsai, C.H.; Martin, R.R.; Dreher, T.W. A tymovirus with an atypical 3′-UTR illuminates the possibilities for 3′-UTR evolution. *Virology* **2009**, *392*, 238–245. [CrossRef] [PubMed]

74. Fan, W.H.; Roberts, A.P.E.; McElwee, M.; Bhella, D.; Rixon, F.J.; Lauder, R. The large tegument protein pUL36 is essential for formation of the capsid vertex-specific component at the capsid-tegument interface of herpes simplex virus 1. *J. Virol.* **2015**, *89*, 1502–1511. [CrossRef] [PubMed]

75. Lakshman, D.K.; Jian, J.; Tavantzis, S.M. A double-stranded RNA element from a hypovirulent strain of *Rhizoctonia solani* occurs in DNA form and is genetically related to the pentafunctional AROM protein of the shikimate pathway (RNA-dependent RNA polymerase). *Microbiology* **1998**, *95*, 6425–6429. [CrossRef]

76. Deng, F.; Xu, R.; Boland, G.J. Hypovirulence-Associated Double-Stranded RNA from *Sclerotinia homoeocarpa* Is Conspecific with *Ophiostoma novo-ulmi* Mitovirus 3a-Ld. *Phytopathology* **2003**, *93*, 1407–1414. [CrossRef] [PubMed]

77. Lee, K.M.; Yu, J.; Son, M.; Lee, Y.W.; Kim, K.H.; Arkowitz, R.A. Transmission of *Fusarium boothii* Mycovirus via Protoplast Fusion Causes Hypovirulence in Other Phytopathogenic Fungi. *PLoS ONE* **2011**, *6*, e21629. [CrossRef] [PubMed]

viruses

MDPI

Communication

Molecular Characterization and Geographic Distribution of a Mymonavirus in the Population of *Botrytis cinerea*

Fangmin Hao [1,2], Mingde Wu [1,2,*] and Guoqing Li [1,2]

1   The State Key Laboratory of Agricultural Microbiology, Huazhong Agricultural University,
    Wuhan 430070, China; haofangmin@163.com (F.H.); guoqingli@mail.hzau.edu.cn (G.L.)
2   The Key Laboratory of Plant Pathology of Hubei Province, Huazhong Agricultural University,
    Wuhan 430070, China
*   Correspondence: mingde@mail.hzau.edu.cn

Received: 24 July 2018; Accepted: 14 August 2018; Published: 15 August 2018

Abstract: Here, we characterized a negative single-stranded (−ss)RNA mycovirus, Botrytis cinerea mymonavirus 1 (BcMyV1), isolated from the phytopathogenic fungus *Botrytis cinerea*. The genome of BcMyV1 is 7863 nt in length, possessing three open reading frames (ORF1–3). The ORF1 encodes a large polypeptide containing a conserved mononegaviral RNA-dependent RNA polymerase (RdRp) domain showing homology to the protein L of mymonaviruses, whereas the possible functions of the remaining two ORFs are still unknown. The internal cDNA sequence (10-7829) of BcMyV1 was 97.9% identical to the full-length cDNA sequence of Sclerotinia sclerotiorum negative stranded RNA virus 7 (SsNSRV7), a virus-like contig obtained from *Sclerotinia sclerotiorum* metatranscriptomes, indicating BcMyV1 should be a strain of SsNSRV7. Phylogenetic analysis based on RdRp domains showed that BcMyV1 was clustered with the viruses in the family *Mymonaviridae*, suggesting it is a member of *Mymonaviridae*. BcMyV1 may be widely distributed in regions where *B. cinerea* occurs in China and even over the world, although it infected only 0.8% of tested *B. cinerea* strains.

Keywords: *Botrytis cinerea*; Botrytis cinerea mymonavirus 1; *Mymonaviridae*

1. Introduction

Botrytis spp., a group of widespread plant pathogenic fungi, can infect more than 1400 plant species, causing gray mold disease on many economically important crops [1]. Besides being pathogens of many plants, *Botrytis* spp. are also ideal hosts for viruses. Among sequenced Botrytis viruses, most positive single-stranded (+ss)RNA viruses were classified into five families—*Alphaflexiviridae*, *Gammaflexiviridae*, *Hypoviridae*, *Narnaviridae*, and a recent proposed family *Fusariviridae*—while most double-stranded (ds)RNA viruses were assigned into three families—*Endornaviridae*, *Partitiviridae*, and *Totiviridae*—and the genus *Botybirnavirus* [2–5]. In addition, a few sequenced Botrytis viruses, including Botrytis cinerea RNA virus 1 [6], Botrytis ourmia-like virus [7], and Botrytis cinerea negative-stranded RNA virus 1 (BcNSRV1) [8], remained unclassified.

Compared with +ssRNA and dsRNA viruses, (−ss)RNA viruses are rarely reported in *Botrytis cinerea* as well as in other fungi [1]. Mononegaviruses are a group of nonsegmented (−ss)RNA viruses with the genomes of 8.9–19 kb in length, although there are some exceptions [9]. Most reported mononegaviruses infect invertebrates, vertebrates, and plants, whereas only few have been shown to infect fungi [10–13]. Mononegaviruses are divided into eight families—*Bornaviridae*, *Filoviridae*, *Paramyxoviridae*, *Rhabdoviridae*, *Pneumoviridae*, *Sunviridae*, *Nyamiviridae*, and *Mymonaviridae* [12]—of which *Mymonaviridae* (genus *Sclerotimonavirus*, type species Sclerotinia sclerotiorum negative-stranded RNA virus 1 (SsNSRV1)) is a newly established viral family that accommodates mononegaviruses infecting fungi in the order *Mononegavirales* [14]. In addition, eight other viruses/virus-like contigs, including soybean leaf-associated negative-stranded RNA viruses 1–4 (SlaNSRV1–4), Sclerotinia sclerotiorum negative stranded viruses 2–4 (SsNSRV2–4), and Fusarium graminearum negative-stranded RNA virus 1 (FgNSRV1), are phylogenetically closer to SsNSRV1, and may also belong to *Mymonaviridae* [15–17].

In the present study, we describe the genome of a (−ss)RNA virus infecting the fungus *B. cinerea*, namely Botrytis cinerea mymonavirus 1 (BcMyV1). Genomic and phylogenetic analysis indicates that BcMyV1 was most closely related to Sclerotinia sclerotiorum negative-stranded RNA virus 7 (SsNSRV7) [18] and also showed homology to other fungal mononegaviruses. In addition, we also determined the incidence and geographic distribution of BcMyV1 in the population of *B. cinerea* in China.

2. Materials and Methods

2.1. Fungal Strains, Culture Conditions, and Biological Characterization

B. cinerea strains Ecan17-2 was originally obtained through single conidium isolation from diseased oilseed rape (*Brassica napus*) stem in Shiyan, Hubei Province, China, and strain B05.10 of *B. cinerea* was used as a control [3]. In addition, 508 *B. cinerea* strains from 40 counties/cities in 11 provinces of China were used for testing the presence of BcMyV1. All strains were stored at 4 °C and working culture was established [19]. The mycelial growth was determined on potato dextrose agar (PDA) in petri dishes [3], while the pathogenicity of *B. cinerea* strains was determined on detached *Nicotiana benthamiana* leaves [3,19,20].

2.2. dsRNA Extraction and Purification

dsRNAs from *B. cinerea* mycelia was extracted and purified as described previously [20] and was further confirmed based on resistance to DNase I and S1 nuclease (Promega, Madison, WI, USA). The extracted dsRNA was fractionated by agarose gel (1%, *w/v*) electrophoresis and visualized by staining with ethidium bromide (1.5 μg/L) and viewing on a UV transilluminator.

2.3. cDNA Cloning and Sequencing

After separation by agarose gel electrophoresis, the dsRNA segment (BcMyV1 replication intermediates, approximately 10 kb in size based on the DNA marker) was gel-purified by using AxyPrepTM DNA Gel Extraction Kit (Axygen Scientific, Inc.; Union City, CA, USA) as described by Wu et al. [21]. The cDNAs of BcMyV1 were produced using a random-primer-mediated PCR amplification protocol [6] and were then sequenced [21]. The terminal sequences of the dsRNA were cloned through ligating the 3′-terminus for each strand of the dsRNA with the 5′-terminus of the 110A adaptor (Table S1) using T4 RNA ligase (Promega Corporation, 2800 Woods Hollow Road, Madison, WI, USA) at 16 °C for 18 h and were then reverse transcribed using primer RC110A (Table S1). The cDNA strands were then used as template for PCR amplification of the 5′- and 3′-terminal sequences with the primer RC110A and the corresponding sequence specific primers (Figure S1 and Table S1). Cloning of the 3′- or 5′-terminal sequences of the dsRNA was performed for three rounds. In addition, the internal region of the BcMyV1 genome was also amplified through RT-PCR with four sequence-specific primer pairs (Figure S1 and Table S1). All these amplicons were detected by agarose gel electrophoresis, gel-purified, and cloned into *Escherichia coli* DH5α and sequenced as previously described [19]. All cDNA sequences were assembled to obtain the full-length cDNA sequence of BcMyV1.

2.4. Sequence Analysis

Open reading frames (ORFs) in the full-length cDNA sequences of BcMyV1 in strain Ecan17-2 of *B. cinerea* were deduced using the ORF Finder program on the website of the National Center for Biotechnology Information (NCBI, http://www.ncbi.nlm.nih.gov/gorf/gorf.html). The homologous sequences searching for the full-length cDNA sequences and deduced polypeptides of BcMyV1 were carried out at the NCBI database by using the BlastN and BlastP programs, respectively. CDD database (http://www.ncbi.nlm.nih.gov/Structure/cdd/wrpsb.cgi) searching predicted the domains present in the polypeptide sequence. Multiple alignments of the sequences of mononegaviral RNA-dependent RNA polymerase (RdRp) domains in the polypeptides encoded by BcMyV1 and other mononegaviruses were performed using the ClustalW program in MEGA 7.0 [22]. Phylogenetic trees based on the sequences of RdRp domains were constructed using the neighbor-joining (NJ) method and tested with a bootstrap of 1000 replicates to ascertain the reliability of a given branch pattern in MEGA 7.0. Putative transmembrane helices sequences were predicted using the TMHMM server version 2.0 (http://www.cbs.dtu.dk/services/TMHMM/) [23].

2.5. Detection of BcMyV1 in B. cinerea Population

The total RNAs of 508 *B. cinerea* strains were extracted using the TRIzol® reagent (Invitrogen Corp, Carlsbad, CA, USA) as described previously [19], and the presence of BcMyV1 was determined by using RT-PCR with primer pairs M-RT-F/R (Table S1), which was designed to amplify a specific band of 728 bp in size.

3. Results

3.1. B. cinerea Strain Ecan17-2 Exhibits Hypovirulence Traits

After cultivation on a PDA plate for 9 days, strain Ecan17-2 formed colonies with no production of sclerotia, whereas strain B05.10 produced massive sclerotia on the colony (Figure 1A). In addition, the radial mycelial growth of Ecan17-2 on PDA, averaging 2.9 mm/day, was significantly slower than that of strain B05.10 (14.8 mm/day). The virulence assay on detached *N. benthamiana* leaves revealed that the average lesion diameter (6.9 mm) caused by strain Ecan17-2 was significantly smaller than that (29.3 mm) of strain B05.10 (Figure 1B).

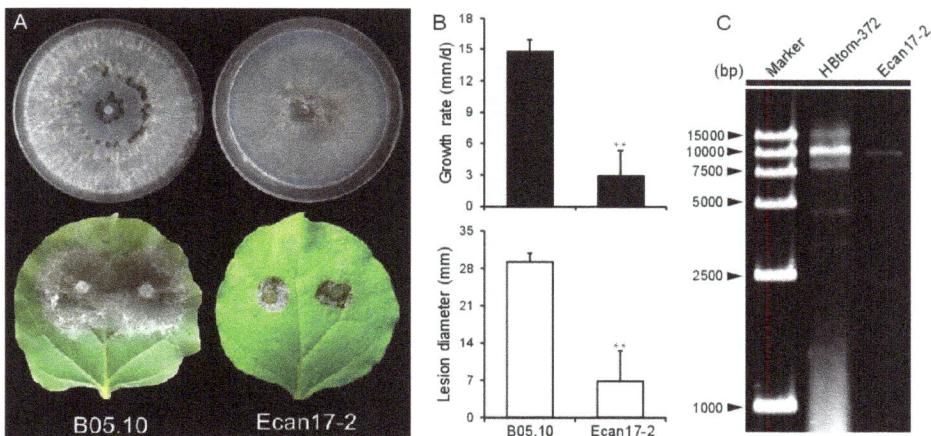

Figure 1. (**A**) Culture morphology (upper, 20 °C, 9 days) and pathogenicity assay (lower, 20 °C, 3 days) of *Botrytis cinerea* strains Ecan17-2 and B05.10 on potato dextrose agar (PDA) and detached *N. benthamiana* leaves, respectively. (**B**) Radial mycelial growth rate (20 °C, upper) on PDA and lesion diameter (20 °C, 72 h, lower) on detached *N. benthamiana* leaves of strains Ecan17-2 and B05.10. "**"** indicates a significant difference ($p < 0.01$) between strains Ecan17-2 and B05.10 in both pathogenicity and radial mycelial growth rate. (**C**) Agarose gel electrophoresis of dsRNAs extracted from the mycelia of *B. cinerea* strains Ecan17-2 and HBtom-372.

3.2. Genome Analysis of BcMyV1

After DNase I and S1 nuclease digestion, a major dsRNA segment was detected through electrophoresis in the mycelium of *B. cinerea* strain Ecan17-2 with the size of approximately 10.0 kb, which was slightly smaller than the dsRNA-B (Botrytis cinerea fusarivirus 1 (BcFV1), 8411 bp) detected in strain HBtom-372 [3]. The coding strand (GenBank accession no. MH648611) of BcMyV1 was 7863 nt long, with a GC content of 41.6%, possessing three ORFs (ORF1–3) and two short untranslated regions (UTRs) of 76 nt and 472 nt in length at the 5′- and 3′-terminus, respectively. The ORF1 was predicted to encode a putative large polypeptide of 1968 amino acid (aa) residues (Figure 2A), which contains a putative mononegaviral RdRp domain and a mononegaviral mRNA-capping region V (Figure 2A). The ORF2 and ORF3 encoded two proteins of 169 aa and 250 aa in length, respectively. In addition, the 21 nt long repeated sequence 3′-UAAAUUUCUUUGAUCCUCUAU-5′ was detected in the two UTRs between the three ORFs (Figure 2B).

The results of the Blast search showed that the nucleotide sequence of the internal region (10-7829) of the BcMyV1 genome was 97.9% identical to the full-length nucleotide sequence of a contig obtained from metatranscriptomes of *Sclerotinia sclerotiorum* isolates, SsNSRV7 [18]. In addition, the polypeptide encoded by BcMyV1 ORF1 was almost 99.9% identical to the protein L of SsNSRV7 and also showed homology to the protein L of SsNSRV1 (33.1% aa identity), FgNSRV1 (33.8% aa identity), and several (−ss)RNA viruses identified through deep sequencing (Table 1). Four conserved motifs, I–IV, found in mononegaviruses were also identified in the protein L encoded by BcMyV1 ORF1 (Figure 2C). Unlike SsNSRV1, transmembrane (TM) domains (Figure S2) were found at the C-proximal protein L of BcMyV1. However, the proteins encoded by ORF2 and ORF3 showed no significant sequence similarity with proteins in the database of NCBI by using BlastP search.

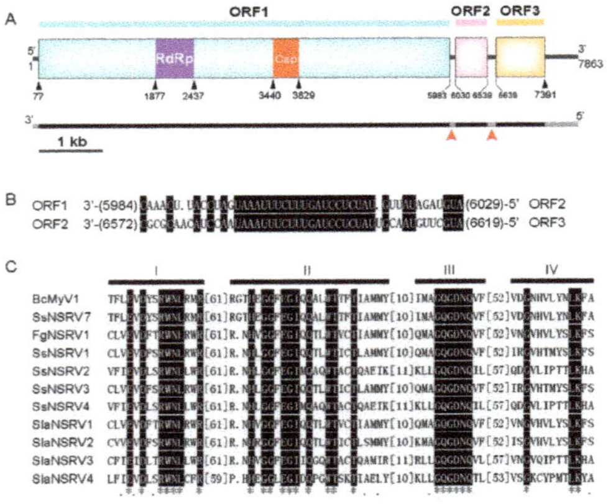

Figure 2. (**A**) Schematic diagram of the genome organization of Botrytis cinerea mymonavirus 1 (BcMyV1). The coding strand of BcMyV1 is 7863 nt long and contains three Open reading frames (ORFs), and the ORF1 encode a protein L of 1968 amino acids (aa), possessing a mononegaviral RNA-dependent RNA polymerase (RdRp) domain and a mRNA-capping region V (Cap) domain. ORF2 and ORF3 encode two proteins of 169 aa and 250 aa, respectively. The black bars indicate the coding regions, and the gray bars represent the untranslated regions (UTRs) on the genome of BcMyV1. Two red arrowheads point out the positions of a 21 nt repeat region on the two UTRs, and the detailed sequence information are listed in (**B**). The numbers in the parentheses indicate the nt positions nearby the parentheses. (**C**) Multiple alignment of the amino acid sequences of RdRp in the protein L encoded by BcMyV1 and other (−ss)RNA viruses. "*" indicates identical amino acid residues; and "." indicate low chemically similar amino acid residues. The abbreviations of virus names are listed in Table 1.

3.3. Phylogenetic Analysis of BcMyV1

To define the phylogenetic relationship of BcMyV1 with other viruses in *Monoregavirales* (Table 1), a phylogenetic tree was established based on the mononegaviral RdRp domain. BcMyV1 firstly formed a tight cluster with SsNSRV7 and then clustered with (−ss)RNA mycoviruses from *S. sclerotiorum*, *F. graminearum*, and other viral-like contigs, forming an independent clade of *Mymonaviridae* with the bootstrap support of 99%. In addition, other viruses from *Bornaviridae*, *Sunviridae*, *Filoviridae*, *Rhabdoviridae*, *Paramyxoviridae*, *Pneumoviridae*, and *Nyamiviridae* also formed the corresponding viral family clades (Figure 3). Therefore, we suppose that BcMyV1 should be a member in the virial family *Mymonaviridae*.

Table 1. Percentage of sequence identities between Botrytis cinerea mymonavirus 1 and other mononegaviruses according to the multiple alignments of the full-length protein L and the RNA-dependent RNA polymerase domain.

Family	Virus	Acronym	aa Identity (%)		Accession no.
			Full Sequence	RdRp	
Mymonaviridae	Sclerotinia sclerotiorum negative-stranded RNA virus 7	SsNSRV7	99.85	100	MF444285
	Sclerotinia sclerotiorum negative-stranded RNA virus 1	SsNSRV1	33.12	56.68	NC_025383.1
	Sclerotinia sclerotiorum negative-stranded RNA virus 2	SsNSRV2	22.28	41.45	KP900931.1
	Sclerotinia sclerotiorum negative-stranded RNA virus 3	SsNSRV3	33.57	56.15	NC_026732.1
	Sclerotinia sclerotiorum negative-stranded RNA virus 4	SsNSRV4	21.67	39.38	KP900930.1
	Soybean leaf-associated negative-stranded RNA virus 1	SlaNSRV1	32.75	59.36	KT598225.1
	Soybean leaf-associated negative-stranded RNA virus 2	SlaNSRV2	33.12	61.5	KT598227.1
	Soybean leaf-associated negative-stranded RNA virus 3	SlaNSRV3	21.43	37.31	KT598228.1
	Soybean leaf-associated negative-stranded RNA virus 4	SlaNSRV4	17.82	34.57	KT598229.1
	Fusarium graminearum negative-stranded RNA virus 1	FgNSRV1	32.75	59.36	MF276904.1
Bornaviridae	Jungle carpet python virus	JCPV	14.84	24.6	MF135780
	Southwest carpet python virus	SWCPV	14.33	21.93	MF135781
	Loveridge's garter snake virus 1	LGSV1	14.28	23.53	KM114265
	Variegated squirrel bornavirus 1	VSBV1	14.16	25.13	LN713681
Rhabdoviridae	Rabies virus	RabV	14.97	23.62	AB517659
	Maize mosaic virus	MMV	15.19	23.12	NC_005975.1
Paramyxoviridae	Newcastle disease virus	NDV	14.94	27.69	JF827026.1
	Measles virus	MV	15.44	30.77	NC_001498.1
Nyamiviridae	Midway nyavirus	MIDMV	15.81	26.42	NC_012702.1
	Nyamanini nyavirus	NYMV	16.01	26.42	NC_012703.1
Filoviridae	Rose rosette virus	RRV	10.57	11.4	HQ871942
	Raspberry leaf blotch virus	RLBV	10.71	10.27	FR823299
Sunviridae	Reptile sunshinevirus 1	RSV-1	14.26	24.26	NC_025345
Pneumoviridae	Human respiratory syncytial virus	HRSV	13.43	22.68	NC_001781
	Pneumonia virus of mice	PVM	14.22	22.8	NC_006579

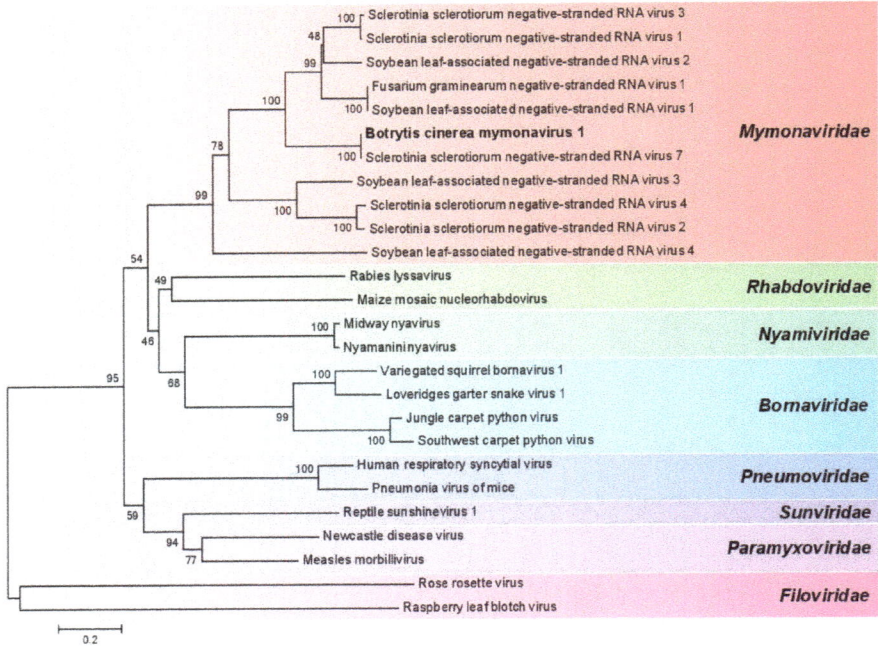

Figure 3. Phylogenetic analysis of Botrytis cinerea mymonavirus 1 (BcMyV1) based on RdRp domain from strain Ecan17-2 of *B. cinerea*.

3.4. Incidence and Distribution of BcMyV1

In order to investigate the incidence and distribution of BcMyV1 in China, 508 *B. cinerea* strains from China were tested for the presence of BcMyV1 by using RT-PCR with the primer pair M-RT-F/R (Table S1). BcMyV1 infection was detected in only 4 out of the 508 (0.8%) tested *B. cinerea* strains (Figures 4 and 5.). In these BcMyV1-infected strains, Bs6-23 and Bs6-33 were collected from the same location (Beijing, China), whereas strain JLaub-11 was collected from Changchun of Jilin Province.

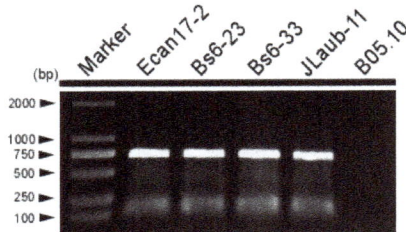

Figure 4. RT-PCR detection of Botrytis cinerea mymonavirus 1 (BcMyV1) in five *B. cinerea* strains. Strains Ecan17-2 and B05.10 served as positive and negative controls, respectively.

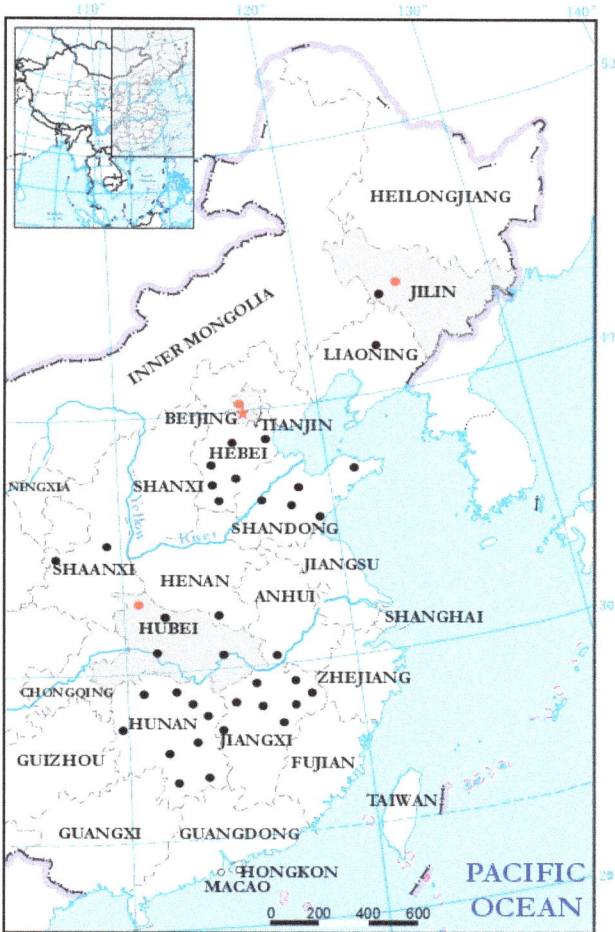

Figure 5. Geographic distribution of Botrytis cinerea mymonavirus 1 (BcMyV1) in 11 provinces of China. The black dots indicate the places where BcMyV1 was not detected, whereas places where BcMyV1 was detected are indicated as red dots and the corresponding provinces are also highlighted in grey on the map.

4. Discussion

In the present study, we characterized the genome of a (−ss)RNA mycovirus, namely, BcMyV1, infecting the hypovirulent strain Ecan17-2 of *B. cinerea*. Notwithstanding that numerous mycoviruses have been reported in *B. cinerea*, only one case of a (−ss)RNA virus (BcNSRV1) had been characterized [8]. BcNSRV1 is phylogenetically related to members of the genus *Emaravirus* in the viral family *Bunyaviridae*. However, phylogenetic analysis based on RdRp domain indicated that BcMyV1 should belong to the viral family *Mymonaviridae* in the order *Mononegavirales*. As high sequence similarity was observed between BcMyV1 and SsNSRV7, BcMyV1 should be a strain of SsNSRV7 [18]. Nonetheless, the nucleotide sequence of BcMyV1 was longer at both 5′- and 3′-termini than those of SsNSRV7, indicating the sequence of SsNSRV7 in the NCBI database might be incomplete.

In *S. sclerotiorum*, SsNSRV1 infection was closely related to the debilitation symptoms of the infected *S. sclerotiorum* strain, including slow growth on PDA, loss of the ability to produce sclerotia, and pathogenicity on oilseed rape [14]. Similarly, *B. cinerea* strain Ecan17-2 carrying BcMyV1 also displayed reduced mycelial growth on PDA and attenuated virulence on *N. benthamiana*, indicating possible negative effects of BcMyV1 on its host. However, a faint dsRNA segment of approximately 2.4 kb in length (Figure 1C) was also detected in strain Ecan17-2, indicating coinfection of other viruses or defective/satellite RNAs with BcMyV1 [3]. Thus, sequencing the 2.4 kb dsRNA in the traditional way is warranted to ascertain the causal agent of hypovirulence in *B. cinerea* strain Ecan17-2. Moreover, deep sequencing [8] may also be an option to determine the full view of the viral infection in strain Ecan17-2. Generally, two aspects of approach have been used for construction of isogenic lines that aim to elucidate the effects of mycoviral infection on their hosts. Firstly, viruses could be introduced into virus-free strains by using the techniques like pairing-culture [3], virion transfection [21], and construction of infectious cDNA clones [24]. On the other hand, some investigations, including sequential hyphal tip isolation [25], protoplasts/small mycelial fragments regeneration [26], treatment of cycloheximide [27,28], or cAMP-rifamycin [29], and single spore isolation [20] were also explored to cure the viruses in their original strains. Therefore, similar experiments will also be carried out to elucidate the role of BcMyV1 on *B. cinerea* biology in the future.

It is of interest that BcMyV1 was detected in two different fungal species, *B. cinerea* and *S. sclerotiorum*. Although the same mycovirus is rarely detected in different fungi, there are still a few exceptions. In addition to the present case, another mycovirus, Botrytis porri botybirnavirus 1 (BpBV1), was also detected in both *B. porri* and *S. sclerotiorum* [18,21]. We suppose that viral interspecific transmission may frequently occur between *B. cinerea* and *S. sclerotiorum*, although they belong to different genera. Some factors may increase the possibility of viral transmission between the two species. Firstly, *B. cinerea* is a close relative of *S. sclerotiorum*, and their genes share 83% aa identity on average between the two fungi [30]. Thus, viruses may adapt a new host more easily when viral transmission occurs from one host to the other. Secondly, both fungi have a broad host range [1,31], and many plant species are hosts for both *B. cinerea* and *S. sclerotiorum*. Therefore, the contact between the two fungi may frequently occur in small niches under field conditions [30]. Thirdly, viral interspecies transmission through anastomosis has also been reported in a few cases, including from *Aspergillus niger* to *A. nidulans* [32], from *S. sclerotiorum* to *S. minor* [33], and from *Cryphonectria parasitica* to *C. nitschkei* [34], suggesting that viral interspecies transmission by anastomosis between *B. cinerea* and *S. sclerotiorum* might also be possible. Finally, recent studies have shown that insects and mites are also potential vectors during mycoviral transmission [35,36], and similar mechanisms may increase the possible viral transmission between *B. cinerea* and *S. sclerotiorum* as well.

Although –ssRNA viruses have been reported in several fungal species [8,14–18], the information of their incidence and distribution remain unclear. Moreover, investigation of other BcMyV1-infected *B. cinerea* strains may also help us to uncover their effects on *B. cinerea*. On the contrary, unlike strain Ecan17-2, two strains of *B. cinerea* (Bs6-33 and JLaub-11) carrying BcMyV1 grew quickly and produced massive sclerotia, and only one strain, Bs6-23, showed similar culture morphology to that of strain Ecan17-2 (data not shown). This suggests that more complex interactions may exist between BcMyV1 and the *B. cinerea* population. Therefore, genome-wide association study [37,38] may be helpful to further elucidate the role of genetic variation in the response of *B. cinerea* to viral infection. Compared with other viruses infecting *B. cinerea*, including Botrytis cinerea endornavirus 1 [2], Botrytis cinerea hypovirus 1, BcFV1 [3], Botrytis cinerea mitovirus 1 [39], Botrytis virus F, and Botrytis virus X [5], the incidence of BcMyV1 in the Chinese *B. cinerea* population was very low, accounting for only 0.8% in the tested strains. The linear distance between Beijing and Shiyan is over 900 km, and Changchun is over 800 km away from Beijing. These results indicate that BcMyV1 might be widely distributed in regions where *B. cinerea* occurs of China, although the infection rate is low. This may be one reason why (−ss)RNA viruses were rarely reported in the population of *B. cinerea*. Despite the low incidence, BcMyV1 still had a wide geographic distribution and may not be only limited to China, as the homologous virus-like contig was obtained from *S. sclerotiorum* strains in Australia. This suggests BcMyV1 may have a global distribution.

Supplementary Materials: The following are available online at http://www.mdpi.com/1999-4915/10/8/432/s1. Table S1: Oligonucleotide primers/adaptor used in this study. Figure S1: Schematic representation of the strategy used for full cDNA sequence cloning of Botrytis cinerea mymonavirus 1 (BcMyV1). Figure S2: Transmembrane domains prediction for ORF1 encoded polypeptide L of BcMyV1.

Author Contributions: F.H., M.W., and G.L. conceived and designed the experiments. F.H. performed the experiments; F.H. and M.W analyzed the data. F.H. and M.W wrote the paper.

Funding: This research was funded by the R & D Special Fund for Public Welfare Industry (Agriculture) of China (grant number 201303025) and the Natural Science Foundation of China (grant number 31401690, 31772212).

Conflicts of Interest: The authors declare that there is no conflict of interest.

References

1. Elad, Y.; Pertot, I.; Marina, A.; Prado, A.M.; Stewart, A. Plant Hosts of *Botrytis* spp. In *Botrytis—The Fungus, the Pathogen and Its Management in Agricultural Systems*; Fillinger, S., Elad, Y., Eds.; Springer: Cham, Switzerland, 2016; pp. 413–486.
2. Hao, F.M.; Zhou, Z.L.; Wu, M.D.; Li, G.Q. Molecular characterization of a novel endornavirus from the phytopathogenic fungus *Botrytis cinerea*. *Arch. Virol.* **2017**, *162*, 313–316. [CrossRef] [PubMed]
3. Hao, F.M.; Ding, T.; Wu, M.D.; Zhang, J.; Yang, L.; Chen, W.D.; Li, G.Q. Two novel hypovirulence-associated mycoviruses in the phytopathogenic fungus *Botrytis cinerea*: Molecular characterization and suppression of infection cushion formation. *Viruses* **2018**, *10*, 254. [CrossRef] [PubMed]
4. Wu, M.D.; Zhang, J.; Yang, L.; Li, G.Q. RNA mycoviruses and their role in *Botrytis* Biology. In *Botrytis—The Fungus, the Pathogen and Its Management in Agricultural Systems*; Fillinger, S., Elad, Y., Eds.; Springer: Cham, Switzerland, 2016; pp. 71–90.
5. Pearson, M.N.; Bailey, A.M. Viruses of botrytis. *Arch. Virol.* **2013**, *86*, 249–272.
6. Yu, L.; Sang, W.; Wu, M.D.; Zhang, J.; Yang, L.; Zhou, Y.J.; Chen, W.D.; Li, G.Q. Novel hypovirulence-associated RNA mycovirus in the plant pathogenic fungus *Botrytis cinerea*. *Appl. Environ. Microbiol.* **2015**, *81*, 2299–2310. [CrossRef] [PubMed]
7. Donaire, L.; Rozas, J.; María, A. Molecular characterization of Botrytis ourmia-like virus, a mycovirus close to the plant pathogenic genus *Ourmiavirus*. *Virology* **2016**, *489*, 158–164. [CrossRef] [PubMed]
8. Donaire, L.; Pagan, I.; Ayllon, M.A. Characterization of *Botrytis cinerea* negative-stranded RNA virus 1, a new mycovirus related to plant viruses, and a reconstruction of host pattern evolution in negative-sense ssRNA viruses. *Virology* **2016**, *499*, 212–218. [CrossRef] [PubMed]
9. Easton, A.J.; Pringle, C.R. Order Mononegavirales. In *Virus Taxonomy: Classification and Nomenclature of Viruses, Ninth Report of the International Committee on Taxonomy of Viruses*; King, M.Q., Adams, M.J., Carstens, E.B., Lefkowitz, E.J., Eds.; Elsevier, Academic Press: London, UK, 2011; pp. 653–657.
10. Dietzgen, R.G.; Kondo, H.; Goodin, M.M.; Kurath, G.; Vasilakis, N. The family *Rhabdoviridae*: mono- and bipartite negative-sense RNA viruses with diverse genome organization and common evolutionary origins. *Virus Res.* **2017**, *227*, 158–170. [CrossRef] [PubMed]
11. Kondo, H.; Chiba, S.; Toyoda, K.; Suzuki, N. Evidence for negative-strand RNA virus infection in fungi. *Virology* **2013**, *435*, 201–209. [CrossRef] [PubMed]
12. Afonso, C.L.; Amarasinghe, G.K.; Bányai, K.; Bao, Y.; Basler, C.F.; Bavari, S.; Bejerman, N.; Blasdell, K.R.; Briand, F.; Briese, T.; et al. Taxonomy of the order *Mononegavirales*: Update. 2016. *Arch. Virol.* **2016**, *161*, 2351–2360. [CrossRef] [PubMed]
13. Walker, P.J.; Firth, C.; Widen, S.G.; Blasdell, K.R.; Guzman, H.; Wood, T.G.; Paradkar. P.N.; Holmes, E.C.; Tesh, R.B.; Vasilakis, N. Evolution of genome size and complexity in the *Rhabdoviridae*. *PLoS Pathog.* **2015**, *11*, 1–25. [CrossRef] [PubMed]
14. Liu, L.; Xie, J.; Cheng, J.; Fu, Y.; Li, G.; Yi, X.; Jiang, D. Fungal negative-stranded RNA virus that is related to bornaviruses and nyaviruses. *Proc. Natl. Acad. Sci. USA* **2014**, *111*, 12205–12210. [CrossRef] [PubMed]
15. Marzano, S.Y.; Domier, L.L. Novel mycoviruses discovered from metatranscriptomics survey of soybean phyllosphere phytobiomes. *Virus Res.* **2016**, *213*, 332–342. [CrossRef] [PubMed]

16. Marzano, L.S.Y.; Nelson, B.D.; Ajayi-Oyetunde, O.; Bradley, C.A.; Hughes, T.J.; Hartman, G.L.; Eastburn, D.M.; Domiera, L.L. Identification of diverse mycoviruses through metatranscriptomics characterization of the viromes of five major fungal plant pathogens. *J. Virol.* **2016**, *90*, 6846–6863. [CrossRef] [PubMed]

17. Wang, L.; He, H.; Wang, S.; Chen, X.; Qiu, D.; Kondob, H.; Guo, L. Evidence for a novel negative-stranded RNA mycovirus isolated from the plant pathogenic fungus *Fusarium graminearum*. *Virology* **2018**, *518*, 232–240. [CrossRef] [PubMed]

18. Mu, F.; Xie, J.T.; Cheng, S.F.; You, M.P.; Barbetti, M.J.; Jia, J.C.; Wang, Q.Q.; Cheng, J.S.; Fu, Y.P.; Chen, T.; et al. Virome characterization of a collection of *Sclerotinia sclerotiorum* from Australia. *Front. Microbial.* **2018**, *8*, 2540. [CrossRef] [PubMed]

19. Wu, M.D.; Zhang, L.; Li, G.Q. Genome characterization of a debilitation-associated mitovirus infecting the phytopathogenic fungus *Botrytis cinerea*. *Virology* **2010**, *406*, 117–126. [CrossRef] [PubMed]

20. Wu, M.D.; Zhang, L.; Li, G.Q.; Jiang, D.H.; Hou, M.S.; Huang, H.C. Hypovirulence and double-stranded RNA in *Botrytis cinerea*. *Phytopathology* **2007**, *97*, 1590–1599. [CrossRef] [PubMed]

21. Wu, M.D.; Jin, F.Y.; Zhang, J.; Yang, L.; Jiang, D.H.; Li, G.Q. Characterization of a novel bipartite double-stranded RNA mycovirus conferring hypovirulence in the phytopathogenic fungus *Botrytis porri*. *J. Virol.* **2012**, *86*, 6605–6619. [CrossRef] [PubMed]

22. Kumar, S.; Stecher, G.; Tamura, K. MEGA7: Molecular Evolutionary Genetics Analysis version 7.0 for bigger datasets. *Mol. Blol. Evol.* **2016**, *33*, 1870–1874. [CrossRef] [PubMed]

23. Krogh, A.; Larsson, B.; von Heijne, G.; Sonnhammer, E.L. Predicting transmembrane protein topology with a hidden Markov model: Application to complete genomes. *J. Mol. Biol.* **2001**, *305*, 567–580. [CrossRef] [PubMed]

24. Choi, G.H.; Nuss, D.L. Hypovirulence of chestnut blight fungus conferred by an infectious viral cDNA. *Science* **1992**, *257*, 800–803. [CrossRef] [PubMed]

25. Van Diepeningen, A.D.; Debets, A.J.M.; Hoekstra, R.F. Dynamics of dsRNA mycoviruses in black *Aspergillus* populations. *Fungal. Genet. Biol.* **2006**, *43*, 446–452. [CrossRef] [PubMed]

26. Kim, J.M.; Jung, J.E.; Park, J.A.; Park, S.M.; Cha, B.J.; Kim, D.H. Biological function of a novel chrysovirus, CnV1-BS122, in the Korean *Cryphonectria nitschkei* BS122 strain. *J. Biosci. Bioeng.* **2013**, *115*, 1–3. [CrossRef] [PubMed]

27. Fulbright, D.W. Effect of eliminating dsRNA in hypovirulent *Endothia parasitica*. *Phytopathology* **1984**, *74*, 722–724. [CrossRef]

28. Elias, K.S.; Cotty, P.J. Incidence and stability of infection by double-stranded RNA genetic elements in *Aspergillus* section *Flavi* and effects on aflatoxigenicity. *Can. J. Bot.* **1996**, *74*, 716–725. [CrossRef]

29. Kwon, Y.C.; Jeong, D.W.; Gim, S.I.; Ro, H.S.; Lee, H.S. Curing viruses in *Pleurotus ostreatus* by growth on a limited nutrient medium containing cAMP and rifamycin. *J. Virol. Methods* **2012**, *185*, 156–159. [CrossRef] [PubMed]

30. Amselem, J.; Cuomo, C.A.; van Kan, J.A.L.; Viaud, M.; Benito, E.P.; Couloux, A.; Coutinho, P.M.; de Vries, R.P.; Dyer, P.S.; Fillinger, S.; et al. Genomic analysis of the necrotrophic fungal pathogens *Sclerotinia sclerotiorum* and *Botrytis cinerea*. *PLoS Genet.* **2011**, *7*, e1002230. [CrossRef] [PubMed]

31. Bolton, M.D.; Thomma, B.P.; Nelson, B.D. *Sclerotinia sclerotiorum* (Lib.) de Bary: Biology and molecular traits of a cosmopolitan pathogen. *Mol. Plant. Pathol.* **2006**, *7*, 1–16. [CrossRef] [PubMed]

32. Coenen, A.; Kevei, F.; Hoekstra, R.F. Factors affecting the spread of double-stranded RNA viruses in *Aspergillus nidulans*. *Genet. Res.* **1997**, *69*, 1–10. [CrossRef] [PubMed]

33. Melzer, M.S.; Ikeda, S.S.; Boland, G.J. Interspecific transmission of double-stranded RNA and hypovirulence from *Sclerotinia sclerotiorum* to *S. minor*. *Phytopathology* **2002**, *92*, 780–784. [CrossRef] [PubMed]

34. Liu, Y.C.; Hillman, B.I.; Linder-Basso, D.; Kaneko, S.; Milgroom, M.G. Evidence for interspecies transmission of viruses in natural populations of filamentous fungi in the genus *Cryphonectria*. *Mol. Ecol.* **2003**, *12*, 1619–1628. [CrossRef] [PubMed]

35. Bouneb, M.; Turchetti, T.; Nannelli, R.; Roversi, P.F.; Paoli, F.; Danti, R.; Simoni, S. Occurrence and transmission of mycovirus Cryphonectria hypovirus 1 from dejecta of *Thyreophagus corticalis* (Acari, Acaridae). *Fungal Biol.* **2016**, *120*, 351–357. [CrossRef] [PubMed]

36. Liu, S.; Xie, J.; Cheng, J.; Li, B.; Chen, T.; Fu, Y. Fungal DNA virus infects a mycophagous insect and utilizes it as a transmission vector. *Proc. Natl. Acad. Sci. USA* **2016**, *113*, 12803–12808. [CrossRef] [PubMed]

37. Korinsak, S.; Tangphatsornruang, S.; Pootakham, W.; Wanchana, S.; Plabpla, A.; Jantasuriyarat, C.; Patarapuwadol, S.; Vanavichit, A.; Toojinda, T. Genome-wide association mapping of virulence gene in rice blast fungus *Magnaporthe oryzae* using a genotyping by sequencing approach. *Genomics* **2018**. [CrossRef] [PubMed]

38. Castiblanco, V.; Marulanda, J.J.; Würschum, T.; Miedaner, T. Candidate gene based association mapping in *Fusarium culmorum* for field quantitative pathogenicity and mycotoxin production in wheat. *BMC Genet.* **2017**, *18*, 49. [CrossRef] [PubMed]

39. Rodríguez-García, C.; Medina, V.; Alonso, A.; Ayllón, M.A. Mycoviruses of *Botrytis cinerea* isolates from different hosts. *Ann. Appl. Biol.* **2013**, *164*, 46–61. [CrossRef]

viruses

MDPI

Article

A Novel Partitivirus in the Hypovirulent Isolate QT5-19 of the Plant Pathogenic Fungus *Botrytis cinerea*

Md Kamaruzzaman [1], Guoyuan He [1], Mingde Wu [1], Jing Zhang [1], Long Yang [1], Weidong Chen [2] and Guoqing Li [1,*]

[1] The Key Laboratory of Plant Pathology of Hubei Province and The State Key Laboratory of Agricultural Microbiology, Huazhong Agricultural University, Wuhan 430070, China; kamaru.m@webmail.hzau.edu.cn (M.K.); hgygh2015@163.com (G.H.); mingde@mail.hzau.edu.cn (M.W.); zhangjing1007@mail.hzau.edu.cn (J.Z.); yanglong@mail.hzau.edu.cn (L.Y.)
[2] U. S. Department of Agriculture, Agricultural Research Service, Washington State University, Pullman, WA 99164, USA; w-chen@wsu.edu
*   Correspondence: guoqingli@mail.hzau.edu.cn

Received: 30 October 2018; Accepted: 28 December 2018; Published: 3 January 2019

Abstract: A pink isolate (QT5-19) of *Botrytis cinerea* was compared with three gray isolates of *B. cinerea* for growth and morphogenesis on potato dextrose agar (PDA), and for pathogenicity on tobacco. A double-stranded (ds) RNA mycovirus infecting QT5-19 was identified based on its genome feature and morphology of the virus particles. The results showed that QT5-19 grew rapidly and established flourishing colonies as the gray isolates did. However, it is different from the gray isolates, as it failed to produce conidia and sclerotia asthe gray isolates did. QT5-19 hardly infected tobacco, whereas the gray isolates aggressively infected tobacco. Two dsRNAs were detected in QT5-19, dsRNA 1 and dsRNA 2, were deduced to encode two polypepetides with homology to viral RNA-dependent RNA polymerase (RdRp) and coat protein (CP), respectively. Phylogenetic analysis of the amino acid sequences of RdRp and CP indicated that the two dsRNAs represent the genome of a novel partitivirus in the genus *Alphapartitivirus*, designated here as Botrytis cinerea partitivirus 2 (BcPV2). BcPV2 in QT5-19 was successfully transmitted to the three gray isolates through hyphal contact. The resulting BcPV2-infected derivatives showed rapid growth on PDA with defects in conidiogenesis and sclerogenesis, and hypovirulence on tobacco. This study suggests that BcPV2 is closely associated with hypovirulence of *B. cinerea*.

Keywords: *Botrytis cinerea*; hypovirulence; partitivirus; conidiogenesis; sclerogenesis

1. Introduction

Mycoviruses (or fungal viruses) are viruses infecting filamentous fungi, yeasts, and oomycetes [1–3]. Previous studies demonstrated that mycoviruses widely exist in all major taxonomic groups of fungi and oomycetes [4]. Most mycoviruses have the genomes of RNA, either double-stranded RNA (dsRNA) or single-stranded RNA (ssRNA), whereas a few mycoviruses have the genomes of single-stranded DNA or ssDNA [5]. The mycoviruses in the families *Chrysoviridae*, *Megabirnaviridae*, *Patitiviridae*, *Reoviridae*, and *Totiviridae* have the dsRNA genomes, which are encapsidated within the coat proteins (CP), thereby forming virus particles. On the other hand, the mycoviruses in the other families such as *Hypoviridae* and *Narnaviridae* have the positive ssRNA (+ssRNA) genomes, which are unencapsidated without formation of virus particles [6,7]. Recently, a few mycoviruses with the genomes of negative ssRNA (−ssRNA) have been identified in a few plant pathogenic fungi, including *Botrytis cinerea* [8,9], *Erysiphe pisi* [10], *Fusarium graminearum* [11], and *Sclerotinia sclerotiorum* [12].

Partitiviruses have been identified in fungi, as well as in plants and protozoa [1,13]. The genome of each partitivirus possesses two dsRNA segments (dsRNA 1 and dsRNA 2), which have the length ranging from 1300 bp to 2500 bp. Each dsRNA segment contains one open reading frame (ORF) on the positive RNA strand. The ORF in dsRNA 1 usually codes for RNA-dependent RNA polymerase (RdRp), whereas the ORF in dsRNA 2 usually codes for coat protein (CP), thereby forming virus particles with the isometric shape of 25 to 40 nm in diameter [13,14].

Botrytis cinerea Pers.: Fr. (teleomorph *Botryotinia fuckeliana* (de Bary) Whetzel) is a ubiquitous plant pathogenic fungus [15,16]. It causes gray mold disease on more than 1400 plant species, including many economically important crops such as cucumber (*Cucumis sativus*), strawberry (*Fragaria* × *ananassa*), table grapes (*Vitis vinifera*), and tomato (*Lycopersicon esculentum*) [17]. Substantial economic losses can be caused by *B. cinerea* in these crops before or after harvest under cool-to-temperate and humid conditions [18]. *B. cinerea* is a typical necrotrophic fungus and it usually uses an arsenal of chemical weapons (i.e., cell-wall-degrading enzymes and phytotoxic metabolites) to infect plant tissues [19]. Thus far, cultivars/varieties highly resistant to infection by *B. cinerea* are not available in the crops mentioned above. Control of *B. cinerea* in these crops largely depends on the repeated use of fungicides in China as well as in many other countries [16]. The chemical control is effective when appropriate fungicides are applied to plant tissues in a timely manner. However, some unintended side effects, including fungicide residues in crop produces, pollution to environments, development of fungicide-resistant individuals of *B. cinerea*, may arise due to repeated use of fungicides, thereby causing public concerns over use of the fungicides for control of *B. cinerea*. Therefore, it is necessary to explore alternative measures to control *B. cinerea*, including biological control using 'natural enemies' such as mycoviruses.

Studies on mycoviruses infecting *B. cinerea* dated back to the middle of the 1990s, when Howitt and colleagues (1995) first observed virus-like particles (VLPs) in *B. cinerea* and detected dsRNA molecules in this fungus [20]. Since then, many researchers showed interests in identification of mycoviruses in different isolates of *B. cinerea* [8,21–30]. Several mycoviruses from *B. cinerea* have been characterized at the genome level, including Botrytis virus F (BVF) [21], Botrytis virus X (BVX) [22], Botrytis cinerea endornavirus 1 (BcEV1) [27], Botrytis cinerea hypovirus 1 (BcHV1) [28], Botrytis cinerea fusarivirus 1 (BcFV1) [28], Botrytis cinerea mitovirus 1 (BcMV1) [24], Botrytis cinerea mymonavirus 1 (BcMyV1) [9], Botrytis cinerea negative-stranded RNA virus 1 (BcNSRV-1) [8], Botrytis ourmia-like virus (BOLV1) [26], Botrytis cinerea partitivirus 1 (BcPV1) [31], Botrytis cinerea RNA virus 1 (BcRV1) [32], Botryotinia fuckeliana partitivirus 1 (BfPV1) [33], and Botryotinia fuckeliana totivirus 1 (BfTV1) [34]. Among these mycoviruses, BcEV1, BcFV1, BcHV1, BcMV1, BcMyV1, BcNSRV-1, BcPV1, BcRV1, and BOLV1 were found to be closely associated with hypovirulence of *B. cinerea* [8,9,23,26,28,31,32]. However, none of these mycoviruses has been found to have potential as a biocontrol agent for control of the diseases caused by *B. cinerea*. Two factors might be responsible for this situation. First, the mycoviruses have limited horizontal transmissibility possibly due to hyphal (vegetative) incompatibility between the mycoviruses-containing and mycoviruses-free individuals. Second, the mycovirus-infected *B. cinerea* isolates usually grow poorly [23,35] and they may have lower competitive ability than the virulent isolates. Therefore, potential mycoviruses for successful control of *B. cinerea* need to overcome these limitations.

A pink-colored isolate (QT5-19) of *B. cinerea* was obtained from a diseased fruit of tomato collected in Shaanxi Province of China. Our preliminary study showed that QT5-19 was almost non-pathogenic on many plants, including tomato, tobacco and oilseed rape (*Brassica napus*), suggesting that this isolate was debilitated in pathogenicity or virulence. It is suspected that some mycoviruses may exist in this particular isolate. Therefore, this study was conducted to characterize the mycovirus in QT5-19.

2. Materials and Methods

2.1. Fungal Isolates and Cultural Conditions

Five fungal isolates of *B. cinerea* (08168, B05.10, QT5-19, RoseBC-3, XN-1) were used in this study. QT5-19 was isolated from a diseased fruit of tomato collected from Shaanxi Province of China. The identity of QT5-19 as *B. cinerea* was confirmed both by PCR detection using the *B. cinerea*-specific PCR primers reported by [36], and by analysis of the sequence of ITS-rDNA [37] region. Four other *B. cinerea* isolates (08168, B05.10, RoseBC-3, XN-1) were obtained from our previous collection and their origin was described in related reports [37–41]. All the fungal isolates were incubated at 20 °C on potato dextrose agar (PDA) in the dark for 2 to 3 days to determine radial growth rates and for 20 days to determine sclerotial production [42].

Production of the pink color in the colonies of QT5-19 suggests that this isolate may produce the pink pigment bikaverin [43]. In order to confirm this hypothesis, expression of six bikaverin biosynthesis-related genes (*bcbik1* to *bcbik6*) in QT5-19 (pink color) and B05.10 (gray color) was detected by RT-PCR (Figure S1) using the total RNA of each isolate as template and the specific PCR primer sets listed in Table S1. To confirm correctness of the *bcbik1* gene in QT5-19, the RT-PCR product of *bcbik1* from QT5-19 was purified from the agarose gel, cloned into *E. coli* DH5α and sequenced, and the resulting cDNA sequence was compared with the DNA sequence of *bcbik1* in *B. cinerea* isolate 1750 (GenBank Acc. No. HE802550) (Figure S2).

2.2. Determination of Pathogenicity

Seeds of tobacco (*Nicotiana benthamiana*) were sown in plastic pots containing Organic Culture Mix containing 2% to 5% of N + P_2O_5 + K_2O (*w:w*, N:P_2O_5:K_2O = 1:1:1) (Zheng Jiang Pei Lei Organic Manure Manufacturing Co., Ltd., Zhengjiang City, Jiangsu Province, China). The pots were placed in a growth room at 20 °C for 40 days under the lighting regime of 12-h light/12-h dark and watered as required. Fully-expanded leaves were detached from the plants and placed on moist paper towels in a plastic tray (45 × 30 × 2.5 cm, length × width × height) with their adaxial surface facing up. Mycelial agar plugs (5 mm in diameter) from the margin area of a two-day-old PDA culture of an investigated isolate of *B. cinerea* were placed on the leaves, one mycelial agar plug per leaf and six leaves (as six replicates) for each isolate. The tray was covered with a 0.1-mm-thick transparent plastic film (Gold Mine Plastic Industry Ltd., Jiang Men City, China) to maintain high humidity and placed in a growth chamber at 20 °C for three days (12-h light/12-h dark). Diameter of the leaf lesion formed around each mycelial agar plug was measured.

Additionally, isolates QT5-19 and B05.10 were tested for infection of apple (*Malus demestica*), table grapes (*Vitis vinifera*), tomato (*Lycopersicon esculentum*), cucumber (*Cucumis sativus*), oilseed rape (*Brassica napus*), and strawberry (*Fragaria × ananassa*) (cultivars unknown for all of the investigated crops). Fruits of apple, table grapes and tomato purchased from a local supermarket, and leaves of cucumber, oilseed rape, and strawberry detached from adult plants were placed on moist paper towels in plastic trays with three replicates for each of the two isolates, three fruits or three leaves per replicate, one agar plug on each fruit or leaf. The trays were individually covered with transparent plastic films and placed in the growth chamber (20 °C) for three days. Lesion diameter on each fruit or leaf around each mycelial agar plug was measured (Figure S3).

2.3. Extraction of dsRNAs

Isolate QT5-19 was inoculated on cellophane films (CF) placed on PDA in Petri dishes as the CF-PDA cultures. The cultures were incubated at 20 °C in the dark for three days. The mycelial mass was harvested from the cultures and ground to a fine powder in liquid nitrogen. DsRNAs were extracted from the mycelial powder using the cellulose (CF-11) chromatography method described by Wu and colleagues (2007, 2010, 2012). The extract was digested with RQ1 RNase-free DNase (Promega, Madison, WI, USA) for elimination of DNA contamination and with S1 nuclease (TaKaRa) for elimination of single-stranded RNAs (ssRNAs) in the extracts. The dsRNAs were detected by agarose gel (0.7% or 1.0%, w/v) and visualized by staining with Biosharp® SuperRed/GelRed (Guangzhou Sai Guo Biotechnol. Co., Ltd., Guangzhou, China).

2.4. cDNA Cloning of the dsRNAs

The dsRNAs extracted from isolate QT5-19 were purified from an agarose gel after electrophoresis using Axygen® DNA gel extraction kit (Axygen® Scientific Inc., Union City, CA, USA). The pure dsRNAs were ligated with the adaptor 110a (Figure S4, Table S1) at the 3′-terminus using T4 RNA ligase (Promega, Madison, WI, USA) at 16 °C for 12 h. The adaptor-ligated dsRNAs were purified using AxyPrep™ PCR cleanup kit (Axygen Scientific Inc., Union City, CA, USA). For synthesis of cDNAs, the adaptor-ligated dsRNAs (10 µL) were mixed with 1 µL oligonucleotide primer RC110a (complementary to the adaptor 110a, Supporting Information Table S1), 1 µL dNTPs (10 mmol/L), 1 µL dimethyl sulfoxide (DMSO) and 2 µL DEPC-H$_2$O. The mixture was denatured at 98 °C and chilled on ice for 120 s. Then, PrimeScript II reverse transcriptase (TaKaRa Biotechnol. Co., Ltd., Dalian, China) and 5× PrimeScript II buffer (TaKaRa) were added to the mixture and the reverse transcription was done at 42 °C for 1.5 h. The full-length cDNAs were PCR amplified in S1000™ Thermal Cycler (BIO-RAD Laboratories Inc., Hercules, CA USA) using the synthetic cDNAs as templates in the presence of the primer RC110a [44]. The cDNA cloning was repeated five times to guarantee sequence accuracy. All the PCR products were separated by 1% (w/v) agarose gel electrophoresis, followed by gel purification of the cDNA molecules and ligation of the cDNAs into the pMD18-T vector (TaKaRa), which was then transformed into *E. coli* DH5α (TaKaRa) for proliferation and sequencing.

2.5. Analysis of the Mycoviral Genome

The open reading frame (ORF) in the full-length cDNA sequence of each mycoviral dsRNA was identified using the ORF Finder program in GenBank at the website of NCBI (https://www.ncbi.nlm.nih.gov/) with the standard codon usage. The conserved domains and motifs in each ORF were identified by comparison with those in related genomes deposited in the public databases of GenBank and PROSITE (http://prosite.expasy.org/). Multiple sequence alignment of the amino acid sequence for RNA-dependent RNA polymerase (RdRp) was performed using the T-Coffee server in the website (http://tcoffee.crg.cat/apps/tcoffee). The gaps were manually removed using the CLUSTAL_W program. Nucleotide (nt) and amino acid (aa) sequences of the previously-reported partitiviruses and related members were retrieved from GenBank (Table S2). They were used for comparative analysis and for inference of the phylogenetic relationship of the mycovirus in QT5-19. Phylogenetic trees were generated based on the amino acid sequences of RNA-dependent RNA polymerase (RdRp) and coat protein (CP) using the Maximum-Likelihood (ML) method in the MEGA software version 7.0 [45]. They were individually tested with the bootstrap value of 1000 to ascertain the reliability of a given pattern of the ML trees. Potential secondary structures at the 5′- and 3′-termini of each dsRNA were predicted using RNA Structure version 5.8.1 [46].

2.6. Purification of Virus Particles

Isolate QT5-19 was incubated on CF-PDA at 20 °C for eight days. The mycelial mass (~8 g) harvested from the CF-PDA cultures was ground to fine powder in liquid nitrogen and the powder was suspended in 200 mL phosphate-buffered saline (PBS) (0.1 mol/L, pH 7) amended with 6 mL Triton-X 100. The slurry was further homogenized in a glass homogenizer for 30 min. The homogenate was transferred to a 200-mL-plastic tube and centrifuged at 10,000× g for 20 min to remove the hyphal debris. The supernatant was transferred to another plastic tube and centrifuged at 119,000× g under 4 °C for 2 h. The supernatant was discarded and the pellet was suspended in 0.1 mol/L PBS and the suspension was centrifuged at 16,000× g for 20 min to remove large particles. The supernatant was loaded in a centrifuge tube containing sucrose with the concentration gradient ranging from 10% to 50% (w/v) and centrifuged at 70,000× g under 4 °C for 2 h. The fractions were carefully collected by punching the centrifuge tube with a clean sterilized syringe and separately measured for the existence of the virus partials by detection of the abundance of the dsRNAs by agarose electrophoresis. The fraction containing the virus particles was suspended in 50 μL PBS (50 mmol/L, pH 7.0). The virus particles were stained with phosphotungstic acid (20 g/L, *w/v*, pH 7.4) and observed under transmission electron microscope (TEM). The dsRNAs from the viral particles were extracted with phenol chloroform isoamyl alcohol (25:24:1, *v:v:v*), and detected by agarose gel electrophoresis. Meanwhile, the purified virus particles in PBS were boiled and denatured at 100 °C for 5 min, and the solution was centrifuged at 12,000 r/min for 2 min. The resulting supernatant was mixed with 5× Tris-glycerol loading buffer (250 mM Tris·HCl, 10% SDS, 30% glycerol, 10 mM DTT, 0.05% bromophenol blue, pH 6.8) at the volume ratio of 1:5. The mixture was the loaded in a 10% SDS-PAGE gel. The pre-stained protein ladder, PageRuler™ (Thermo Fisher Scientific Inc., Watham, MA USA) was loaded in the gel as the molecular weight standard. After electrophoresis (120 V, 2.5 h), the gel was stained with Coomassie brilliant blue R-250.

2.7. RT-PCR Detection of the Mycovirus in QT5-19 and Northern Blotting

Total RNA was extracted from the mycelia of the investigated isolates of *B. cinerea* using TaKaRa RNAiso Plus Total RNA Extraction Reagents (TaKaRa). It was used in reverse transcription-PCR (RT-PCR) for detection of the mycovirus in QT5-19 and other isolates with the specific primer pairs BcPVRd-F/BcPVRd-R and BcPVCP-F/BcPVCP-R targeting the ORFs for RdRp and CP, respectively (Table S1). Meanwhile, a specific primer pair Act-F/Act-R targeting the actin gene of *B. cinerea* was used as a reference gene in RT-PCR [32].

Northern blotting was carried out to confirm authenticity of the full-length cDNAs from the dsRNAs in isolate QT5-19. It was done using the procedure described in our previous studies [23,35]. Two DNA probes targeting the two ORFs of the dsRNAs coding for RdRp and CP were generated by PCR with the two cDNAs as templates and the specific primer pairs RdRp-F/RdRp-R and CP-F/CP-R (Figure S5, Table S1).

2.8. Horizontal Transmission of the Mycovirus in QT5-19

Horizontal transmission of the mycovirus from QT5-19 (donor) to *B. cinerea* isolates 08168, B05.10, RoseBc-3, and XN-1 (recipients) was done on PDA in Petri dishes (9 cm in diameter) using the pair culturing technique described in our previous study [44]. Three derivative isolates were obtained from the colonies of each recipient in the four-day-old pair-cultures. These derivatives were easily distinguished from QT5-19, as they formed gray-colored colonies rather than pink-colored colonies like those of QT5-19. Five representative derivatives (08168T, B05.10T, RoseBc-3T, and XN-1T from 08168, B05.10, RoseBc-3, and XN-1 in the pair-cultures, respectively) were tested for the presence of the mycovirus by dsRNA profiling and RT-PCR with the mycovirus-specific primer pairs RdRp-f1/RdRp-r1 and CP-f1/CP-r1 (Table S1). Meanwhile, these derivatives, as well as their parents, were tested for growth rates and production of conidia/sclerotia on PDA, and for pathogenicity on tobacco leaves using the methods described above.

2.9. Elimination of the Mycovirus in QT5-19

Four trials (protoplast regeneration, hyphal tipping, thermal treatment, and ribavirin treatment) were done to eliminate the mycovirus in isolate QT5-19. In the protoplast regeneration trial, the two-day-old mycelial mass of QT5-19 from the CF-PDA cultures was harvested and blended to hyphal fragments in sterile distilled water (10 g fresh mycelia in 100 mL water). The resulting hyphal fragment suspension (HFS) was inoculated in a 250-mL-Erlenmeyer flask containing PDB at the volume ratio of 1:20 (HFS:PDB), and incubated overnight at 150 rpm and 20 °C. The culture was centrifuged at 2000× *g* and 4 °C for 10 min to collect the mycelial pellet, which was then suspended in 1 mol/L sorbitol amended with 1% (*w/v*) Lysing Enzyme L2265 (Sigma-Aldrich Chemical Company Ltd., St. Louis, MO, USA) and 0.1% snailase (*w/v*) (Solarbio® Life Sciences, Beijing, China). The suspension was gently shaken at 30 °C for 3 h to release the protoplasts and the resulting mixture was filtered through double-layered filter papers to remove the hyphal fragments. The protoplasts were collected by centrifugation under 4 °C at 2000× g for 20 min, washed twice with 1 mol/L D-sorbitol, and suspended in STC buffer (1 mol/L D-sorbitol, 50 mmol/L Tris-HCl at pH 7.5, 50 mmol/L CaCl$_2$). The suspension was plated on TB3 medium (2 g casein hydrolysate, 5 g yeast extract, 25 g sucrose, 15 g agar, 1000 mL water) and the culture was incubated at 20 °C for seven days for protoplast regeneration and hyphal growth. Fifty-one emerged fungal colonies were individually transferred to fresh PDA, incubated at 20 °C in the dark for 15 days and the colony morphology (color, production of conidia and sclerotia) was observed.

In the hyphal-tipping trial, QT5-19 was incubated on water agar at 20 °C for three days. The cultures were placed under a dissecting light microscope. Hyphal tips (approximately 80 μm in length) in the colony margin were individually cut using a fine needle and transferred to PDA in Petri dishes, one hyphal tip per dish. The resulting cultures were incubated at 20 °C for 15 days and the colony morphology was observed.

In the thermal treatment, QT5-19 was incubated on PDA at 30 °C for 30 days. Mycelial agar plugs (~3 mm in diameter) were removed from the margin area of the cultures. They were individually transferred to fresh PDA in Petri dishes, one mycelial agar plug per dish. The resulting cultures were incubated at 30 °C for 30 days and each culture was sub-cultured again on PDA at 30 °C for 30 days. Finally, the cultures were transformed to PDA and incubated at 20 °C for 15 days, and the colony morphology of each subculture was observed.

In the ribavirin treatment, QT5-19 was inoculated in PDA amended with ribavirin at 0.5, 1.0, or 1.5 mg/mL. The cultures were incubated at 20 °C for seven days. Then, mycelial agar plugs were removed from the cultures of each ribavirin treatment and transferred to new PDA containing the same concentrations of ribavirin and the subcultures were incubated at 20 °C also for seven days. This process was repeated two more times. Finally, the subcultures were incubated on PDA alone at 20 °C for 15 days and the colony morphology of each subculture was observed.

Seventy out of the 111 derivative isolates of QT5-19 from the above-mentioned trials were arbitrarily selected (15, 15, 20, and 20 isolates from hyphal tipping, thermotherapy, chemotherapy and protoplast regeneration, respectively). They were incubated on PDA at 20 °C in the dark for observation of the colony morphology and for detection of the mycovirus by dsRNA profiling and RT-PCR with the specific primers RdRp-f1/RdRp-r1 and CP-f1/CP-r2 (Table S1, Figure S6).

2.10. Statistical Analysis

Data in each experiment or assay were analyzed using the procedure of Analysis of Variance (ANOVA) in the SAS software (SAS Institute, Cary, NC, USA, v. 8.0, 1999). Treatment means about mycelial growth rates on PDA in the colony morphology-characterization trials, lesion diameters on leaves of tobacco in the pathogenicity tests were separated using the least significant difference (LSD) test at $\alpha = 0.05$. Means about mycelial growth rates and numbers of sclerotia per dish on PDA, and leaf lesion diameters on tobacco and between each BcPV2-transfected derivative (08168T, B05.10T, RoseBc-3T, XN-1T) and its parental isolate (08168, B05.10, RoseBc-3, XN-1) were compared using Student's t test at $\alpha = 0.05$ or 0.01.

3. Results

3.1. Cultural Characteristics

Isolate QT5-19 grew at an average rate of 1.2 cm/d on PDA at 20 °C and colonized the entire Petri dishes (9 cm in diameter) after incubation for four days. This growth rate was slightly lower than those (1.5 cm/d) in three other isolates of *B. cinerea* (08168, B05.10, XN-1) under the same cultural conditions (Figure 1A,B). QT5-19 formed pink-colored colonies without production of any conidia and sclerotia even after incubation for 20 days. These morphological features differed from those of isolates 08168, B05.10 and XN-1, which formed gray-colored colonies with production of conidia and sclerotia (Figure 1A).

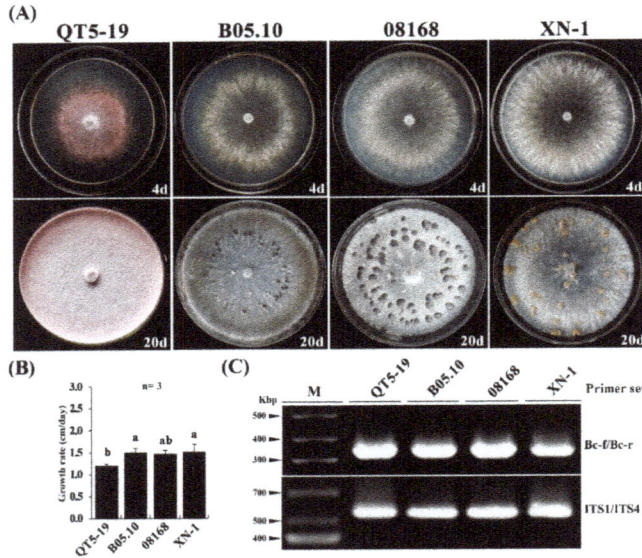

Figure 1. Cultural characteristics of *Botrytis cinerea* isolates on potato dextrose agar. (**A**) Four-day-old (top) and 20-day-old cultures (bottom) of isolates QT5-19, 08168, B05.10, and XN-1 (20 °C). Note difference in colony color, pink for QT5-19, whereas gray for the other three isolates. Note also difference in sclerotial color, orange sclerotia produced by isolate XN-1, whereas black sclerotia produced by isolates 08168 and B05.10; (**B**) A histogram showing average mycelial growth rates of the four isolates. Means ± S.D. ($n = 4$) labeled with the same letters are not significantly different at $p > 0.05$; (**C**) PCR-based identification of QT5-19 using the *B. cinerea*-specific primer set Bc-f/Bc-r and the universal primer set ITS1/ITS4. Isolates 08168, B05.10, and XN-1 of *B. cinerea* were used as reference.

Despite the lack of typical cultural and morphological characteristics, QT5-19 was identified as *B. cinerea* based on two molecular features. First, QT5-19 showed the *B. cinerea*-specific marker, a 327-bp DNA fragment (Figure 1C), in PCR with the *B. cinerea*-specific primers Bc-f/Bc-r reported by Fan and colleagues [36]. Second, the nucleotide sequence of the internal transcribed spacer (ITS)-rDNA region of QT5-19 (GenBank Acc. No. KX822693) is 100% identical to the ITS sequence of *B. cinerea* B05.10 (GenBank Acc. No. CP009808) and 99% identical to the ITS sequence of *B. cinerea* XN-1 (GenBank Acc. No. KT266229).

The pink pigment produced by QT5-19 is possibly bikaverin, as all the six bikaverin biosynthesis-related genes (*bcbik1* to *bcbik6*) were expressed in this isolate (Figure S1). The RT-PCR product of *bcbik1* coding for polyketide synthase (PKS) was cloned from QT5-19 and sequenced. The 679-bp cDNA fragment (GenBank Acc. No. MH747471) was 100% identical to the DNA sequence of *bcbik1* in the bikaverin-producing isolate INRA 1750 of *B. cinerea* (GenBank Acc. No. HE802550) (Figure S2). On the other hand, isolate B05.10 of *B. cinerea* failed to produce bikaverin on PDA. It had expression of five of these six bikaverin biosynthesis-related genes (*bcbik2* to *bcbik6*), but had no expression of *bcbik1* (Figure S1), as it lacks this gene in its genome [43].

3.2. Hypovirulence

Results of the pathogenicity test showed that isolate QT5-19 caused no infection or produced tiny lesions on leaves of tobacco, as well as on leaves of cucumber, oilseed rape, and strawberry, and on fruits of apple, table grapes and tomato at 3 days post-inoculation (dpi) under 20 °C with average lesion diameters ranging from 0.0 to 0.3 cm (Figure 2A,B, Figure S3). In contrast, B05.10, 08168 and XN-1 caused large necrotic lesions on the tobacco leaves with average lesion diameters of 3.1, 3.8, and 4.0 cm, respectively (Figure 2). B05.10 caused large necrotic lesions on apple, cucumber, oilseed rape, strawberry, table grapes, tobacco, and tomato with average lesion diameters ranging from 1.8 cm to 3.0 cm (Figure S3).

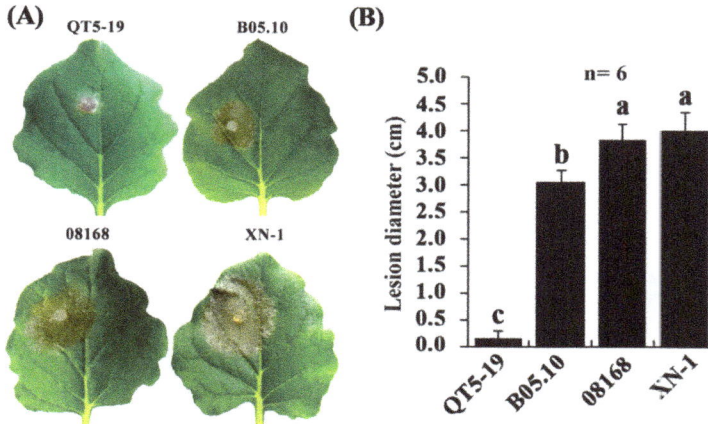

Figure 2. Pathogenicity of *Botrytis cinerea* isolates on tobacco (*Nicotiana benthamiana*). (**A**) Four tobacco leaves inoculated with the mycelia of the isolates QT5-19, 08168, B05.10, and XN-1, respectively (20 °C, 3 dpi). Note no visible lesion formation on the leaf inoculated with isolate QT5-19, whereas formation of large necrotic lesions on the leaves inoculated with the other three isolates; (**B**) A histogram showing average leaf lesion diameters caused by the four isolates. Means ± S.D. (*n* = 6) labeled with the same letters are not significantly different at *p* > 0.05.

3.3. The dsRNA Elements

Two dsRNA molecules (dsRNA 1 and dsRNA 2) of ~2 kb in size were detected in the extracts of nucleic acids from the mycelia of QT5-19 (Figure 3). In contrast, no dsRNAs were detected in the virulent isolates 08168, B05.10, RoseBc-3, and XN-1 of *B. cinerea*. To determine persistency of the two dsRNAs in QT5-19, repeated attempts using four different approaches (protoplast regeneration, thermotherapy, hyphal tipping, chemotherapy) were made to eliminate the dsRNAs from QT5-19. A total of 111 derivatives were obtained, 15 from hyphal tipping, 15 from thermotherapy, 30 from chemotherapy, and 51 from protoplast regeneration. All these derivatives of QT5-19 formed colonies with the morphology similar to that of the parent QT5-19 (pink color, no production of conidia and sclerotia) on PDA at 20 °C. Seventy derivatives (15 from hyphal tipping, 15 from thermotherapy, 20 from chemotherapy, and 20 from protoplast regeneration) were randomly selected and individually tested for the presence of the dsRNAs by dsRNA profiling and RT-PCR using the total RNA as template and RdRp-f1/RdRp-r1 and CP-f1/CP-r1 as specific primers (Table S1). The results showed that all the selected derivatives contained two dsRNAs with the same size as those in QT5-19 (Figure S6).

3.4. Virus Particles

Virus particles were isolated from the mycelia of QT5-19 and purified by sucrose density-gradient centrifugation. They are isometric in shape under transmission electronic microscope (TEM) with an average diameter of 36 nm (Figure 4A). They contain two dsRNAs with the same size as those extracted from the mycelia of QT5-19 (Figure 4B) and a protein with the molecular weight of ~68 kDa (Figure 4C).

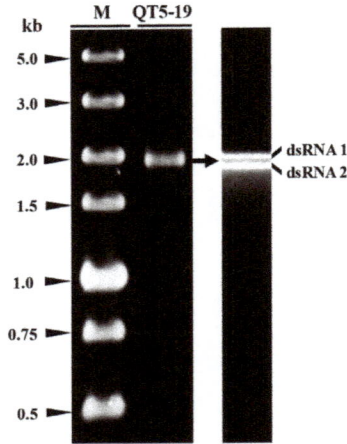

Figure 3. Double-stranded (ds) RNAs in isolate QT5-19 of *Botrytis cinerea*. The electrophoregram on the left was carried out in a 1.0% agarose gel after 1.5-h electrophoresis under room temperature. The electrophoregram on the right was carried out in a 0.7% agarose gel after 15-h electrophoresis at 4 °C. The dsRNAs were treated with RNase-free DNase I and S1 nuclease before electrophoresis. M = DL5000 dsDNA marker (TaKaRa).

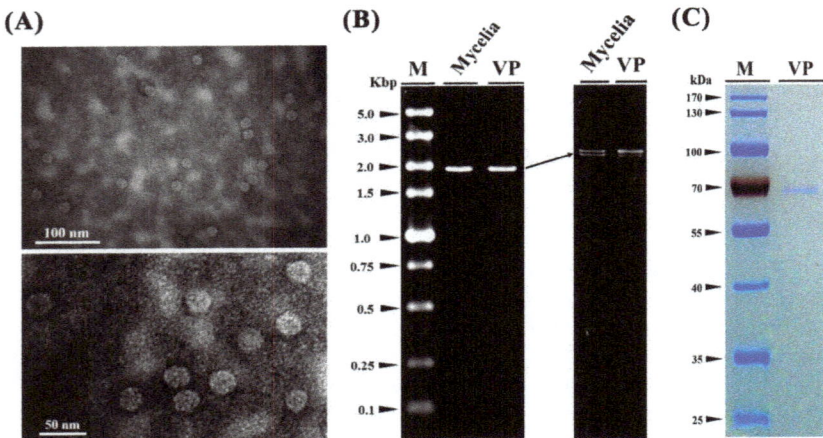

Figure 4. Virus particles isolated from the mycelia of isolate QT5-19 of *Botrytis cinerea*. (**A**) Two transmission electron microscope (TEM) graphs showing the shape and size of the virus particles of the mycovirus in QT5-19; (**B**) Electrophoregrams showing the two dsRNAs extracted from the virus particles (VP) and the mycelia of QT5-19 (Mycelia). The electrophoregram on the right was created in a 1% agarose gel after 1.5-h electrophoresis under room temperature. The electrophoregram on the left was created in a 0.7% agarose gel after 15-h electrophoresis at 4 °C. M = DL5000 marker (TaKaRa); (**C**) An SDS-PAGE (10%) electrophoregram showing the band of the structural protein (coat protein) extracted from the virus particles of the mycovirus in QT5-19. The gel was stained with Coomassie brilliant blue R-250. M = PageRuler™ Prestained Protein Ladder.

3.5. Identity of the Two dsRNAs

The dsRNAs in QT5-19 were purified from agarose gels after electrophoresis. They were ligated with the adaptor sequence 110a (Figure S4) at the 3′-terminus of each dsRNA. The adaptor-dsRNA molecules were reverse transcribed to cDNAs, which were then used as templates in PCR for cDNA amplification with the primer RC110a complementary to the adaptor 110a (Table S1). The resulting DNA products were separated by agarose gel electrophoresis and the purified fragments were cloned into the pMD18-T vector, which was then transformed into *Escherichia coli* DH5α and sequenced. The full-length cDNAs for dsRNA 1 (GenBank Acc. No. MG011707) and dsRNA 2 (GenBank Acc. No. MG011708) are 1909 and 1883 bp long, respectively. Northern blotting confirmed authenticity of the cDNAs from the dsRNAs in QT5-19 (Figure S5).

Analysis of the nucleotide sequences of the two dsRNAs showed that the coding strand of each dsRNA contains one open reading frame (ORF) flanked by two untranslated regions (UTRs) at 3′- and 5′-termini (Figure 5A). The two dsRNAs shares 63.9% nucleotide identity at the 3′-terminus (72 bp long) and 96.1% nucleotide identity at the 5′-terminus (76 bp long) (Figure 5B). Additionally, the 5′- and 3′-UTRs of each dsRNA were predicted to be able to form stem-loop structures (Figure 5C).

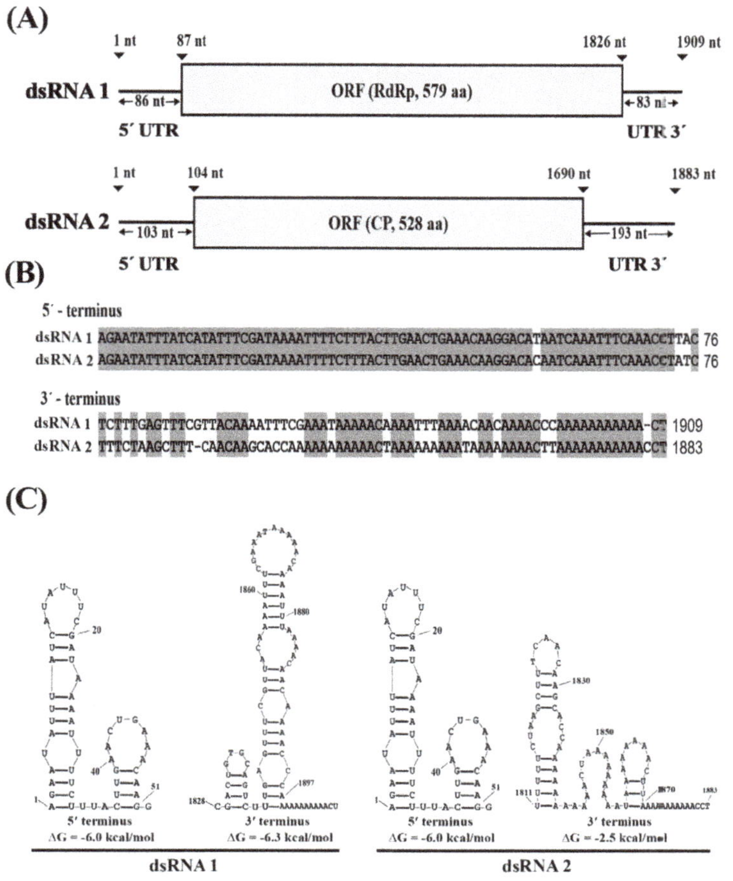

Figure 5. Genome structure of BcPV2. (**A**) A schematic diagram illustrating the segmented dsRNA genome. Each dsRNA has one open reading frame (ORF) flanked by two untranslated regions (UTR) at 3′- and 5′-termini; (**B**) Alignments of the partial nucleotide sequences of the 3′- and 5′-UTRs of dsRNA 1 and dsRNA 2. The identical nucleotides were highlighted in gray color. "-", missing nucleotides; (**C**) Predicted secondary structures for the terminal regions of dsRNA 1 and dsRNA 2

The ORF in dsRNA 1 was predicted to encode a polypeptide containing 579 amino acid (aa) residues with the estimated molecular weight of 67.08 kDa. Homology searches in the database of GenBank showed that the 579-aa protein is identical by 13%–76% to the RNA-dependent RNA polymerases (RdRp) encoded by ten partitiviruses, including Botryotinia fuckeliana partitivirus 1 (BfPV1, 13% identity), Ceratobasidium partitivirus (CPV, 76% identity), Flammulina velutipes isometric virus (FvIV, 42% identity), Fusarium solani partitivirus 2 (FsPV2, 41% identity), Grapevine partitivirus (GPV, 53% identity), Powdery mildew-associated partitivirus (PmAPV, 51% identity), Rhizoctonia solani partitivirus 1 (RsPV1, 43% identity), Rosellinia necatrix partitivirus 5 (RnPV5, 40% identity), Sclerotinia sclerotiorum partitivirus S (SsPV-S, 46% identity), and Soybean leaf-associated partitivirus 2 (SLAPV2, 58% identity) (Table 1). It contains six motifs (Motif III to Motif VIII) (Figure 6A), which are also present in the RdRps of some other partitiviruses, such as Beet cryptic virus 1 (BCV1), CPV, Dill cryptic virus 1 (DCV1), PmAPV, Rosellinia necatrix partitivirus 2 (RsPV2), Vicia cryptic virus (VCV), and white clover cryptic virus 1 (WCCV1) (Table 1).

The ORF in dsRNA 2 was predicted to encode a polypeptide containing 528 aa residues with estimated molecular weight of 58.84 kDa. The polypeptide is identical by 10–60% to the coat proteins (CP) encoded by seven partitiviruses, including Arabidopsis halleri partitivirus 1 (AhPV1, 22% identity), BfPV1, Botrytis cinerea partitivirus 1 (BcPV1, 14% identity), CPV (60% identity), PmAPV (31% identity), RnPV2 (23% identity), and SLAPV2 (34% identity), and by 24–31% to the CPs encoded by three plant cryptic viruses, including carrot cryptic virus (CCV, 26% identity), Diuris pendunculata cryptic virus (DpCV, 31% identity) and spinach cryptic virus 1 (SCV, 24% identity) (Table 1). The aa sequence of CP found in this study and its homologs in CCV, CPV, DpCV, PmAPV, and SLAPV2 appeared to be highly divergent (Figure 6B). Therefore, the two dsRNAs in QT5-19 represent the genome of a novel partitivirus in the family *Partitiviridae*, which is hereafter designated as Botrytis cinerea partitivirus 2 (BcPV2).

The phylogeny of BcPV2 was inferred based on the aa sequences of RdRp and CP of this mycovirus and the known members in *Partitiviridae* (Table S2) using the Maximum-likelihood method. The results showed that BcPV2 belongs to the genus *Alphapartitivirus* (Figure 7A,B). It is closely related to CPV, PmAPV, and SLAPV2, but is distantly related to BcPV1 and BfPV1, which belong to the genera *Betapartitivirus* and *Gammapartitivirus*, respectively.

Table 1. Summary of the results in the BLASTp search for RNA-dependent RNA polymerase (RdRp) and coat protein (CP) of *Botrytis cinerea* partitivirus 2 (BcPV2).

Viruses	GenBank Acc. No.	Size (aa)	Identity (%)	Overlap (Positions)	Bit Score	E-Value
RdRp						
Ceratobasidium partitivirus (CPV)	AOX47571	579	76	440/578	917	0.0
Soybean leaf-associated partitivirus 2 (SLAPV2)	ALM62247	602	58	341/583	681	0.0
Grapevine partitivirus (GPV)	AFX73022	584	53	311/587	631	0.0
Powdery mildew-associated partitivirus (PmAPV)	YP_009272944	584	51	301/587	605	0.0
Sclerotinia sclerotiorum partitivirus S (SsPV-S)	YP_003082248	580	46	263/566	472	1×10^{-157}
Rhizoctonia solani partitivirus 1 (RsPV1)	AND83003	603	43	234/549	431	4×10^{-141}
Flammulina velutipes isometric virus (FvIV)	BAH08700	587	42	252/594	436	2×10^{-143}
Fusarium solani partitivirus 2 (FsPV2)	BAQ36631	608	41	242/587	402	5×10^{-130}
Rosellinia necatrix partitivirus 5 (RnPV5)	BAM36403	647	40	230/568	414	6×10^{-134}
Botryotinia fuckeliana partitivirus 1 (BfPV1) *	YP_001686789	540	13 *	40/293	-**	-
CP						
Ceratobasidium partitivirus (CPV)	AOX47604	370	60	221/370	473	1×10^{-161}
Soybean leaf-associated partitivirus 2 (SLAPV2)	ALM62248	496	34	169/498	272	8×10^{-82}
Powdery mildew-associated partitivirus (PmAPV)	YP_009272945	527	31	164/528	238	2×10^{-68}
Diuris pendunculata cryptic virus (DpCV)	AFY23215	496	31	36/116	58	2×10^{-05}
Carrot cryptic virus (CCV)	ACL93279	490	26	38/147	45.1	0.33
Spinach cryptic virus 1 (SCV)	APX42422	488	24	44/181	48.5	0.023
Rosellinia necatrix partitivirus 2 (RnPV2)	YP_007419078	483	23	77/333	47.4	0.057
Arabidopsis halleri partitivirus 1 (AhPV1)	YP_009273019	487	22	72/330	57	6×10^{-05}
Botrytis cinerea partitivirus 1 (BcPV1) *	AGQ21570	634	14 *	56/389	-	-
Botryotinia fuckeliana partitivirus 1 (BfPV1) *	YP_001686790	436	10 *	41/408	-	-

* The values were calculated using the DNAMAN software (version 7.0), rather than provided in BLASTp search in NCBI, as the identity level between BcPV2 and BcPV1/BfPV1 is too low. ** The values cannot be calculated in the analysis with DNAMAN.

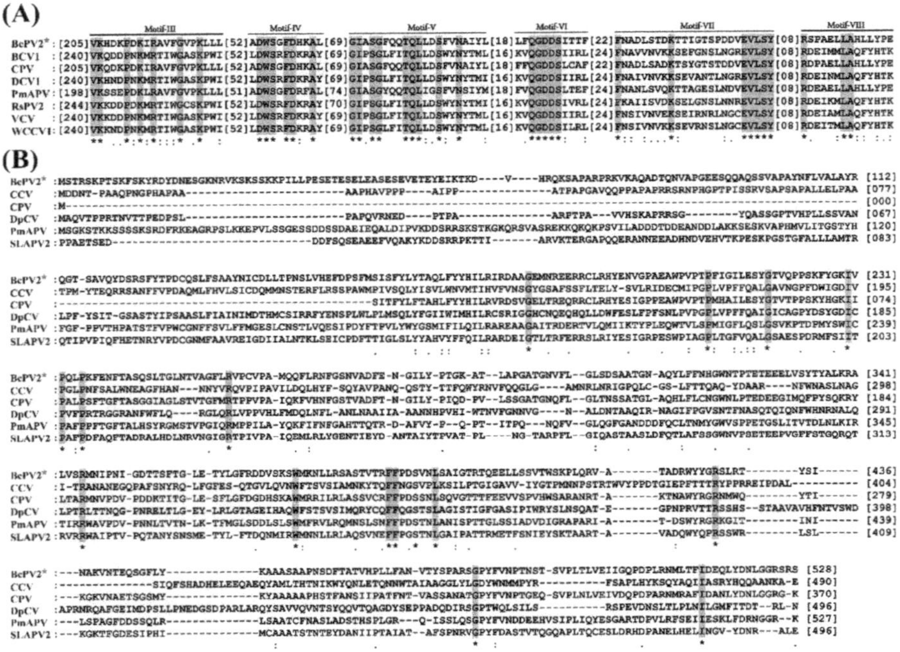

(A)

(B)

Figure 6. Multiple alignments of the amino acid sequences of RNA-dependent RNA polymerase and coat protein of BcPV2 with other selected members in *Alphapartitivirus*. (**A**) Amino acid sequences of RNA-dependent RNA polymerase (RdRp); and (**B**) amino acid sequences of coat protein (CP). "*", identical amino acids, ":" and "." high and low chemically similar amino acids, respectively. Note the five conserved motifs (Motif-III to Motif-VIII) in RdRp and diversified CP sequences among the compared viruses. Abbreviations: BCV1, Beet cryptic virus 1; BcPV2, Botrytis cinerea partitivirus 2; CCV, carrot cryptic virus; CPV, Ceratobasidium partitivirus; DCV1, Dill cryptic virus 1; DpCV, Diuris pendunculata cryptic virus; PmAPV, Powdery mildew-associated partitivirus; RsPV2, Rhizoctonia solani partitivirus 2; SLAPV2, Soybean leaf-associated partitivirus 2; **VCV**, Vicia cryptic virus; WCCV1, white clover cryptic virus 1. The GenBank accession numbers for RdRp and CP of the above-mentioned viruses are listed in Table S2.

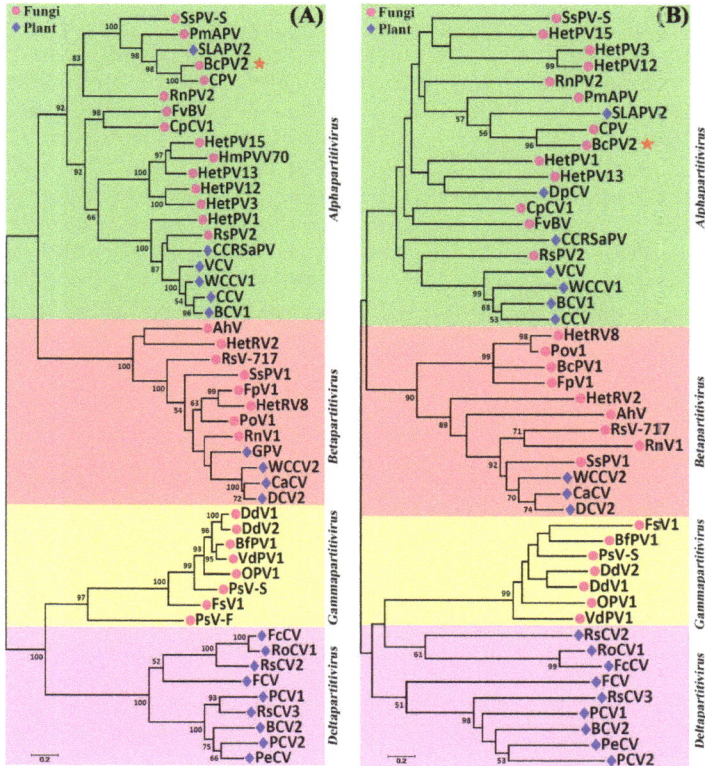

Figure 7. Two phylogenetic trees showing the relationship of BcPV2 with other partitiviruses. The trees were inferred based on amino acid sequences of RdRp (**A**) and CP (**B**) using the Maximum-Likelihood method with bootstrap values determined by 1000 replicates. Bootstrap values higher than 50% are shown in the graphs. The scale bars represent substitutions per nucleotide position. Abbreviations: AhV, Atkinsonella hypoxylon partitivirus; BcPV1, Botrytis cinerea Partitivirus 1; BcPV2, Botrytis cinerea Partitivirus 2; BCV1, Beet cryptic virus 1; BCV2, Beet cryptic virus 2; BfPV1, Botryotinia fuckeliana partitivirus 1; CaCV, Cannabis cryptic virus; CCRSaPV, Cherry chlorotic rusty spot associated partitivirus; CpCV1, Chondrostereum purpureum cryptic virus 1; CPV, Ceratobasidium partitivirus; DCV1, Dill clover cryptic virus 1; DCV2, Dill cryptic virus 2; DdV1, Discula destructiva virus 1; DdV2, Discula destructiva virus 2; FcCV, Fragaria chiloensis cryptic virus; FCV, Fig cryptic virus; FpV1, Fusarium poae virus 1; FsV1, Fusarium solani virus 1; FvBV, Flammulina velutipes browning virus; GPV, Grapevine partitivirus; HetPV1, Heterobasidion partitivirus 1; HetPV12, Heterobasidion partitivirus 12; HetPV13, Heterobasidion partitivirus 13; HetPV15, Heterobasidion partitivirus 15; HetPV3, Heterobasidion partitivirus 3; HetRV2, Heterobasidion partitivirus 2; HetRV8, Heterobasidion partitivirus 8; HmPVV70, Helicobasidium mompa partitivirus V70; OPV1, Ophiostoma partitivirus 1; PCV1, Pepper cryptic virus 1; PCV2, Pepper cryptic virus 2; PeCV, Persimmon cryptic virus; PmAPV, Powdery mildew-associated partitivirus; PoV1, Pleurotus ostreatus virus 1; PsV-F, Penicillium stoloniferum virus F; PsV-S, Penicillium stoloniferum virus S; RnPV2, Rosellinia necatrix partitivirus 2; RnV1, Rosellinia necatrix partitivirus 1; RoCV1, Rose cryptic virus 1; RsCV2, Raphanus sativus cryptic virus 2; RsCV3, Raphanus sativus cryptic virus 3; RsPV2, Rhizoctonia solani partitivirus 2; RsV-717, Rhizoctonia solani virus 717; SLAPV2, Soybean leaf-associated partitivirus 2; SsPV1, Sclerotinia sclerotiorum partitivirus 1; SsFV-S, Sclerotinia sclerotiorum partitivirus S; VCV, Vicia cryptic virus; VdPV1, Verticillium dahliae partitivirus 1; WCCV1, White clover cryptic virus 1; WCCV2, White clover cryptic virus 2. The GenBank accession number of the above-mentioned viruses are listed in Table S2.

3.6. Horizontal Transmission of BcPV2

The pair-culturing technique [23,24,35] was used (Figure 8A) to test horizontal transmission of BcPV2 from the donor isolate QT5-19 to each of the four recipient isolates of *B. cinerea* (08168, B05.10, XN-1, RoseBc-3) on PDA at 20 °C (Figure 8B). Three derivatives (one from each pair culture) were obtained from the colonies of each recipient in the 4-day-old pair cultures of QT5-19 and that recipient. Four representative derivatives (08168T, B05.10T, RoseBc-3T, and XN-1T from 08168, B05.10, RoseBc-3, and XN-1, respectively) were selected (Figure 8C) and compared with their parents for the accumulation of BcPV2 in mycelia, mycelial growth rates and production of conidia/sclerotia on PDA (20 °C), and for pathogenicity on tobacco leaves. Results showed that B05.10T, 08168T, and XN-1T had BcPV2 accumulation, whereas RoseBc-3T did not have any detectable BcPV2 accumulation (Figure 9; Table 2). Therefore, horizontal transmission of BcPV2 was successful from QT5-19 to three of the four investigated isolates, but was unsuccessful from QT5-19 to RoseBc-3. Results also showed that B05.10T, 08168T, and XN-1T grew at the average rate of 1.5 cm/d, which did not significantly differ ($p > 0.05$) from those of their parents. However, they failed to produce conidia and sclerotia after incubation at 20 °C for 20 days, and these features differed from their parents (Table 2). As expected, RoseBc-3T showed the same phenotype as its parental isolate RoseBc-3 both in growth rate and in the production of conidia and sclerotia (Table 2). On tobacco leaves, B05.10T, 08168T and XN-1T caused average leaf lesion diameters of 2.2, 2.3, and 2.9 cm (20 °C, 3 dpi), respectively, and the values were significantly ($p < 0.05$) smaller than those of B05.10, 08168 and XN-1, which caused lesion diameters of 3.1, 3.8, and 4.0 cm, respectively (Table 2, Figure 10). However, RoseBc-3T caused an average lesion diameter of 2.2 cm, not significantly different ($p > 0.05$) from the average leaf lesion caused by RoseBc-3 (2.3 cm in diameter) (Table 2, Figure 10). Therefore, the introduction of BcPV2 from QT5-19 to B05.10, 08168, and XN-1 caused debilitation of pathogenicity and deficiency in formation of conidia and sclerotia.

Table 2. Horizontal transmission of BcPV2 from QT5-19 to isolates of *B. cinerea* (B05.10, 08168, RoseBc-3, XN-1) and effect of BcPV2 infection on mycelial growth, conidial production, sclerotial production, and pathogenicity of the recipients.

Isolate	BcPV2 [x]	Mycelial Growth Rate (cm/day) [y]	Conidial Production [y]	No. Sclerotia per Dish [y]	Lesion Diameter (cm) [z]
QT5-19	+	1.2	-	0	0.2
B05.10	-	1.5	+	36	3.1
B05.10T	+	1.5	-	0 **	2.2 *
08168	-	1.5	+	38	3.8
08168T	+	1.5	-	0 **	2.3 *
XN-1	-	1.5	+	37	4.0
XN-1T	+	1.5	-	0 **	2.9 *
RoseBc-3	-	1.4	+	25	2.3
RoseBc-3T	-	1.4	+	22	2.2

[x] BcPV2 was detected by dsRNA profiling and RT-PCR using the specific primer pairs RdRp-f1/RdRp-r1 (dsRNA 1) and CP-f1/CP-r1 (dsRNA-2). "+", the presence of BcPV2; "-", the absence of BcPV2. [y] Mycelial growth rate, conidial production, and sclerotial production were determined on PDA at 20 °C. Each value is an average of 5 replicates. "+" with conidia, "-" without conidia. "**" indicates significant difference at $p < 0.01$ between each recipient and its transfected derivative according to Student's *t* test. [z] Pathogenicity was determined on detached tobacco leaves (20 °C, 72 h). Each value of lesion diameter is an average of seven replicates. "*" indicates significant difference at $p < 0.05$ between each recipient and its transfected derivative isolate according to Student's *t* test.

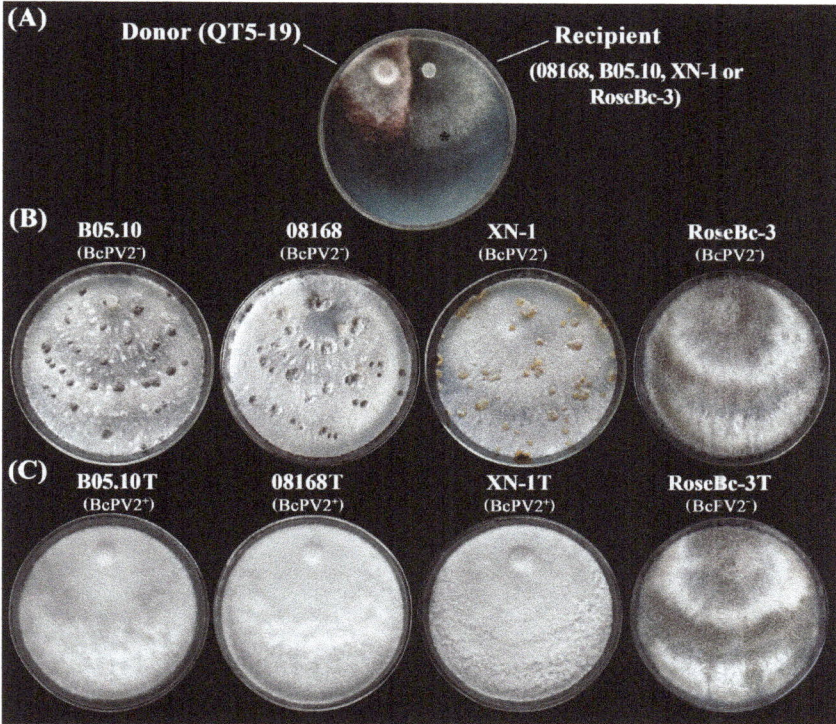

Figure 8. Horizontal transmission of BcPV2 through hyphal contact in pair cultures on PDA. (**A**) A pair culture on PDA with the colonies of isolate QT5-19 (BcPV2 donor, pink color) and a recipient (gray color). The symbol (*) in the colony indicated the area where a mycelial agar plug was removed and transferred to PDA for establishing a BcPV2-transmitted derivative; (**B**) Four PDA cultures (20 °C, 20 day) of the *B. cinerea* isolates free of infection by BcPV2 (BcPV2$^-$); (**C**) Four PDA cultures (20 °C, 20 day) of the transfected derivatives of *B. cinerea*. BcPV2$^+$, infected by BcPV2, BcPV2$^-$, not infected by BcPV2.

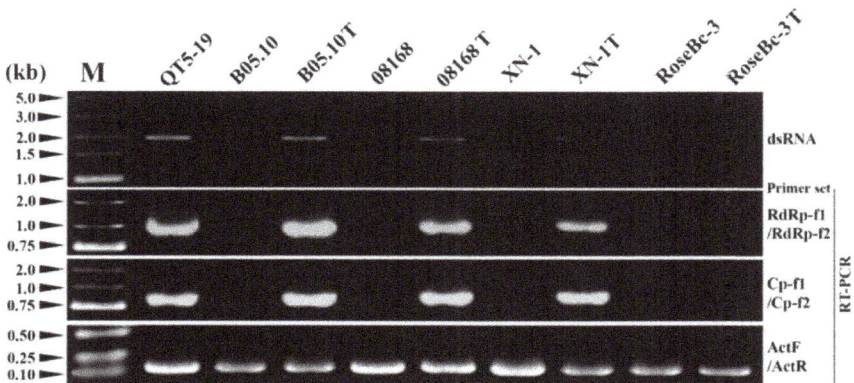

Figure 9. Detection of BcPV2 in different isolates of *Botrytis cinerea* by dsRNA profiling and RT-PCR with specific primers. M = DL5000 DNA marker (TaKaRa). Isolates B05.10T, 08168T, XN-1T, and RoseBc-3T were derived from B05.10, 08168, XN-1 and RoseBc-3 in the pairing cultures of QT5-19/B05.10, QT5-19/08168, QT5-19/XN-1 and QT5-19/RoseBc-3, respectively.

Figure 10. Pathogenicity of different isolates of *Botrytis cinerea* on detached leaves of *Nicotiana benthamiana* (20 °C, 72 h). Note smaller leaf lesion size caused by B05.10T, 08168T, and XN-1T than that caused by B05.10, 08168 and XN-1, whereas similar leaf lesion size caused by RoseBc-3T and RoseBc-3. BcPV$^+$ = infected by BcPV2; BcPV2$^-$ = not infected by BcPV2.

4. Discussion

This study demonstrated that the pink isolate QT5-19 of *B. cinerea* is a hypovirulent isolate on apple, cucumber, oilseed rape, strawberry, table grapes, tobacco, and tomato, compared to the gray isolates 08168, B05.10, and XN-1of *B. cinerea*, which aggressively infected tobacco. Botrytis cinerea partitivirus 2 (BcPV2) was identified in QT5-19, whereas BcPV2 and other RNA mycoviruses were not detected in 08168, B05.10 and XN-1. Moreover, isolates 08168T, B05.10T and XN-1T infected by BcPV2 in the horizontal transmission experiment showed a significant ($p < 0.05$) decrease in pathogenicity on tobacco, compared to their parents (08168, B05.10, XN-1). This result suggests that BcPV2 is closely associated with hypovirulence of *B. cinerea*. Previous studies have reported several mycoviruses-infected hypovirulent isolates in B. cinerea, including CanBc-1 (infested with BcMV1), CCg378 (infested with BfPV1), BerBc-1 (infested with BcRV1), Ecan17-2 (infested with BcMyV1), and HBtom-372 (infested with BcFV1 and BcHV1) [9,23,27,28,31,32]. QT5-19 grew almost as rapidly as 08168, B05.10 and XN-1 on PDA at 20 °C, but failed to produce conidia and sclerotia. These cultural features differ from those in CanBc-1, Ecan17-2, and HBtom-372, which grew slowly on PDA [23], and from those of BerBc-1, which grew rapidly and formed sclerotia on PDA [32]. Moreover, the BcPV2-transfected isolates 08168T, B05.10T, and XN-1T also showed vigorous growth on PDA and the defect in production of conidia and sclerotia. These results suggest that BcPV2 in QT5-19 as well as in 08168T, B05.10T, and XN-1T may have no negative effect on vegetative growth of these isolates of *B. cinerea*; however, it has attenuation effects on pathogenicity, as well as on production of conidia and sclerotia. Further transcriptomic analysis of the isogenic BcPV2-free and BcPV2-infected isolates is necessary to determine the BcPV2-responsive genes associated with (or responsible for) pathogenicity, and production of conidia and sclerotia in *B. cinerea*.

The viruses in the family *Partitiviridae* have been reported in fungi as well as in plants and protozoa [13]. There are at least five genera in this family, namely *Alphapartitivirus*, *Betapartitivirus*, *Cryspovirus*, *Gammapartitivirus* and *Deltapartitivirus* according to the recent report in ICTV (https://talk.ictvonline.org). Previous studies showed that most mycoviruses in *Partitiviridae* have no deleterious effects on their fungal hosts [4] and only a few partitiviruses were found to be associated with abnormal phenotypes in fungi. For example, Rhizoctonia solani partitivirus 2 (RsPV2) is closely associated with hypovirulence in *Rhizoctonia solani*, the causal agent of rice sheath blight [47]. Sclerotinia sclerotiorum partitivirus 1 (SsPV1) can confer hypovirulence in *S. sclerctiorum* and related species [48]. Two partitiviruses, namely Botrytis cinerea partitivirus 1 (BcPV1) and Botryotinia fuckeliana partitivirus 1 (BfPV1), have been reported in *B. cinerea* [31,34]. BcPV1 was found to be associated with hypovirulence of *B. cinerea* [31], whereas the effect of BfPV1 on pathogenicity of *B. cinerea* remains unknown. Moreover, Xiao and colleagues (2014) reported that introduction of the virus particles of SsPV1 to *B. cinerea* caused decrease in mycelial growth and pathogenicity of *B. cinerea*, but caused increase in conidial production by this fungus [48]. Phylogenetic analysis in this study showed that BcPV2 is a unique member in *Alphapartitivirus* and it is phylogenetically distant from BcPV1 in *Betapartitivirus*, SsPV1 in *Betapartitivirus*, and BfPV1 in *Gammapartitivirus*. The close association between BcPV2 infection and hypovirulence observed in this study provides an example of *Alphapartitivirus*-mediated hypovirulence in *B. cinerea*.

As mentioned above, QT5-19 infected with BcPV2 hardly infected tobacco and six other crops. BcPV2 in QT5-19 was successfully transmitted to three other isolates of *B. cinerea* (08168, B05.10, XN-1) through hyphal contact (possibly through anastomosis). The resulting derivatives B05.10T, XN-1T, and 08168T infested with BcPV2 grew rapidly on PDA at 20 °C, but failed to produce conidia and sclerotia. These results suggest that BcPV2 in B05.10T, XN-1T and 08168T has a significant attenuation effect on the formation of conidia and sclerotia. However, the pathogenicity test showed that B05.10T, XN-1T, and 08168T could infect tobacco and caused formation of necrotic leaf lesions (Figure 10), although the average leaf lesion diameters caused by these BcPV2-infected derivatives were significantly ($p < 0.05$) smaller than those caused by their parents (reduced by 27% to 39%). This result suggests that BcPV2 in QT5-19 may have a larger pathogenicity-attenuation effect than it does in B05.10T, XN-1T, and 08168T. Different genetic backgrounds in isolates QT5-19, B05.10T, XN-1T and 08168T might be responsible for this phenomenon.

Previous studies showed that it is difficult to use the RNA mycoviruses such as BcMV1, BcFV1, BcHV1, BcRV1, and BcMyV1 in the corresponding hypovirulent isolates to control virulent isolates of *B. cinerea* [9,23,28,32]. There are two possible reasons: (*i*) limited horizontal transmissibility of the mycoviruses possibly due to hyphal incompatibility between the mycoviruses-infected hypovirulent isolates and virulent isolates; and (*ii*) low competitiveness of the mycoviruses-infected hypovirulent isolates due to poor mycelial growth. The hypovirulent isolate QT5-19 grew almost as rapidly as the virulent isolates 08168, B05.10, and XN-1 on PDA, suggesting that it may have a strong competitive ability. Meanwhile, BcPV2 in QT5-19 (donor) was successfully transmitted to four of the five investigated recipient isolates, suggesting that BcPV2 in QT5-19 may have a wider horizontal transmission spectrum than the mycoviruses reported in previous studies [9,23,28,32]. Therefore, QT5-19 with BcPV2 may have potential to be exploited as a biocontrol agent against virulent individuals of *B. cinerea*. Additional studies on evaluation of the biocontrol efficacy of QT5-19 are warranted.

In conclusion, this study found that the hypovirulent isolate QT5-10 of *B. cinerea* has distinct cultural characteristics (rapid mycelial growth, but with defects in formation of conidia and sclerotia). A novel partitivirus in *Alphapartitivirus*, Botrytis cinerea partitivirus 2 (BcPV2), was identified in QT5-19. BcPV2 was successfully transmitted from QT5-19 to virulent isolates 08168, B05.10, and XN-1, and the resulting BcPV2-transfected derivatives were attenuated both in pathogenicity and in production of conidia and sclerotia.

Supplementary Materials: The following are available online at http://www.mdpi.com/1999-4915/11/1/24/s1, Figure S1: Phenotypic and molecular characterization of bikaverin biosynthesis in the pink-colored isolate QT5-19 of *Botrytis cinerea* (A) Difference between QT5-19 and B05.10 in accumulation of the pink pigment on potato dextrose agar (20 °C) (B) RT-PCR detection of expression of the genes (*bcbik1* to *bcbik6*) responsible for biosynthesis of the pink pigment bikaverin. The primers for related genes used in RT-PCR are listed in Table S1. (C) A schematic diagram showing the bikaverin-biosynthesis gene cluster in isolates QT5-19 and B05.10. The gray-colored region indicates the absence of *bcbik1* in B05.10. Arrows indicate the direction of gene transcription. Figure S2: Alignment of the partial nucleotide sequences of *bcbik1* coding for polyketide synthase (PKS) in isolates QT5-19 and 1750 of *Botrytis cinerea*, and the homologs of *bcbik1* in *Fusarium oxysporum* and *Fusarium anthophilum*. Figure S3: Pathogenicity of isolates QT5-19 and B05.10 of *B. cinerea* on fruits of tomato, table grape and apple, and on leaves of cucumber, rapeseed and strawberry. Figure S4: A schematic diagram showing the strategy for cDNA cloning of the full-length sequences of the two dsRNAs in Botrytis cinerea partitivirus 2 (BcPV2). Figure S5: Northern-blotting detection of the dsRNAs extracted from the mycelia of *Botrytis cinerea* QT5-19. Figure S6: Detection of BcPV2 in different isolates of *Botrytis cinerea* by dsRNA profiling and RT-PCR. Table S1: List of the oligonucleotide primers or adaptor used in this study. Table S2: List of partitiviruses for multiple sequence alignment and phylogenetic analysis.

Author Contributions: M.K. conceived and designed the experiments; M.K. and G.H. performed the experiments; M.K., M.W., J.Z., G.L., and L.Y. analyzed the data; M.K., G.L., and W.C. wrote the paper.

Funding: This study was financially supported by the National Natural Science Foundation of China (Grant No. 31772212) and the Special Fund for Agro-Scientific Research in the Public Interest (Grant No. 201303025).

Conflicts of Interest: The authors declare no conflict of interest.

References

1. Ghabrial, S.A.; Suzuki, N. Viruses of plant pathogenic fungi. *Annu. Rev. Phytopathol.* **2009**, *47*, 353–384. [CrossRef] [PubMed]

2. Pearson, M.N.; Beever, R.E.; Boine, B.; Arthur, K. Mycoviruses of filamentous fungi and their relevance to plant pathology. *Mol. Plant Pathol.* **2009**, *10*, 115–128. [CrossRef] [PubMed]

3. Herrero, N.; Márquez, S.S.; Zabalgogeazcoa, I. Mycoviruses are common among different species of endophytic fungi of grasses. *Arch. Virol.* **2009**, *154*, 327–330. [CrossRef] [PubMed]

4. Ghabrial, S.A.; Castón, J.R.; Jiang, D.; Nibert, M.L.; Suzuki, N. 50-plus years of fungal viruses. *Virology* **2015**, *479*, 356–368. [CrossRef] [PubMed]

5. Xie, J.; Jiang, D. New insights into mycoviruses and exploration for the biological control of crop fungal diseases. *Annu. Rev. Phytopathol.* **2014**, *52*, 45–68. [CrossRef] [PubMed]

6. Wang, S.; Kondo, H.; Liu, L.; Guo, L.; Qiu, D. A novel virus in the family *Hypoviridae* from the plant pathogenic fungus *Fusarium graminearum*. *Virus Res.* **2013**, *174*, 69–77. [CrossRef] [PubMed]

7. Martínez-Álvarez, P.; Vainio, E.J.; Botella, L.; Hantula, J.; Diez, J.J. Three mitovirus strains infecting a single isolate of *Fusarium circinatum* are the first putative members of the family *Narnaviridae* detected in a fungus of the genus *Fusarium*. *Arch. Virol.* **2014**, *159*, 2153–2155. [CrossRef]

8. Donaire, L.; Pagán, I.; Ayllón, M.A. Characterization of *Botrytis cinerea* negative-stranded RNA virus 1, a new mycovirus related to plant viruses, and a reconstruction of host pattern evolution in negative-sense ssRNA viruses. *Virology* **2016**, *499*, 212–218. [CrossRef]

9. Hao, F.; Wu, M.; Li, G. Molecular characterization and geographic distribution of a Mymonavirus in the population of *Botrytis cinerea*. *Viruses* **2018**, *10*, 432. [CrossRef]

10. Kondo, H.; Chiba, S.; Toyoda, K.; Suzuki, N. Evidence for negative-strand RNA virus infection in fungi. *Virology* **2013**, *435*, 201–209. [CrossRef]

11. Wang, L.; He, H.; Wang, S.; Chen, X.; Qiu, D.; Kondo, H.; Guo, L. Evidence for a novel negative-stranded RNA mycovirus isolated from the plant pathogenic fungus *Fusarium graminearum*. *Virology* **2018**, *518*, 232–240. [CrossRef] [PubMed]

12. Liu, L.; Xie, J.; Cheng, J.; Fu, Y.; Li, G.; Yi, X.; Jiang, D. Fungal negative-stranded RNA virus that is related to bornaviruses and nyaviruses. *Proc. Natl. Acad. Sci. USA* **2014**, *111*, 12205–12210. [CrossRef] [PubMed]

13. Nibert, M.L.; Ghabrial, S.A.; Maiss, E.; Lesker, T.; Vainio, E.J.; Jiang, D.; Suzuki, N. Taxonomic reorganization of family *Partitiviridae* and other recent progress in partitivirus research. *Virus Res.* **2014**, *188*, 128–141. [CrossRef] [PubMed]

14. Nibert, M.L.; Tang, J.; Xie, J.; Collier, A.M.; Ghabrial, S.A.; Baker, T.S.; Tao, Y.J. 3D structures of fungal partitiviruses. *Adv. Virus Res.* **2013**, *86*, 59–85. [PubMed]

15. Dean, R.; Van Kan, J.A.; Pretorius, Z.A.; Hammond-Kosack, K.E.; Di Pietro, A.; Spanu, P.D.; Rudd, J.J.; Dickman, M.; Kahmann, R.; Ellis, J. The top 10 fungal pathogens in molecular plant pathology. *Mol. Plant Pathol.* **2012**, *13*, 414–430. [CrossRef] [PubMed]
16. Williamson, B.; Tudzynski, B.; Tudzynski, P.; van Kan, J.A. *Botrytis cinerea*: The cause of grey mould disease. *Mol. Plant Pathol.* **2007**, *8*, 561–580. [CrossRef] [PubMed]
17. Elad, Y.; Pertot, I.; Marina, A.; Prado, A.M.; Stewart, A. Plant Hosts of *Botrytis* spp. In *Botrytis-The Fungus, the Pathogen and Its Management in Agricultural Systems*; Fillinger, S., Elad, Y., Eds.; Springer: Cham, Switzerland, 2016; pp. 413–486.
18. Fernández-Ortuño, D.; Chen, F.; Schnabel, G. Resistance to pyraclostrobin and boscalid in *Botrytis cinerea* isolates from strawberry fields in the Carolinas. *Plant Dis.* **2012**, *96*, 1198–1203. [CrossRef]
19. Choquer, M.; Fournier, E.; Kunz, C.; Levis, C.; Pradier, J.M.; Simon, A.; Viaud, M. *Botrytis cinerea* virulence factors: New insights into a necrotrophic and polyphageous pathogen. *FEMS Microbiol. Lett.* **2007**, *277*, 1–10. [CrossRef]
20. Howitt, R.L.; Beever, R.E.; Pearson, M.N.; Forster, R.L. Presence of double-stranded RNA and virus-like particles in *Botrytis cinerea*. *Mycol. Res.* **1995**, *99*, 1472–1478. [CrossRef]
21. Howitt, R.L.; Beever, R.E.; Pearson, M.N.; Forster, R.L. Genome characterization of Botrytis virus F, a flexuous rod-shaped mycovirus resembling plant 'potex-like' viruses. *J. Gen. Virol.* **2001**, *82*, 67–78. [CrossRef]
22. Howitt, R.; Beever, R.; Pearson, M.; Forster, R. Genome characterization of a flexuous rod-shaped mycovirus, Botrytis virus X, reveals high amino acid identity to genes from plant 'potex-like' viruses. *Arch. Virol.* **2006**, *151*, 563–579. [CrossRef] [PubMed]
23. Wu, M.; Zhang, L.; Li, G.; Jiang, D.; Hou, M.; Huang, H.C. Hypovirulence and double-stranded RNA in *Botrytis cinerea*. *Phytopathology* **2007**, *97*, 1590–1599. [CrossRef] [PubMed]
24. Wu, M.; Zhang, L.; Li, G.; Jiang, D.; Ghabrial, S.A. Genome characterization of a debilitation-associated mitovirus infecting the phytopathogenic fungus *Botrytis cinerea*. *Virology* **2010**, *406*, 117–126. [CrossRef] [PubMed]
25. Rodríguez-García, C.; Medina, V.; Alonso, A.; Ayllón, M. Mycoviruses of *Botrytis cinerea* isolates from different hosts. *Ann. Appl. Biol.* **2014**, *164*, 46–61. [CrossRef]
26. Donaire, L.; Rozas, J.; Ayllón, M.A. Molecular characterization of Botrytis ourmia-like virus, a mycovirus close to the plant pathogenic genus *Ourmiavirus*. *Virology* **2016**, *489*, 158–164. [CrossRef] [PubMed]
27. Hao, F.; Zhou, Z.; Wu, M.; Li, G. Molecular characterization of a novel encornavirus from the phytopathogenic fungus *Botrytis cinerea*. *Arch. Virol.* **2017**, *162*, 313–316. [CrossRef] [PubMed]
28. Hao, F.; Ding, T.; Wu, M.; Zhang, J.; Yang, L.; Chen, W.; Li, G. Two novel hypovirulence-associated mycoviruses in the phytopathogenic fungus *Botrytis cinerea*: Molecular characterization and suppression of infection cushion formation. *Viruses* **2018**, *10*, 254. [CrossRef] [PubMed]
29. Castro, M.; Kramer, K.; Valdivia, L.; Ortiz, S.; Castillo, A. A double-stranded RNA mycovirus confers hypovirulence-associated traits to *Botrytis cinerea*. *FEMS Microbiol. Lett.* **2003**, *228*, 87–91. [CrossRef]
30. Castro, M.; Kramer, K.; Valdivia, L.; Ortiz, S.; Benavente, J.; Castillo, A. A new double-stranded RNA mycovirus from *Botrytis cinerea*. *FEMS Microbiol. Lett.* **1999**, *175*, 95–99. [CrossRef]
31. Potgieter, C.A.; Castillo, A.; Castro, M.; Cottet, L.; Morales, A. A wild-type *Botrytis cinerea* strain co-infected by double-stranded RNA mycoviruses presents hypovirulence-associated traits. *Virol. J.* **2013**, *10*, 220–228. [CrossRef]
32. Yu, L.; Sang, W.; Wu, M.D.; Zhang, J.; Yang, L.; Zhou, Y.J.; Chen, W.D.; Li, G.Q. Novel hypovirulence-associated RNA mycovirus in the plant-pathogenic fungus *Botrytis cinerea*: Molecular and biological characterization. *Appl. Environ. Microbiol.* **2015**, *81*, 2299–2310. [CrossRef] [PubMed]
33. De Guido, M.; Minafra, A.; Santomauro, A.; Pollastro, S.; De Miccolis Angelini, R.; Faretra, F. Molecular characterization of mycoviruses from *Botryotinia fuckeliana*. *J. Plant Pathol.* **2005**, *87*, 293.
34. De Guido, M.; Minafra, A.; Santomauro, A.; Faretra, F. Molecular characterization of mycoviruses in *Botryotinia fuckeliana*. In Proceedings of the 14th International Botrytis Symposium Abstract Book, Cape Town, South Africa, 2007; p. 40.
35. Wu, M.; Jin, F.; Zhang, J.; Yang, L.; Jiang, D.; Li, G. Characterization of a novel bipartite double-stranded RNA mycovirus conferring hypovirulence in the pytophathogenic fungus *Botrytis porri*. *J. Virol.* **2012**, *86*, 6605–6619. [CrossRef]

36. Fan, X.; Zhang, J.; Yang, L.; Wu, M.; Chen, W.; Li, G. Development of PCR-based assays for detecting and differentiating three species of *Botrytis* infecting broad bean. *Plant Dis.* **2015**, *99*, 691–698. [CrossRef]
37. Zhou, Y.; Li, N.; Yang, J.; Yang, L.; Wu, M.; Chen, W.; Li, G.; Zhang, J. Contrast between orange and black colored sclerotial isolates of *Botrytis cinerea*: Melanogenesis and ecological fitness. *Plant Dis.* **2018**, *102*, 428–436. [CrossRef]
38. Büttner, P.; Koch, F.; Voigt, K.; Quidde, T.; Risch, S.; Blaich, R.; Brückner, B.; Tudzynski, P. Variations in ploidy among isolates of *Botrytis cinerea*: Implications for genetic and molecular analyses. *Curr. Genet.* **1994**, *25*, 445–450. [CrossRef] [PubMed]
39. Li, G.; Wang, D.; Huang, H.; Zhou, Q. Polymorphisms of *Sclerotinia sclerotiorum* isolated from eggplant in Jiamusi, Heilongjiang Province. *Acta Pharmacol. Sin.* **1996**, *26*, 237–242.
40. Fan, L.; Jing, Z.; Long, Y. *Agrobacterium tumefaciens*-mediated transformation of *Botrytis cinerea* strain RoseBc-3. *J. Huazhong Agri. Uni.* **2013**, *32*, 30–35.
41. Yu, Q.Y. Functional Analysis of Autophage-Related Genes *Bcatg26*, *Bcatg17* and *Bcatg14* in *Botrytis cinerea*. Master's Thesis, Huazhong Agricultural University, Wuhan, China, 2007.
42. Zhang, L.; Wu, D.M.; Li, G.Q.; Jiang, D.H.; Huang, H.C. Effect of mitovirus infection on formation of infection cushions and virulence of *Botrytis cinerea*. *Physiol. Mol. Plant Pathol.* **2010**, *75*, 71–80. [CrossRef]
43. Schumacher, J.; Gautier, A.; Morgant, G.; Studt, L.; Ducrot, P.-H.; Le Pecheur, P.; Azeddine, S.; Fillinger, S.; Leroux, P.; Tudzynski, B.; et al. A functional bikaverin biosynthesis gene cluster in rare strains of *Botrytis cinerea* is positively controlled by VELVET. *PLoS ONE* **2013**, *8*, e53729. [CrossRef]
44. Wu, M.; Deng, Y.; Zhou, Z.; He, G.; Chen, W.; Li, G. Characterization of three mycoviruses co-infecting the plant pathogenic fungus *Sclerotinia nivalis*. *Virus Res.* **2016**, *223*, 28–38. [CrossRef] [PubMed]
45. Kumar, S.; Stecher, G.; Tamura, K. MEGA7: Molecular evolutionary genetics analysis version 7.0 for bigger datasets. *Mol. Biol. Evol.* **2016**, *33*, 1870–1874. [CrossRef] [PubMed]
46. Mathews, D.H.; Disney, M.D.; Childs, J.L.; Schroeder, S.J.; Zuker, M.; Turner, D.H. Incorporating chemical modification constraints into a dynamic programming algorithm for prediction of RNA secondary structure. *Proc. Natl. Acad. Sci. USA* **2004**, *101*, 7287–7292. [CrossRef] [PubMed]
47. Zheng, L.; Zhang, M.; Chen, Q.; Zhu, M.; Zhou, E. A novel mycovirus closely related to viruses in the genus *Alphapartitivirus* confers hypovirulence in the phytopathogenic fungus *Rhizoctonia solani*. *Virology* **2014**, *456*, 220–226. [CrossRef] [PubMed]
48. Xiao, X.; Cheng, J.; Tang, J.; Fu, Y.; Jiang, D.; Baker, T.S.; Ghabrial, S.A.; Xie, J. A novel partitivirus that confers hypovirulence on plant pathogenic fungi. *J. Virol.* **2014**, *88*, 10120–10133. [CrossRef] [PubMed]

viruses

MDPI

Article

Two Novel Hypovirulence-Associated Mycoviruses in the Phytopathogenic Fungus *Botrytis cinerea*: Molecular Characterization and Suppression of Infection Cushion Formation

Fangmin Hao [1,2], Ting Ding [1,2], Mingde Wu [1,2,*], Jing Zhang [1,2], Long Yang [1,2], Weidong Chen [3] and Guoqing Li [1,2]

[1] The State Key Laboratory of Agricultural Microbiology, Huazhong Agricultural University, Wuhan 430070, China; haofangmin@163.com (F.H.); dingting19910216@sina.com (T.D.); zhangjing1007@mail.hzau.edu.cn (J.Z.); yanglong@mail.hzau.edu.cn (L.Y.); guoqingli@mail.hzau.edu.cn (G.L.)

[2] The Key Laboratory of Plant Pathology of Hubei Province, Huazhong Agricultural University, Wuhan 430070, China

[3] U.S. Department of Agriculture, Agricultural Research Service, Washington State University, Pullman, WA 99164, USA; w-chen@wsu.edu

* Correspondence: mingde@mail.hzau.edu.cn

Received: 13 April 2018; Accepted: 9 May 2018; Published: 13 May 2018

Abstract: *Botrytis cinerea* is a necrotrophic fungus causing disease on many important agricultural crops. Two novel mycoviruses, namely Botrytis cinerea hypovirus 1 (BcHV1) and Botrytis cinerea fusarivirus 1 (BcFV1), were fully sequenced. The genome of BcHV1 is 10,214 nt long excluding a poly-A tail and possesses one large open reading frame (ORF) encoding a polyprotein possessing several conserved domains including RNA-dependent RNA polymerase (RdRp), showing homology to hypovirus-encoded polyproteins. Phylogenetic analysis indicated that BcHV1 may belong to the proposed genus Betahypovirus in the viral family *Hypoviridae*. The genome of BcFV1 is 8411 nt in length excluding the poly A tail and theoretically processes two major ORFs, namely ORF1 and ORF2. The larger ORF1 encoded polypeptide contains protein domains of an RdRp and a viral helicase, whereas the function of smaller ORF2 remains unknown. The BcFV1 was phylogenetically clustered with other fusariviruses forming an independent branch, indicating BcFV1 was a member in Fusariviridae. Both BcHV1 and BcFV1 were capable of being transmitted horizontally through hyphal anastomosis. Infection by BcHV1 alone caused attenuated virulence without affecting mycelial growth, significantly inhibited infection cushion (IC) formation, and altered expression of several IC-formation-associated genes. However, wound inoculation could fully rescue the virulence phenotype of the BcHV1 infected isolate. These results indicate the BcHV1-associated hypovirulence is caused by the viral influence on IC-formation-associated pathways.

Keywords: *Botrytis cinerea*; hypovirus; fusarivirus; hypovirulence; infection cushion

1. Introduction

Fungi in the genus of *Botrytis* are able to infect more than 1400 species of cultivated plants, and are responsible for heavy losses of many important agricultural crops [1]. Among *Botrytis* spp., *B. cinerea* has the widest distribution and broadest host range, and has received most attention due to its high economic impact. The control of *B. cinerea* mostly relies on chemical fungicides [2]. However, the evolution of fungicide resistance increases the difficulty of only using fungicides for *B. cinerea* management. Therefore, some alternative methods, such as biological control, were developed for the control of *B. cinerea*. Mycoviruses, as a biocontrol agent, have been successfully used for the control of chestnut blight in Europe [3]. This consequently inspired further research of mycoviruses, and many mycoviruses were constantly reported in different groups of plant pathogenic fungi [4], although the complex vegetative compatibility groups (VCGs) limits the use of hypovirus for the control of chestnut blight in Northern America [5–7].

Mycoviruses are also commonly reported in the population of *Botrytis* spp., and most of these infect *B. cinerea*. Some mycoviruses infecting *Botrytis* spp. were assigned to viral families *Gammaflexiviridae, Alphaflexiviridae, Narnaviridae, Endornaviridae, Partitiviridae,* and *Totiviridae* [8,9], or the new established genus *Botybirnavirus* [10], whereas the remaining are still unassigned [8,11–13]. Among sequenced mycoviruses infecting *Botrytis* spp., such as Botrytis cinerea mitovirus 1 (BcMV1) [14,15], Botrytis porri botybirnavirus 1 (previously known as Botrytis porri RNA virus 1) [10], Botrytis cinerea RNA virus 1 [11] and Botrytis cinerea CCg378 virus 1 [16] were determined to be capable of attenuating virulence of *Botrytis*. Nevertheless, mycoviruses such as Botrytis ourmia-like virus [12], Botrytis cinerea negative-stranded RNA virus 1 [13], Botrytis virus F [17] and Botrytis virus X [18], seem to have no significant effects on the pathogenicity of *B. cinerea*. Recently, deep sequencing has also been employed for investigating virus diversity in *Botrytis* [19].

Hypoviruses are a group of positive single-stranded RNA (+ssRNA) mycoviruses, 9–13 kb in length excluding a poly A tail, possessing one or two open reading frames (ORFs) on their coding strands, without formation of true virions [20]. In addition to their potential for the biological control of chestnut blight, hypoviruses have also been developed as a tool for investigating interactions between viruses and host fungi [21]. Based on the phylogenetic analysis and genomic characteristics, the viral family *Hypoviridae* was proposed to be divided into two (Alphahypovirus and Betahypovirus) [22–24] or three genera (Alphahypovirus, Betahypovirus, and Gammahypovirus) [25].

The viral family Fusariviridae is a newly proposed +ssRNA viral family [26], probably encompassing eight viral members including Fusarium graminearum virus-DK21 (FgV-DK21) [27], Sclerotinia sclerotiorum fusarivirus 1 (SsFV1) [28], Penicillium roqueforti ssRNA mycovirus 1 (PrRV1), Rosellinia necatrix fusarivirus 1 (RnFV1) [26], Pleospora typhicola fusarivirus 1 (PtFV1) [29], Penicillium aurantiogriseum fusarivirus 1 (PaFV1) [29], Macrophomina phaseolina single-stranded RNA virus 1 (MpRV1) and Alternaria brassicicola fusarivirus 1 (AbFV1) [30]. The genomes of fusariviruses are 6–8 kb in length excluding a poly A tail, and contains two or four ORFs. The large ORF encoded polypeptides by fusariviruses usually contain an RNA dependent RNA polymerase (RdRp) domain and a viral helicase domain [30].

Although numerous mycoviruses have been reported in species of *Botrytis*, no hypovirus or fusarivirus has been documented in the population of *Botrytis*. In this study, three mycoviruses, including a hypovirus, namely Botrytis cinerea hypovirus 1 (BcHV1), a fusarivirus, namely Botrytis cinerea fusarivirus 1 (BcFV1), and an endornavirus, namely Botrytis cinerea endornavirus 1 (BcEV1), were detected in a hypovirulent *B. cinerea* strain HBtom-372. Besides these three mycoviruses, four smaller dsRNAs with the length ranging from 1.0 kb to 4.0 kb were also detected in the mycelium of strain HBtom-372. As the genome organization of BcEV1 was described previously [31], the objectives of this study were: (i) to determine the full length sequences of BcHV1, BcFV1 and four smaller dsRNAs co-infecting *B. cinerea* strain HBtom-372 in China; (ii) to investigate the biological effects of the three mycoviruses and smaller four dsRNAs on *B. cinerea*; and (iii) to elucidate the underling mechanism that may be responsible for the virus-induced hypovirulence of *B. cinerea*.

2. Materials and Methods

2.1. Fungal Strains and Culture Conditions

B. cinerea strains HBtom-372 and HBtom-459 were originally isolated from diseased tomato fruits in Jingmen County and Yichang County, Hubei Province, China. Strain B05.10 of *B. cinerea* (whole genome sequence available) was isolated from diseased table grape (*Vitis vinifera*) in Germany.

All strains were stored as described previously [15], and working culture for each strain was established through transferring the stored mycelial plugs onto the PDA plates and subsequently incubated at 20 °C for 3 days. Isolates Z1, Z3, Z26 and Z33 were derived from strain HBtcm-459 via hyphae anastomosis with HBtom-372.

2.2. dsRNA Extraction and Purification

Extraction and purification of dsRNA from *B. cinerea* mycelia was performed as described previously [14], the dsRNA nature was further confirmed based on resistance to DNase I and S1 nuclease (Promega, Madison, WI, USA). The extracted dsRNA was fractionated by agarose gel (1%, w/v) electrophoresis and visualized by staining with ethidium bromide (1.5 μg/L) and viewing on a UV trans-illuminator.

2.3. cDNA Cloning and Sequencing

The dsRNAs extracted from strain HBtom-372 of *B. cinerea* were separated by agarose gel electrophoresis. Each dsRNA band was excised and purified from the agarose gel using AxyPrep^{TM} DNA Gel Extraction Kit (Axygen Scientific, Inc.; Union City, CA, USA). The cDNA library of each dsRNA (dsRNA-A2, dsRNA-B, dsRNA-C, dsRNA-D and dsRNA-E) was produced using a random primer-mediated PCR amplification protocol [11] and sequenced as previously described [10]. The terminal sequences of each dsRNA were cloned through ligating the 3′-terminus for each strand of each dsRNA with the 5′-terminus of the 110A adaptor (Table S1) using T4 RNA ligase (Promega Corporation, 2800 Woods Hollow Road, Madison, WI, USA) at 16 °C for 18 h. and then reverse transcribed using primer RC110A (Table S1). The cDNA strands were then used as template for PCR amplification of the 5′- and 3′-terminal sequences with primer RC110A and corresponding sequence specific primer for each dsRNA segment (Figures S1 and S2, and Table S1). Cloning of the 3′- or 5′-terminal sequences of the dsRNA was performed on three separate occasions (Figures S1 and S2). The gaps between the cDNA contigs among different dsRNAs were amplified by RT-PCR with sequence specific primer pairs (Figures S1 and S2, Table S1). The dsRNA-F was agarose gel-purified, ligated with the 110A adaptor, reverse transcribed to cDNA with the primer RC110A, and the cDNA was then used as template in PCR to amplify the full length cDNA sequence of dsRNA-F directly with the primer RC110A (Figure S2) [32]. Cloning of the full-length sequence of dsRNA-F was repeated three times. All these amplicons were detected by agarose gel electrophoresis, gel-purified, and cloned into *E. coli* DH5α and sequenced as previously described [15]. All partial cDNA sequences were assembled to obtain the full-length cDNA sequence of BcHV1, BcFV1 and the other three dsRNAs (dsRNA-C, dsRNA-D, and dsRNA-E).

2.4. Nucleotide Sequences and Amino Acid Residues Sequences Analysis

ORFs in the full-length cDNA sequences of the dsRNAs in strain HBtom-372 of *B. cinerea* were deduced using the ORF Finder program in the website of the National Center for Biotechnology Information (NCBI, http://www.ncbi.nlm.nih.gov/gorf/gorf.html). The BlastN and BlastP programs in the public database at NCBI were used for searching the full-length cDNA sequences and deduced polypeptides of each dsRNA, respectively. CDD database (http://www.ncbi.nlm.nih.gov/Structure/cdd/wrpsb.cgi) searching deduced the domains present in the polypeptide sequence. Multiple alignment of the sequences of conserved domains in the polypeptides encoded by different mycoviruses were performed using the MUSCLE program in MEGA 5.0 [33]. Phylogenetic trees based on the sequences of conserved domains of BcHV1 and BcFV1 were constructed using the neighbor-joining (NJ) method and tested with a bootstrap of 1000 replicates to ascertain the reliability of a given branch pattern in MEGA 5.0. Putative transmembrane helices sequences were predicted using the TMHMM server version 2.0 (http://www.cbs.dtu.dk/services/TMHMM/) [34].

2.5. Northern Hybridization

Northern hybridization was performed to confirm the authenticity of the cDNA sequences generated from BcHV1 and BcFV1 in strain HBtom-372 of *B. cinerea*. Two DNA probes, nt positions 5514–6234 for Probe 1 and nt positions 2586–3336 for Probe 2, were designed based on full-length cDNA sequences of BcHV1 and BcFV1, respectively. The gel-purified dsRNA-A2 and dsRNA-B were separated in 1% (*w*/*v*) agarose gel and transferred to positively charged nylon membranes (Millipore, Bedford, MA, USA) [10] by the capillary transfer method using 20 × SSC as transfer buffer [35]. Probe 1 and Probe 2 were pre-labeled with the enzyme as described by the manufacturers (GE Healthcare, Little Chalfont, United Kingdom)) for hybridization with the denatured dsRNAs blotted on two membranes, respectively. The chemiluminescent signals of the probe-RNA hybrids were detected using a CDP-Star kit (GE Healthcare).

2.6. Biological Properties of Botrytis cinerea Strain HBtom-372

MAPs (mycelium agar plugs, 6 mm in diameter) removed from the colony margin of a 2–4-day-old culture of each strain or isolate were placed on PDA in petri dishes (9 cm in diameter), one plug per dish. The dishes were incubated at 20 °C for determination of the mycelial growth rate and for observation of the colony morphology. Lesion diameter on rapeseed (*Brassica napus* L.) (20 °C, 72 h) and tomato (*Lycopersicon esculentum* Mill.) leaves (20 °C, 48 h), and radial mycelial growth rate on PDA (20 °C, in the dark) was determined using the procedures described in our previous studies [14,15].

To rule out the possibility that the derivative isolates were contaminated by the donor strain, the genetic backgrounds of all the derived isolates, along with strains HBtom-372 and HBtom-459, were profiled by randomly amplified polymorphic DNA (RAPD) using the 10-mer primer OPC-04 as described previously [36] (Table S1).

2.7. Viral Horizontal Transmission and Detection of Mycoviruses by RT-PCR

Horizontal transmission refers to the transmission of hypovirulence-associated dsRNAs from hypovirulent to virulent fungal strains through hyphal anastomosis or contact [37]. The experiment was carried out by using the pairing culture technique as previously described [14,15]. In each pairing culture (9 cm in diameter), the dsRNA-harboring hypovirulent strain HBtom-372 served as the donor, whereas strains HBtom-459 served as the recipient. Derivative isolates were obtained from the recipient strain HBtom-459 in the contact cultures using the method described by Wu et al. [14]. All derivative isolates were subjected to test for the presence of three mycoviruses and other dsRNAs through RT-PCR with primer pairs H-RT-F/H-RT-R, F-RT-F/F-RT-R, C-RT-F/C-RT-R, and RT-F-F/RT-F-R, respectively (Table S1). Strains HBtom-372 and HBtom-459 were included as controls in this experiment. Four derivative isolates, namely Z1, Z3, Z26 and Z33, were selected and individually tested for the pathogenicity on both intact and wounded detached rapeseed leaves (20 °C, 72 h), mycelial growth rate on PDA (20 °C), production of conidia and sclerotia, and the presence of dsRNA in mycelia. In addition, 98 strains of *B. cinerea* from different places of China (Table S2) were subjected to test for the presence of BcHV1 and BcFV1.

2.8. Stereomicroscopic Observation of Infection Cushions

The MAPs (6 mm diameter) of strains HBtom-372 and HBtom-459 and their derivative isolates of *B. cinerea* were inoculated on onion bulb scales, one MAP per each bulb scale, three scales for each isolate/strain. All inoculated onion scales were placed on moistened paper towels in plastic trays and covered with transparent plastic films to maintain high humidity. After incubation at 20 °C for 9 h, MAPs and the bulb scales of onion were stained with methyl blue and examined for formation of ICs under a stereomicroscope. The number of ICs formed by each *B. cinerea* isolate/strain around the MAPs was counted.

2.9. Quantitative Real-Time PCR

The mycelia of strain HBtom-459 and isolate Z33 were harvested from PDA plate and onion bulb scales (9 h post inoculation), respectively, and the total RNA was extracted from the harvested mycelia with TRIzol® reagent (Invitrogen Corp, Carlsbad, CA, USA) using the procedures recommended by the manufacturer. The extracted RNA was then used for detection the expression of IC formation-associated genes by RT-PCR and qRT-PCR with the primer sets listed in Table S3. The calculation of the relative expression level of each gene was done using the procedures described previously [38]. The experiment was repeated two more times.

3. Results

3.1. Botrytis cinerea Strain HBtom-372 Exhibits Hypovirulence Traits

After cultivation on potato dextrose agar (PDA) for 15 days, strain HBtom-372 formed abnormal colonies with no production of conidia and sclerotia, and was unable to cover the entire Petri dishes (Figure 1A). In contrast, strain B05.10 formed normal colonies with the formation of conidia and sclerotia (Figure 1A). It is notable that strain HBtom-372 also failed to produce conidia and sclerotia on PDA even after 30 days. The virulence assay on detached rapeseed leaves revealed that the average lesion diameter (0.5 mm) caused by strain HBtom-372 was significantly smaller than that (18.0 mm) of strain B05.10 (Figure 1A,B, Table S6). Compared with the virus free strain B05.10, the radial mycelial growth of HBtom-372 on PDA was significantly slower. The average radial mycelial growth rate of strain HBtom-372 was 2.0 mm/day, which was significantly slower than that of strain B05.10 (15.1 mm/day). Virulence assay on detached tomato leaves also indicated that the virulence of strain HBtom-372 was significantly reduced compared with that of strain HBtom-459 (Figure S3). After DNase I and S1 nuclease digestion, multiple dsRNA segments were detected through electrophoresis in the mycelium of HBtom-372 of *B. cinerea* with the sizes ranging from 13.5 kb to 1.5 kb, named from largest to smallest as dsRNA-A1, dsRNA-A2, dsRNA-B, dsRNA-C, dsRNA-D, dsRNA-E and dsRNA-F (Figure 1C), whereas no dsRNA was detected in HBtom-459 (Figure S3). A faint dsRNA segment close to dsRNA-B was also observed. However, no other sequences except BcFV1 (dsRNA-B) were obtained through the sequencing of the cDNA library (about 60 clones) constructed by the gel-purified dsRNA. Therefore, we suppose this band may also be derived from dsRNA-B. Among these dsRNAs, dsRNA-A1 was previously determined to be BcEV1. It is notable that the dsRNA of BcEV1 was purified from the agarose gel and then detected through electrophoresis once more, thus only one dsRNA segment was shown in our previous study [31].

3.2. Full-Length cDNA Sequences of BcHV1 and BcFV1

The assembled full length genome sequence of BcHV1 (dsRNA-A2) is 10,214 nt, with a GC content of 44.7%, excluding the poly A tail (GenBank accession No. MG554632). The genome of BcHV1 was hypothesized to contain one large ORF with two 395 nts and 924 nts long untranslated regions (UTR) located at both 5′- and 3′- terminus of the positive strand of BcHV1, respectively. The large ORF, namely ORF L, was predicted to encode a putative polypeptide of 2964 amino acid (aa) residues with a deduced molecular mass of 336 kDa (Figure 2A). The BLASTp analysis of the polypeptide showed that ORF L-encoded polypeptide is closely related to Sclerotinia sclerotiorum hypovirus 1 (SsHV1, 67.4% identity), Cryphonectria hypovirus 3 (CHV3, 58.31% identity), Phomopsis longicolla hypovirus 1 (PlHV1, 57.84% identity) and Valsa ceratosperma hypovirus 1 (VcHV1, 56.48% identity) (Table S4). Therefore, we supposed that BcHV1 should be a member in the viral family *Hypoviridae*.

The full length cDNA sequence of BcFV1 (dsRNA-B) was 8411 bp long with a GC content of 46.4%, excluding the poly A tail (GenBank accession No. MG554633). The whole genome of BcFV1 possesses two large ORFs and two 510-nt and 364-nt long untranslated regions (UTR) located at both 5′- and 3′- terminus of the positive strand of BcFV1, respectively (Figure 2A). A 456 nts long UTR was located between the two ORFs. The ORF 1 and ORF 2 were predicted to encode a 186 kDa and an

81.7 kDa putative polypeptide of 1644 and 734 aa residues in length, respectively. The BLASTp analysis of ORF 1-encoded polypeptide of BcFV1 showed that it was related to polypeptides encoded by AbFV1 (25.58% identity), SsFV1 (24.3% identity) and PtFV1 (23.84% identity). In addition, the polypeptide encoded by ORF 1 of BcFV1 also showed low sequence similarity to viruses of *Hypoviridae*, *Potyviridae*, *Poxviridae*, *Iflaviridae* and *Secoviridae* (Table S5). However, the polypeptide encoded by ORF 2 showed no significant sequence similarity with proteins in the database of NCBI by using BlASTp search. Therefore, we suppose that BcFV1 might be a novel member in the proposed virial family Fusariviridae. Moreover, the viral sequences of both BcHV1 and BcFV1 were confirmed through the Northern hybridization analysis (Figure 2B). Considering both viruses possess +ssRNA genomes, the observed dsRNAs are likely intermediates during viral replication.

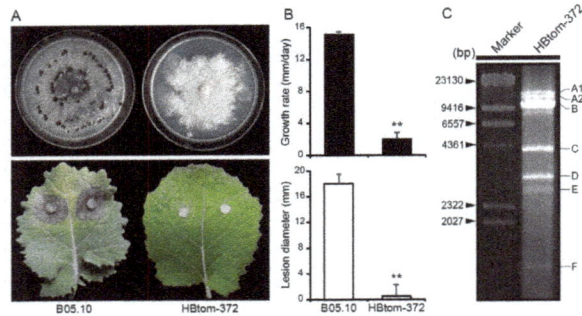

Figure 1. Biological properties and dsRNA detection of *Botrytis cinerea* strain HBtom-372 and B05.10. (**A**) Colony morphology (upper, 20 °C, 15 days) and pathogenicity assay (lower, 20 °C, 3 days) of strains HBtom-372 and B05.10 on potato dextrose agar (PDA) and detached rapeseed leaves, respectively. (**B**) Radial mycelial growth rate (20 °C, upper) on PDA and lesion diameter (20 °C, 72 h, lower) on detached rapeseed leaves of strains HBtom-372 and B05.10. "**" indicates a significant difference ($p < 0.01$) between strains HBtom-372 and B05.10 in both pathogenicity and radial mycelial growth rate. (**C**) Agarose gel electrophoresis of dsRNAs extracted from the mycelium of *Botrytis cinerea* strain HBtom-372. Marker, λ-*Hind* III digest DNA marker.

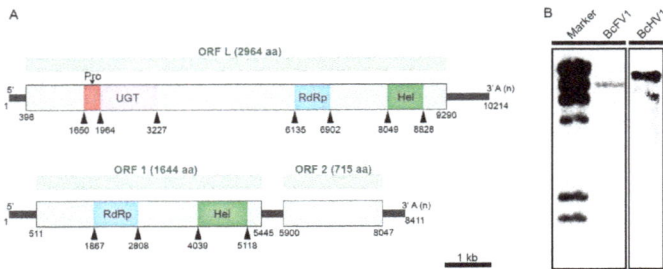

Figure 2. (**A**) Schematic diagram of the genome organization of Botrytis cinerea hypovirus 1 (BcHV1) and Botrytis cinerea fusarivirus 1 (BcFV1). The coding strand of BcHV1 is 10,214 nt long and contains a large ORF encoding a polyprotein of 2964 aa, possessing conserved domains: Prot, papain-like protease; UGT, UDP glucose/sterol glucosyltransferase; RdRp, RNA-dependent RNA polymerase; Hel, viral helicase superfamily. The coding strand of BcFV1 is 8411 nt in length and contains two major ORFs, ORF1 and ORF2, encoding two polypeptides of 1644 aa and 715aa, respectively. ORF1-encoded polypeptide possesses two conserved domains, RdRp and Hel. (**B**) Northern blotting detection of BcHV1 and BcFV1 dsRNAs extracted from the mycelium of *B. cinerea* strain HBtom-372. Marker, λ-Hind III digest DNA marker.

3.3. Putative Polyprotein Encoded by BcHV1 and BcFV1

CDD database search of the BcHV1-encoded polypeptide in the database of NCBI revealed that it contained a putative papain-like protease (Prot) domain, a UDP glucose/sterol glucosyltransferase (UGT) domain, an RdRp domain and a viral RNA Helicase (Hel) domain (Figure 2A). The predicted RdRp domain was located between the UGT domain and Hel domain, including eight conserved motifs (I–VIII) (Figure 3) as described in other hypoviruses [39]. The RdRp domain of BcHV1 was closely related to SsHV1 (85.94% identity) and PlHV1 (85.16% identity) (Table S4). A typical Prot domain with conserved predicted autoproteolytic catalytic site (at positions Cys^{426} and His^{473}) and a putative polyprotein cleavage site (at position Gly^{523}) was detected in the polyprotein encoded by BcHV1 based on sequence alignment with other hypoviruses (Figure 2A, Table S4, and Figure S4). CDD database search showed that BcHV1 also contained a conserved UGT domain as reported in SsHV1 and other hypoviruses [40] (Figure 2A and Figure S4). Three characteristic motifs, namely GKST box, DExH box and QRxGR box, of the predicted Hel domain of BcHV1 were also detected through multiple sequence alignment (Figure S4).

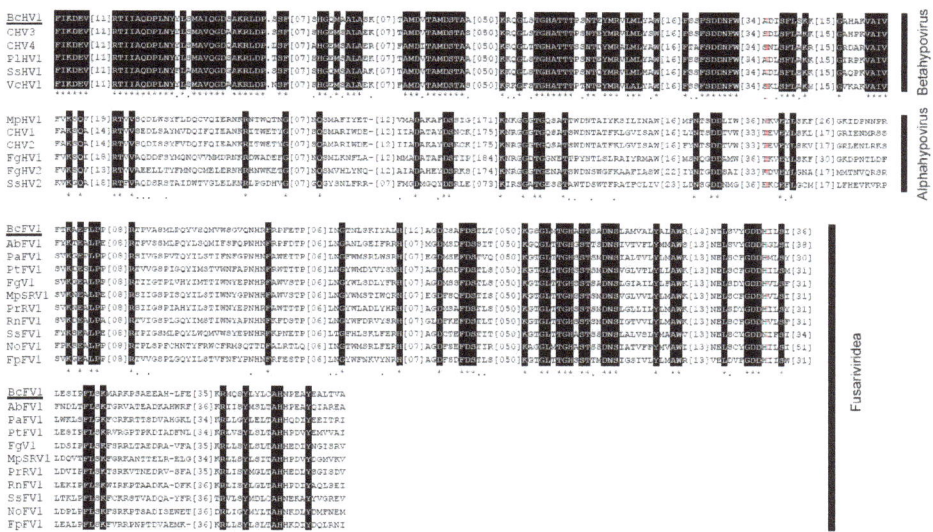

Figure 3. Multiple alignment of the amino acid sequences of RNA-dependent RNA polymerase domains in the polyprotein encoded by Botrytis cinerea hypovirus 1 (BcHV1) and Botrytis cinerea fusarivirus 1 (BcFV1), respectively. "*" indicates identical amino acid residues; and "." indicate low chemically similar amino acid residues. The abbreviations of virus names are listed in Tables S4 and S5. The names of BcHV1 and BcFV1 are underlined.

The BcFV1 ORF 1-encoded polypeptide contained a putative RdRp domain and a viral Hel domain (Figure 2A). The predicted RdRp domain included eight conserved motifs (I–VIII) and was closely related to RdRp domains of other fusariviruses, especially AbFV1 (47.17% identity) and SsFV1 (46.98% identity) (Table S5 and Figure 3). Six conserved motifs in the viral Hel of fusariviruses were also detected in the Hel of BcFV1 through multiple sequence alignment (Table S5 and Figure S4). However, no protein showed significant similarity to the ORF 2 encoded protein through BLASTp search in NCBI database. Similar to other fusariviruses, transmembrane (TM) domains (Figure S5) were also found at the *N*-proximal ORF 1-coded protein of BcFV1 [34].

3.4. Phylogenetic Analysis of BcHV1 and BcFV1

To define the phylogenetic relationship of BcHV1 and BcFV1 with other mycoviruses (Table S4), a phylogenetic tree was established based on the RdRp-Hel region including the RdRp domain, Hel domain and the aa sequence between the two domains. Three major clades, namely Alphahypovirus, Betahypovirus, and Fusariviridae, were observed in the RdRp-Hel phylogenetic tree with bootstrap of 100% for each clade (Figure 4). BcHV1 appeared to be mostly close related to SsHV1 with bootstrap of 81%, and then clustered with other betahypoviruses forming an independent Betahypovirus clade. Although BcFV1 was mostly homologous to AbFV1 through BLAST search, BcFV1 did not cluster with AbFV1, instead of forming an independent branch, and then clustered with other fusariviruses forming the Fusariviridae clade (Figure 4).

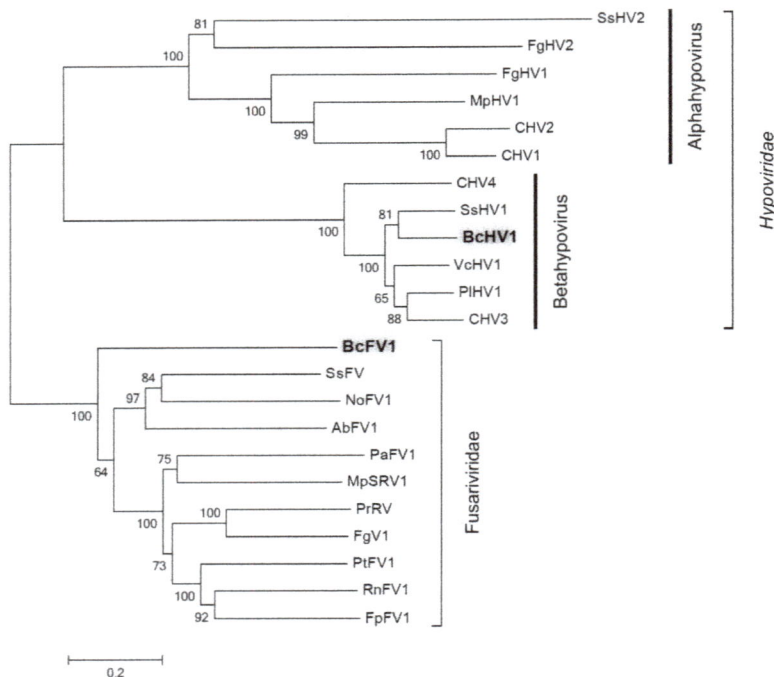

Figure 4. Phylogenetic analysis of Botrytis cinerea hypovirus 1 (BcHV1) and Botrytis cinerea fusarivirus 1 (BcFV1) based on RdRp-Hel region from strain HBtom-372 of *B. cinerea*. The abbreviations of virus names for constructing the phylogenetic tree are listed in Tables S4 and S5.

3.5. Nucleotide Sequence of other dsRNAs

The full-length cDNA sequences of remaining four dsRNAs, namely dsRNA-C, dsRNA-D, dsRNA-E and dsRNA-F, were also determined. Sequence analysis indicated that dsRNA-C (GenBank accession No. MG554634) and dsRNA-F (GenBank accession No. MG554637) were 3912 bp and 1375 bp in length excluding the Poly A tail, respectively. Alignment of the cDNA sequences of dsRNA-C and dsRNA-F showed that the nucleotide (nt) sequence of dsRNA-F (1-1375) was 100%, 99.57% and 98.91% identical to three regions of dsRNA-C, namely 1–171, 2451–3142 and 3401–3912, respectively (Figure S6). Thus, dsRNA-F might be the defective RNA of dsRNA-C. Both the 5′- and 3′-terminal sequences of dsRNA-C were closely related to those of BcHV1. Sequence alignment showed that of the first 298 nts sequences at 5′-termini of SsHV1 and dsRNA-C were 88.93% identical (Figure 5), and their last 202 nts at the 3′-termini were 83.65% identical (Figure 5). However, no significant sequence similarity was detected in the middle region between dsRNA-C and BcHV1 at nt level. DsRNA-C was deduced to encode a protein of 670 aa with an approximate molecular mass of 73.8 kDa. However, no homologous protein and conserved domain were detected through BLASTp and CDD search with the deduced aa sequence of dsRNA-C.

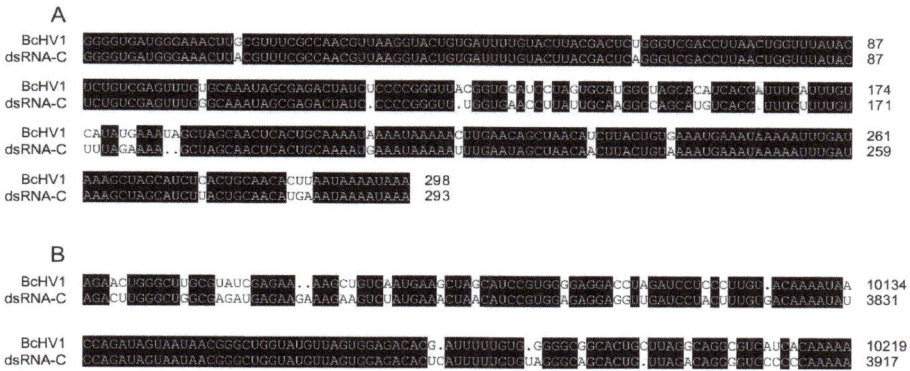

Figure 5. Alignment of the 5′ (**A**) and 3′-terminal (**B**) sequences of the coding strands of Botrytis cinerea hypovirus 1 (BcHV1) and dsRNA-C. An identical nucleotide is highlighted with black shading. The black dot indicates a missing nucleotide.

Sequence analysis indicated that the full length cDNA of dsRNA-D (GenBank accession No. MG554635) and dsRNA-E (GenBank accession No. MG554636) were 3482 bp and 3253 bp in length, respectively. The nt sequences of dsRNA-D and dsRNA-E were almost 100% identical to two regions, namely nt 1–3480 and 1–3247, of BcFV1, respectively, except few nt insertions at both 5′-termini and middle region of the two dsRNAs (Figure S6). Therefore, we suppose that dsRNA-D and dsRNA-E may be two defective RNAs of BcFV1.

3.6. Horizontal Transmission of Hypovirulence-Associated dsRNAs

To determine the transmission capacity of BcHV1, BcFV1 and BcEV1, strain HBtom-372 was dually cultured with strain HBtom-459 on the same PDA plate, while the inhibition of mycelial growth on the margin of HBtom-459 was observed after seven days (Figure 6A and Figure S7). Several mycelial plugs were picked out from the abnormal margin for establishing the derivative isolates, and 33 isolates were obtained. The presences of three mycoviruses, dsRNA-C and dsRNA-F in all 33 derivative isolates were detected through reverse transcription (RT)-PCR with the specific primers (Table S1). The results showed that isolate Z33 only contained one virus, BcHV1, and isolate Z26 contained all three viruses, BcEV1, BcHV1 and BcFV1, while the other isolates including Z1 and Z3 were infected by both BcHV1 and BcFV1 (Figure 7A). The dsRNA-C and dsRNA-F were only detected in isolate Z26, while not detected in isolates Z33, Z1 and Z3 (Figure 7A). All four derived isolates showed same RAPD profiles as those of their recipient HBtom-459, indicating that the derived isolates were not from contamination (Figure S8). Four isolates, Z1, Z3, Z26 and Z33, were selected for further biological characterization.

Isolate Z33 grew fast on PDA plate with the average radial growth rate of 15.8 mm/day, and was comparable to strain HBtom-459 with the radial growth rate of 14.6 mm/day (Figure 7B and Table S6). The other strain or isolates grew slower than both isolate Z33 and strain HBtom-459 with the average radial growth rate ranging from 2.6 mm/day to 9.4 mm/day (Figure 7B and Table S6). Although both isolates Z1 and Z3 were infected by BcFV1 and BcHV1, isolate Z3 grew slightly faster than isolate Z1. The culture morphology of isolate Z33 was similar to that of strain HBtom-459 without the formation of mycelial sectors, and both strains colonized the entire plate within three days (Figure 6B).

However, isolates Z1 and Z3 were unable to colonize the entire plate within three days and formed many mycelial sectors after 5 days (Figure 6B). Isolate Z26 was severely debilitated with the formation of many mycelial sectors, and could not colonize the entire plate within seven days (Figure 6B). The production of conidia and sclerotia varied in different derivative strains (Table 1). Compared with strain HBtom-459, the yields of conidia were slightly decreased in isolate Z33, Z1 and Z3, whereas the yields of sclerotia were significantly increased in isolate Z33, Z1 and Z3, with 79, 66 and 115 sclerotia per dish, respectively, but with a smaller size (Table 1). However, similar to strain HBtom-372, isolate Z26 formed no conidia and sclerotia after 30 days (Figure 6B).

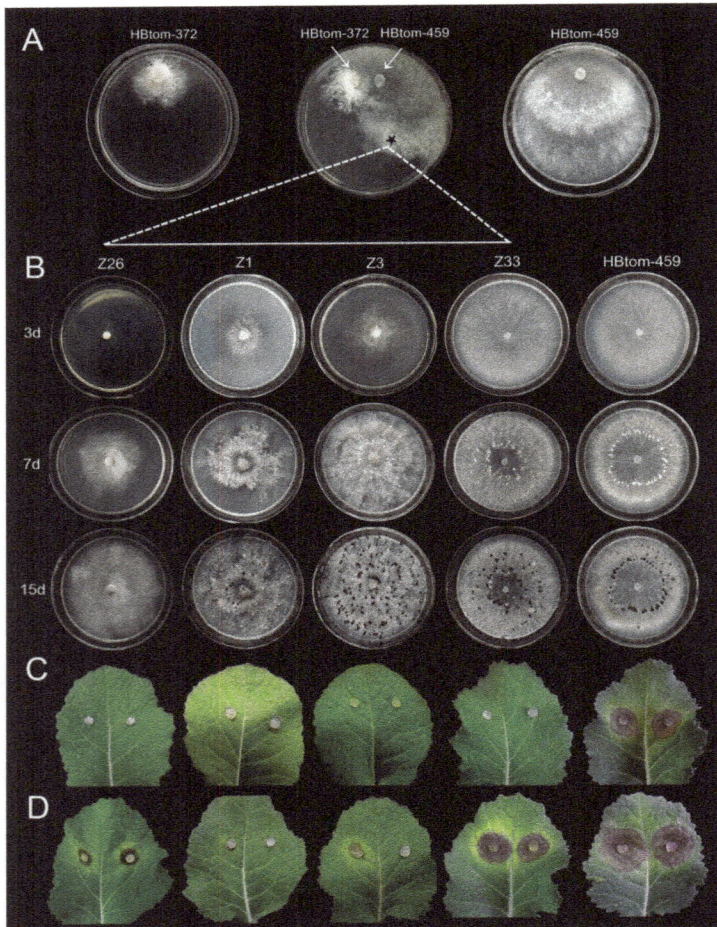

Figure 6. Horizontal transmission of three mycoviruses and biological properties of derivative *Botrytis cinerea* isolates. (**A**) Colony morphology of virus-infected strain HBtom-372, virus-free strain HBtom-459 and dual culture of these two strains allowing horizontal transmission of viruses from strain HBtom-372 to strain HBtom-459 through hyphal anastomosis on potato dextrose agar (PDA) (20 °C, seven days). (**B**) Colony morphology among strain HBtom-459 and derivative *B. cinerea* isolates and on PDA at 20 °C for 3, 7 and 15 days, respectively. Pathogenicity assay on intact (**C**) and wounded (**D**) detached rapeseed leaves (*Brassica napus*) of derivative *B. cinerea* isolates and strain HBtom-459. The star in the dual culture indicates the area where mycelial agar plugs were taken for generation of derivative isolates.

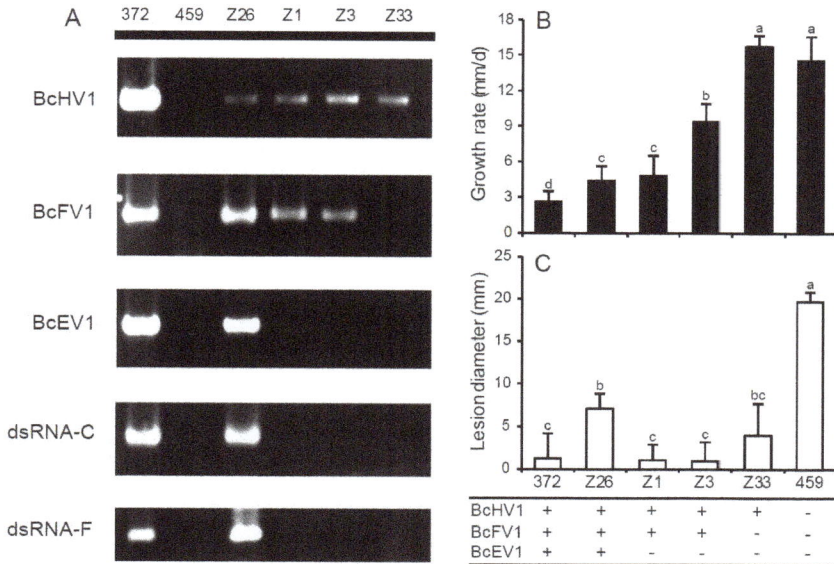

Figure 7. The radial mycelial growth and virulence of *B. cinerea* isolates/strains infected with different mycoviruses. (**A**) The reverse transcription (RT)-PCR detection of Botrytis cinerea endornavirus 1 (BcEV1), Botrytis cinerea hypovirus 1(BcHV1), Botrytis cinerea fusarivirus 1 (BcFV1). dsRNA-C and dsRNA-F in *B. cinerea* strains with primer pairs listed in Table S1. The radial growth rate on potato dextrose agar at 20 °C (**B**) and lesion diameter on detached leaves of rapeseed leaves (20 °C, 72 h) (*Brassica napus*) (**C**) for *B. cinerea* strains/isolates. The names of strains HBtom-372 and HBtom-459 are abbreviated here as 372 and 459, respectively. Bars represent arithmetic mean ± S.E.M. In each histogram, bars labeled with the same letters are not significantly different ($p > 0.05$) according to the least-significant-difference test. The symbols "+" and "−" indicate the presence and absence of corresponding virus, respectively.

Table 1. Yield of conidium and sclerotium, and size of sclerotium produced by different strains of *Botrytis cinerea*.

Strain	Pathogenicity [1]	Conidia [2]	Sclerotium	
		Log$_{10}$ Conidia/Dish ($n = 5$)	Sclerotia/Dish ($n = 3$)	Size (mm) ($n = 100$)
HBtom-459	V	6.95 a [3]	38 c	3.6 × 2.5
Z33	HV	6.64 b	79 b	2.8 × 2.3
Z1	HV	6.3 c	66 b	2.2 × 1.9
Z3	HV	6.36 b,c	115 a	3.5 × 2.4
Z26	HV	0 d	0 d	NS [4]
HBtom-372	HV	0 d	0 d	NS

[1] V = Virulent; HV = Hypovirulent. [2] Conidia and sclerotia were collected from 30-day-old cultures grown on potato dextrose agar (20 °C). [3] Numbers in the same column followed by the same letters are not significantly different ($p > 0.05$). [4] NS = no sclerotial formation.

The virulence assay on intact detached rapeseed leaves showed that the virulence of all derivative isolates were significantly impaired (Figure 6C). The average lesion diameters on rapeseed leaves were 4.7 mm, 0.4 mm, 0.0 mm, and 3.8 mm for isolates Z26, Z1, Z3, and Z33, respectively, which were apparently reduced in comparison with strain HBtom-459 of the average lesion diameter of 16.5 mm (Figures 6C and 7C, and Table S6).

3.7. Formation of Infection-Cushions

After inoculation on onion bulb epidermis for 9 h, lots of infection cushions (ICs) were formed around the mycelial agar plugs (MAPs) of strain HBtom-459 with the average number of 195 (Figure 8A,B, Figure S9, and Table S6). In contrast, the numbers of ICs formed by four derivative isolates were dramatically decreased in varying degrees, and the average numbers of ICs formed by isolates Z1, Z3, Z26 and Z33, were 11, 20, 15, and 55, respectively (Figure 8B and Table S6). However, no IC was observed around the MAPs of strain HBtom-372. To test the association of hypovirulence to the decreased formation of ICs, the MAPs of each strain/isolate were inoculated on both intact and wounded rapeseed leaves. The results showed that the virulence on wounded rapeseed leaves was significantly enhanced compared with that on intact rapeseed leaves for most *B. cinerea* strains/isolates (Figures 6D and 8C). It is notable that the virulence of isolate Z33 was almost fully recovered on wounded leaves, as the average lesion diameter on wounded leaves (17.4 mm) was dramatically increased in comparison with that on intact leaves (3.8 mm) (Figure 8C and Table S6). More interestingly, the virulence of isolate Z33 on wounded leaves was also comparable to that of strain HBtom-459 on either intact leaves (16.5 mm) or wounded leaves (19.4 mm) (Figures 6D and 8C, and Table S6).

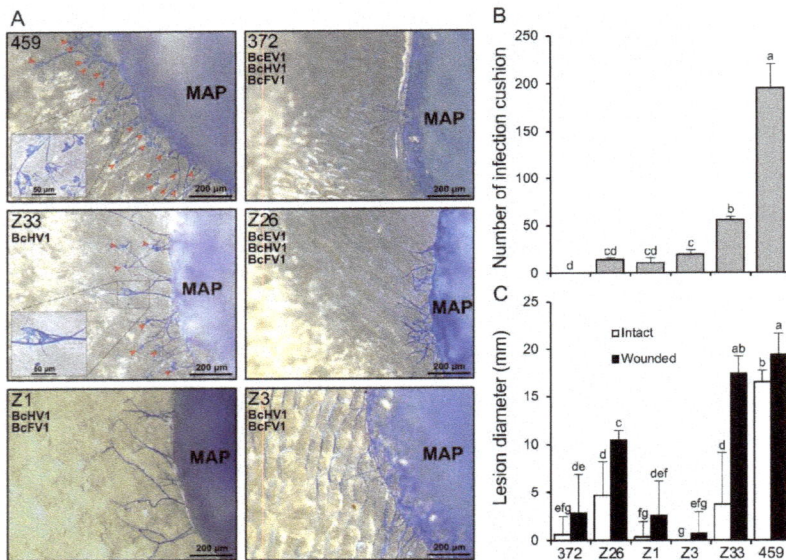

Figure 8. Infection cushion formation (IC) and virulence of *Botrytis cinerea* virus-infected strain HBtom-372 (372), virus-free strain HBtom-459 (459) and virus-infected isolates Z1, Z3, Z26 and Z33 derived from HBtom-459; (**A**) Comparison of IC (red arrowheads) formation on epidermis of onion bulbs after staining with methyl blue. The mycoviruses harbored in each isolate/strain are indicated. The MAP = mycelial agar plug. (**B**) Number of ICs formed around the mycelial agar plugs (6 mm diam). (**C**) Lesion diameters formed on intact and wounded detached rapeseed leaves (*Brassica napus*). (**B,C**) Bars represent arithmetic mean ± S.E.M. In each histogram, bars labeled with the same letters are not significantly different ($p > 0.05$) according to the least-significant-difference test.

3.8. Transcripts of Infection Cushion Formation-Associated Genes

The functions of six infection cushion formation associated genes are summarized in Table 2. Among the six genes, three genes, *Bciqg1* [41], *Bcpdi1* [42] and *Bcmsb2* [43], were previously characterized in *B. cinerea*, whereas the other three genes are homologs of IC-associated genes in the closely related species *S. sclerotiorum* [44–46]. Bciqg1 is a scaffold that mediates interaction of the catalytic subunits with the regulator BcNoxR, which is involved in the MAP kinase- and calcium-dependent signaling pathways [41]. The Bcpdi1 is the essential endoplasmic reticulum (ER) protein as an interaction partner of the NoxA complex, and affects the redox homeostasis in *B. cinerea* [42]. The function of Bcmsb2 is likely to sense hard surfaces for germ tubes and hyphae that triggers the formation of appressoria or ICs via the activation of the BMP1 MAP kinase pathway [43]. The Bcsac1 is the homolog of *S. sclerotiorum* adenylate cyclase (sac1), probably participating in the cAMP-signaling pathway in *B. cinerea*. The homolog of Bcrgb1 in *S. sclerotiorum* (rgb1) was determined to be the regulatory B subunit of the Type 2A phosphoprotein phosphatase (PP2A) involved in several cellular signal-transduction pathways [46]. Sscaf1, the homolog of *B. cinerea* Bccaf1, is a secretory protein and possesses a putative Ca^{2+}-binding EF-hand motif [44].

Table 2. Summary of function analysis of six infection cushion formation associated genes in *Botrytis cinerea* or *Sclerotinia sclerotiorum*.

Gene	Description	D [1]	V [2]	IC [3]	CATs [4]	References
Bciqg1	IQGAP homolog	+	+	+	+	[41]
Bcmsb2	Sensor protein	−	+	+	unknow	[43]
Bcpdi1	ER protein	+	+	+	+	[42]
Sssac1	Adenylate cyclase	+	+	+	unknow	[45]
Ssrgb1	B regulatory 55-kDa R2 subunit	+	+	+	unknow	[46]
Sscaf1	Putative Ca^{2+}-binding protein with an EF-hand motif	+	+	+	unknow	[44]

[1] D: differentiation (includes all tested growth characteristics). [2] V: virulence altered in respective deletion mutant (+), not altered (−). [3] IC: infection cushions. [4] CATs: conidial anastomosis tubes.

The transcriptions of six IC formation-associated genes were investigated using RT-PCR and quantitative real-time PCR (qRT-PCR) (Figure 9). Compared with strain HBtom-459, only one gene, *Bccaf1*, showed a decreased transcription in isolate Z33 on onion bulb scale, while the transcriptions of two genes, *Bcmsb2* and *Bcsac1*, were increased. Three genes including *Bciqg1*, *Bcpdi1* and *Bcrgb1* transcribed in isolate Z33 at almost same levels as in strain HBtom-459 on onion bulb scale. On PDA plate, the transcriptions of genes *Bccaf1* and *Bcrgb1* were down-regulated, whereas transcriptions of genes *Bcmsb2* and *Bciqg1* were up-regulated in isolate Z33 compared with strain HBtom-459.

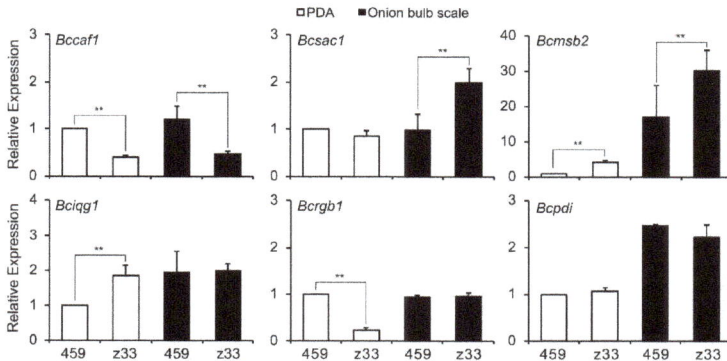

Figure 9. Comparison of relative transcript levels of six infection cushion formation-associated genes including *Bccaf1*, *Bcsac1*, *Bcmsb2*, *Bciqg1*, *Bcrgb1*, and *Bcpdi1* in strain HBtom-459 (459) and isolate Z33 of *B. cinerea* on PDA or onion bulb scale. The results in each histogram are expressed as arithmetic means with the standard errors of the means. "**" indicates a significant difference ($p < 0.01$) between strain HBtom-459 and isolate Z33 in relative transcript level according to the Student *t* test.

3.9. Incidence and Distribution of BcHV1 and BcFV1

Both BcHV1 and BcFV1 were widely distributed in populations of *B. cinerea*, as about 29.6% and 14.3% of *B. cinerea* strains carried BcHV1 and BcFV1 as detected by RT-PCR, respectively (Table S2). It is notable that 13 BcFV1-infected *B. cinerea* strains were co-infected with BcHV1, while only one strain of *B. cinerea* was infected by BcFV1 without BcHV1. The geographic distributions of the two viruses were vast, of which BcHV1 was detected in all five provinces, while BcFV1 was detected in three of five provinces.

4. Discussion

In the present study, we described the molecular and biological properties of two mycoviruses, BcHV1 and BcFV1, co-infecting the hypovirulent strain HBtom-372 of *B. cinerea* with BcEV1. Notwithstanding numerous mycoviruses have been reported in *Botrytis* spp., no hypovirus or fusarivirus has been described in species of *Botrytis* [8].

Sequence and phylogenetic analyses of the putative polyprotein encoded by BcHV1 strongly suggested that BcHV1 belonged to the viral family *Hypoviridae*. Phylogenetic tree based on the RdRp-Hel region clearly supported that the viral family *Hypoviridae* could be divided into two proposed genera Alphahypovirus and Betahypovirus, wherein BcHV1 was clustered with SsHV1, CHV3, CHV4, PlHV1 and VcHV1 forming an independent Betahypovirus clade (bootstrap support 100%). Excluding BcHV1, pairwise sequence analysis of the rest hypoviruses in the Betahypovirus clade showed that they shared 49.9–64.4% nt and 47.6–68.6% aa sequence identity with each other. The pairwise identities of BcHV1 with the other five betahypoviruses range from 49.5% to 65.8% and from 46.8 to 67.4% at levels of the nt sequence and aa sequence, respectively. Considering the virus host, phylogenetic analysis, and sequence identity with other hypoviruses, BcHV1 might be a novel hypovirus isolated from *B. cinerea*. Although BcFV1 clustered with other fusariviruses through phylogenetic analysis, BcFV1 only showed relatively low sequence similarity with those fusariviruses (Table S5). Thus, BcFV1 should be considered as a novel viral member in the proposed viral family Fusariviridae isolated from *B. cinerea*. BLAST analysis also indicated that BcFV1 (nt region 1807–2822) showed 70.96% nt and 83.08% aa identity with those of a virus-like contig (grapevine associated mycovirus-2, GaMV2, GU108600) associated with grapevine plants [47]. Therefore, GaMV2 might be a strain or relative of BcFV1. As *B. cinerea* is a common fungal pathogen of grape, GaMV2 may also be carried by *B. cinerea* isolates infecting grape.

Cryphonectria hypovirus 1 (CHV1), the type species of the genus *Hypovirus*, has been successfully used for the control of chestnut blight caused by *C. parasitica* [40] in Europe. Although the complex VCGs limiting the use of hypovirus for the control of *C. parasitica* [5–7], successful engineering super mycovirus donor strains makes the control of *C. parasitica* with hypovirus in Northern America more hopefully in future [48]. Not only used for biocontrol, the CHV1-*C. parasitica* system has also be used as tools for deciphering molecular mechanisms involved in interaction between viruses and hosts [21]. In addition to *B. cinerea*, hypoviruses have been reported in eight other plant pathogenic fungal species, including *C. parasitica* [49], *S. sclerotiorum* [40], *Fusarium graminearum* [24], *F. poae* [50], *F. langsethiae* [51], *Valsa ceratosperma* [22], *Macrophomina phaseolina* [52] and *Phomopsis longicolla* [53]. The effects of hypovirus infection varied in different virus–host systems. The infections of FgHV1, CHV4/SR2 and VcHV1/MVC86 have no significant effect on their hosts [22,54,55], whereas CHV1/EP713, CHV2/NB58, CHV3/GH2 and FgHV2 infections caused or were associated with the hypovirulence of their hosts [24,39,56–59]. In *S. sclerotiorum*, it was shown that co-infection of SsHV1/SZ150 and its satellite-like small dsRNA, not SsHV1/SZ150 alone, caused the hypovirulence [40]. As only 5'- and 3'- terminal sequences of dsRNA-C were homologous to those of BcHV1 and the hypothetical protein of dsRNA-C possessed no RdRp domain, dsRNA-C was believed to be unable to replicate independently and should be the satellite RNA of BcHV1. However, the infection of BcHV1 alone without the satellite RNA dsRNA-C (in strain Z33) significantly reduced virulence without affecting mycelial growth (Figure 7B, Table S6). Moreover, isolates Z1, Z3 and Z33 of *B. cinerea* lacking the satellite RNAs (dsRNA-C and dsRNA-F) also showed the hypovirulence trait. Therefore, the presence of BcHV1, not the satellite RNAs, is associated with the hypovirulence of strain HBtom-372.

Among seven reported fusariviruses, SsFV1, RnFV1, AbFV1, PtFV1, PaFV1, FpFV1 and FgV1, only FgV1 showed a significant association with pronounced morphological changes, like impaired mycelial growth, increased pigmentation, reduced mycotoxin production and attenuated virulence of its host [60]. The other fusariviruses, like PtFV1, FpFV1 and PaFV1, lack the information of their biological properties [29,50]. SsFV1 and RnFV1 were able to be horizontally transmitted via anastomosis, but their infection did not cause any symptoms [26,28]. In addition, AbFV1 was co-infected with Alternaria brassicicola endornavirus 1 in *Alternaria brassicicola* strain 817-14. As AbFV1-cured strain showed comparable biological properties with AbFV1-infected strain, AbFV1 was believed to have no conspicuous impact on its host [30]. In the present study, we obtained a series derivative isolates of *B. cinerea*, unfortunately, no derivative isolate was infected by BcFV1 alone. Compared with isolate Z33 (only carrying BcHV1), the derivative isolates Z1 and Z3 (carrying both BcHV1 and BcFV1) showed more severe debilitation. Isolate Z33 only showed impaired virulence, whereas isolates Z1 and Z3 also showed slower mycelial growth and abnormal colony morphology besides impaired virulence (Figure 6). Moreover, the derivative isolate Z26 (carrying BcEV1, BcHV1 and BcFV1), showed almost the same phenotype of the donor strain HBtom-372 (Figure 6). Thus, the co-infection of BcHV1 with BcFV1, or with both BcEV1 and BcFV1, may exacerbate the debilitation of virus-infected strain at varying degrees. Since no isolate solely infected by BcFV1 or BcEV1 was obtained, the effect of BcFV1 on *B. cinerea* biology remains to be investigated in the future. In addition, most derivative isolates were co-infected with BcHV1 and BcFV1 or infected by BcHV1 alone, indicating the faster transmission of BcHV1 and BcFV1 in comparison with BcEV1 during hyphal anastomosis. In addition, only BcHV1 solely infected isolate was detected, also suggesting BcHV1 might be transmitted faster than BcFV1 through hyphal anastomosis.

Although hypovirulence-associated mycoviruses were recorded in many phytopathogenic fungi, the underling mechanism responsible for virus-mediated hypovirulence remains unclear. The impact of CHV1/EP713 on *C. parasitica* seems to be global, besides hypovirulence, multiple phenotypical changes including colony pigmentation, ascospore production and female sterility were also observed [37], accompanying with the transcriptional alternation of hundreds of genes involved in over a dozen pathways [61,62]. In addition, comparative proteomics analysis also indicated dozens of vesicle proteins were expressed differently between virus-infected and virus-free *C. parasitica* strains [63]. Therefore, the influence of CHV1/EP713 on *C. parasitica* is not only limited to its pathogenesis.

Infection structures, mainly including appressoria and infection cushions (ICs), play crucial role during the infection process for many plant pathogenic fungi with the function of direct penetration of host plants [41]. ICs, also described as "claw-like" structures [64], are observed when plants are inoculated with mycelia of *B. cinerea* or *S. sclerotiorum* [65,66]. Inhibition or deficiency of IC formation lead to severely debilitation of pathogenicity for both two fungi, and wound inoculation could restore the pathogenicity caused by IC deficiency in varying degrees [43–45,67,68]. The inhibition of IC formation was probably responsible for the BcMV1-associated hypovirulence in *B. cinerea* [69]. Nevertheless, the vegetative growth of BcMV1-infected strain CanBc-1 was severely impaired. Moreover, wound inoculation was not able to restore the pathogenicity of strain CanBc-1 [69]. Thus, deficiency in formation of ICs is probably not the only factor responsible for BcMV1-associated hypovirulence in *B. cinerea*. Interestingly, similar to fungal strains infected by BcMV1, isolate Z33 also showed significantly decreased formation of ICs. Nonetheless, the vegetative growth of isolate Z33 was normal and comparable to strain HBtom-459; more importantly, wound inoculation could fully recover the virulence of isolate Z33. Thus, inhibition of IC formation may be the major factor responsible for the hypovirulence of isolate Z33. In addition, co-infection by more viruses was accompanied with more phenotypical changes, such as more severe inhibition of IC formation, reduced mycelial growth, and decreased production of conidia and sclerotia. Moreover, wound inoculation was unable to fully recover the virulence of other derivative isolates. These results indicate more cellular pathways may be influenced by the co-infection of BcFV1 and BcEV1, or by the combined effects of the two or three viruses.

To investigate the underling mechanism responsible for the impaired formation of IC, six IC formation-associated genes were selected and tested for their relative transcriptional levels in strain HBtom-459 and isolate Z33 on both PDA and onion bulb scale. All six genes were required for the formation of IC, as their knock-out mutants were unable to form ICs and showed significant reduced virulence [41–46]. Consequently, the decreased transcript levels of *Bccaf1* (Figure 9) in isolate Z33 suggest it is related to the reduced formation of IC in isolate Z33. This is consistent with the *Sscaf1* knockout mutant in *S. sclerotiorum*, as wound inoculation could also restore the virulence of *Sscaf1* knockout mutant [44]. Besides participating in the IC formation process of *S. sclerotiorum*, *Sscaf1* may be also involved in sclerotium formation, as the *Sscaf1* knockout mutant produced fewer but larger sclerotia than the wild-type strain [44]. However, isolate Z33 produced more but smaller sclerotia than strain HBtom-459, which is opposite to the case in *S. sclerotiorum*. Some factors may be responsible for the differences, like the transcription of *Bccaf1* was decreased, not null in Z33; other pathways may be affected by viral infection; or the two fungi may have evolved different pathways for sclerotium formation. Therefore, the infection of BcHV1 may affect the expression of certain IC formation-associated genes including *Bccaf1* directly or indirectly, causing the decreased formation of IC, resulting in the hypovirulence of *B. cinerea*. Transcriptome analysis may be helpful to uncover the overall view of gene expressions altered by the infection of BcHV1 in future.

Besides Hubei Province where strain HBtom-372 was originally isolated, BcHV1 was detected in strains from four other provinces, namely Hunan, Jiangxi, Jilin, and Shangdong, occupying a total area of about 907,800 square kilometers for all five provinces (Table S2 and Figure S10). The absence of BcFV1-infected *B. cinerea* strain in Hunan and Jiangxi is most likely due to the small population size (one or two strains) from these two provinces considering their proximity to Hubei province. It is interesting that, in most cases, BcFV1 was detected to be co-infected with BcHV1, as only 1 out of 14 strains showed the BcFV1 infection alone. This is also coincidence with the horizontal transmission of BcHV1 and BcFV1, as most derivative isolates (32 out of 33 isolates) were co-infected by both BcHV1 and BcFV1. Whether BcFV1 has some inner connections with BcHV1 is worth to be tested in further studies. These results indicate that both BcHV1 and BcFV1 have a wide geographic distribution and can spread in natural *B. cinerea* populations.

Supplementary Materials: The following are available online at http://www.mdpi.com/1999-4915/10/5/254/s1, Table S1: Oligonucleotide primers/adaptor used for cDNA cloning and revere transcription (RT)-PCR detection of different mycoviruses and dsRNAs in this study. Table S2: The presence of Botrytis cinerea hypovirus 1 (BcHV1) and Botrytis cinerea fusarivirus 1 (BcFV1) in the population of *Botrytis cinerea* isolated from Hubei, Shangdong, Hunan, Jiangxi and Jilin Province, China. Table S3: Oligonucleotide primers/adaptor used for quantitative reverse transcription (qRT)-PCR detection of infection formation related gene in *Botrytis cinerea* strains. Table S4: Sequence identities of Botrytis cinerea hypovirus 1 (BcHV1) to other viruses through multiple alignments of the polyprotein sequences and the amino acid (aa) residue sequences of different domains. Table S5: Sequence identities of Botrytis cinerea fusarivirus 1 (BcFV1) to other viruses through multiple alignments of the polyprotein sequences and the amino acid (aa) residue sequences of different domains. Table S6: Data used to generate histogram. Figure S1: Schematic representation of the strategy used for full cDNA sequence cloning of BcHV1 and BcFV1. PCR primers and the 3′-adaptor used for the cDNA cloning are shown on the diagram. The sequences of the primers and the 3′-adaptor are listed in Table S1. Figure S2: Schematic representation of the strategy used for full cDNA sequence cloning of dsRNA-C, dsRNA-D, dsRNA-E and dsRNA-F. PCR primers and the 3′-adaptor used for the cDNA cloning are shown on the diagram. The sequences of the primers and the 3′-adaptor are listed in Table S1. Figure S3: Pathogenicity assay (**A**); and lesion diameter (**B**) on detached tomato leaves of strains HBtom-372 and HBtom-459 (20 °C, 48 h, lower). "*****" indicates a significant difference ($p < 0.01$) between strains HBtom-372 and HBtom-459 in pathogenicity. (**C**) Agarose gel electrophoresis of dsRNAs extracted from the mycelium of *Botrytis cinerea* strains HBtom-372 and HBtom-459. Figure S4: Multiple alignment of amino acid sequences of conserved domains including papain-like protease domain (Prot) and UDP glucose/sterol glucosyltransferase domain (UGT) of BcHV1, and viral RNA Helicase domain (Hel) of both BcHV1 and BcFV1. Motifs in the corresponding domains are indicated by the roman numerals I–VIII. "*" indicates identical amino acid residues; and "." indicates low chemically similar amino acid residues. Figure S5: Transmembrane domains prediction for ORF1 and ORF2 encoded protein of BcFV1, and the ORF1 encoded protein of RnFV1, SsFV1 and PrRV1. Figure S6: A diagram showing the structure of dsRNA-C, dsRNA-D, dsRNA-E and dsRNA-F and their relationship with BcHV1 and BcFV1. The dashed line rectangle represents the corresponding deletion range on the genome BcHV1, BcFV1 and dsRNA-C. The arrows and sequences in the dashed line frame indicate the extra-nucleotides on dsRNA-D and dsRNA-E in comparison with BcHV1. The gray shades linked both 5′- and 3′-termini of BcHV1 and dsRNA-C indicates the homologous region between BcHV1 and dsRNA-C. Figure S7: Enlarged colony morphology from Figure 6A, red arrowheads indicate the inhibition of mycelial growth of HBtom-459 in duel culture during viral horizontal transmission (20 °C, 7 days). Figure S8: RAPD profiling for different strains of *Botrytis cinerea* using random primer OPC-04. Figure S9: IC formation (enlarged from Figure 8) of different strains/_solates on epidermis of onion bulbs after staining with methyl blue. Figure S10: Geographic distribution of Botrytis cinerea hypovirus 1 (BcHV1) and Botrytis cinerea fusarivirus 1 (BcFV1) in provinces Hubei, Hunan, Jiangxi, Shandong and Jilin, China. The five provinces are highlighted in grey on the map. The numbers of *B. cinerea* strains are labeled as "number of total strains/number of BcHV1 infected strains/number of BcFV1 infected strains" in the parentheses for each province.

Author Contributions: F.H., M.W. and G.L. conceived and designed the experiments. F.H. and T.D. performed the experiments; F.H., M.W., J.Z., L.Y., W.C. and G.L. analyzed the data. F.H., M.W. and W.C. wrote the paper.

Acknowledgments: This research was supported by the R & D Special Fund for Public Welfare Industry (Agriculture) of China (grant number 201303025) and the Natural Science Foundation of China (grant number 31401690). We greatly appreciate the assistance of Jiejing Tang of Huazhong Agricultural University for providing some primers used for the qRT-PCR analysis.

Conflicts of Interest: The authors declare that they have no conflict of interest.

References

1. Elad, Y.; Pertot, I.; Marina, A.; Prado, A.M.; Stewart, A. Plant Hosts of *Botrytis* spp. In *Botrytis—The Fungus, the Pathogen and Its Management in Agricultural Systems*; Fillinger, S., Elad, Y., Eds.; Springer: Cham, Switzerland, 2016; pp. 413–486.

2. Phillips, M.W.A.; McDougall, J. Crop protection market trends and opportunities for new active ingredients. In *Abstracts of Papers of the American Chemical Society*; American Chemical Society: Washington, DC, USA, 2012; p. 244.

3. Heiniger, U.; Rigling, D. Biological control of chestnut blight in Europe. *Annu. Rev. Phytopathol.* **1994**, *32*, 581–599. [CrossRef]

4. Ghabrial, S.A.; Castón, J.R.; Jiang, D.; Nibert, M.L.; Suzuki, N. 50-plus years of fungal viruses. *Virology* **2015**, *479*, 356–368. [CrossRef] [PubMed]

5. Boland, G.J. Fungal viruses, hypovirulence, and biological control of *Sclerotinia* species. *Can. J. Plant Pathol.* **2004**, *26*, 6–18. [CrossRef]

6. Brasier, C.M. Inter-mycelial Recognition Systems in Ceratocystis ulmi: Their Physiological Properties and Ecological Importance. In *The Ecology and Physiology of the Fungal Mycelium*; Jennings, D.H., Rayner, A.D.M., Eds.; Cambridge University Press: Cambridge, UK, 1984; pp. 451–497.

7. Anagnostakis, S.L.; Hau, B.; Kranz, J. Diversity of vegetative incompatibility groups of *Cryphonectria parasitica* in Connecticut and Europe. *Plant Dis.* **1986**, *70*, 36–538. [CrossRef]

8. Wu, M.D.; Zhang, J.; Yang, L.; Li, G.Q. RNA mycoviruses and their role in *Botrytis* Biology. In *Botrytis—The Fungus, the Pathogen and Its Management in Agricultural Systems*; Fillinger, S., Elad, Y., Eds.; Springer: Cham, Switzerland, 2016; pp. 71–90.

9. Pearson, M.N.; Bailey, A.M. Viruses of *Botrytis*. In *Advances In Virus Research*; Ghabrial, S.A., Ed.; Academic Press: Cambridge, MA, USA, 2013; pp. 249–272.

10. Wu, M.D.; Jin, F.Y.; Zhang, J.; Yang, L.; Jiang, D.H.; Li, G.Q. Characterization of a novel bipartite double-stranded RNA mycovirus conferring hypovirulence in the phytopathogenic fungus *Botrytis porri*. *J. Virol.* **2012**, *86*, 6605–6619. [CrossRef] [PubMed]

11. Yu, L.; Sang, W.; Wu, M.D.; Zhang, J.; Yang, L.; Zhou, Y.J.; Chen, W.D.; Li, G.Q. Novel hypovirulence-associated RNA mycovirus in the plant pathogenic fungus *Botrytis cinerea*. *Appl. Environ. Microbiol.* **2015**, *81*, 2299–2310. [CrossRef] [PubMed]

12. Donaire, L.; Rozas, J.; María, A. Molecular characterization of Botrytis ourmia-like virus, a mycovirus close to the plant pathogenic genus *Ourmiavirus*. *Virology* **2016**, *489*, 158–164. [CrossRef] [PubMed]

13. Donaire, L.; Pagan, I.; Ayllon, M.A. Characterization of *Botrytis cinerea* negative-stranded RNA virus 1, a new mycovirus related to plant viruses, and a reconstruction of host pattern evolution in negative-sense ssRNA viruses. *Virology* **2016**, *499*, 212–218. [CrossRef] [PubMed]

14. Wu, M.D.; Zhang, L.; Li, G.Q.; Jiang, D.H.; Hou, M.S.; Huang, H.C. Hypovirulence and double-stranded RNA in *Botrytis cinerea*. *Phytopathology* **2007**, *97*, 1590–1599. [CrossRef] [PubMed]

15. Wu, M.D.; Zhang, L.; Li, G.Q. Genome characterization of a debilitation-associated mitovirus infecting the phytopathogenic fungus *Botrytis cinerea*. *Virology* **2010**, *406*, 117–126. [CrossRef] [PubMed]

16. Potgieter, C.A.; Castillo, A.; Castro, M.; Cottet, L.; Morales, A. A wild-type *Botrytis cinerea* strain co-infected by double-stranded RNA mycoviruses presents hypovirulence-associated traits. *Virol. J.* **2013**, *10*, 220–228. [CrossRef] [PubMed]

17. Howitt, R.L.; Beever, R.E.; Pearson, M.N.; Forster, R.L. Genome characterization of Botrytis virus F, a flexuous rod-shaped mycovirus resembling plant 'potex-like' viruses. *J. Gen. Virol.* **2001**, *82*, 67–78. [CrossRef] [PubMed]

18. Howitt, R.L.; Beever, R.E.; Pearson, M.N.; Forster, R.L. Genome characterization of a flexuous rod-shaped mycovirus, Botrytis virus X, reveals high amino acid identity to genes from plant 'potex-like' viruses. *Arch. Virol.* **2006**, *151*, 563–579. [CrossRef] [PubMed]

19. Donaire, L.; Ayllón, M.A. Deep sequencing of mycovirus-derived small RNAs from *Botrytis* species. *Mol. Plant Pathol.* **2016**, *18*, 1127–1137. [CrossRef] [PubMed]

20. Suzuki, N.; Ghabrial, S.A.; Kim, K.H.; Pearson, M.; Marzano, S.Y.L.; Yaegashi, H.; Xie, J.T.; Guo, L.H.; Kondo, H.; Koloniuk, L.; et al. ICTV Virus Taxonomy Profile: Hypoviridae. *J. Gen. Virol.* **2018**. [CrossRef] [PubMed]

21. Dawe, A.L.; Nuss, D.L. Hypovirus molecular biology: From Koch's postulates to host self-recognition genes that restrict virus transmission. *Adv. Virus Res.* **2013**, *86*, 109–147. [PubMed]

22. Yaegashi, H.; Kanematsu, S.; Ito, T. Molecular characterization of a new hypovirus infecting a phytopathogenic fungus, *Valsa ceratosperma*. *Virus Res.* **2012**, *165*, 143–150. [CrossRef] [PubMed]

23. Khalifa, M.E.; Pearson, M.N. Characterisation of a novel hypovirus from *Sclerotinia sclerotiorum* potentially representing a new genus within the *Hypoviridae*. *Virology* **2014**, *464*, 441–449. [CrossRef] [PubMed]

24. Li, P.F.; Zhang, H.L.; Chen, X.G.; Qiu, D.W.; Guo, L.H. Molecular characterization of a novel hypovirus from the plant pathogenic fungus *Fusarium graminearum*. *Virology* **2015**, *481*, 151–160. [CrossRef] [PubMed]

25. Hu, Z.; Wu, S.; Cheng, J.; Fu, Y.; Jiang, D.; Xie, J. Molecular characterization of two positive-strand RNA viruses co-infecting a hypovirulent strain of *Sclerotinia sclerotiorum*. *Virology* **2014**, *464*, 450–459. [CrossRef] [PubMed]

26. Zhang, R.; Liu, S.; Chiba, S.; Kondo, H.; Kanematsu, S.; Suzuki, N. A novel single-stranded RNA virus isolated from a phytopathogenic filamentous fungus, *Rosellinia necatrix*, with similarity to hypo-like viruses. *Front. Microbiol.* **2014**, *5*, 1–12. [CrossRef] [PubMed]

27. Kwon, S.J.; Lim, W.S.; Park, S.H.; Park, M.R.; Kim, K.H. Molecular characterization of a dsRNA mycovirus, *Fusarium graminearum* virus-DK21, which is phylogenetically related to hypoviruses but has a genome organization and gene expression strategy resembling those of plant potex-like viruses. *Mol. Cells* **2007**, *23*, 304–315. [CrossRef] [PubMed]

28. Liu, R.; Cheng, J.; Fu, Y.; Jiang, D.; Xie, J. Molecular Characterization of a Novel Positive-Sense, Single-Stranded RNA Mycovirus Infecting the Plant Pathogenic Fungus *Sclerotinia sclerotiorum*. *Viruses* **2015**, *7*, 2470–2484. [CrossRef] [PubMed]

29. Nerva, L.; Ciuffo, M.; Vallino, M.; Margaria, P.; Varese, G.C.; Gnavi, G.; Turina, M. Multiple approaches for the detection and characterization of viral and plasmid symbionts from a collection of marine fungi. *Virus Res.* **2015**, *219*, 22–38. [CrossRef] [PubMed]

30. Zhong, J.; Shang, H.H.; Zhu, C.X.; Zhu, J.Z.; Zhu, H.J.; Hu, Y.; Gao, B.D. Characterization of a novel single-stranded RNA virus, closely related to fusariviruses, infecting the plant pathogenic fungus *Alternaria brassicicola*. *Virus Res.* **2015**, *217*, 1–7. [CrossRef] [PubMed]

31. Hao, F.M.; Zhou, Z.L.; Wu, M.D.; Li, G.Q. Molecular characterization of a novel endornavirus from the phytopathogenic fungus *Botrytis cinerea*. *Arch. Virol.* **2016**, *162*, 313–316. [CrossRef] [PubMed]

32. Wu, M.D.; Deng, Y.; Zhou, Z.L.; He, G.Y.; Chen, W.D.; Li, G.Q. Characterization of three mycoviruses co-infecting the plant pathogenic fungus *Sclerotinia nivalis*. *Virus Res.* **2016**, *223*, 28–38. [CrossRef] [PubMed]

33. Tamura, K.; Peterson, D.; Peterson, N.; Stecher, G.; Nei, M.; Kumar, S. MEGA5: Molecular evolutionary genetics analysis using maximum likelihood, evolutionary distance, and maximum parsimony methods. *Mol. Biol. Evol.* **2011**, *28*, 2731–2739. [CrossRef] [PubMed]

34. Krogh, A.; Larsson, B.; von Heijne, G.; Sonnhammer, E.L. Predicting transmembrane protein topology with a hidden Markov model: Application to complete genomes. *J. Mol. Biol.* **2001**, *305*, 567–580. [CrossRef] [PubMed]

35. Jiang, D.H.; Ghabrial, S.A. Molecular characterization of *Penicillium chrysogenum* virus: Reconsideration of the taxonomy of the genus *Chrysovirus*. *J. Gen. Virol.* **2004**, *85*, 2111–2121. [CrossRef] [PubMed]

36. Fan, X.; Zhang, J.; Yang, L.; Wu, M.D.; Chen, W.D.; Li, G.Q. Development of PCR-based Assays for Detecting and Differentiating Three Species of Botrytis Infecting Broad Bean. *Plant Dis.* **2015**, *99*, 691–698. [CrossRef]

37. Nuss, D.L. Hypovirulence: Mycoviruses at the fungal-plant interface. *Nat. Rev. Microbiol.* **2005**, *3*, 632–642. [CrossRef] [PubMed]

38. Zeng, L.M.; Zhang, J.; Han, Y.C.; Yang, L.; Wu, M.D.; Jiang, D.H.; Chen, W.D.; Li, G.Q. Degradation of oxalic acid by the mycoparasite *Coniothyrium minitans* plays an important role in interacting with *Sclerotinia sclerotiorum*. *Environ. Microbiol.* **2014**, *16*, 2591–2610. [CrossRef] [PubMed]

39. Hillman, B.I.; Halpern, B.T.; Brown, M.P. A viral dsRNA element of the chestnut blight fungus with a distinct genetic organization. *Virology* **1994**, *201*, 241–250. [CrossRef] [PubMed]

40. Xie, J.; Xiao, X.; Fu, Y.; Liu, H.; Cheng, J.; Ghabrial, S.A.; Li, G.; Jiang, D. A novel mycovirus closely related to hypoviruses that infects the plant pathogenic fungus *Sclerotinia sclerotiorum*. *Virology* **2011**, *418*, 49–56. [CrossRef] [PubMed]

41. Marschall, R.; Tudzynski, P. BcIqg1, a fungal IQGAP homolog, interacts with NADPH oxidase, MAP kinase and calcium signaling proteins and regulates virulence and development in *Botrytis cinerea*. *Mol. Microbiol.* **2016**, *101*, 281–298. [CrossRef] [PubMed]

42. Marschall, R.; Tudzynski, P. The Protein Disulfide Isomerase of *Botrytis cinerea*: An ER Protein Involved in Protein Folding and Redox Homeostasis Influences NADPH Oxidase Signaling Processes. *Front. Microbiol.* **2017**, *8*, 1–15. [CrossRef] [PubMed]

43. Leroch, M.; Mueller, N.; Hinsenkamp, I.; Hand, M. The signalling mucin Msb2 regulates surface sensing and host penetration via BMP1 MAP kinase signalling in *Botrytis cinerea*. *Mol. Plant Pathol.* **2015**, *16*, 787–798. [CrossRef] [PubMed]

44. Xiao, X.Q.; Xie, J.T.; Cheng, J.C.; Li, G.Q.; Yi, X.H.; Jiang, D.H.; Fu, Y.P. Novel Secretory Protein Ss-Caf1 of the Plant-Pathogenic Fungus *Sclerotinia sclerotiorum* Is Required for Host Penetration and Normal Sclerotial Development. *Mol. Plant Microbe Interact.* **2014**, *27*, 40–55. [CrossRef] [PubMed]

45. Jurick, W.M.; Rollins, J.A. Deletion of the adenylatecyclase (*sac1*) gene affects multiple developmental pathways and pathogenicity in *Sclerotinia sclerotiorum*. *Fungal Genet. Biol.* **2007**, *44*, 521–530. [CrossRef] [PubMed]

46. Erental, A.; Harel, A.; Yarden, O. Type 2A phosphoprotein phosphatase is required for asexual development and pathogenesis of *Sclerotinia sclerotiorum*. *Mol. Plant Microbe Interact.* **2007**, *20*, 944–954. [CrossRef] [PubMed]

47. Al Rwahnih, M.; Daubert, S.; Urbez-Torres, J.R.; Cordero, F.; Rowhani, A. Deep sequencing evidence from single grapevine plants reveals a virome dominated by mycoviruses. *Arch. Virol.* **2011**, *156*, 397–403. [CrossRef] [PubMed]

48. Zhang, D.X.; Nuss, D.L. Engineering super mycovirus donor strains of chestnut blight fungus by systematic disruption of multilocus vic genes. *Proc. Natl. Acad. Sci. USA* **2016**, *113*, 2062–2067. [CrossRef] [PubMed]

49. Shapira, R.; Choi, G.H.; Nuss, D.L. Virus-like genetic organization and expression strategy for a double-stranded RNA genetic element associated with biological control of chestnut blight. *EMBO J.* **1991**, *10*, 731–739. [PubMed]

50. Osaki, H.; Sasaki, A.; Nomiyama, K.; Tomioka, K. Multiple virus infection in a single strain of *Fusarium poae* shown by deep sequencing. *Virus Genes* **2016**, *52*, 835–847. [CrossRef] [PubMed]

51. Li, P.F.; Chen, X.G.; He, H.; Qiu, D.; Guo, L.H. Complete genome sequence of a novel hypovirus isolated from the phytopathogenic fungus *Fusarium langsethiae*. *Genome Announc.* **2017**, *5*, 1–2. [CrossRef] [PubMed]

52. Marzano, S.Y.; Nelson, B.D.; Oyetunde, Q.A.; Bradley, C.A.; Hughes, T.J.; Hartman, G.L. Identification of diverse mycoviruses through metatranscriptomics characterization of the viromes of five major fungal plant pathogens. *J. Virol.* **2016**, *90*, 6846–6863. [CrossRef] [PubMed]

53. Koloniuk, I.; El-Habbak, M.H.; Petrzik, K.; Ghabrial, S.A. Complete genome sequence of a novel hypovirus infecting *Phomopsis longicolla*. *Arch. Virol.* **2014**, *159*, 1861–1863. [CrossRef] [PubMed]

54. Wang, S.; Kondo, H.; Liu, L.; Guo, L.; Qiu, D. A novel virus in the family *Hypoviridae* from the plant pathogenic fungus *Fusarium graminearum*. *Virus Res.* **2013**, *174*, 69–77. [CrossRef] [PubMed]

55. Linder-Basso, D.; Dynek, J.N.; Hillman, B.I. Genome analysis of *Cryphonectria hypovirus* 4, the most common hypovirus species in North America. *Virology* **2005**, *337*, 192–203. [CrossRef] [PubMed]

56. Hillman, B.I.; Shapira, R.; Nuss, D.L. Hypovirulence-associated suppression of host functions in *Cryphonectria parasitica* can be partially relieved by high light intensity. *Phytopathology* **1990**, *80*, 950–956. [CrossRef]

57. Hillman, B.I.; Tian, Y.; Bedker, P.J.; Brown, M.P. A North American hypovirulent isolate of the chestnut blight fungus with European isolated-related dsRNA. *J. Gen. Virol.* **1992**, *73*, 681–686. [CrossRef] [PubMed]

58. Fulbright, D.W. Effect of eliminating dsRNA in hypovirulent *Endothia parasitica*. *Phytopathology* **1984**, *74*, 722–724. [CrossRef]

59. Smart, C.D.; Yuan, W.; Foglia, R.; Nuss, D.L.; Fulbright, D.W.; Hillman, B.I. Cryphonectria hypovirus 3, a virus species in the family Hypoviridae with a single open reading frame. *Virology* **1999**, *265*, 66–73. [CrossRef] [PubMed]

60. Chu, Y.M.; Jeon, J.J.; Yea, S.J.; Kim, Y.H.; Yun, S.H. Double-stranded RNA mycovirus from *Fusarium graminearum*. *Appl. Environ. Microbiol.* **2002**, *68*, 2529–2534. [CrossRef] [PubMed]

61. Allen, T.D.; Dawe, A.L.; Nuss, D.L. Use of cDNA Microarrays To Monitor Transcriptional Responses of the Chestnut Blight Fungus *Cryphonectria parasitica* to Infection by Virulence-Attenuating Hypoviruses. *Eukaryot. Cell* **2003**, *2*, 1253–1265. [CrossRef] [PubMed]

62. Allen, T.D.; Nuss, D.L. Specific and common alterations in host gene transcript accumulation following infection of the chestnut blight fungus by mild and severe hypoviruses. *J. Virol.* **2004**, *78*, 4145–4155. [CrossRef] [PubMed]

63. Wang, J.Z.; Wang, F.Z.; Feng, Y.J.; Mi, K.; Chen, Q.; Shang, J.J.; Chen, B.S. Comparative vesicle proteomics reveals selective regulation of protein expression in chestnut blight fungus by a hypovirus. *J. Proteom.* **2013**, *78*, 221–230. [CrossRef] [PubMed]

64. Kunz, C.; Vandelle, E.; Rolland, S.; Poinssot, B.; Bruel, C.; Cimerman, A.; Zotti, C.; Moreau, E.; Vedel, R.; Pugin, A.; et al. Characterization of a new, nonpathogenic mutant of *Botrytis cinerea* with impaired plant colonization capacity. *New Phytol.* **2006**, *170*, 537–550. [CrossRef] [PubMed]

65. Backhouse, D.; Willets, H.J. Development and structure of infection cushions of *Botrytis cinerea*. *Trans. Br. Mycol. Soc.* **1987**, *89*, 89–95. [CrossRef]

66. Lumsden, R.D.; Dow, R.L. Histopathology of *Sclerotinia sclerotiorum* infection of bean. *Phytopathology* **1973**, *63*, 708–715. [CrossRef]

67. Bashi, T.; Shovman, O.; Fridkin, M.; Volkov, A.; Barshack, I.; Blank, M.; Shoenfeld, Y. Novel therapeutic compound tuftsin-phosphorylcholine attenuates collagen-induced arthritis. *Clin. Exp. Immunol.* **2016**, *184*, 19–28. [CrossRef] [PubMed]

68. Feng, H.Q.; Li, G.H.; Du, S.W.; Yang, S.; Li, X.Q.; de Figueiredo, P.; Qin, Q.M. The septin protein Sep4 facilitates host infection by plant fungal pathogens via mediating initiation of infection structure formation. *Environ. Microbiol.* **2017**, *19*, 1730–1749. [CrossRef] [PubMed]

69. Zhang, L.; Wu, M.D.; Li, G.Q.; Jiang, D.H.; Huang, H.C. Effect of Mitovirus on formation of infecfion cushions and production of some virulence factors by *Botrytis cinerea*. *Physiol. Mol. Plant Pathol.* **2010**, *75*, 71–80. [CrossRef]

viruses

MDPI

Article

Sclerotinia minor Endornavirus 1, a Novel Pathogenicity Debilitation-Associated Mycovirus with a Wide Spectrum of Horizontal Transmissibility

Dan Yang [1], Mingde Wu [1], Jing Zhang [1], Weidong Chen [2], Guoqing Li [1] and Long Yang [1,*]

[1] The State Key Laboratory of Agricultural Microbiology and Hubei Key Laboratory of Plant Pathology, Huazhong Agricultural University, Wuhan 430070, China; danyang@mail.hzau.edu.cn (D.Y.); mingde@mail.hzau.edu.cn (M.W.); zhangjing1007@mail.hzau.edu.cn (J.Z.); guoqingli@mail.hzau.edu.cn (G.L.)
[2] U.S. Department of Agriculture, Agricultural Research Service, Washington State University, Pullman, WA 99164, USA; w-chen@wsu.edu
* Correspondence: yanglong@mail.hzau.edu.cn

Received: 18 September 2018; Accepted: 26 October 2018; Published: 27 October 2018

Abstract: Sclerotinia minor is a phytopathogenic fungus causing sclerotinia blight on many economically important crops. Here, we have characterized the biological and molecular properties of a novel endornavirus, Sclerotinia minor endornavirus 1 (SmEV1), isolated from the hypovirulent strain LC22 of *S. minor*. The genome of SmEV1 is 12,626 bp long with a single, large open reading frame (ORF), coding for a putative protein of 4020 amino acids. The putative protein contains cysteine-rich region (CRR), viral methyltransferase (MTR), putative DEXDc, viral helicase (Hel), and RNA-dependent RNA polymerase (RdRp) domains. The putative protein and the conserved domains are phylogenetically related to endornaviruses. SmEV1 does not contain a site-specific nick characteristic of most previously described endornaviruses. Hypovirulence and associated traits of strain LC22 and SmEV1 were readily cotransmitted horizontally via hyphal contact to isolates of different vegetative compatibility groups of *S. minor*. Additionally, SmEV1 in strain LC22 was found capable of being transmitted vertically through sclerotia. Furthermore, mycelium fragments of hypovirulent strain LC22 have a protective activity against attack by *S. minor*. Taken together, we concluded that SmEV1 is a novel hypovirulence-associated mycovirus with a wide spectrum of transmissibility, and has potential for biological control (virocontrol) of diseases caused by *S. minor*.

Keywords: Sclerotinia minor; endornavirus; hypovirulence; transmissibility; biological control

1. Introduction

Mycoviruses or fungal viruses are viruses that can infect fungi and Oomycetes, and can replicate in these organisms [1]. They exist in all major taxonomic groups of fungi and Oomycetes [2]. As intracellular and molecular parasites, mycoviruses completely depend on their hosts for replication and transmission. Therefore, the mycoviruses have evolved various subtle and well-regulated relationships with their hosts. Previous studies showed that most mycoviruses have little or no detectable effects on the morphology and/or physiology of their hosts, and this kind of infection is usually called cryptic (or latent) infection [1]. However, a few mycoviruses do have evident effects on the morphology and/or physiology of their hosts. The beneficial effects on the hosts include increase of virulence (hypervirulence) in some plant pathogenic fungi or oomycetes [3,4], and improvement of competitive ability of some yeasts by producing killer proteins [5].

The detrimental effects on their hosts include virus diseases on cultivated mushrooms [6] and reduction of virulence (hypovirulence). So far, mycovirus conferring (or associated with)

hypovirulence has been reported in many plant pathogenic fungi, including *Botrytis* spp., *Cryphonectria* (=*Endothia*) *parasitica*, *Fusarium graminearum*, *Ophiostoma novo-ulmi*, *Rhizoctonia solani*, *Rosellinia necatrix*, and *Sclerotinia* spp. [7]. Some mycoviruses can be potentially exploited for controlling plant fungal diseases. In this aspect, a well-known example is the use of Cryphonectria hypovirus 1 (CHV1) to control chestnut blight caused by *C. parasitica* [8,9]. Recently, Yu and colleagues [10] reported that Sclerotinia sclerotiorum hypovirulence-associated DNA virus 1 (SsHADV-1) in strain DT-8 of *S. sclerotiorum* can effectively suppress Sclerotinia stem rot of oilseed rape (*Brassica napus*) caused by virulent strains of *S. sclerotiorum* when the hyphal fragments of DT-8 or the virus particles of SsHADV-1 were applied on the plants. These successful examples have inspired many other researchers to search for mycoviruses with a promising biocontrol potential.

Previous studies showed that efficient transmission of the hypovirulence-associated mycoviruses appears to be very important in use of the mycoviruses to control plant diseases [8,9]. Transmission of the mycoviruses occurs vertically to offspring in asexual or sexual spores, and horizontally between individuals by hyphae-hyphae fusion or hyphal anastomosis [11]. It seems that the DNA viruses differ from the RNA viruses in horizontal transmission. SsHADV-1 can be transmitted in two ways, namely hyphal anastomosis and contagious infection by the virus particles [10,12]. Moreover, Liu and colleagues [13] found that the mycophageous insect *Lycoriella ingenua* can act as a vector for transmission of SsHADV-1 from DT-8 to virulent strains of *S. sclerotiorum*. On the contrary, there is currently no good evidence for extracellular transmission of the RNA mycoviruses. Vegetative incompatibility between donor and recipient individuals usually blocks the horizontal transmission of the RNA viruses. It is actually a phenomenon of non-self allorecognition, which is characterized by formation of vacuoles, cytoplasm shrinkage, cell collapse and death [14]. Vegetative incompatibility is genetically controlled by heterokaryon genes (*het*) or vegetative incompatibility genes (*vic*) [15,16]. Two individuals sharing the same alleles at all *het* (or *vic*) loci are compatible and the RNA mycoviruses are transmissible between the two individuals [11]. In contrast, two individuals with different alleles at all or some *het* (or *vic*) loci are incompatible and the RNA mycoviruses can be blocked for the horizontal transmission. By disruption of the *vic* genes in *C. parasitica*, Zhang and Nuss [17] developed super CHV1-donor strains, which showed more efficient transmission of CHV1 than the wild-type CHV-donor strains. As a consequence, the genetically-modified strains were enhanced for suppression of chestnut blight, compared to the wild-type hypovirulent strains [17]. Recently, Wu and colleagues [18] reported that Sclerotinia sclerotiorum mycoreovirus 4 (SsMYRV4) can suppress the non-self-recognition in interaction between the SsMYRV4-donor strain and a mycelially incompatible mycovirus-free strain, thereby facilitating the horizontal transmission of other RNA viruses, such as Sclerotinia sclerotiorum debilitation-associated RNA virus (SsDRV).

Most mycoviruses have the genomes of RNA and a few have the genomes of DNA [1, 7]. The mycoviruses in the families *Alphaflexiviridae*, *Gammaflexiviridea*, *Barnaviridae*, *Hypoviridae*, and *Narnaviridae* have the genomes of positive single-stranded RNA (+ssRNA). Recently, a few mycoviruses with the genomes of negative single-stranded RNA (−ssRNA) have been identified in *Botrytis cinerea* [19] and *Sclerotinia sclerotiorum* [20]. On the other hand, the mycoviruses in the families *Chrysoviridae*, *Endornaviridae*, *Megabirnaviridae*, *Patitiviridae*, *Quadriviridae*, *Reoviridae*, and *Totiviridae* have the genomes of double-stranded RNA (dsRNA). The members in families *Chrysoviridae*, *Megabirnaviridae*, *Patitiviridae*, *Quadriviridae*, *Reoviridae*, and *Totiviridae* produce coat proteins to accommodate their dsRNA genomes, whereas the members in *Endornaviridae* have no coat proteins and their dsRNA genomes are unencapsidated.

The endornaviruses were first reported in plants [21,22]. Later studies showed that they exist not only in plants [23], but also in fungi and oomycetes [24–28]. Each endornavirus has a non-segmented dsRNA genome, coding for a large polypeptide with the conserved RNA-dependent RNA polymerase domain [29]. Some endornaviruses have a site-specific nick in the positive RNA strand at the 5′-terminus [30,31].

Sclerotinia minor Jagger, a homothallic fungus, exhibits two mating type alleles (Inv+ and Inv−) that are mitotically stable [32] and are useful markers for tracking strains in special situations. *S. minor* is an important plant pathogenic fungus that infects many economically important crops, including Chinese cabbage (*Brassica rapa* subsp. *pekinensis*), lettuce (*Lactuca sativa*), oilseed rape (*Brassica napus*), peanut (*Arachis hypogea*), and sunflower [33–36]. It causes substantial economic losses in these crops in many countries, including Canada, China, and USA [37–39]. So far, management of *S. minor* on lettuce and peanut largely depends on application of fungicides, as highly resistant commercial cultivars against *S. minor* are not available in the two crops [40,41]. However, this chemical control often causes public concerns over the fungicide residues on plant products and environmental pollution. Therefore, it is imperative to develop alternative measures for control of *S. minor*, including biological control using mycoviruses.

Melzer and Boland [38] first reported transmissible dsRNA elements (possibly RNA viruses) in *S. minor*. However, none of the dsRNA elements in *S. minor* has ever been sequenced and identified. Our previous study identified a hypovirulent strain (LC22) of *S. minor*, which carries a dsRNA element of approximately 13 kb in size [39], but very little is known about the dsRNA element. This study was done to fulfill the following objectives: (i) to identify the dsRNA element in strain LC22 by cDNA cloning and sequence analysis; (ii) to characterize transmission of the dsRNA element among different strains of *S. minor*; and (iii) to evaluate the biocontrol potential of the dsRNA element.

2. Materials and Methods

2.1. Fungal Strains and Cultural Media

A total of 26 fungal strains were used in this study (Supporting Information Table S1). Twenty-six strains, representing 11 mycelial compatibility groups (MCGs), belong to *S. minor*. Twenty-three strains of *S. minor*, including LC22 and LC41, were isolated in 2012 from lettuce (*Lactuca sativa*) and a few weeds in Lichuan County and Xianning County of Hubei Province, China [38]. The remaining three isolates of *S. minor* (W1, W26, P13) were kindly provided by Dr. Barbara Shew of North Carolina State University in USA. Two cultural media, namely potato dextrose agar (PDA), and potato dextrose broth (PDB) were used in this study. Both PDA and PDB were prepared with peeled potato tubers.

2.2. Determination of Mycelial Growth Rate, Sclerotial Production, and Pathogenicity

Mycelial agar plugs of the strains were transferred to PDA in Petri dishes (9 cm diameter), one agar plug per dish and five dishes (replicates) for each strain. The diameter of each colony was measured at day 1 and day 2, and the diameter difference between the two measurements was used to calculate the radial mycelial growth rates [42]. These two measurements were used for growth rate calculation because some fast growing isolates could reach to the edge of the plates in three days. The cultures were further incubated for 20 days. Colony morphology was observed and number of the sclerotia formed in each dish was counted. The experiment was repeated one more time.

Pathogenicity of the strains was tested on detached leaves of oilseed rape (*Brassica napus* cultivar "Zhongshuang No. 9") from 45-day-old plants. Mycelial agar plugs were inoculated on detached leaves placed on moisturized paper towels in plastic trays (45 × 30 × 2.5 cm, length × width × height), two agar plugs per leaf and 3 leaves (replicates) for each strain. The trays were individually covered with 1-mm-thick transparent films to maintain high humidity. Diameter of each necrotic lesion was measured two days after inoculation. The test was repeated once.

2.3. Extraction and Identification of dsRNA

DsRNA molecules in the mycelia were extracted, purified using the same procedures as those used in our previous studies [42–45]. Presence of dsRNA viruses was detected by agarose gel (1%, *w/v*) electrophoresis. The nature of the dsRNA was confirmed by digestion with RNase A (TaKaRa Biotechnology Co., Ltd., Dalian, China), RQ1 RNase-free DNase (Promega, Madison, WI, USA) and S1 nuclease (TaKaRa) [42,43]. The molecules that can be digested by RNase A, but not by DNase and S1 nuclease were considered to be dsRNA.

2.4. cDNA Cloning and Sequencing of SmEV1

The dsRNA was used as a template to generate cDNA fragments following the procedure used by Wu and colleagues [42–44]. The amplified DNA fragments were cloned, sequenced, and used to assemble the full-length cDNA sequence. Every base pair in the assembled sequence was ascertained by sequencing at least three clones.

2.5. Sequence Analysis

Open reading frame (ORF) in the full-length cDNA sequence of SmEV1 and polypeptides encoded by the ORF was deduced using the ORF Finder program in NCBI (http://www.ncbi.nlm.nih.gov/gorf/) with the standard codon usages. The sequences of previously reported endornaviruses and related outgroup viruses were retrieved from the NCBI GenBank database (http://www.ncbi.nlm.nih.gov/genomes) and used for comparative analysis. Multiple sequence alignment was carried out using DNAMAN software (V6.0, Lynnon Corporation, San Ramon, CA, USA) to determine the conserved motifs for the domains of MTR, Hel 1, and RdRp. Phylogenetic trees were constructed based on the amino acid sequences of MTR, Hel 1, and RdRp using the neighbor-joining (NJ) method in MEGA 5.0 [46] and tested with a bootstrap of 1000 replicates.

2.6. Northern Hybridization

For Northern blotting, 1 μg viral dsRNA was loaded in 1.2% (w/v) agarose gel in MOPS buffer (20 mM MOPS, 5 mM sodium acetate, 2 mM EDTA, pH 7.0) containing 2% formaldehyde (v/v). After electrophoresis, the gel was transferred to a nylon Zeta-Probe membrane (Bio-Rad, Hercules, CA, USA) by capillary blotting for 16 h using 20× SSC (3.0 M NaCl, 0.3 M sodium citrate, pH 7.0) as transfer buffer. Two probes, namely Probe 1 (800 bp long, nt 11335–12134) and Probe 2 (351 bp long, nt 40–390), were used in Northern hybridization. Two *E. coli* clones harboring the cDNA fragments of SmEV1 were used for generation of the probes by PCR with the primer pairs Probe 1F/1R and Probe 2F/2R (Supporting Information Figure S1, Table S2). The probes were labeled with digoxigenin (DIG). Northern blotting was done following the method described by Streit and colleagues [47]. The chemiluminescent signals in the probe DNA-RNA hybrids were detected using the reagents in the CDO-Star kit (GE Healthcare Life Sciences, Pittsburgh, PA, USA).

2.7. Horizontal Transmission of SmEV1 in S. minor

Strain LC22 of *S. minor* (MCG 5, *MAT* Inv+) was used as a donor of SmEV1 and 25 other strains of *S. minor* (MCGs 1 to 11) were used as recipients of SmEV1 (Table 1). Transmission of SmEV1 from strain LC22 to each of the recipient strains was tested using the pair culture technique described by Wu and colleagues [42–44]. The pair cultures on PDA were incubated at 20 °C for seven days. A mycelial agar plug from the margin area of the recipient colony distant from the inoculation point was transferred to a new PDA dish to establish a derivative of that recipient, designated by suffixing the recipient strain with a "V". The transmission for each recipient strain was repeated three times. To ascertain that the derivative isolates were indeed due to transmission of SmEV1 from the donor strain LC22 to the recipient and not due to any possibility of contamination of LC22, the *MAT* alleles of the derivative isolates XN01V1, XN01V2, XN01V3, along with the donor strain LC22 (Inv+), and the recipient strain XN01 (Inv−), were detected by using PCR with the previously developed primers [32,39]. All of the 75 derivative isolates were individually tested for the presence of SmEV1 by RT-PCR using specific primer pair RdRp F/R (Supporting Information Table S2). Meanwhile, mycelial growth rate, sclerotial production, and pathogenicity of these derivatives were compared with their progenitors, using methods as described above.

Table 1. Endornavirus SmEV1 transmission from the donor strain LC22 (MCG 5) to other strains of Sclerotinia minor (MCGs 1 to MCG 11), and effect of SmEV1 introduction on mycelial growth, sclerotial yield, and pathogenicity of the recipients.

Recipient Strain (MCG)	SmEV1 [1]		Growth Rate (cm/d) [2]		Sclerotia Per Dish [2]		Leaf Lesion Diameter (cm) [3]	
	Before	After	Before	After	Before	After	Before	After
LC53 (MCG1)	−	+	1.8	0.6 **	1806	1259 **	4.5	1.6 **
XN19 (MCG1)	−	+	1.8	0.7 **	2286	835 **	3.3	2.2 **
XN21 (MCG1)	−	+	1.9	0.5 **	1881	674 **	3.6	0.4 **
XN35 (MCG1)	−	+	1.6	0.2 **	2550	1200 *	2.9	0.5 **
DY02 (MCG2)	−	+	1.5	1.0 **	1978	1411 **	3.7	1.6 **
LC02 (MCG2)	−	+	1.7	1.1 **	2300	843 **	3.6	1.6 **
LC20 (MCG2)	−	+	1.5	0.7 **	1719	734 **	3.1	1.0 **
LC28 (MCG2)	−	+	1.4	1.1 **	2416	1278 *	3.8	0.9 **
LC41 (MCG2)	−	+	1.8	1.2 **	2198	1196 **	3.2	0.5 **
LC11 (MCG3)	−	+	1.6	0.3 **	2224	575 **	3.3	1.0 **
LC19 (MCG3)	−	+	1.7	0.2 **	1742	1014 **	3.2	0.6 **
LC46 (MCG3)	−	+	1.7	0.9 **	2432	889 **	2.6	0.8 **
XN01 (MCG3)	−	+	1.8	0.8 **	2880	1673 **	2.5	0.6 **
XN12 (MCG4)	−	+	2.0	1.2 **	2792	1408 **	3.9	0.2 **
XN13 (MCG4)	−	+	1.6	1.1 **	2556	1000 **	3.5	2.4 **
LC15 (MCG5)	−	+	1.8	1.4 **	2316	1475 **	3.4	1.3 **
LC38 (MCG5)	−	+	1.3	0.3 **	1188	674 **	0.8	0.2 **
XN14 (MCG6)	−	+	1.7	1.4 **	2350	1225 **	2.2	1.3 **
XN34 (MCG6)	−	+	1.8	0.4 **	1661	1124 *	4.0	0.7 **
LC29 (MCG7)	−	+	2.0	0.8 **	2492	1105 **	3.8	0.7 **
LC47 (MCG7)	−	+	2.1	1.2 **	2459	1472 **	3.5	1.1 **
LC36 (MCG8)	−	+	1.5	0.6 **	2691	1152 **	2.9	2.1 **
P13 (MCG9)	−	+	1.9	0.3 **	1029	528 **	3.5	0.6 **
W1 (MCG10)	−	+	1.7	0.2 **	1788	782 **	3.7	0.7 **
W26 (MCG11)	−	+	1.6	0.3 **	1230	956 **	3.4	0.7 **

[1] Based on detection of SmEV1 RdRp domain using RT-PCR tests for each strain three times with independent total RNA extractions. "+" and "−" indicate presence and absence of SmEV1, respectively; "*" and "**" indicate significant difference at $p < 0.05$ and $p < 0.01$, respectively, according to Student's *t* test. [2] Average growth rate and the sclerotial yield values for each *S. minor* strain of 10 replicates (two tests each with 5 replicates) on PDA at 20 °C in Petri dishes (9 cm in diameter). [3] Average of lesion diameter of detached leaves assays of oilseed rape with 12 replicates (two tests each with three leaves, each leaf with two inoculation sites).

2.8. Transmission of SmEV1 through Sclerotia

Sclerotia (a total of 120) of strain LC22 produced on PDA (20 °C, 30 days in the dark) were surface-sterilized with 0.1% HgCl$_2$ solution (*w/v*) for 5 min, followed by rinsing for three times (1 min each) in sterilized water and individually transferred on PDA, one sclerotium per dish. The cultures were incubated at 20 °C for 10 days. The resulting single-sclerotium (SS) cultures were tested for pathogenicity on detached oilseed rape leaves (20 °C, 72 h). Eight SS isolates (S002, S004, S017, S038, S054, S085, S097, S104) were randomly selected and tested for present of SmEV1 by RT-PCR using the specific primer pair RdRp F/R (Supporting information Table S2), as well as for mycelial growth rates and sclerotial production on PDA at 20 °C. In these tests, LC22 and LC41 were used as controls.

2.9. Extraction of the Total RNA and RT-PCR Detection of SmEV1

Total RNA was extracted from 3-day-old mycelia (20 °C) of each strain or isolate using the TRIzol reagent (TaKaRa) following the manufacturer's instructions. It was treated with DNase I to remove DNA contamination. Then, the extract was used as template for reverse transcription to synthesize cDNA using PrimeScript® Reverse Transcriptase (TaKaRa) with the oligo (dT)$_{18}$ primer. Finally, the cDNA was used as template in PCR for amplification of the RdRp region of SmEV1 with the primer pair RdRp F/R. Detection of the actin gene (*Actin*) by RT-PCR using the primer pair Actin qF2/qR4 (Supporting information Table S2) was used as control.

2.10. Biocontrol Assay

Mycelium of strain LC22 from five-day-old shake-culture (150 rpm, 20 °C) in PDB was collected by centrifugation at 5000 rpm for 10 min. The mycelial pellet was weighed and re-suspended in fresh PDB (3 g wet mycelial pellet in 100 mL PDB), and blended to generate a hyphal fragment suspension (HFS). Meanwhile, the detached leaves of oilseed rape were placed in two rows on moisturized paper towels in a plastic tray. The leaves in one row were treated with the HFS at 500 µL HFS per leaf. Then, the leaves in the other row were treated with fresh PDB (control) also at 500 µL per leaf. The leaves were inoculated with the mycelial agar plugs from a two-day-old PDA culture of strain LC41 (20 °C), one agar plug per leaf. The tray was covered with a 1-mm-transparent plastic film to maintain a high level of humidity. It was then placed in a growth chamber (20 °C, 12 h light/12 h dark). The lesion diameter around each inoculated mycelial agar plug was measured at 48, 72, and 96 h after inoculation. The experiment was repeated three times.

2.11. Data Analysis

The procedure UNIVARIATE in the SAS software (SAS Institute, Cary, NC, USA, v. 8.0, 1999) was used to analyze the data on mycelial growth rates, number of sclerotia per dish, sclerotium weight, and leaf lesion diameters for strain LC22, the SmEV1-transfected recipient derivatives and their parental strains in related comparative assays. Data on each parameter between strains LC22 and LC41, between each SmEV1-transfected recipient derivative and its parental recipient were compared using Student's t test at $p < 0.05$ or $p < 0.01$. Meanwhile, the procedure ANOVA (analysis of variance) in the SAS software was used to analyze the data on mycelial growth rates and number of sclerotia per dish produced by LC22, eight representative single-sclerotium isolates of LC22 and LC41. Means of each parameter among the isolate/strains were separated using least significant difference (LSD) test at $p < 0.05$.

3. Results

3.1. Cultural Characteristics and Pathogenicity of Strain LC22

Strain LC22 of *S. minor* grew on PDA significantly slower ($p < 0.01$) than that of strain LC41 of *S. minor* (Figure 1A). Strain LC22 formed colonies with abnormal morphology characterized by producing numerous irregular mycelial sectors in the colony margin. In contrast, LC41 colonized the entire PDA dishes after 3 days without producing any irregular mycelial sectors. LC22 formed fewer sclerotia in 10-day-old PDA cultures (574 sclerotia/dish) than LC41 (1741 sclerotia/dish). Furthermore, strain LC22 was less virulent on detached oilseed rape leaves (Figure 1B). Therefore, compared to LC41, LC22 is attenuated in mycelial growth, sclerotial production, and pathogenicity (or aggressiveness). It is a hypovirulent strain in terms of pathogenicity.

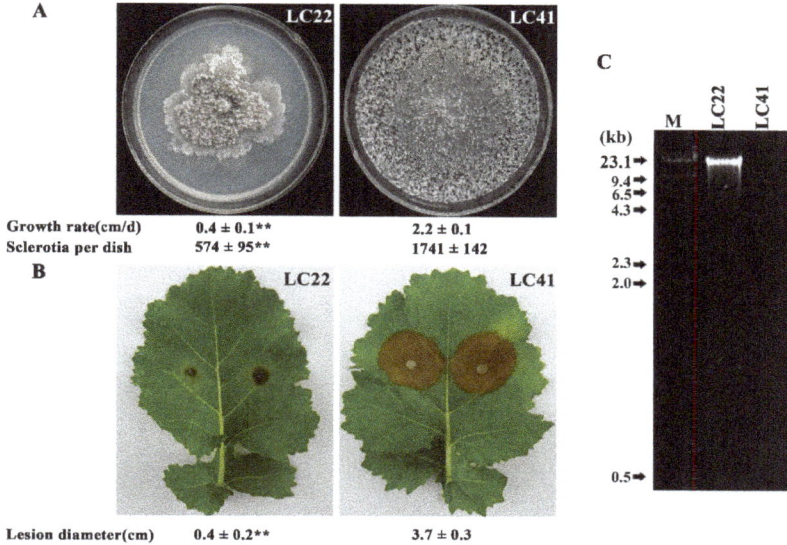

Figure 1. Comparison of strains LC22 and LC41 of Sclerotinia minor in mycelial growth rate, sclerotial production, pathogenicity and presence/absence of dsRNA. (**A**) Colony morphology, growth rate and sclerotial formation on potato dextrose agar (20 °C, 10 d). ** $p < 0.01$ between LC22 and LC41, Student's *t* test; (**B**) Lesions on oilseed rape leaves two days after inoculation (20 °C). ** $p < 0.01$ between LC22 and LC41, Student's *t* test; (**C**) An agarose gel electrophoregram showing the band of the dsRNA extracted from the mycelia of LC22 and LC41. The extracts were treated with S1 nuclease and DNase I before being loaded in the gel for electrophoresis. The DNA marker is λDNA/*Hind* III.

3.2. DsRNA in Strain LC22 and Its Mycoviral Nature

A dsRNA element of ~13 kb in size was detected in the mycelia of LC22, but was not detected in the mycelia of LC41 or any other of the 24 *S. minor* strains (Figure 1C; Supporting information Table S1). The full-length cDNA sequence of 12,626 bps for the dsRNA in LC22 was finally obtained and deposited in the GenBank under the accession number MG255170. The cDNA sequence has the GC content of 47.4%. It was deduced to have one large open reading frame (ORF) starting from nt 502 to nt 12,564 and two untranslated regions (UTR) at 3'- and 5'-termini with the length of 62 and 502 bp, respectively (Figure 2A). The ORF codes for a polyprotein containing 4020 amino acids (aa) with a calculated molecular mass of 448 kDa. BLAST search indicated that the polyprotein contains a viral methyltransferase (MTR) domain, a cysteine-rich region (CRR), two putative viral helicase domains (DExDc and Hel 1) and a domain for RNA-dependent RNA polymerase_2 (RdRp_2) (Figure 2A; Supporting information Figures S3–S6). This genome structure is similar to that of the previously-reported endornaviruses [1]. Therefore, the dsRNA in LC22 represents the genome of an endornavirus, designated thereafter as Sclerotinia minor endornavirus 1/LC22 (SmEV1/LC22). Northern blotting result showed that both probes detected only one dsRNA band (Supporting Information Figure S2), suggesting that no nick exists in SmEV1.

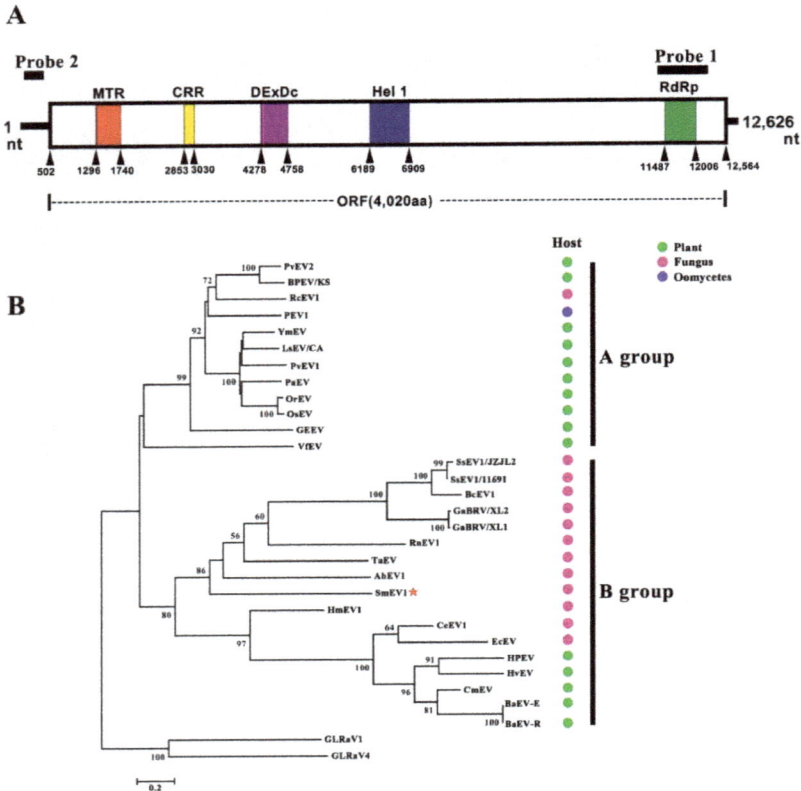

Figure 2. Genome characteristics and phylogeny of SmEV1. (**A**) A schematic diagram showing the genome organization of SmEV1. The coding stand of SmEV1 comprises one large open reading frame (ORF) coding for a polyprotein with the conserved domains of methyltransferase (MTR), cysteine-rich region (CRR), DEXDc, helicase 1 (Hel 1) and RNA-dependent RNA polymerase superfamily_2 (RdRp); (**B**) A maximum-likelihood tree showing the phylogenetic relationship of SmEV1 with 28 other endornaviruses. Grapevine leafroll-associated virus 1 and 4 (GLRaV1 and GLRaV4, GenBank Acc. Nos. ANP22157.1 and NC_016416.1, respectively) were used as outgroups. The tree was inferred from the RNA-dependent RNA polymerase (RdRp) domain. See Table 2 for abbreviation of the endornaviruses. Bootstrap support values lower than 50% are not shown.

3.3. Phylogeny of SmEV1/LC22

SmEV1/LC22 was compared with 29 other endornaviruses in homology of the whole polyprotein and each of the three conserved domains (MTR, Hel 1, RdRp_2). SmEV1/LC22 is identical by 7.60–21.99%, 13.87–34.62%, 15.48–30.20%, and 23.92–47.43% to the other endornaviruses in the whole polyprotein, MTR, Hel 1 and RdRp_2, respectively (Table 2). Phylogenetic analyses were done based on the conserved domains of RdRp_2, MTR, and Hel 1. The endornaviruses formed two groups (A and B). SmEV1 is located in B group in all the three phylogenetic trees. It is most closely related to Sclerotinia sclerotiorum endornavirus 1 (SsEV1), Botrytis cinerea endornavirus 1 (BcEV1), Gremmeniella abietina type B RNA virus (GaBRV), Rosellinia necatrix endornavirus 1 (RnEV1), Tuber aestivum endornavirus (TaEV), and Alternaria brassicicola endornavirus 1 (AbEV1) (Figure 2; Supporting Information Figures S3–S6). Based on currently valid species demarcation criteria for the family *Endornaviridae* (i.e., isolated from a different host species, and less than 75% sequence identity) [48], SmEV1/LC22 should be considered a novel species in the genus *Endornavirus*.

3.4. Horizontal Transmission of SmEV1

Results of the transmission experiments showed that all the recipient-derived strains exhibited attenuated mycelial growth and formation of numerous irregular mycelial sectors in the colony margin at 7 days after pair culturing at 20 °C with the donor strain LC22 on PDA (Figure 3). In order to ascertain the origin of the derivatives from the recipients, not from the donor (LC22), three derivatives, namely XN01V1, XN01V2 and XN01V3, were selected and compared with XN01 and LC22 in the *MAT* alleles (Inv+ and Inv−), which were detected by PCR using specific primers MAT1-1-F/MAT1-1-R (for Inv−) and Type IIF/Type-IIR (for Inv+) [32]. Results showed that XN01V1, XN01V2 and XN01V3 had the same *MAT* allele as XN01 (Inv−), whereas LC22 had the *MAT* allele of Inv+ (Supporting information Figure S7).

All the 25 derivatives were positively for SmEV1 accumulation (Table 1). They were significantly ($p < 0.01$) suppressed for mycelial growth and sclerotial production on PDA, and for infection of leaves of oilseed rape, compared to the virus-free recipient strain (Table 1). Therefore, hypovirulence and decline characteristics of strain LC22 and SmEV1 were successfully transmitted by hyphal contact to isolates of different vegetative compatibility groups of *S. minor*. These results indicated that SmEV1 has a wide spectrum of horizontal transmissibility.

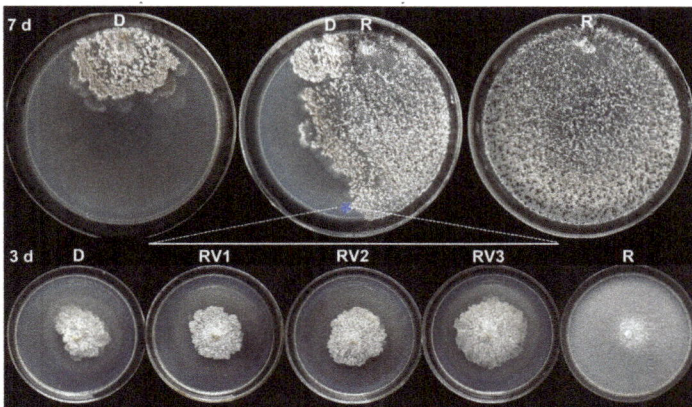

Figure 3. Horizontal transmission of SmEV1 from the donor strain LC22 (D, donor; SmEV1+) to the recipient strain LC41 (R, recipient; SmEV1−) through hyphal contact in a pair culture, with comparison of the strains in single cultures. Top row: two single cultures of the donor strain LC22 and the recipient strain LC41, respectively, and a pair culture of LC22/LC41 (D/R) on PDA (20 °C, seven days). "*" in the pair-culture indicates the area where a mycelial agar plug was removed for generating a derivative of the recipient strain designated as RV. Bottom row, PDA cultures (20 °C, three days) of the donor strain LC22 (D, SmEV1+) and the recipient strain LC41 (R, SmEV1−), and three derivative isolates generated from the recipient strain LC41 in the pair cultures of LC22/LC41 (D/R).

3.5. Transmission of SmEV1 through Sclerotia

Sclerotia produced by *Sclerotinia* species play an important role (survival and reproduction) in their life cycle. Sclerotia of *S. minor* can survive the harsh summer and winter seasons and they usually germinate to produce mycelia, whereas occasionally germinate to produce ascospores [36]. Therefore, transmission of SmEV1 from mycelia to sclerotia is important in terms of biological control using SmEV1. A total of 104 single-sclerotium (SS) isolates were obtained from the 120 sclerotia of strain LC22 formed in PDA cultures. Results showed that the leaf lesion diameters caused by these SS isolates ranged from 0.0 to 2.9 cm, significantly smaller than that of 3.6 cm caused by LC41 (Figure 4A). Eight SS isolates (S002, S004, S017, S038, S054, S085, S097, S104) were randomly selected for determining mycelial growth rates and sclerotial formation on PDA (20 °C) and for the presence of SmEV1 by dsRNA profiling and RT-PCR. All these SS isolates formed colonies with the abnormal morphology (irregular sectors in the colony margin), compared to the colonies formed by LC41 (Figure 4B). They grew at slower growth rates (0.2 to 1.2 cm/d) than LC41 (1.8 cm/d) (Figure 4C) and formed fewer sclerotia (118 to 743 sclerotia/dish) than LC41 (2430 sclerotia/dish) (Figure 4D). SmEV1 was detected in mycelia of each of these SS isolates (Figure 4E). These results suggested that SmEV1 could be transmitted vertically through sclerotia.

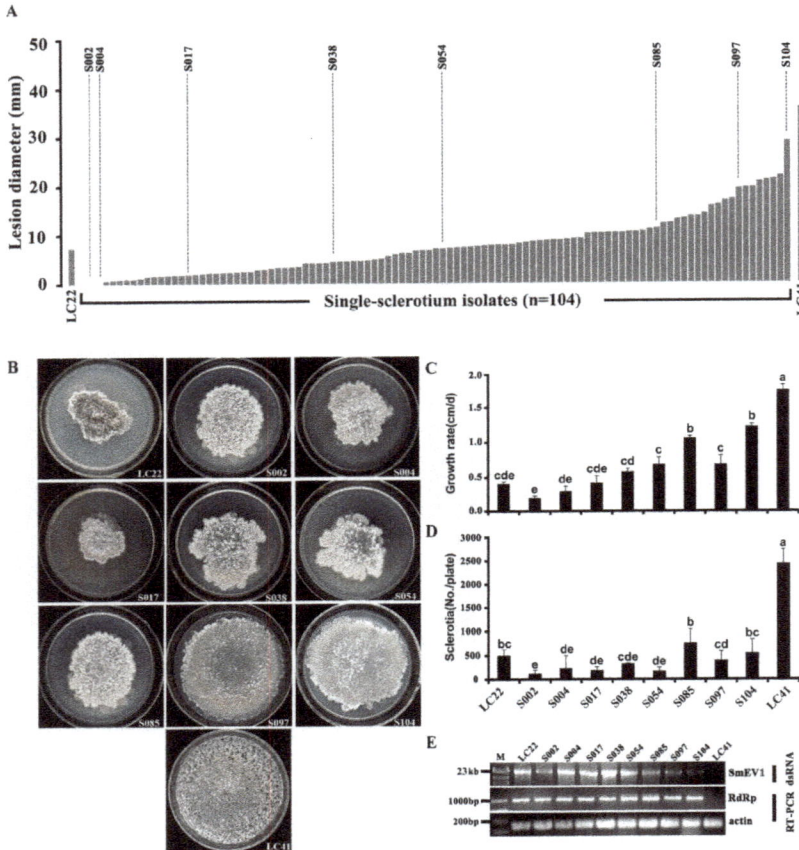

Figure 4. Transmission of SmEV1 in strain LC22 of Sclerotinia minor through sclerotia. (**A**) A histogram showing the average lesion diameters caused by the hypovirulent strain LC22, 104 single-sclerotium (SS) isolates of strain LC22 and the virulent strain LC41 on detached leaves of oilseed rape (20 °C, 72 h). Note reduced average leaf lesion diameters caused by strain LC22 and its SS progenies, compared to the lesion caused by LC41. Eight representative SS isolates selected for further analyses were labeled in the graph; (**B**) Colony morphology of LC22, eight SS isolates and LC41 (PDA, 20 °C, 30 d); (**C**) Average growth rates ($n = 5$) for LC22, eight SS isolates and LC41. Different letters on the bars in each graph indicate significant difference ($p < 0.05$) according to the Least Significant Difference test; (**D**) Average sclerotial yield ($n = 5$) produced by LC22, eight SS isolates and strain LC41 in 30-day-old PDA cultures; (**E**) Detection of SmEV1 in LC22, eight SS isolates and LC41 by dsRNA extraction and RT-PCR.

3.6. Biocontrol Efficacy of SmEV1

An indoor assay on leaves of oilseed rape was done to test efficacy of pre-treatment with the mycelial fragments of LC22 in suppression of the infection by challenge-inoculated virulent strain LC41 (Figure 5A). On the leaves of the control treatment (PDB alone), strain LC41 caused severe infection on the leaves (Figure 5B). In contrast, on the leaves pre-treated with the hyphal fragments of strain LC22 alone without inoculation with strain LC41, no visible symptoms were observed. On the leaves pre-treated with the hyphal fragments of strain LC22 and challenge-inoculated with LC41, only slight infection was observed with formation of tiny restricted necrotic lesions or spots (Figure 5B). Additionally, leaf tissues with the lesions in the LC22 HFS-treated leaves were cut off, surface-sterilized with NaOCl and incubated on PDA for isolation of the fungus in the lesions. Five fungal isolates were obtained. These isolates grew on PDA (20 °C) and formed the colonies with the abnormal colony morphology similar to the colonies formed by LC22. Therefore, LC22 can effectively suppress infection of leaves of oilseed rape by virulent *S. minor*.

Figure 5. Effect of pre-treatment of oilseed rape leaves with the hyphal fragment suspension (HFS) of the hypovirulent strain LC22 of Sclerotinia minor on infection by the virulent strain LC41 of *S. minor*. (**A**) Lesions caused by strain LC41 at 96 h post inoculation (20 °C) on leaves of different treatments. PDB + Agar, the leaf was pre-treated with potato dextrose broth (PDB) alone and challenge-inoculated with a fresh PDA plug; HFS + Agar, the leaf was pre-treated with the HFS of LC22 alone and challenge-inoculated with a fresh PDA agar; PDB + LC41, the leaf was pre-treated with PDB alone and challenge-inoculated with a mycelial agar plug of LC41; HFS + LC41, the leaf was pre-treated with the HFS of LC22 and challenge-inoculated with a mycelial agar plug of LC41; (**B**) A histograph showing average leaf lesion diameters ($n = 18$) of different treatments. ** $p < 0.01$, Student's t test.

Table 2. Percent identity of amino acid sequences between SmEV1 and other endornaviruses determined by multiple alignments of the full-length polyprotein sequence and the conserved domains coding for methyltransferase (MTR), helicase 1 (Hel 1), RNA-dependent RNA polymerase (RdRp).

Virus	Host	Genome Length	Identity (%)				Acc. No.	Presence of Nick
			ORF	MTR	Hel	RdRp		
Botrytis cinerea endornavirus 1 (BcEV1)	F	11,557 bp	20.37	33.33	24.39	45.45	KU923747	–
Alternaria brassicicola endornavirus 1 (AbEV1)	F	10,290 bp	15.09	30.90	24.28	32.41	NC_026136	ND
Tuber aestivum endornavirus (TaEV)	F	9760 bp	15.87	34.62		41.18	NC_014904	ND
Sclerotinia sclerotiorum endornavirus 1/JZJL2 (SsEV1/JZJL2)	F	10,770 bp	21.81	33.33	23.48	47.43	NC_021706	–
Sclerotinia sclerotiorum endornavirus 1/11691 (SsEV1/11691)	F	10,513 bp	21.53	32.90	23.89	47.04	NC_023893	–
Gremmeniella abietina type B RNA virus XL2 (GaBRV/XL2)	F	10,374 bp	21.72	33.19	30.20	46.64	DQ399290	–
Gremmeniella abietina type B RNA virus XL1 (GaBRV/XL1)	F	10,375 bp	21.99	33.19	29.80	45.45	NC_007920	–
Rhizoctonia cerealis endornavirus 1 (RcEV1)	F	17,486 bp	9.29	15.15	17.74	25.39	NC_022619	–
Rhizoctonia solani endornavirus RS002 (RsEV/RS002)	F	14,694 bp	9.78	18.10	18.43		KC792590	ND
Yerba mate endornavirus (YmEV)	P	13,954 bp	11.67		17.32	26.56	NC_024455	ND
Persea americana endornavirus (PaEV)	P	13,459 bp	10.75		15.48	24.22	NC_016648	ND
Oryza rufipogon endornavirus (OrEV)	P	13,936 bp	11.71		15.69	25.00	NC_007649	+
Bell pepper endornavirus (BPEV)	P	14,728 bp	12.01	17.32	17.74	27.73	NC_015781	ND
Oryza sativa endornavirus (OsEV)	P	13,952 bp	18.68		17.65	24.61	D32136	+
Vicia faba endornavirus (VfEV)	P	17,635 bp	8.91		15.63	26.95	AJ000929	+
Phytophthora endornavirus 1 (PEV1)	O	13,883 bp	12.59		16.47	26.95	AJ877914	+
Lagenaria siceraria endornavirus-California (LsEV-CA)	P	15,088 bp	10.04			23.92	NC_023641	ND
Phaseolus vulgaris endornavirus 1 (PvEV1)	P	13,908 bp	11.99		20.00	26.17	AB719397	+
Phaseolus vulgaris endornavirus 2 (PvEV2)	P	14,820 bp	11.16	19.05	20.40	26.56	AB719398	+
Helicobasidium mompa endornavirus 1 (HmEV1)	F	16,614 bp	9.88		17.94	26.07	AB218287	+
Grapevine endophyte endornavirus (GEEV)	P	12,154 bp	11.89		17.53	24.71	NC_019493	ND
Chalara endornavirus (CeEV1)	F	11,602 bp	12.48		19.22	26.95	GQ494150	ND
Basella alba endornavirus-Eclipse (BaEV-E)	P	14,027 bp	7.86		18.50	24.61	AB844264	+
Basella alba endornavirus-Rubra (BaEV-R)	P	14,027 bp	7.86		18.50	24.22	AB844265	+
Erysiphe cichoracearum endornavirus (EcEV)	F	11,908 bp	8.71		18.18	24.61	KT38810	ND
Rosellinia necatrix endornavirus (RnEV1)	F	9639 bp	14.45	24.03	27.98	37.65	LC076696	–
Hordeum vulgare endornavirus (HvEV)	P	14,243 bp	7.60		18.55	25.78	KT721705	ND
Cucumis melo endornavirus (CmEV)	P	15,078 bp	9.33			25.10	KT727022	ND
Hot pepper endornavirus (HPEV)	P	14,729 bp	8.6	13.87		26.95	JN019858	+

Note: F = Fungus, P = Plant, O = Oomycete; ORF = Open reading frame; MTR = Methyltransferase, Hel = Viral RNA helicase, RdRp = RNA-dependent RNA polymerase; "–" = domain not present or no nick; "+" = nicked, ND = not determined.

4. Discussion

In this study, we characterized a novel mycovirus (SmEV1) in a hypovirulent strain LC22 of *S. minor*. Based on phylogenetic analysis of RdRp and genome organization, SmEV1 belongs to the subclade B group in the clade of endornaviruses. The SmEV1 horizontal transmission experiments showed that hypovirulence and associated decline characteristics of strain LC22 and SmEV1 were successfully cotransmitted to strains belonging to different vegetative compatibility groups of *S. minor*. As far as we know, this is the first report of a pathogenicity debilitation-associated endornavirus in *S. minor*. With one exception of Helicobasidium mompa endornavirus 1 (HmEV-1) [28], endornavirus infection usually does not cause any visible abnormal symptoms for host fungi [26]. SmEV1 associated debilitation symptoms are similar to previous studies of hypovirulence caused by other mycoviruses in strains of *S. sclerotiorum* [12,49–51] and other plant pathogenic fungi [8,52–54]. Although SmEV1 cannot be tested for virion transfection without virus particles, the infectious viral cDNA method [55] might be used to determine the causal relationship between SmEV1 infection and host hypovirulence in future studies.

So far, 29 *Endornavirus* genomes have been fully sequenced and characterized, 15 infecting plants, 13 infecting fungi, and one infecting oomycetes (Table 2). One feature of most endornaviruses is that there is a site-specific nick in the coding strand at 5′ terminus [48]. In previous reports, 15 endornaviruses were investigated for presence or absence of the nick in the coding strand. Nine endornaviruses belonging to *Alphaendornavirus* have the nick in the coding strand [24,28,31,53–59]. Other six endornaviruses infecting fungi do not have the site-specific nicks in its dsRNA genomes [25–27,60,61], five of them belong to *Betaendornavirus*, with the exception of Rhizoctonia cerealis endornavirus 1 (RcEV1), which belongs to *Alphaendornavirus*. Our results also showed that SmEV1 in *Betaendornavirus* does not have a nick in 5′-terminus of the coding strand. However, the biological significance of the site-specific nicks in any of these endornaviruses remains unknown. They may play roles in regulation of replication or transcription of endornaviruses.

Endornaviruses usually cause a persistent infection in hosts [23,29]. The endornavirus members in plant can transmit vertically by pollen grains and seeds, but can hardly transmit by contact [62]. In fungi, only one endornavirus (i.e., HmEV1-670) was reported to be able to transmit through hyphal anastomosis [63]. In our study, hyphal incompatibility did not restrict horizontal transmission of SmEV1 in *S. minor*. Previous studies have shown that the two mycoviruses (SsHADV-1 and SsPV1) could easily and efficiently transmit between two vegetative incompatible strains in *S. sclerotiorum* [12,51]. Furthermore, SsMYRV4 was found capable of suppression of vegetative incompatibility-mediated programmed cell death (PCD), thus facilitating horizontal transmission of other mycoviruses among vegetative incompatible strains of *S. sclerotiorum* [18]. Whether the horizontal transmission of SmEV1 is achieved by inhibiting vegetative incompatibility reaction remains unknown and needs to be further characterized. Another possible explanation is that the vegetative incompatibility reaction in *S. minor* is not strong enough to restrict the horizontal transmission of SmEV1.

SmEV1 persistently exist in its hosts. This phenomenon is similar to that reported for other endornaviruses [64] and some other mycoviruses, such as SsPV1 [51], and BcRV1 [45]. We had tried to eliminate SmEV1 from strain LC22 using different methods, including hyphal-tip culturing, protoplast regeneration and plant-inoculation and re-isolation. However, all the methods failed to generate a virus-free strain. Further studies are warranted on curing *S. minor* strain LC22 of SmEV1 by single ascospore isolation. Our previous study has demonstrated that the population characteristics of *S. minor* were simple with a few mycelial compatibility groups [39]. *S. minor* usually germinate myceliogenically. Persistence of SmEV1 in mycelia and sclerotia might be useful for future exploitation of SmEV1 to control sclerotinia diseases caused by *S. minor*. The ability of the hypovirulent strain LC22 in suppressing plant infection by virulent strain LC41 of *S. minor* could be due to SmEV1 horizontal transmission via hyphal anastomosis from LC22 to LC41 and caused hypovirulence. Another reason may be the hypovirulent strain LC22 induces plant resistance to the virulent strain LC41 of *S. minor*.

In conclusion, we isolated and characterized a novel pathogenicity debilitation-associated endornavirus (SmEV1) in *S. minor*, and demonstrated SmEV1 has a wide spectrum of transmissibility. SmEV1 has the ability to spread horizontally among isolates regardless of vegetative incompatibility barriers between different VCGs in *S. minor*. The hypovirulent strain LC22 has a protective activity against attack by virulent *S. minor* according to the detached leaf assays in this study. Further evaluation of using hypha and sclerotia to transmit SmEV1 as a biological control agent to control *S. minor* under field conditions is needed.

Supplementary Materials: The following are available online at http://www.mdpi.com/1999-4915/10/11/589/s1, Figure S1. A schematic diagram showing the strategy used for full-length cDNA cloning of the SmEV1 genome. The PCR primers and the 3'-adaptor used for the cDNA cloning are shown. See Table S2 for the oligonucleotide sequences of the primers/adaptor; Figure S2. (A) A schematic diagram showing positions of Probe 1 and Probe 2 used in Northern blotting; (B) Banding pattern of SmEV1 dsRNA (left), and Northern blotting with the two DNA probes (right); Figure S3. (A) Multiple alignment of the amino acid sequences of the methyltransferase (MTR) domain of SmEV1 and other endornaviruses. "I" to "IV" represent motifs I to IV. "*" and "." represent identical and chemically-similar amino acids, respectively; (B) A maximum-likelihood tree showing the phylogeny of SmEV1 and other endornaviruses. The tree was inferred from the MTR domain. Rehmannia mosaic virus (RMV, GenBank acc. no. JX575184), ngewotan virus (NV, GenBank acc. no. AFY98072.1) and plum bark necrosis stem pitting-associated virus (PBNSPaV, GenBank acc. no. CDM63857.1) were used as outgroups. See Table 1 for abbreviations of the endornaviruses in the phylogenetic tree. Bootstrap support values lower than 50% are not shown; Figure S4. Multiple alignment of the amino acid sequences of the cysteine-rich region (CRR) of SmEV1 and other endornaviruses. Cysteine in the sequences were highlighted; Figure S5. (A) Multiple alignment of the amino acid sequences of the helicase 1 domain (Hel 1) of SmEV1 and other endornaviruses. "I" to "VI" represent motifs I to VI. "*" and "." represent the identical and chemically similar amino acids, respectively. (B) Phylogeneny of SmEV1 and other endornaviruses based on the Hel 1 domain. Grapevine leafroll-associated virus 1 (GLRaV1, GenBank acc. no. ANP22157.1) was used as outgroup. See Table 1 for abbreviation of the virus. Bootstrap support values lower than 50% are not shown; Figure S6. Multiple alignment of the amino acid sequences of the domain for RNA-dependent RNA polymerase superfamily_2 (RdRp) of SmEV1 and other endornaviruses. "I" to "VIII" represent motifs I to VIII. "*" and "." represent the identical and chemically similar amino acids, respectively; Figure S7. Agarose gel electrophoregrams showing the genetic markers for the *MAT* alleles (Inv− and Inv+) in the donor strain LC22, the recipient strain XN01, and the three derivative isolates XN01V1, XN01V2, and XN01V3. M, DNA marker; ITS, Internal Transcribed Spacer, Table S1. List of Sclerotinia minor strains used in the study; Table S2. Primers/adaptor used in this study and their oligonucleotide sequences.

Author Contributions: D.Y., G.L., and L.Y. conceived and designed the experiments; D.Y. performed the experiments; D.Y., M.W., J.Z., G.L., and L.Y. analyzed the data; and D.Y., G.L., W.C., and L.Y. wrote the paper.

Funding: This research was funded by the National Natural Science Foundation of China (NSFC) (grant no. 31701832) and the National Key Research and Development program of China (2017YFD0201100).

Conflicts of Interest: The authors declare that there is no conflict of interest.

References

1. Ghabrial, S.A.; Suzuki, N. Viruses of plant pathogenic fungi. *Annu. Rev. Phytopathol.* **2009**, *47*, 353–384. [CrossRef] [PubMed]

2. Pearson, M.N.; Beever, R.E.; Boine, B.; Arthur, K. Mycoviruses of filamentous fungi and their relevance to plant pathology. *Mol. Plant Pathol.* **2009**, *10*, 115–128. [CrossRef] [PubMed]

3. Jian, J.; Lakshman, D.K.; Tavantzis, S.M. Association of distinct double-stranded RNAs with enhanced or diminished virulence in *Rhizoctonia solani* infected potato. *Mol. Plant-Microbe Interact.* **1997**, *10*, 1002–1009. [CrossRef]

4. Ahn, I.P.; Lee, Y.H. A viral double-stranded RNA up regulates the fungal virulence of *Nectria radicicola*. *Mol. Plant-Microbe Interact.* **2001**, *14*, 496–507. [CrossRef] [PubMed]

5. Schmitt, M.J.; Breinig, F. Yeast viral killer toxins: Lethality and self-protection. *Nat. Rev. Microbiol.* **2006**, *4*, 212–221. [CrossRef] [PubMed]

6. Hollings, M. Viruses associated with a die-back disease of cultivated mushroom. *Nature* **1962**, *196*, 962–965. [CrossRef]

7. Xie, J.T.; Jiang, D.H. New insights into mycoviruses and exploration for the biological control of crop fungal diseases. *Annu. Rev. Phytopathol.* **2014**, *52*, 45–68. [CrossRef] [PubMed]

8. Anagnostakis, S.L. Biological control of chestnut blight. *Science* **1982**, *215*, 466–471. [CrossRef] [PubMed]

9. Milgroom, M.G.; Cortesi, P. Biological control of chestnut blight with hypovirulence: A critical analysis. *Annu. Rev. Phytopathol.* **2004**, *42*, 311–338. [CrossRef] [PubMed]

10. Yu, X.; Li, B.; Fu, Y.P.; Xie, J.T.; Cheng, J.S.; Ghabrial, S.A.; Li, G.Q.; Yi, X.H.; Jiang, D.H. Extracellular transmission of a DNA mycovirus and its use as a natural fungicide. *Proc. Natl. Acad. Sci. USA* **2013**, *110*, 1452–1457. [CrossRef] [PubMed]

11. Nuss, D.L. Hypovirulence: Mycoviruses at the fungal-plant interface. *Nat. Rev. Microbiol.* **2005**, *3*, 632–642. [CrossRef] [PubMed]

12. Yu, X.; Li, B.; Fu, Y.P.; Jiang, D.H.; Ghabrial, S.A.; Li, G.Q.; Peng, Y.L.; Xie, J.T.; Cheng, J.S.; Huang, J.B.; et al. A geminivirus-related DNA mycovirus that confers hypovirulence to a plant pathogenic fungus. *Proc. Natl. Acad. Sci. USA* **2010**, *107*, 8387–8392. [CrossRef] [PubMed]

13. Liu, S.; Xie, J.T.; Cheng, J.S.; Li, B.; Chen, T.; Fu, Y.P.; Li, G.Q.; Wang, M.Q.; Jin, H.N.; Wan, H.; et al. Fungal DNA virus infects a mycophagous insect and utilizes it as a transmission vector. *Proc. Natl. Acad. Sci. USA* **2016**, *113*, 12803–12808. [CrossRef] [PubMed]

14. Biella, S.; Smith, M.L.; Cortesi, P.; Milgroom, M.G. Programmed cell death correlates with virus transmission in a filamentous fungus. *Proc. Biol. Sci.* **2002**, *269*, 2269–2276. [CrossRef] [PubMed]

15. Leslie, J.F. Fungal vegetative compatibility. *Annu. Rev. Phytopathol.* **1993**, *31*, 127–150. [CrossRef] [PubMed]

16. Cortesi, P.; Milgroom, M.G. Genetics of vegetative incompatibility in *Cryphonectria parasitica*. *Appl. Environ. Microbiol.* **1998**, *64*, 2988–2994. [PubMed]

17. Zhang, D.X.; Nuss, D.L. Engineering super mycovirus donor strains of chestnut blight fungus by systematic disruption of multilocus *vic* genes. *Proc. Natl. Acad. Sci. USA* **2016**, *113*, 2062–2067. [CrossRef] [PubMed]

18. Wu, S.S.; Cheng, J.S.; Fu, Y.P.; Chen, T.; Jiang, D.H.; Ghabrial, S.A.; Xie, J.T. Virus-mediated suppression of host non-self recognition facilitates horizontal transmission of heterologous viruses. *PLoS Pathog.* **2017**, *13*, e1006234. [CrossRef] [PubMed]

19. Donaire, L.; Pagán, I.; Ayllón, M.A. Characterization of Botrytis cinerea negative-stranded RNA virus 1, a new mycovirus related to plant viruses, and a reconstruction of host pattern evolution in negative-sense ssRNA viruses. *Virology* **2016**, *499*, 212–218. [CrossRef] [PubMed]

20. Liu, L.J.; Xie, J.T.; Cheng, J.S.; Fu, Y.P.; Li, G.Q.; Yi, X.H.; Jiang, D.H. Fungal negative-stranded RNA virus that is related to bornaviruses and nyaviruses. *Proc. Natl. Acad. Sci. USA* **2014**, *111*, 12205–12210. [CrossRef] [PubMed]

21. Valverde, R.A.; Nameth, S.; Abdallha, O.; Al-Musa, O.; Desjardins, P.; Dodds, A. Indigenous double-stranded RNA from pepper (*Capsicum annuum*). *Plant Sci.* **1990**, *67*, 195–201. [CrossRef]

22. Wakarchuk, D.A.; Hamilton, R.I. Partial nucleotide sequence from enigmatic dsRNA in *Phaseolus vulgaris*. *Plant Mol. Biol.* **1990**, *14*, 637–639. [CrossRef] [PubMed]

23. Fukuhara, T.; Koga, R.; Aoki, N.; Yuki, C.; Yamamoto, N.; Oyama, N.; Udagawa, T.; Horiuchi, H.; Miyazaki, S.; Higashi, Y.; et al. The wide distribution of endornaviruses, large double-stranded RNA replicons with plasmid-like properties. *Arch. Virol.* **2006**, *151*, 995–1002. [CrossRef] [PubMed]

24. Hacker, C.V.; Brasier, C.M.; Buck, K.W. A double-stranded RNA from a *Phytophthora* species is related to the plant endornaviruese and contains a putative UDP glycosyltransferase gene. *J. Gen. Virol.* **2005**, *85*, 1561–1570. [CrossRef] [PubMed]

25. Hao, F.M.; Zhou, Z.L.; Wu, M.D.; Li, G.Q. Molecular characterization of a novel endornavirus from the phytopathogenic fungus *Botrytis cinerea*. *Arch. Virol.* **2017**, *162*, 313–316. [CrossRef] [PubMed]

26. Khalifa, M.E.; Pearson, M.N. Molecular characterisation of an endornavirus infecting the phytopathogen *Sclerotinia sclerotiorum*. *Virus Res.* **2014**, *189*, 303–309. [CrossRef] [PubMed]

27. Li, W.; Zhang, T.; Sun, H.; Deng, Y.; Zhang, A.; Chen, H.; Wang, K. Complete genome sequence of a novel endornavirus in the wheat sharp eyespot pathogen *Rhizoctonia cerealis*. *Arch. Virol.* **2014**, *159*, 1213–1216. [CrossRef] [PubMed]

28. Osaki, H.; Nakamura, H.; Sasaki, A.; Matsumoto, N.; Yoshida, K. An endornavirus from a hypovirulent strain of the violet root rot fungus, *Helicobasidium mompa*. *Virus Res.* **2006**, *118*, 143–149. [CrossRef] [PubMed]

29. Roossinck, M.J.; Sabanadzovic, S.; Okada, R.; Valverde, R.A. The remarkable evolutionary history of endornaviruses. *J. Gen. Virol.* **2011**, *92*, 2674–2678. [CrossRef] [PubMed]

30. Okada, R.; Kiyota, E.; Sabanadzovic, S.; Moriyama, H.; Fukuhara, T.; Saha, P.; Roossinck, M.J.; Severin, A.; Valverde, R.A. Bell pepper endornavirus: Molecular and biological properties, and occurrence in the genus Capsicum. *J. Gen. Virol.* **2011**, *92*, 2664–2673. [CrossRef] [PubMed]

31. Okada, R.; Kiyota, E.; Moriyama, H.; Fukuhara, T.; Valverde, R.A. A new endornavirus species infecting Malabar spinach (*Basella alba* L.). *Arch. Virol.* **2014**, *159*, 807–809. [CrossRef] [PubMed]

32. Chitrampalam, P.; Pryor, B.M. Characterization of mating type (*MAT*) alleles differentiated by a natural inversion in *Sclerotinia minor*. *Plant Pathol.* **2015**, *64*, 911–920. [CrossRef]

33. Gaetán, S.A.; Madia, M. Occurrence of sclerotinia stem rot on canola caused by *Sclerotinia minor* in Argentina. *Plant Dis.* **2008**, *92*, 172. [CrossRef]

34. Li, M.; Zhang, Y.Y.; Wang, K.; Hou, Y.G.; Zhou, H.Y.; Jin, L.; Chen, W.D.; Zhao, J. First report of sunflower white mold caused by *Sclerotinia minor* Jagger in Inner Mongolia region, China. *Plant Dis.* **2016**, *100*, 211. [CrossRef]

35. Lyu, A.; Zhang, J.; Yang, L.; Li, G.Q. First report of *Sclerotinia minor* on *Brassica rapa* subsp. *pekinensis* in central China. *Plant Dis.* **2014**, *98*, 992. [CrossRef]

36. Melzer, M.S.; Smith, E.A.; Boland, G.J. Index of plant hosts of *Sclerotinia minor*. *Can. J. Plant Pathol.* **1997**, *19*, 272–280. [CrossRef]

37. Abawi, G.S.; Grogan, R.G. Epidemiology of diseases caused by *Sclerotinia* species. *Phytopathology* **1979**, *69*, 899–904. [CrossRef]

38. Melzer, M.S.; Boland, G.J. Transmissible hypovirulence in *Sclerotinia minor*. *Can. J. Plant Pathol.* **1996**, *18*, 19–28. [CrossRef]

39. Yang, D.; Zhang, J.; Wu, M.D.; Chen, W.D.; Li, G.Q.; Yang, L. Characterization of the mycelial compatibility groups and mating type alleles in populations of *Sclerotinia minor* in central China. *Plant Dis.* **2016**, *100*, 2313–2318. [CrossRef]

40. Hubbard, J.C.; Subbarao, K.V.; Koike, S.T. Development and significance of dicarboximide resistance in *Sclerotinia minor* isolates from commercial lettuce fields in California. *Plant Dis.* **1997**, *81*, 148–153. [CrossRef]

41. Damicone, J.P.; Jackson, K.E. Effects of application method and rate on control of *Sclerotinia* blight of peanut with iprodione and fluazinam. *Peanut Sci.* **2001**, *28*, 28–33. [CrossRef]

42. Wu, M.D.; Zhang, L.; Li, G.Q.; Jiang, D.H.; Hou, M.S.; Huang, H.C. Hypovirulence and double-stranded RNA in *Botrytis cinerea*. *Phytopathology* **2007**, *97*, 1590–1599. [CrossRef] [PubMed]

43. Wu, M.D.; Zhang, L.; Li, G.Q.; Jiang, D.H.; Ghabrial, S.A. Genome characterization of a debilitation-associated mitovirus infecting the phytopathogenic fungus *Botrytis cinerea*. *Virology* **2010**, *406*, 117–126. [CrossRef] [PubMed]

44. Wu, M.D.; Jin, F.Y.; Zhang, J.; Yang, L.; Jiang, D.H.; Li, G.Q. Characterization of a novel bipartite double-stranded RNA mycovirus conferring hypovirulence in the phytopathogenic fungus *Botrytis porri*. *J. Virol.* **2012**, *86*, 6605–6619. [CrossRef] [PubMed]

45. Yu, L.; Sang, W.; Wu, M.D.; Zhang, J.; Yang, L.; Zhou, Y.J.; Chen, W.D.; Li, G.Q. Novel hypovirulence-associated RNA mycovirus in the plant-pathogenic fungus *Botrytis cinerea*: Molecular and biological characterization. *Appl. Environ. Microbiol.* **2015**, *81*, 2299–2310. [CrossRef] [PubMed]

46. Tamura, K.; Peterson, D.; Peterson, N.; Stecher, G.; Nei, M.; Kumar, S. MEGA5: Molecular evolutionary genetic analysis using maximum likelihood, evolutionary distance, and maximum parsimony methods. *Mol. Biol. Evol.* **2011**, *28*, 2731–2739. [CrossRef] [PubMed]

47. Streit, S.; Michalski, C.W.; Erkan, M.; Kleeff, J.; Friess, H. Northern blot analysis for detection and quantification of RNA in pancreatic cancer cells and tissues. *Nat. Protoc.* **2009**, *4*, 37–43. [CrossRef] [PubMed]

48. Fukuhara, T.; Gibbs, M.J. Family *Endornaviridae*. In *Virus Taxonomy: Classification and Nomenclature of Viruses*; King, A.M.Q., Adams, M.J., Carstens, E.B., Lefkowitz, E.J., Eds.; Ninth Report of the International Committee on Taxonomy of Viruses; Elsevier Academic Press: London, UK, 2012; pp. 519–521.

49. Xie, J.T.; Ghabrial, S.A. Molecular characterization of two mitoviruses co-infecting a hypovirulent isolate of the plant pathogenic fungus *Sclerotinia sclerotiorum*. *Virology* **2012**, *428*, 77–85. [CrossRef] [PubMed]

50. Khalifa, M.E.; Pearson, M.N. Molecular characterisation of three mitoviruses co-infecting a hypovirulent isolate of *Sclerotinia sclerotiorum* fungus. *Virology* **2013**, *441*, 22–30. [CrossRef] [PubMed]

51. Xiao, X.Q.; Chen, J.S.; Tang, J.H.; Fu, Y.P.; Jiang, D.H.; Baker, T.S.; Ghabrial, S.A.; Xie, J.T. A novel partitivirus that confers hypovirulence on plant pathogenic fungi. *J. Virol.* **2014**, *88*, 10120–10133. [CrossRef] [PubMed]

52. Brasier, C.M. A cytoplasmically transmitted disease of *Ceratocysis ulmi*. *Nature* **1983**, *305*, 220–223. [CrossRef]

53. Chu, Y.M.; Jeon, J.J.; Yea, S.J.; Kim, Y.H.; Yun, S.H.; Lee, Y.W.; Kim, K.H. Double-stranded RNA mycovirus from *Fusarium graminearum*. *Appl. Environ. Microbiol.* **2002**, *68*, 2529–2534. [CrossRef] [PubMed]

54. Chiba, S.; Salaipeth, L.; Lin, Y.-H.; Sasaki, A.; Kanematsu, S.; Suzuki, N. A novel bipartite double-stranded RNA mycovirus from the white root rot fungus *Rosellinia necatrix*: Molecular and biological characterization, taxonomic considerations, and potential for biological control. *J. Virol.* **2009**, *83*, 12801–12812. [CrossRef] [PubMed]

55. Choi, G.H.; Nuss, D.L. Hypovirulence of chestnut blight fungus conferred by an infectious viral cDNA. *Science* **1992**, *257*, 800–803. [CrossRef] [PubMed]

56. Moriyama, H.; Nitta, T.; Fukuhara, T. Double-stranded RNA in rice: A novel RNA replicon in plants. *Mol. Gen. Genet.* **1995**, *248*, 364–369. [CrossRef] [PubMed]

57. Pfeiffer, P. Nucleotide sequence, genetic organization and expression strategy of the double-stranded RNA associated with the '447' cytoplasmic male sterility trait in *Vicia faba*. *J. Gen. Virol.* **1998**, *79*, 2349–2358. [CrossRef] [PubMed]

58. Moriyama, H.; Horiuchi, H.; Nitta, T.; Fukuhara, T. Unusual inheritance of evolutionarily-related double-stranded RNAs in interspecific hybrid between rice plants *Oryza sativa* and *Oryza rufipogon*. *Plant Mol. Biol.* **1999**, *39*, 1127–1136. [CrossRef] [PubMed]

59. Okada, R.; Young, C.K.; Valverde, R.A.; Sabanadzovic, S.; Aoki, N.; Hotate, S.; Kiyota, E.; Moriyama, H.; Fukuhara, T. Molecular characterization of two evolutionarily distinct endornaviruses co-infecting common bean (*Phaseolus vulgaris*). *J. Gen. Virol.* **2013**, *94*, 220–229. [CrossRef] [PubMed]

60. Tuomivirta, T.T.; Kaitera, J.; Hantula, J. A novel putative virus of *Gremmeniella abietina* type B (Ascomycota: Helotiaceae) has a composite genome with endornavirus affinities. *J. Gen. Virol.* **2009**, *90*, 2299–2305. [CrossRef] [PubMed]

61. Yaegashi, H.; Kanematsu, S. Natural infection of the soil-borne fungus *Rosellinia necatrix* with novel mycoviruses under greenhouse conditions. *Virus Res.* **2016**, *219*, 83–91. [CrossRef] [PubMed]

62. Horiuchi, H.; Moriyama, H.; Fukuhara, T. Inheritance of *Oryza sativa endornavirus* in F1 and F2 hybrids between *japonica* and *indica* rice. *Genes. Genet. Syst.* **2003**, *78*, 229–234. [CrossRef] [PubMed]

63. Ikeda, K.; Nakamura, H.; Matsumoto, N. Hypovirulent strain of the violet root rot fungus *Helicobasidium mompa*. *J. Gen. Plant Pathol.* **2003**, *69*, 385–390. [CrossRef]

64. Roossinck, M.J. Lifestyles of plant viruses. *Phil. Trans. R. Soc. B* **2010**, *365*, 1899–1905. [CrossRef] [PubMed]

![viruses logo] *viruses*

MDPI

Article

Transcriptional and Small RNA Responses of the White Mold Fungus *Sclerotinia sclerotiorum* to Infection by a Virulence-Attenuating Hypovirus

Shin-Yi Lee Marzano [1,2,*], Achal Neupane [1] and Leslie Domier [3,*]

[1] Department of Biology and Microbiology, South Dakota State University, Brookings, SD 57006, USA; achal.neupane@sdstate.edu (A.N.)

[2] Department of Agronomy, Horticulture, and Plant Science, South Dakota State University, Brookings, SD 57006, USA

[3] United States Department of Agriculture, Agricultural Research Service, Department of Crop Sciences, University of Illinois, Urbana, IL 61801, USA

[*] Correspondence: shinyi.marzano@sdstate.edu (S.M.); leslie.domier@ars.usda.gov (L.D.); Tel.: +01-605-688-5469 (S.M.); +01-217-333-0510 (L.D.)

Received: 3 October 2018; Accepted: 10 December 2018; Published: 14 December 2018

Abstract: Mycoviruses belonging to the family *Hypoviridae* cause persistent infection of many different host fungi. We previously determined that the white mold fungus, *Sclerotinia sclerotiorum*, infected with Sclerotinia sclerotiorum hypovirus 2-L (SsHV2-L) exhibits reduced virulence, delayed/reduced sclerotial formation, and enhanced production of aerial mycelia. To gain better insight into the cellular basis for these changes, we characterized changes in mRNA and small RNA (sRNA) accumulation in *S. sclerotiorum* to infection by SsHV2-L. A total of 958 mRNAs and 835 sRNA-producing loci were altered after infection by SsHV2-L, among which >100 mRNAs were predicted to encode proteins involved in the metabolism and trafficking of carbohydrates and lipids. Both *S. sclerotiorum* endogenous and virus-derived sRNAs were predominantly 22 nt in length suggesting one dicer-like enzyme cleaves both. Novel classes of endogenous small RNAs were predicted, including phasiRNAs and tRNA-derived small RNAs. Moreover, *S. sclerotiorum* phasiRNAs, which were derived from noncoding RNAs and have the potential to regulate mRNA abundance in trans, showed differential accumulation due to virus infection. tRNA fragments did not accumulate differentially after hypovirus infection. Hence, in-depth analysis showed that infection of *S. sclerotiorum* by a hypovirulence-inducing hypovirus produced selective, large-scale reprogramming of mRNA and sRNA production.

Keywords: hypovirus; small RNA; tRFs; mycovirus

---

1. Introduction

Fungal viruses (mycoviruses) are highly diverse, infect pathogenic and nonpathogenic fungi, and can significantly reduce the virulence of pathogenic fungi [1]. Fungal–virus interactions involve the interplay between gene expression networks, some of which are influenced by small noncoding RNAs [2]. Central to those interactions are RNA-mediated antiviral defenses that are activated by double-stranded RNA (dsRNA) in a process termed RNA silencing [3–5]. Specificity of the defense is imparted by short (21–24 nt) interfering RNAs (siRNAs) produced from viral dsRNA by RNase III-type enzymes called dicer-like (DCL) proteins [6]. One strand of each siRNA is combined with Argonaute (AGO) proteins into RNA-induced silencing complexes and directs degradation of complementary viral RNA sequences [3].

While RNA silencing genes are highly conserved in animals and plants, they are much less conserved in fungi and oomycetes [7]. For example, the genomes of *Saccharomyces cerevisiae* and

Ustilago maydis lack homologs of canonical AGO and DCL genes, while they are present in the genomes of *Saccharomyces castellii* and *Ustilago hordei* [8]. In addition, some fungi, e g., *Candida albicans*, have evolved novel DCL proteins that differ significantly from those of higher eukaryotes [9], making it challenging to ascertain the importance of fungal endogenous small RNA processing in response to virus infection.

Endogenous small RNAs (sRNAs) play important roles in gene regulation in development and responses to biotic and abiotic stresses in animals and plants [10], but their roles are not well defined in phytopathogenic fungi. It remains elusive whether filamentous fungi have functional microRNAs (miRNAs), typically 21–24 nt in length, generated by DCL processing of short hairpin structures and regulate gene expression directly through interactions with AGO and indirectly through the production of secondary phased siRNAs (phasiRNAs) from long noncoding RNAs [11–13]. PhasiRNAs are a class of secondary sRNAs triggered by miRNAs; the substrate of phasiRNAs is produced by host RNA-dependent RNA polymerase (RDR). The presence of phasiRNAs is an indication of functional miRNAs.

While most plants share a common set of miRNAs [10], fungal species analyzed so far however express diverse suites of miRNAs [14,15]. A few studies have attempted to identify the roles fungal miRNAs play in regulating gene expression. Small RNA and high-throughput rapid amplification of cDNA ends (HT-RACE) to identify cleavage targets for *Fusarium oxysporum* microRNA-like RNAs (milRNAs) did not identify any milRNAs that were present in the current miRNA database, and none of the milRNAs were predicted to trigger cleavage of *F. oxysporum* mRNAs [16]. Analysis of differentially expressed sRNA loci and miRNA accumulation in *Aspergillus flavus* under different growth conditions suggested that miRNAs play important roles in cellular functions including mycotoxin biosynthesis and mycelial growth [17]. The predicted targets for sRNAs from the wheat stripe rust fungus (*Puccinia striiformis*) were enriched for kinases and small secreted proteins suggesting that development-related signaling pathways are regulated by sRNAs in *P. striiformis* [14].

Another source of abundant sRNAs is derived from tRNAs. In plants and animals, mature tRNAs have been confirmed to be sources of functional sRNAs produced in a DCL-independent manner that have been implicated in post-transcriptional and epigenetic regulation of gene expression and repression of retrotransposons [18–20]. In the plant-pathogenic fungus *Magnaporthe oryzae*, tRNA-derived RNA fragments (tRFs) were more abundant in appressoria—specialized-infection tissues—than in vegetative mycelia [21], suggesting that small RNAs in fungi also play active roles in the regulation of growth and development as in higher eukaryotes.

Virus-derived siRNAs (vsiRNAs) have been shown to direct the cleavage of host RNAs in plants [22–25]. Also, another new class of endogenous small RNAs were shown to be activated in *Arabidopsis* plants in response to infection by cucumber mosaic virus that were termed virus-activated siRNAs (vasiRNAs) that lead to the silencing of a broad set of host genes and establishment of an antiviral state [26]. Clearly, small RNA are not just products of RNA silencing but instead exhibit biological functions to down regulate gene expression.

Changes in fungal gene expression in response to virus infection vary greatly depending on the nature of the host–virus interaction that hinges on the activity of antiviral RNA silencing pathway. Saccharomyces cerevisiae LA virus, a member of the Totiviridae, and its M1 satellite impart a beneficial killer phenotype to *S. cerevisiae*, which lacks RNA silencing genes, and produces relatively small changes in gene expression [27]. In contrast, infection of plant pathogenic fungi with robust RNA silencing by mycoviruses that alter fungal virulence have much more pronounced effects on fungal gene expression. For example, infection of *Cryphonectria parasitica* with Cryphonectria hypovirus 1 (CHV1) significantly altered the expression of more than 13% of the analyzed transcripts [28]. Proteomic analysis of proteins secreted by CHV1-infected *C. parasitica* identified 99 proteins with differential accumulation relative to virus-free cultures [29]. Similar proteomic analysis of infection of *Fusarium graminearum* by Fusarium graminearum virus DK21, which perturbs development and attenuates the virulence of the fungus, altered the accumulation of nearly 150 proteins [30].

High-throughput (HT) sequencing of different RNA species provides an approach to functionally characterize how small RNAs interact with potential targets to determine whether milRNAs and siRNAs repress gene expression by mRNA cleavage in fungi. The combination of three types of sequencing data—small RNA-seq, HT-5′-RACE (degradome), and mRNA-seq—allows a global analysis of small RNA function. By analyzing predicted milRNA sequences together with mapped degradome reads, high confidence cleavage events can be identified. Predicted cleavage events can be supported by observations of downregulation of predicted mRNA targets in RNA-seq analysis. Confirmation of predictions is especially important in fungi because, unlike plant or animal miRNAs, fungal pre-milRNAs are do not have well-defined secondary structures and the thermodynamics of the interactions between miRNAs and targets is not well defined, which can lead to false positive cleavage site predictions that cannot be confirmed empirically.

Because mycoviruses can reduce fungal virulence to animal and plant hosts, we aimed to examine the effect of mycovirus infection on gene expression and different classes of sRNA accumulation and gather functional evidence for the existence of miRNAs in the white mold fungus *Sclerotinia sclerotiorum*, a devastating plant pathogen. Previously, we determined that Sclerotinia sclerotiorum hypovirus 2-L (SsHV2-L) reduced/delayed development in sclerotia, enhanced production of aerial mycelia, and induced hypovirulence. Now, we report the use of high throughput sequencing to (1) show that infection by SsHV2-L changes in mRNA and sRNA accumulation in *S. sclerotiorum*; (2) identify new classes of fungal sRNAs; and (3) examine whether predicted sRNA cleavage events correspond to changes in gene expression. Supported by HT-RACE analysis, we show that endogenous *S. sclerotiorum* sRNAs were capable of directing the cleavage of coding and noncoding RNAs, the latter leading to the production of phasiRNAs, demonstrating the evidence of miRNA activity.

2. Materials and Methods

2.1. Preparation of Sclerotinia sclerotiorum Cultures and RNA Extraction

Virus-free (VF) and virus-transfected (VT) cultures of *S. sclerotiorum* strain SsDK3 were produced as described previously [31]. Two sequencing runs were performed: Trial #1 and Trial #2. For the sequencing run of 4 libraries (Trial #1), fungal cultures were grown in potato dextrose broth (PDB) at 25 °C for 10 d. For the separate sequencing run of 10 libraries (Trial #2), fungal cultures were grown on potato dextrose agar (PDA) at 21 °C for 4 d. Total RNAs were extracted from VF and VT cultures using RNeasy Mini (Qiagen, Valencia, CA, USA). Small RNAs were extracted from two biological replications each for VF and VT cultures using mirVana miRNA isolation kits (ThermoFisher Scientific, Waltham, MA, USA) following the manufacturer's instructions.

2.2. Analysis of S. sclerotiorum Transcriptome

Trial #1 of four RNA-Seq libraries (two VF and two VT) was prepared using the TruSeq Stranded mRNA Library Prep Kit (Illumina, San Diego, CA, USA). The RNA-Seq libraries were barcoded and sequenced as single-end 100-nt reads in one lane on an Illumina HiSeq2500. Trial #2 of ten RNA-Seq libraries (five VF and five VT) were prepared using the same kit and barcoded/sequenced as single-end 100-nt reads on an Illumina HiSeq4000. All RNA-Seq reads were assembled de novo using Trinity [32] and used to identify novel coding regions in the *S. sclerotiorum* genome using BRAKER1 [33]. The abundance of reads aligning to predicted coding regions was estimated using RSEM [34]. Differentially expressed coding regions were identified using the DESeq2 Bioconductor package [35]. Differentially expressed coding regions were classified using PANTHER [36]. The relative abundance of transcripts from *S. sclerotiorum* coding regions was determined using reads per kilobase of coding sequence per million mapped reads (RPKM) [37].

2.3. Analysis of S. sclerotiorum Small RNA Populations

Four sRNA libraries (two VF and two VT) were prepared using the TruSeq Small RNA Library Prep Kit (Illumina). The four sRNA libraries were barcoded and sequenced in one lane as 50-nt single-end reads on an Illumina HiSeq2500. Loci with significantly different accumulations of sRNAs between VF and VT samples were identified using DESeq2. Candidate miRNAs were predicted with six different programs, miRDeep2 [38], MiRDeep* [39], miRDeep-P [40], miReap (http://sourceforge. net/projects/mireap/), MiRPlant [41], and ShortStack [42], using the default settings for each program. Small RNA sequences with fewer than 20 reads, and sequences that aligned to coding regions or ribosomal RNA or more than 10 locations in the *S. sclerotiorum* genome were excluded. Loci producing phased sRNAs were identified with Shortstack [43]. Plots of phased siRNAs were prepared using the equation described by Howell et al. [44]. Candidate tRNA genes in the *S. sclerotiorum* genome were predicted using the tRNAscan-SE server [45].

2.4. High-Throughput RNA Ligase-Mediated Rapid Amplification of cDNA Ends (HT-RACE) Analysis

Four HT-RACE (degradome) libraries (two VF and two VT) were constructed as described by Li et al. [46]. Briefly, the Illumina sRNA-seq 5′ adapter was ligated to the purified polyadenylated RNAs from two VF and two VT cultures using T4 RNA ligase. The 5′ termini of most intact fungal mRNAs are blocked by an m7GTP cap structure, while cleaved RNAs have a 5′ terminal phosphate that is available for ligation with the adapter. The 5′-adapter-ligated RNAs were purified, fragmented, treated with phosphatase, purified, and ligated to the Illumina sRNA-seq 3′ adapter. The dual-adapter ligated RNAs were reverse transcribed using the Illumina sRNA-seq RT primer and amplified with the Illumina Gx1 and GX2 sRNA-seq PCR primers. The four barcoded libraries were pooled and sequenced in one lane for 50-nt single-end reads. Potential targets for candidate miRNAs and cluster sequences were identified using the HT-RACE sequence data and the predicted *S. sclerotiorum* coding sequences with the CleaveLand4 pipeline [47]. Predicted miRNA targets with *P*-values of less than 0.05 and degradome categories 0 and 1 hits were selected for downstream cluster analysis.

3. Results

3.1. Changes in mRNA Accumulation Associated with Infection of S. sclerotiorum by SsHV2-L

Two sequencing runs of the transcriptome were performed. Trial #1 had two biological replications each, and Trial #2 had five biological replications each. The accession numbers for the infected and virus free treatments from the Trial #1 are SRR8306347 and SRR8306348, respectively, while SRR8305679 and SRR8305680 are the accession numbers for the infected and virus free treatments from the Trial #2, respectively.

RNA-Seq analysis of Trial #1 produced a total of 2.11×10^8 100-nt reads from the four libraries. On average, 89.9% and 87.1% of the reads from the virus-free and SsHV2-L-infected libraries aligned to the *S. sclerotiorum* genome sequence [48], respectively (Table S1a), and assembled into 23,968 contigs with an N50 of 2834 nt. RNA-Seq analysis of Trial #2 produced a total of 3.99×10^3 100-nt reads from the ten libraries with 89.0% and 83.7% of the reads from the virus free and SsHV2-L infected libraries aligned to the *S. sclerotiorum* genome sequence, respectively (Table S1b). The *S. sclerotiorum* genome is predicted to contain 14,053 coding regions [48], but re-annotation of the *S. sclerotiorum* genome using the RNA-Seq data identified an additional 174 putative coding regions with an average size of 680 nt in this study. Overall, in the RNAs from the infected samples, an average of 1.45% and 5.7% of the sequence reads were derived from the SsHV2-L genome from Trial #1 and Trial #2, respectively.

The RNA-Seq data analyzed from Trial #1 when the cultures are 10-d old in PDB at 25 °C, among the proteins differentially expressed primarily involved in carbohydrate metabolism, *Sclerotinia sclerotiorum* hexose transporter 1 (Sshxt1; SS1G_04841), and a predicted invertase (SS1G_07184; beta-fructofuranosidase) were upregulated 37.8-fold and 18.8-fold, respectively, in SsHV2-L-infected cultures. The expression of *S. sclerotiorum* hexose transporter 2 (SS1G_13734) was upregulated 89.3-fold by SsHV2-L infection. In Trial #1, 958 coding regions differed significantly; 471 were upregulated and 487 were downregulated (Table S2a). The same trial also found that SsHV2-L infection significantly enhanced the accumulation of mRNAs predicted to encode a DEAD-box ATP-dependent RNA helicase (SS1G_04322) and a mitochondria-localized pentatricopeptide repeat-containing protein (SS1G_08638) (Table S2a). Similar to plant gene expression changes induced by plant virus infection [49], mycovirus SsHV2-L infection also increased transcription levels of retrotransposon-related coding sequences in the fungus. SsHV2-L enhanced expression of coding regions for ubiquitin-related proteins (SS1G_02395 and SS1G_00267), aldo/keto reductase (SS1G_00727), and a heat shock protein (SS1G_12897). SsHV2-L infection altered the accumulation of mRNAs encoding 10 cytochrome P450s and 13 methyltransferases. SsHV2-L downregulated SS1G_01530 which is a putative SPX-domain protein in a manner similar to the *S. cerevisiae* protein SPX-domain protein Syg1 that interacts with the β subunit of G-proteins to suppress the lethality of α subunit deficiency [50].

The two trials of transcriptome sequencing runs from *S. sclerotiorum* grown on different media, for different times and at different temperatures generated very different lists of differentially expressed genes in response to SsHV2-L infection. Both sequencing runs showed that infection by SsHV2-L did not significantly alter the expression of *S. sclerotiorum* homologs of AGO (SS1G_00334 and SS1G_11723), DCL (SS1G_13747 and SS1G_10369), or RDR (SS1G_03377, SS1G_13161 and SS1G_09915). With the ten libraries of Trial #2 RNA-Seq analysis from 4-d old cultures grown on PDA at 21 °C, infection of *S. sclerotiorum* with SsHV2-L significantly ($p < 0.05$) altered the abundance of transcripts from 1319 coding regions; 779 were upregulated and 540 were downregulated. Panther analysis of the functional classifications of the annotated *S. sclerotiorum* coding regions found that sequences predicted to be involved in rRNA metabolic processes ($P = 6.4 \times 10^{-20}$) and cellular component biogenesis ($P = 4.6 \times 10^{-13}$) were significantly over represented in the SsHV2-L-infected cultures. rRNA-processing protein 7 homolog A-related protein (SS1G_00995) was downregulated by 2.8-fold. Ribosomal U3 small nucleolar ribonucleoprotein proteins IMP3—SS1G_05011 and SS1G_05012—were both downregulated ~2.5-fold. SS1G_03709 and SS1G_00849, which are predicted to be virulence effectors when infecting soybean [51], were downregulated by 5.6-fold upon SsHV2-L infection. Several methyltransferases were downregulated, including SS1G_12790 by 4.9-fold and SS1G_11246 and SS1G_01682 both by 4-fold (Table S2b). Inactive methyltransferase may lead to terminal modification of small RNAs that have been found to trigger phasiRNA production in plants [52].

3.2. Small RNA Accumulation in Healthy and SsHV2-L-infected S. sclerotiorum Cultures

Sequence analysis of the four sRNA libraries produced a total of 9.04×10^7 reads of 17 to 36 nt, of which, 63.5% aligned to the *S. sclerotiorum* genome sequence (Table S3). The accession numbers are SRR8306349 and SRR8306350 for the virus-infected and virus-free treatments, respectively. On average, 16.5% of the sRNA reads aligned to the 192 *S. sclerotiorum* tRNAs predicted by tRNAscan-SE [45], 12.9% to ribosomal RNAs, 5.9% to retrotransposons, and similar to *Fusarium oxysporum* and *N. crassa*, 12.9% of the *S. sclerotiorum* sRNA reads aligned to the *S. sclerotiorum* mitochondrial genome [16,53,54]. In the RNAs from the two infected samples 9.0% of the sequence reads were derived from the SsHV2-L genome on average.

In all four samples, 22-nt sRNAs were the most abundant with a strong preference for uracil at the 5′ position (Figure 1B). Among the reads that aligned to unique positions in the *S. sclerotiorum* genome, there was a second peak at 27-nt (Figure 1A). Most of the 27-nt sRNAs aligned to *S. sclerotiorum* ribosomal RNA genes (Figure 1D). Few of the 27-nt sRNAs aligned to coding regions or sequences with homology to retrotransposons (Figure 1C). The density with which reads aligned to the genome sequence was highly variable. Twenty of the 25 most abundant nonribosomal RNA and noncoding sequences were of 27- to 32-nt in length. Most aligned to three to six loci in the *S. sclerotiorum* genome and many were conserved in the genome of *B. cinerea*. For example, the most abundant sequence (5′-UCCGAAUUAGUGUAGGGGUUAACAUAACUC-3′) with over 4.5×10^6 reads (5.0% of total reads) aligned to one locus each on *S. sclerotiorum* chromosomes 4 and 7, and two loci on chromosome 16 that corresponded to the locations of predicted glutamine tRNAs (Figure 1E). The sequence, designated as tRF5-Glu(GAA), also aligned to predicted glutamine tRNAs on *B. cinerea* chromosomes 2, 8, and 14 and two loci on chromosome 16 (data not shown). Hence the highly abundant sRNAs were likely derived from mature tRNAs and probably represented tRNA halves that are produced by endonucleolytic cleavage of mature tRNAs, sometimes in response to biotic and abiotic stresses [18]. However, the abundance of sRNAs derived from tRNAs was on average 1.4-fold lower in infected-SsHV2-L samples than virus-free controls.

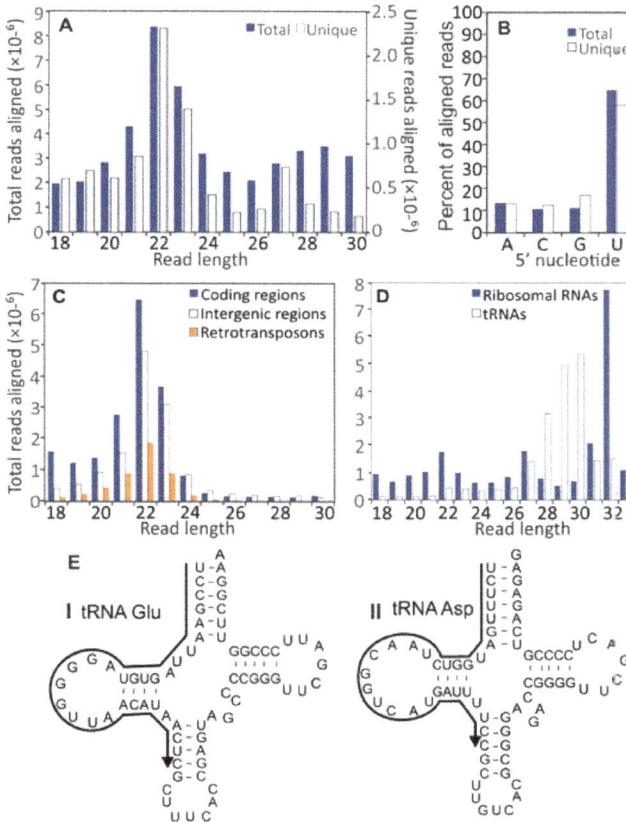

Figure 1. Size distributions of small RNA sequences that aligned to the *S. sclerotiorum* genome. (**A**) Size distribution of small RNA libraries from combined virus-free and hypovirus-infected *S. sclerotiorum* cultures. White and black columns represent unique and total reads of the sRNAs, respectively. (**B**) Frequency of 5′ terminal nucleotides from pooled small RNA samples. Size distribution of small RNA reads aligning to (**C**) coding regions, intergenic regions, retrotransposon sequences, and (**D**) ribosomal RNA and tRNA sequences. (**E**) Mature tRNA structures predicted by tRNAscan-SE with sequences of the two most abundant small RNA sequences that resembled stress-induced tRNA halves. I: 4.5×10^6 reads (5.0% of total reads); tRNA Glu-derived tRF5-Glu(GAA) on *S. sclerotiorum* chromosomes 4, 7, and 16 (two copies); and *B. cinerea* chromosomes 2, 8, 14, and 16 (two copies). II: 1.6×10^6 reads (1.9%); tRNA Asp on *S. sclerotiorum* chromosomes 1, 5, 11, 12, and 14; and *B. cinerea* chromosomes 9, 10, and 13 (two copies).

Similar to sRNA reads derived from the *S. sclerotiorum* genome, the predominant size of sRNAs derived from the SsHV2-L genome were 22 nt in length (Figure 2A) and nearly 90% of the aligned reads contained a 5′-terminal U residue (Figure 2B). As has been reported for other virus infections, the sRNA reads were derived nonuniformly from both RNA strands (Figure 2C).

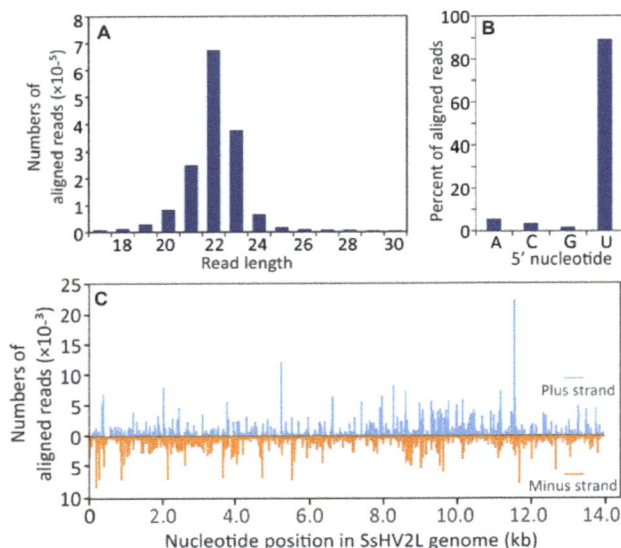

Figure 2. Size distribution of small RNA sequences that aligned to the Sclerotinia sclerotiorum hypovirus 2 L (SsHV2L) genome sequence. (**A**) Size distribution and (**B**) frequency of 5′ terminal nucleotides of small RNAs that aligned to the SsHV2L genome. (**C**) Distribution of small RNA reads that aligned to the SsHV2L genome. Bars above zero indicate alignment to the positive strand, and bars below zero indicate alignment to the negative strand.

3.3. Identification of S. sclerotiorum Loci Producing microRNA-Like RNAs

Cluster analysis in ShortStack identified 8617 intergenic loci that produced at least 100 sRNA reads. Nearly 10% (835) of the intergenic loci producing sRNAs responded significantly to SsHV2-L infection. Because of the variability in RNA silencing-related genes in fungi, candidate miRNAs were predicted from the sRNA data using six different programs. The analysis predicted a total of 459 candidate miRNAs in noncoding and nonribosomal regions of the *S. sclerotiorum* genome with at least 20 reads in the combined data set. Notably, the numbers of candidate miRNAs predicted differed for each of the ten programs used because of the evolution position of fungi between plants and animals. For example, miRDeep2 [38], MiRDeep* [39], miRDeep-P [40], miReap (http://sourceforge.net/projects/mireap/), MiRPlant [41], and ShortStack [42], predicted 46, 173, 175, 27, 139, and 0 candidate miRNA, respectively (Table S4). Just 18% of the candidate miRNAs were predicted by more than one of the programs, among which the predicted structures of candidate miRNAs 0159, 0714, and 1287 contained asymmetrical bulges (Figure 3A).

Because animals, plants, and fungi produce lineage-specific and species-specific miRNAs [10,55,56], the conservation of the candidate miRNA loci of *S. sclerotiorum* was analyzed in the family *Sclerotiniaceae*. Among the candidate miRNA loci, just 18.6% were conserved in at least one other member of the family *Sclerotiniaceae*. For example, candidate miRNAs 0386 and 0437 were conserved in the genomes of *Botrytis cinerea* and *Sclerotinia homoeocarpa* (Figure 3B).

A

candidate_miRNA_0854

```
GC         AU    UCAA      GUUU    C
  UCUCCGGCUUU  GA-AUGA    CUGA        CUA  C
UA AGAGGUCGAAA G CU UACU CUCA GACU A CU GAU A C
```

candidate_miRNA_0159

```
AA GAAAAAGUCAA GUUGAAGCA U GAGAAC-UC AA    C
 UA UUUUUU UAGUU-CAACUUCGU AAG CUCUCU A AG CU
```

candidate_miRNA_1287

```
C GGAAUUGGG GCGGAG GGACA GAAUU U CAAU  C  UUUAG AUCAU CUUG CU GU GU C
  A CUUUAAUCCUCGCUUU  A  CUUAA C GUUA ACUU AAAUC  C  GAAU U CA A A
```

candidate_miRNA_0714

```
UA UUUU A AAA AU GG A AAUGGC AA GUAA CC UCACCAG ACAU AGAUAUGA GA CUGUUGAG A GU A U U U
 CA AAAG CUUU CU CC A UUGCCG A CGUU AU AGUGGUC AU UC G AUACU AUG GAUGACUC CGA A C A G
```

B

candidate_miRNA_0386

S. homoeocarpa	CCUCGUGGCACUUUCGAAGCUUUGGCGACCCUUUUCUCAAUGGUCUCUCACAGUCAAAAAGUAUGCCCAG
B. cinerea	UGUAGUGGCACUUUCGAAGCUUUGGCGACCUAUUUCUCAAUGGUCUCUCACAGUCAAAAAGUAUGCCCAG
S. sclerotiorum	UGAAGUGGCACUUUCGAAGCUUUGGCGACCCUUUUCCCAAUGGUCUCUCACAGUCAAAAAGUAUGCCCAG

candidate_miRNA_0437

S. homoeocarpa	UCUUGGAACACUUGUGAGGGCUUCGAAGAAGUGGGAAAUCUUCUACCUAAGCGUGACAAAUACGGCUCCGGAG
B. cinerea	UCUUGGAACACUUGUGAGGGCUUCGGAGAAGUGGGAAAUCUUCUACCUAAGCGUGACAAAUACGGCUCCGGAG
S. sclerotiorum	UCUUGGAGCACUUGUGAGGGCUUCGAAGAAGUGGGAAACCUUCUACCUAAGCGUGACAAAUACGGCUCCGAAA

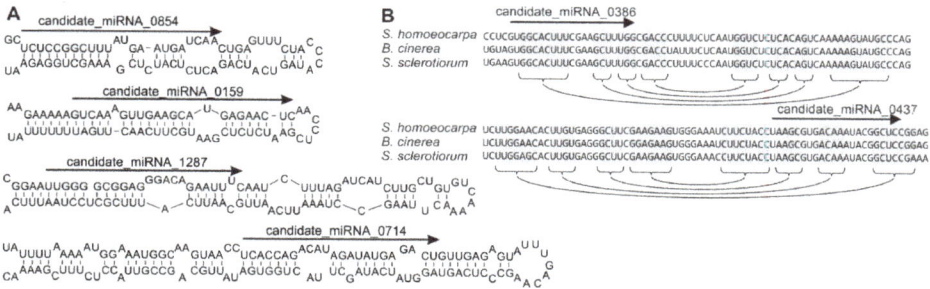

Figure 3. Examples of sRNA producing loci in the *S. sclerotiorum* genome (**A**) capable of folding into structures similar to pre-microRNA and (**B**) conserved in the genomes of other members of the family Sclerotiniaceae. Arrows indicate the positions of mature microRNA-like sequences. Connected shaded boxes indicate regions of conserved base pairing in predicted stem-and-loop structures.

3.4. Potential Targets for S. sclerotiorum sRNAs

Because in silico prediction of miRNA targets alone can produce a large number of false positives in fungi [57], HT-RACE or degradome was used to identify mRNA with uncapped 5′ termini that could represent cleavage events. The abundance of HT-RACE reads were mapped on *S. sclerotiorum* coding sequences relative to the positions of peak cluster sequences and candidate miRNAs sequences with CleaveLand [47] to detect possible sRNA-mediated slicing sites. CleaveLand separated potential cleavage sites into five categories depending on the number of HT-RACE reads that aligned relative to the sRNA. In category 0 cleavage sites, the sRNA aligned over the most abundantly detected 5′ terminus in the coding sequence. Eighteen candidate miRNAs and 22 sequences from cluster analysis had degradome *P* values of less than 0.05 and category 0 or category 1 slicing sites (Table S5). As has been reported in other systems [58], one of the milRNAs, which was also identified by the cluster analysis, was predicted to direct cleavage of transcripts of SS1G_00334, which is predicted to encode an Argonaute 2 homolog. Other predicted targets included an ammonium transporter (SS1G_04502), an extracellular membrane protein (SS1G_12056), a pachytene checkpoint protein (SS1G_12812) and a retrovirus-related polymerase (SS1G_12103) (Table S5).

3.5. Detection of Phased siRNAs from S. sclerotiorum Noncoding RNAs

Most (78.8%) of the *S. sclerotiorum* sRNAs were derived from intergenic regions, and de novo assembly of the *S. sclerotiorum* transcriptome data identified 8761 noncoding transcripts. Analysis of the distribution of phasing of sRNA reads in the *S. sclerotiorum* genome identified 164 loci producing 21-nt or 22-nt phasiRNAs with phasing scores greater than 25 (Figure 4A). Thirty-six loci showed significantly different (*p* < 0.05) accumulations of sRNAs between mock-inoculated and SsHV2-L-infected samples (Figure 4A). Seventeen showed reduced accumulation and 19 showed enhanced accumulation of sRNA sequences. Plots of phasing scores of sliding 10-cycle windows showed characteristic repeating patterns for phased loci (Figure 4B). The detection of large numbers of loci producing phasiRNAs suggests that *S. sclerotiorum* produces sRNAs that function similar to miRNAs in higher eukaryotes.

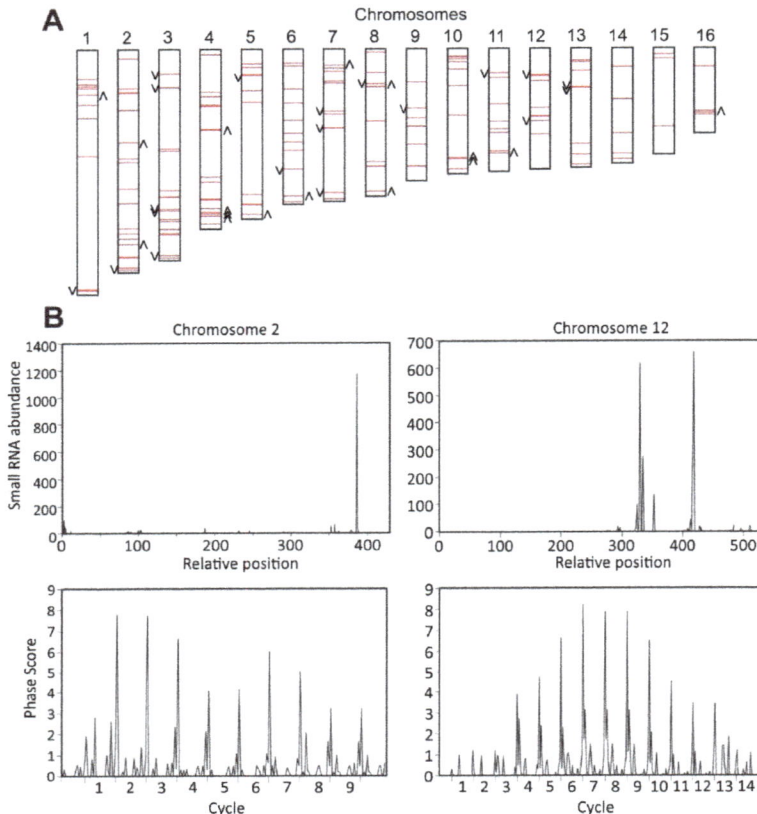

Figure 4. Genome-wide identification of loci producing phased small RNAs in *S. sclerotiorum*. (**A**) Distribution of loci producing phased small RNAs (red horizontal bars) on *S. sclerotiorum* chromosomes. Loci marked with asterisks showed significantly different accumulation of small RNAs between mock-inoculated and SsHV2L-infected samples. Asterisks on the left of the bars indicate reduced small RNA accumulation; bars to the right indicate increased small RNA accumulation. (**B**) Small RNA abundances and phasing score distributions across two loci producing phased small RNAs in the *S. sclerotiorum* genome.

4. Discussion

In this study, we analyzed changes in mRNA and sRNA accumulation in *S. sclerotiorum* in response to persistent hypovirus infection and found that infection by SsHV2-L altered the accumulation of nearly 10% of coding mRNAs and sRNAs. The ability to experimentally inoculate *S. sclerotiorum* with SsHV2-L using in vitro transcripts allowed us to fine map virus-induced changes in mRNA and sRNA accumulation during hypovirus infection. Related studies examined changes in gene expression induced by hypovirus infection in *C. parasitica* and *F. graminearum*. In *C. parasitica*, CHV1 infection altered the expression of mRNAs predicted to encode proteins involved in carbon metabolism, stress responses, and regulation of transcription [28], which was supported by metabolomics analyses [59]. Even though FgHV1 does not induce hypovirulence, FgHV1 infection of *F. graminearum* also significantly altered the accumulation of mRNAs predicted to encode proteins involved in sugar and carbohydrate metabolism [60]. Metabolism of mannitol and other sugars are important in compatible fungal–host interactions [61]. Hence, persistent SsHV2-L infections may be metabolically similar to latent viral infections where carbon is directed away from the TCA cycle for glycolysis to reduce apoptosis [62].

Some of the transcripts differentially expressed in this study provided insights on evolutionarily conserved basal defense systems. Infection of *S. sclerotiorum* by SsHV2-L induced expression of mRNAs predicted to encode stress-responsive proteins including a putative mitochondria-localized pentatricopeptide repeat-containing protein and a DEAD-box RNA helicase. In *Arabidopsis*, the stress-induced mitochondria-localized pentatricopeptide repeat protein, PGN, is involved in defense against necrotrophic fungi and tolerance to abiotic stresses [63]. In addition to their deeply conserved roles in stress responses [64], host DEAD-box RNA helicases enhance and sometimes are required for the replication of animal and plant RNA viruses [65,66]. Hence, it is possible that *S. sclerotiorum* DEAD-box RNA helicases could enhance SsHV2-L accumulation.

Infection of *S. sclerotiorum* by SsHV2-L did not alter the accumulation of mRNAs encoding RNA silencing-related proteins. In contrast, infection of *C. parasitica* by CHV1 upregulated *DCL2* mRNA accumulation up to 20-fold [67]. One of the four *C. parasitica* AGO genes, *AGO2*, which is required for antiviral defense, was also upregulated by CHV1 infection [68,69]. Similarly, one DCL, one AGO, and two RDR genes were upregulated in *Rosellinia necatrix* by infection with Rosellinia necatrix mycoreovirus 3 or Rosellinia necatrix megabirnavirus 1 [70]. Similar to infection of *S. sclerotiorum* by SsHV2-L, infection of *R. necatrix* by Rosellinia necatrix partitivirus 1, Rosellinia necatrix quadrivirus 1, and Rosellinia necatrix victorivirus 1 [70] did not significantly alter the expression of AGO, DCL, or RDR homologs [70].

In animals and plants, the presence of conserved core sets of miRNAs helped define the thermodynamic properties for the biogenesis and activity of miRNAs in those systems [10]. However, unlike animals and plants, several recent studies have found little conservation of miRNA genes across fungal taxa [71–74]. Also, the enzymes involved in RNA silencing are much less conserved in fungi than in higher eukaryotes [75]. For example, neither of the two putative *S. sclerotiorum* DCLs (SS1G_13747 and SS1G_10369) contained canonical PAZ domains, which could mean that parameters appropriate for identification of pre-miRNAs in *S. sclerotiorum* may differ markedly from those used for plants. As previously reported [76], algorithms selected for identification of *S. sclerotiorum* miRNAs and miRNA targets significantly impacted the sequences identified. For example, Zhou et al. [71] described 42 milRNAs in *S. sclerotiorum* and confirmed that a subset of the milRNAs was differentially expressed during sclerotial development. However, none of the milRNAs reported by Zhou et al. [71] were identified in the present study. This may be due to differences in a priori assumptions of milRNA predictions; 58.5% of the milRNA sequences reported by Zhou et al. [71] were derived from exons, but coding sequences were excluded from our analyses.

Plants and animals possess ancient and recent, i.e., species-specific, miRNAs [77]. Ancient miRNAs have conserved sequences and targets, while species-specific miRNAs have targets that are more diverse. A small subset of the candidate *S. sclerotiorum* miRNAs was conserved in closely related fungal species as reported for oomycetes in the genus *Phytophthora* [56]. The most broadly conserved small RNA sequences were derived from predicted *S. sclerotiorum* tRNAs. Small RNAs derived from tRNAs have been shown to regulate gene expression in humans, oomycetes, and plants, and could be important for RNA silencing and gene regulation in *S. sclerotiorum* [20,78–80]. The detection of phasiRNAs suggests that *S. sclerotiorum* miRNAs are capable of cleaving long noncoding RNAs that act as templates for the production of phasiRNAs to regulate other genes in trans [13].

Infection of *S. sclerotiorum* by SsHV2-L was associated with enhanced production of siRNAs from 437 loci that may be analogous to *Arabidopsis* vasiRNAs that have been associated with widespread silencing of host genes and the establishment of a broad-spectrum antiviral state [25]. In *S. sclerotiorum*, 19 of the loci that showed enhanced sRNA accumulation in SsHV2-L-infected samples produced phasiRNAs. In plants, the production of phasiRNAs is initiated by miRNA cleavage of an RNA template, which functions as a substrate for siRNA production at discrete 21 or 22-nt intervals through the combined action of DCL4 and RDR6 [81]. In *Arabidopsis*, phasiRNAs act as negative regulators of nucleotide-binding, leucine-rich repeat plant defense and other genes that can be triggered by a miRNA [82]. While in *Drosophila*, phasiRNAs are expressed primarily in germline cells and suppress retrotransposons and may serve similar functions in *S. sclerotiorum* [81].

Recent studies revealed that fungi are capable of exporting sRNA that can play an important role in pathogenesis [83] and the host can export small RNA to silence fungal genes involved in pathogenesis [84]. The wheat-infecting fungus *P. striiformis* is predicted to use sRNAs to target resistance gene homologs in wheat [14], which is consistent with findings in *B. cinerea* [85] that sRNAs serve as fungal virulence factors. Some *S. sclerotiorum* 22-nt sRNAs from the present study were complementary to mRNAs from *B. napus*, *G. max*, and *H. annuus* mRNAs, including mRNAs predicted to encode ferritin-1. In plants, sRNAs of 22-nt can trigger production of secondary siRNAs and rapid silencing of the targeted mRNA [86]. It remains a question whether a subset of the 22-nt RNAs produced by *S. sclerotiorum* are exported to plant hosts and trigger production of secondary siRNA to downregulate host defense genes, such as ferritin-1 in plants.

We previously demonstrated that SsHV2L infection significantly reduced virulence, changed mycelial growth patterns in 3-day old cultures, and delayed and reduced sclerotia production [31]. In a previous study, we observed that most of the SsHV2L-derived vsiRNAs originated from antisense RNA [87]. However in this study, SsHV2L-derived vsiRNAs were produced from both sense and antisense RNAs, which could be related to the different growth conditions used for the two studies. In summary, we predicted novel classes of sRNAs, showed that infection by a mycovirus induced the expression of sRNAs from both coding and noncoding RNAs, and showed that *S. sclerotiorum* produced large amounts of tRNA-derived siRNAs and phasiRNAs, the latter of which may be capable of regulating gene expression in trans. The roles of these new classes of fungal sRNAs in gene regulation remain to be determined. Future experiments to dissect fungal RNA silencing pathways by introducing mycoviruses may allow us to assign functions to each putative *S. sclerotiorum* RNA silencing gene. A better understanding of small RNA processing will allow us to use the obtained knowledge to compromise fungal antiviral defenses to better control fungal pathogens.

Supplementary Materials: The following are available online at http://www.mdpi.com/1999-4915/10/12/713/s1. Table S1: RNA-Seq analysis of virus-free and hypovirus-transfected *Sclerotinia sclerotiorum*. Table S2: Coding regions differentially expressed between virus-free and hypovirus-infected cultures of *Sclerotinia sclerotiorum*. Table S3: Number of aligning small RNA sequence reads from virus-free and hypovirus-transfected *Sclerotinia sclerotiorum*. Table S4: Candidate microRNA sequence predicted for *Sclerotinia sclerotiorum* small RNA sequence data by give miRNA prediction programs. Table S5: Targets predicted for *Sclerotinia sclerotiorum* small RNAs by degradome sequencing. All the data have been deposited under the NCBI BioProject PRJNA379694.

Author Contributions: S.M. and L.D. conceived, designed, and performed the experiments and wrote the paper. A.N. and L.D. analyzed the data.

Funding: This study was supported in part by National Sclerotinia Initiative Grant SA1800330. This work was partially supported by Project number SD00H606-16/project accession no. 1009451 from the USDA National Institute of Food and Agriculture (to S.M.). This study was also funded by the United States Department of Agriculture/Agricultural Research Service (to L.D.).

Conflicts of Interest: The authors declare no conflict of interest.

References

1. Ghabrial, S.A.; Castón, J.R.; Jiang, D.; Nibert, M.L.; Suzuki, N. 50-plus years of fungal viruses. *Virology* **2015**, *479–480*, 356–368. [CrossRef]

2. Nuss, D.L. Mycoviruses, RNA silencing, and viral RNA recombination. *Adv. Virus Res.* **2011**, *80*, 25–48.

3. Baulcombe, D. RNA silencing. *Trends Biochem. Sci.* **2005**, *30*, 290–293. [CrossRef]

4. Waterhouse, P.M.; Wang, M.B.; Lough, T. Gene silencing as an adaptive defence against viruses. *Nature* **2001**, *411*, 834–842. [CrossRef]

5. Li, Y.; Lu, J.F.; Han, Y.H.; Fan, X.X.; Ding, S.W. RNA interference functions as an antiviral immunity mechanism in mammals. *Science* **2013**, *342*, 231–234. [CrossRef]

6. Baulcombe, D. RNA silencing in plants. *Nature* **2004**, *431*, 356–363. [CrossRef]

7. Nakayashiki, H.; Kadotani, N.; Mayama, S. Evolution and diversification of RNA silencing proteins in fungi. *J. Mol. Evol.* **2006**, *63*, 127–135. [CrossRef]
8. Laurie, J.D.; Ali, S.; Linning, R.; Mannhaupt, G.; Wong, P.; Guldener, U.; Munsterkotter, M.; Moore, R.; Kahmann, R.; Bakkeren, G.; et al. Genome comparison of barley and maize smut fungi reveals targeted loss of RNA silencing components and species-specific presence of transposable elements. *Plant Cell* **2012**, *24*, 1733–1745. [CrossRef]
9. Bernstein, D.A.; Vyas, V.K.; Weinberg, D.E.; Drinnenberg, I.A.; Bartel, D.P.; Fink, G.R. *Candida albicans* Dicer (CaDcr1) is required for efficient ribosomal and spliceosomal RNA maturation. *Proc. Natl. Acad. Sci. USA* **2012**, *109*, 523–528. [CrossRef]
10. Axtell, M.J.; Westholm, J.O.; Lai, E.C. Vive la différence: Biogenesis and evolution of microRNAs in plants and animals. *Genome Biol.* **2011**, *12*, 221. [CrossRef]
11. Song, X.W.; Li, P.C.; Zhai, J.X.; Zhou, M.; Ma, L.J.; Liu, B.; Jeong, D.H.; Nakano, M.; Cao, S.Y.; Liu, C.Y.; et al. Roles of DCL4 and DCL3b in rice phased small RNA biogenesis. *Plant J.* **2012**, *69*, 462–474. [CrossRef]
12. Arikit, S.; Zhai, J.X.; Meyers, B.C. Biogenesis and function of rice small RNAs from non-coding RNA precursors. *Curr. Opin. Plant Biol.* **2013**, *16*, 170–179. [CrossRef]
13. Allen, E.; Xie, Z.X.; Gustafson, A.M.; Carrington, J.C. MicroRNA-directed phasing during trans-acting siRNA biogenesis in plants. *Cell* **2005**, *121*, 207–221. [CrossRef]
14. Mueth, N.A.; Ramachandran, S.R.; Hulbert, S.H. Small RNAs from the wheat stripe rust fungus (*Puccinia striiformis* f.sp. *tritici*). *BMC Genom.* **2015**, *16*, 718. [CrossRef]
15. Lin, R.M.; He, L.Y.; He, J.Y.; Qin, P.G.; Wang, Y.R.; Deng, Q.M.; Yang, X.T.; Li, S.C.; Wang, S.Q.; Wang, W.M.; et al. Comprehensive analysis of microRNA-Seq and target mRNAs of rice sheath blight pathogen provides new insights into pathogenic regulatory mechanisms. *DNA Res.* **2016**, *23*, 415–425. [CrossRef]
16. Chen, R.; Jiang, N.; Jiang, Q.Y.; Sun, X.J.; Wang, Y.; Zhang, H.; Hu, Z. Exploring microRNA-like small RNAs in the filamentous fungus *Fusarium oxysporum*. *PLoS ONE* **2014**, *9*, e104956. [CrossRef]
17. Bai, Y.H.; Lan, F.X.; Yang, W.Q.; Zhang, F.; Yang, K.L.; Li, Z.G.; Gao, P.L.; Wang, S.H. SRNA profiling in *Aspergillus flavus* reveals differentially expressed miRNA-like RNAs response to water activity and temperature. *Fungal Genet. Biol.* **2015**, *81*, 113–119. [CrossRef]
18. Thompson, D.M.; Parker, R. Stressing out over tRNA cleavage. *Cell* **2009**, *138*, 215–219. [CrossRef]
19. Chen, Q.; Yan, M.H.; Cao, Z.H.; Li, X.; Zhang, Y.F.; Shi, J.C.; Feng, G.H.; Peng, H.Y.; Zhang, X.D.; Zhang, Y.; et al. Sperm tsRNAs contribute to intergenerational inheritance of an acquired metabolic disorder. *Science* **2016**, *351*, 397–400. [CrossRef]
20. Martinez, G.; Choudury, S.G.; Slotkin, R.K. tRNA-derived small RNAs target transposable element transcripts. *Nucleic Acids Res.* **2017**, *45*, 5142–5152. [CrossRef]
21. Nunes, C.C.; Gowda, M.; Sailsbery, J.; Xue, M.; Chen, F.; Brown, D.E.; Oh, Y.; Mitchell, T.K.; Dean, R.A. Diverse and tissue-enriched small RNAs in the plant pathogenic fungus, Magnaporthe oryzae. *BMC Genom.* **2011**, *12*, 288. [CrossRef]
22. Wang, M.B.; Bian, X.Y.; Wu, L.M.; Liu, L.X.; Smith, N.A.; Isenegger, D.; Wu, R.M.; Masuta, C.; Vance, V.B.; Watson, J.M.; et al. On the role of RNA silencing in the pathogenicity and evolution of viroids and viral satellites. *Proc. Natl. Acad. Sci. USA* **2004**, *101*, 3275–3280. [CrossRef]
23. Qi, X.; Bao, F.S.; Xie, Z. Small RNA deep sequencing reveals role for *arabidopsis thaliana* RNA-dependent RNA polymerases in viral siRNA biogenesis. *PLoS ONE* **2009**, *4*, e4971. [CrossRef]
24. Shimura, H.; Pantaleo, V.; Ishihara, T.; Myojo, N.; Inaba, J.; Sueda, K.; Burgyan, J.; Masuta, C. A viral satellite RNA induces yellow symptoms on tobacco by targeting a gene involved in chlorophyll biosynthesis using the RNA silencing machinery. *PLoS Pathog.* **2011**, *7*, e1002021. [CrossRef]
25. Smith, N.A.; Eamens, A.L.; Wang, M.B. Viral small interfering RNAs target host genes to mediate disease symptoms in plants. *PLoS Pathog.* **2011**, *7*, e1002022. [CrossRef]
26. Cao, M.J.; Du, P.; Wang, X.B.; Yu, Y.Q.; Qiu, Y.H.; Li, W.X.; Gal-On, A.; Zhou, C.Y.; Li, Y.; Ding, S.W. Virus infection triggers widespread silencing of host genes by a distinct class of endogenous siRNAs in *Arabidopsis*. *Proc. Natl. Acad. Sci. USA* **2014**, *111*, 14613–14618. [CrossRef]
27. McBride, R.C.; Boucher, N.; Park, D.S.; Turner, P.E.; Townsend, J.P. Yeast response to la virus indicates coadapted global gene expression during mycoviral infection. *FEMS Yeast Res.* **2013**, *13*, 162–179. [CrossRef]

28. Allen, T.D.; Dawe, A.L.; Nuss, D.L. Use of cdna microarrays to monitor transcriptional responses of the chestnut blight fungus *cryphonectria parasitica* to infection by virulence-attenuating hypoviruses. *Eukaryot. Cell* **2003**, *2*, 1253–1265. [CrossRef]

29. Wang, J.Z.; Shi, L.M.; He, X.P.; Lu, L.D.; Li, X.P.; Chen, B.S. Comparative secretome analysis reveals perturbation of host secretion pathways by a hypovirus. *Sci. Rep.* **2016**, *6*, 34308. [CrossRef]

30. Kwon, S.J.; Cho, S.Y.; Lee, K.M.; Yu, J.; Son, M.; Kim, K.H. Proteomic analysis of fungal host factors differentially expressed by *Fusarium graminearum* infected with fusarium graminearum virus-DK21. *Virus Res.* **2009**, *144*, 96–106. [CrossRef]

31. Marzano, S.L.; Hobbs, H.A.; Nelson, B.D.; Hartman, G.L.; Eastburn, D.E.; McCoppin, N.K.; Domier, L.L. Transfection of *Sclerotinia sclerotiorum* with *in vitro* transcripts of a naturally occurring interspecific recombinant of Sclerotinia sclerotiorum hypovirus 2 significantly reduces virulence of the fungus. *J. Virol.* **2015**, *89*, 5060–5071. [CrossRef]

32. Grabherr, M.G.; Haas, B.J.; Yassour, M.; Levin, J.Z.; Thompson, D.A.; Amit, I.; Adiconis, X.; Fan, L.; Raychowdhury, R.; Zeng, Q.; et al. Full-length transcriptome assembly from RNA-Seq data without a reference genome. *Nat. Biotechnol.* **2011**, *29*, 644–652. [CrossRef]

33. Hoff, K.J.; Lange, S.; Lomsadze, A.; Borodovsky, M.; Stanke, M. Braker1: Unsupervised RNA-Seq-based genome annotation with GeneMark-ET and AUGUSTUS. *Bioinformatics* **2016**, *32*, 767–769. [CrossRef]

34. Li, B.; Dewey, C.N. Rsem: Accurate transcript quantification from RNA-Seq data with or without a reference genome. *BMC Bioinform.* **2011**, *12*, 323. [CrossRef]

35. Love, M.I.; Huber, W.; Anders, S. Moderated estimation of fold change and dispersion for RNA-Seq data with DESeq2. *Genome Biol.* **2014**, *15*, 550. [CrossRef]

36. Mi, H.Y.; Muruganujan, A.; Casagrande, J.T.; Thomas, P.D. Large-scale gene function analysis with the PANTHER classification system. *Nat. Protoc.* **2013**, *8*, 1551–1566. [CrossRef]

37. Mortazavi, A.; Williams, B.A.; Mccue, K.; Schaeffer, L.; Wold, B. Mapping and quantifying mammalian transcriptomes by RNA-Seq. *Nat. Methods* **2008**, *5*, 621–628. [CrossRef]

38. Mackowiak, S.D. Identification of novel and known miRNAs in deep-sequencing data with miRDeep2. *Curr. Protoc. Bioinform.* **2011**, *36*, 12.10.1–12.10.15.

39. An, J.Y.; Lai, J.; Lehman, M.L.; Nelson, C.C. miRDeep*: An integrated application tool for miRNA identification from RNA sequencing data. *Nucleic Acids Res.* **2013**, *41*, 727–737. [CrossRef]

40. Jones-Rhoades, M.W. Prediction of plant miRNA genes. In *Plant MicroRNAs: Methods and Protocols*; Humana Press: New York, NY, USA, 2010; pp. 19–30.

41. An, J.Y.; Lai, J.; Sajjanhar, A.; Lehman, M.L.; Nelson, C.C. miRPlant: An integrated tool for identification of plant miRNA from RNA sequencing data. *BMC Bioinform.* **2014**, *15*, 275. [CrossRef]

42. Axtell, M.J. Shortstack: Comprehensive annotation and quantification of small RNA genes. *RNA* **2013**, *19*, 740–751. [CrossRef] [PubMed]

43. Johnson, N.R.; Yeoh, J.M.; Coruh, C.; Axtell, M.J. Improved placement of multi-mapping small RNAs. *G3* **2016**, *6*, 2103–2111. [CrossRef] [PubMed]

44. Howell, M.D.; Fahlgren, N.; Chapman, E.J.; Cumbie, J.S.; Sullivan, C.M.; Givan, S.A.; Kasschau, K.D.; Carrington, J.C. Genome-wide analysis of the RNA-DEPENDENT RNA POLYMERASE6/DICER-LIKE4 pathway in *Arabidopsis* reveals dependency on miRNA- and tasiRNA-directed targeting. *Plant Cell* **2007**, *19*, 926–942. [CrossRef] [PubMed]

45. Lowe, T.M.; Chan, P.P. tRNAscan-SE on-line: Integrating search and context for analysis of transfer RNA genes. *Nucleic Acids Res.* **2016**, *44*, W54–W57. [CrossRef] [PubMed]

46. Li, F.; Baker, B. Preparation of cDNA Library for dRNA-Seq. *Bio-protocol* **2012**, *2*, e302. [CrossRef]

47. Addo-Quaye, C.; Miller, W.; Axtell, M.J. Cleaveland: A pipeline for using degradome data to find cleaved small RNA targets. *Bioinformatics* **2009**, *25*, 130–131. [CrossRef]

48. Amselem, J.; Cuomo, C.A.; van Kan, J.A.L.; Viaud, M.; Benito, E.P.; Couloux, A.; Coutinho, P.M.; de Vries, R.P.; Dyer, P.S.; Fillinger, S.; et al. Genomic analysis of the necrotrophic fungal pathogens *Sclerotinia sclerotiorum* and *Botrytis cinerea*. *PLoS Genet.* **2011**, *7*, e1002230. [CrossRef]

49. Whitham, S.A.; Yang, C.L.; Goodin, M.M. Global impact: Elucidating plant responses to viral infection. *Mol. Plant-Microbe Interact.* **2006**, *19*, 1207–1215. [CrossRef]

50. Spain, B.H.; Koo, D.; Ramakrishnan, M.; Dzudzor, B.; Colicelli, J. Truncated forms of a novel yeast protein suppress the lethality of a G-protein α subunit deficiency by interacting with the β subunit. *J. Biol. Chem.* **1995**, *270*, 25435–25444. [CrossRef]

51. Wei, W. Transcriptomic Characterization of Soybean–Sclerotinia Sclerotiorum Interaction at Early Infection Stages. Ph.D. Thesis, University of Illinois at Urbana–Champaign, Champaign, IL, USA, 2017.

52. Fei, Q.; Yu, Y.; Liu, L.; Zhang, Y.; Baldrich, P.; Dai, Q.; Chen, X.; Meyers, B.C. Biogenesis of a 22-nt microRNA in Phaseoleae species by precursor-programmed uridylation. *Proc. Natl. Acad. Sci. USA* **2018**, *115*, 8037–8042. [CrossRef]

53. Hammond, T.M.; Spollen, W.G.; Decker, L.M.; Blake, S.M.; Springer, G.K.; Shiu, P.K.T. Identification of small RNAs associated with meiotic silencing by unpaired DNA. *Genetics* **2013**, *194*, 279–284. [CrossRef] [PubMed]

54. Xu, Z.; Huang, G.; Song, N.; Wang, J.; Cao, L.; Jiang, H.; Ding, T. Complete mitochondrial genome sequence of the phytopathogenic fungi *Sclerotinia sclerotiorum* JX-21. *Mitochondrial DNA Part B* **2016**, *1*, 656–657. [CrossRef]

55. Zhu, R.S.; Li, X.; Chen, Q.S. Discovering numerical laws of plant microRNA by evolution. *Biochem. Biophys. Res. Commun.* **2011**, *415*, 313–318. [CrossRef] [PubMed]

56. Fahlgren, N.; Bollmann, S.R.; Kasschau, K.D.; Cuperus, J.T.; Press, C.M.; Sullivan, C.M.; Chapman, E.J.; Hoyer, J.S.; Gilbert, K.B.; Grunwald, N.J.; et al. *Phytophthora* have distinct endogenous small RNA populations that include short interfering and microRNAs. *PLoS ONE* **2013**, *8*, e77181. [CrossRef] [PubMed]

57. Pinzon, N.; Li, B.; Martinez, L.; Sergeeva, A.; Presumey, J.; Apparailly, F.; Seitz, H. microRNA target prediction programs predict many false positives. *Genome Res.* **2017**, *27*, 234–245. [CrossRef]

58. Vaucheret, H.; Vazquez, F.; Crete, P.; Bartel, D.P. The action of argonaute1 in the miRNA pathway and its regulation by the miRNA pathway are crucial for plant development. *Gene Dev.* **2004**, *18*, 1187–1197. [CrossRef] [PubMed]

59. Dawe, A.L.; Van Voorhies, W.A.; Lau, T.A.; Ulanov, A.V.; Li, Z. Major impacts on the primary metabolism of the plant pathogen *Cryphonectria parasitica* by the virulence-attenuating virus CHV1-EP713. *Microbiology* **2009**, *155*, 3913–3921. [CrossRef]

60. Wang, S.C.; Zhang, J.Z.; Li, P.F.; Qiu, D.W.; Guo, L.H. Transcriptome-based discovery of *Fusarium graminearum* stress responses to FgHV1 infection. *Int. J. Mol. Sci.* **2016**, *17*, 1922. [CrossRef]

61. Meena, M.; Prasad, V.; Zehra, A.; Gupta, V.K.; Upadhyay, R.S. Mannitol metabolism during pathogenic fungal-host interactions under stressed conditions. *Front. Microbiol.* **2015**, *6*, 12. [CrossRef]

62. Delgado, T.; Carroll, P.A.; Punjabi, A.S.; Margineantu, D.; Hockenbery, D.M.; Lagunoff, M. Induction of the Warburg effect by Kaposi's sarcoma herpesvirus is required for the maintenance of latently infected endothelial cells. *Proc. Natl. Acad. Sci. USA* **2010**, *107*, 10696–10701. [CrossRef]

63. Laluk, K.; AbuQamar, S.; Mengiste, T. The Arabidopsis mitochondria-localized pentatricopeptide repeat protein PGN functions in defense against necrotrophic fungi and abiotic stress tolerance. *Plant Physiol.* **2011**, *156*, 2053–2068. [CrossRef]

64. Zhu, M.K.; Chen, G.P.; Dong, T.T.; Wang, L.L.; Zhang, J.L.; Zhao, Z.P.; Hu, Z.L. SlDEAD31, a putative DEAD-box RNA helicase gene, regulates salt and drought tolerance and stress-related genes in tomato. *PLoS ONE* **2015**, *10*, e0133849. [CrossRef] [PubMed]

65. Li, C.; Ge, L.L.; Li, P.P.; Wang, Y.; Sun, M.X.; Huang, L.; Ishag, H.; Di, D.D.; Shen, Z.Q.; Fan, W.X.; et al. The DEAD-box RNA helicase DDX5 acts as a positive regulator of Japanese encephalitis virus replication by binding to viral 3′ UTR. *Antivir. Res.* **2013**, *100*, 487–499. [CrossRef] [PubMed]

66. Kovalev, N.; Pogany, J.; Nagy, P.D. A co-opted dead-box RNA helicase enhances tombusvirus plus-strand synthesis. *PLoS Pathog.* **2012**, *8*, e1002537. [CrossRef] [PubMed]

67. Chiba, S.; Suzuki, N. Highly activated RNA silencing via strong induction of dicer by one virus can interfere with the replication of an unrelated virus. *Proc. Natl. Acad. Sci. USA* **2015**, *112*, E4911–E4918. [CrossRef]

68. Zhang, X.M.; Segers, G.C.; Sun, Q.H.; Deng, F.Y.; Nuss, D.L. Characterization of hypovirus-derived small RNAs generated in the chestnut blight fungus by an inducible DCL-2-dependent pathway. *J. Virol.* **2008**, *82*, 2613–2619. [CrossRef] [PubMed]

69. Sun, Q.; Choi, G.H.; Nuss, D.L. A single argonaute gene is required for induction of RNA silencing antiviral defense and promotes viral RNA recombination. *Proc. Natl. Acad. Sci. USA* **2009**, *106*, 17927–17932. [CrossRef] [PubMed]

70. Yaegashi, H.; Shimizu, T.; Ito, T.; Kanematsu, S. Differential inductions of RNA silencing among encapsidated double-stranded RNA mycoviruses in the white root rot fungus *Rosellinia necatrix*. *J. Virol.* **2016**, *90*, 5677–5692. [CrossRef] [PubMed]

71. Zhou, J.H.; Fu, Y.P.; Xie, J.T.; Li, B.; Jiang, D.H.; Li, G.Q.; Cheng, J.S. Identification of microRNA-like RNAs in a plant pathogenic fungus *Sclerotinia sclerotiorum* by high-throughput sequencing. *Mol. Genet. Genom.* **2012**, *287*, 275–282. [CrossRef] [PubMed]

72. Lau, S.K.P.; Chow, W.N.; Wong, A.Y.P.; Yeung, J.M.Y.; Bao, J.; Zhang, N.; Lok, S.; Woo, P.C.Y.; Yuen, K.Y. Identification of microRNA-like RNAs in mycelial and yeast phases of the thermal dimorphic fungus *Penicillium marneffei*. *PLoS Negl. Trop. Dis.* **2013**, *7*, e2398. [CrossRef]

73. Kang, K.; Zhong, J.S.; Jiang, L.; Liu, G.; Gou, C.Y.; Wu, Q.; Wang, Y.; Luo, J.; Gou, D.M. Identification of microRNA-like RNAs in the filamentous fungus *Trichoderma reesei* by Solexa sequencing. *PLoS ONE* **2013**, *8*, e76288. [CrossRef] [PubMed]

74. Dahlmann, T.A.; Kuck, U. Dicer-dependent biogenesis of small RNAs and evidence for microRNA-like RNAs in the penicillin producing fungus *Penicillium chrysogenum*. *PLoS ONE* **2015**, *10*, e0125989. [CrossRef] [PubMed]

75. Chang, S.S.; Zhang, Z.Y.; Liu, Y. RNA interference pathways in fungi: Mechanisms and functions. *Annu. Rev. Microbiol.* **2012**, *66*, 305–323. [CrossRef] [PubMed]

76. Williamson, V.; Kim, A.; Xie, B.; McMichael, G.O.; Gao, Y.; Vladimirov, V. Detecting miRNAs in deep-sequencing data: A software performance comparison and evaluation. *Brief. Bioinform.* **2013**, *14*, 36–45. [CrossRef] [PubMed]

77. Waterhouse, P.M.; Hellens, R.P. Coding in non-coding RNAs. *Nature* **2015**, *520*, 41–42. [CrossRef]

78. Wang, Q.H.; Li, T.T.; Xu, K.; Zhang, W.; Wang, X.L.; Quan, J.L.; Jin, W.B.; Zhang, M.X.; Fan, G.J.; Wang, M.B.; et al. The tRNA-derived small RNAs regulate gene expression through triggering sequence-specific degradation of target transcripts in the oomycete pathogen *Phytophthora sojae*. *Front. Plant Sci.* **2016**, *7*, 1938. [CrossRef] [PubMed]

79. Haussecker, D.; Huang, Y.; Lau, A.; Parameswaran, P.; Fire, A.Z.; Kay, M.A. Human tRNA-derived small RNAs in the global regulation of RNA silencing. *RNA* **2010**, *16*, 673–695. [CrossRef] [PubMed]

80. Keam, S.P.; Hutvagner, G. tRNA-derived fragments (tRFs): Emerging new roles for an ancient RNA in the regulation of gene expression. *Life* **2015**, *5*, 1638–1651. [CrossRef] [PubMed]

81. Komiya, R. Biogenesis of diverse plant phasiRNAs involves an miRNA-trigger and Dicer-processing. *J. Plant Res.* **2017**, *130*, 17–23. [CrossRef]

82. Fei, Q.L.; Xia, R.; Meyers, B.C. Phased, secondary, small interfering RNAs in posttranscriptional regulatory networks. *Plant Cell* **2013**, *25*, 2400–2415. [CrossRef]

83. Da Silva, R.P.; Puccia, R.; Rodrigues, M.L.; Oliveira, D.L.; Joffe, L.S.; Cesar, G.V.; Nimrichter, L.; Goldenberg, S.; Alves, L.R. Extracellular vesicle-mediated export of fungal RNA. *Sci. Rep.* **2015**, *5*, 7763. [CrossRef] [PubMed]

84. Cai, Q.; Qiao, L.; Wang, M.; He, B.; Lin, F.-M.; Palmquist, J.; Huang, H.-D.; Jin, H. Plants send small RNAs in extracellular vesicles to fungal pathogen to silence virulence genes. *Science* **2018**, *360*, 1126–1129. [CrossRef] [PubMed]

85. Weiberg, A.; Wang, M.; Lin, F.M.; Zhao, H.W.; Zhang, Z.H.; Kaloshian, I.; Huang, H.D.; Jin, H.L. Fungal small RNAs suppress plant immunity by hijacking host RNA interference pathways. *Science* **2013**, *342*, 118–123. [CrossRef] [PubMed]

86. Chen, H.M.; Chen, L.T.; Patel, K.; Li, Y.H.; Baulcombe, D.C.; Wu, S.H. 22-nucleotide RNAs trigger secondary siRNA biogenesis in plants. *Proc. Natl. Acad. Sci. USA* **2010**, *107*, 15269–15274. [CrossRef] [PubMed]

87. Mochama, P.; Jadhav, P.; Neupane, A.; Marzano, S.-Y.L. Mycoviruses as Triggers and Targets of RNA Silencing in White Mold Fungus Sclerotinia sclerotiorum. *Viruses* **2018**, *10*, 214. [CrossRef] [PubMed]

viruses

MDPI

Article

Mycoviruses as Triggers and Targets of RNA Silencing in White Mold Fungus *Sclerotinia sclerotiorum*

Pauline Mochama [1], Prajakta Jadhav [1], Achal Neupane [1] and Shin-Yi Lee Marzano [1,2,*]

[1] Department of Biology and Microbiology, South Dakota State University, Brookings, SD 57007, USA; pauline.mochama@sdstate.edu (P.M.); prajakta.jadhav@sdstate.edu (P.J.); achal.neupane@sdstate.edu (A.N.)

[2] Department of Horticulture, Agronomy, and Plant Sciences, South Dakota State University, Brookings, SD 57007, USA

[*] Correspondence: shinyi.marzano@sdstate.edu or shinyileemarzano@gmail.com; Tel.: +1-605-688-5469

Received: 5 March 2018; Accepted: 20 April 2018; Published: 22 April 2018

Abstract: This study aimed to demonstrate the existence of antiviral RNA silencing mechanisms in *Sclerotinia sclerotiorum* by infecting wild-type and RNA-silencing-deficient strains of the fungus with an RNA virus and a DNA virus. Key silencing-related genes were disrupted to dissect the RNA silencing pathway. Specifically, dicer genes (*dcl-1*, *dcl-2*, and both *dcl-1*/*dcl-2*) were displaced by selective marker(s). Disruption mutants were then compared for changes in phenotype, virulence, and susceptibility to virus infections. Wild-type and mutant strains were transfected with a single-stranded RNA virus, SsHV2-L, and copies of a single-stranded DNA mycovirus, SsHADV-1, as a synthetic virus constructed in this study. Disruption of *dcl-1* or *dcl-2* resulted in no changes in phenotype compared to wild-type *S. sclerotiorum*; however, the double dicer mutant strain exhibited significantly slower growth. Furthermore, the Δ*dcl-1*/*dcl-2* double mutant, which was slow growing without virus infection, exhibited much more severe debilitation following virus infections including phenotypic changes such as slower growth, reduced pigmentation, and delayed sclerotial formation. These phenotypic changes were absent in the single mutants, Δ*dcl-1* and Δ*dcl-2*. Complementation of a single dicer in the double disruption mutant reversed viral susceptibility to the wild-type state. Virus-derived small RNAs were accumulated from virus-infected wild-type strains with strand bias towards the negative sense. The findings of these studies indicate that *S. sclerotiorum* has robust RNA silencing mechanisms that process both DNA and RNA mycoviruses and that, when both dicers are silenced, invasive nucleic acids can greatly debilitate the virulence of this fungus.

Keywords: RNA silencing; gemycircularvirus; mycovirus; antiviral; dicer

---

1. Introduction

RNA-directed gene silencing down-regulates gene expression at the transcriptional and post-transcriptional level. RNA silencing or RNA interference is a mechanism involving the recognition of dsRNA by an RNase III domain containing Dicer enzyme which processes the dsRNA into small RNA (sRNA) duplexes of 18–30-nt in length. These sRNA duplexes are separated into two strands with one of the strands being loaded onto Argonaute proteins to target complementary nucleic acids in a sequence-specific manner.

There are two main biological functions of RNA silencing: the first is endogenous gene regulation in development, stress response, and suppression of transposons and repetitive elements to maintain genome integrity. The second role is to confer defense against invasive nucleic acids including viruses [1–3]. Endogenous gene regulation through RNA silencing has been confirmed in plants and animals but is still debatable for fungi because RNA-silencing gene disruption mutants often do not suffer lethal effects as in plants or animals. However, it is when these mutants are challenged with viruses that the antiviral role of RNA silencing genes becomes evident [4,5]. Therefore, the most noticeable role of RNA silencing in fungi has been identified as an adaptive defense function [5,6]. Although the canonical RNA silencing pathway is deeply conserved, the presence of RNA silencing genes is less uniform in Kingdom Fungi. For instance, *Saccharomyces cerevisiae* has lost all the RNA silencing genes required to internalize a dsRNA mycovirus, L-A: a killer virus that produces a toxin which kills uninfected neighbor cells and leaves the infected cells immune to the toxin (reviewed in [7]). Within the same genera, one fungal species may be predicted to encode RNA silencing genes but another species may not (e.g., *Ustillago hordei* vs. *U. maydis*) [8]. It could be circumstantial that endogenous gene regulation in fungi does not involve RNA silencing mechanisms, but this could partially be due to the existence of unidentified domains producing miRNAs that carry out this function.

The cellular components of RNA silencing have been elucidated in the model fungus *Neurospora crassa*. Two dicer orthologs were identified as DCL-1 and DCL-2 that were shown to play a redundant role in transgene silencing [6]. However, efforts to demonstrate a role for RNA silencing in antiviral defense are lacking due to the absence of a mycovirus experimental system for this fungus. Although it has been determined that *dcl-2* is responsible for antiviral RNA silencing in the ascomycete, *Cryphonectria parasitica* [5], and *dcl-1* has been found to play the antiviral defense role in another ascomycete *Colletotrichum higginsianum* [4], there are currently no reports of evolutionarily conserved dicer homolog specific targets in fungi. Furthermore, no canonical PAZ (Piwi-Argonaute-Zwille) domain has been found in these fungal dicers which is atypical for Class III enzymes that are considered to be RNA silencing initiators in model organisms such as *Drosophila* (reviewed in [9]). Clearly, more studies are needed to dissect the roles of RNA silencing in fungi.

Sclerotinia sclerotiorum is phylogenetically related to *N. crassa* and *C. parasitica* under phylum Ascomycota but in a different class, and its genome has been sequenced and annotated [10]. DNA transformation of *S. sclerotiorum* is straightforward. Moreover, *S. sclerotiorum* has been shown to support the replication of members of more than ten virus families including uniquely, a single stranded (ss)DNA virus, Sclerotinia sclerotiorum hypovirulence-associated DNA virus (SsHADV-1). This virus belongs to a new family, *Genomoviridae*, and has been associated with several infections caused by unknown agents (reviewed in [11]). Previously, a reverse genetics system was developed for a member of the *Hypoviridae* virus family, Sclerotinia sclerotiorum hypovirus 2-lactuca (SsHV2-L) [12]. This diversity in mycoviruses that infect *S. sclerotiorum* allows for an examination of the effect of RNA silencing on viruses with a range of replication strategies in the same host. Antiviral RNA silencing protects an organism against virus infection, however, an outstanding question remains whether the core features against RNA and DNA viruses differ in fungi. In addition, a recent study demonstrated that by simultaneously silencing *dcl-1* and *dcl-2* genes in *Botrytis cinerea*, a close relative of *S. sclerotiorum*, the virulence of *B. cinerea* is greatly hampered due to the reduction in small RNA mediated cross-kingdom RNAi [13]. The two fungal dicer genes are redundant in generating pathogen small RNA effectors that hijack plant immunity [14]. As *Sclerotinia sclerotiorum* is closely related to *B. cinerea* [10], it is intriguing whether corresponding dicer gene(s) have the same effects on *S. sclerotiorum* virulence, small RNA processing, and antiviral defense. We now report the use of the *S. sclerotiorum* experimental system to investigate the role of antiviral RNA silencing in fungi.

2. Materials and Methods

2.1. Fungal Strains and Culture Conditions

Cultures of *Sclerotinia sclerotiorum* wild-type strain DK3 and dicer mutant strains were grown on potato dextrose agar (Sigma, St. Louis, MO, USA) at 20–22 °C. The Δ*dcl-1* and Δ*dcl-2* mutant strains were maintained on PDA supplemented with 100 μg/mL hygromycin B (Alfa Aesar, Haverhill,

MA, USA) and the *Δdcl-1/dcl-2* strain was maintained on PDA supplemented with 100 μg/mL hygromycin and 250 μg/mL Geneticin (G418) [15].

2.2. Construction of dcl-1, dcl-2 and dcl-1/dcl-2 Null Alleles

Sclerotinia sclerotiorum dicer genes (Ss1G_13747 and Ss1G_10369, respectively) were predicted based on homology to those identified in *Neurospora crassa* [8]. Deletion of dicer genes was accomplished using the split marker recombination method which requires two DNA constructs for each gene deletion. To generate the *Δdcl-1* disruption mutant, an 814 bp long upstream region of the gene was amplified using primers F1-DCL1 and F2-DCL2 and a 663 bp long downstream region of the gene was amplified using primers F3-DCL1 and F4-DCL1. F2 and F3 primers include 26–32 bp of complementary sequence to the *Aspergillus nidulans* trpC promoter and terminator respectively. Plasmid pCSN43 containing the hygromycin B resistance (*hph*) gene flanked by the *Aspergillus nidulans* TrpC promoter and terminator [16], obtained from Fungal Genetics Stock Center (Manhattan, KS, USA), was used to amplify the marker gene and promoter and terminator sequences. Primers PtrpC-F and HY-R were used to amplify a 1.2 kb region of the marker gene including the promoter and primers YG-F and TrpC-R were used to amplify a 1.3 kb region of the gene including the terminator. Both amplicons represent roughly two thirds of the marker gene and contain 400 bp of overlapping sequence. The F1–F2 amplicon was then fused to the PrtpC-HY amplicon and the F3–F4 amplicon was fused to the YG-TrpC amplicon using the overlap extension PCR protocol described by Fitch et al. [17]. In the final round of PCR, nested primers were used to give the final gene deletion constructs representing 600 bp of upstream homologous sequence fused to two-thirds of the *hph* gene in the first construct and 600 bp of downstream sequence fused to two-thirds of the *hph* gene in the second construct. Disruption of the *Dcl-2* gene was accomplished with constructs generated as described above using a separate set of primers (Table S1). Final *dcl-2* gene deletion constructs included 830 bp of sequence homologous to the upstream region of the gene and 1 kb of downstream homologous sequence.

The *Δdcl-1/dcl-2* mutant was generated by knocking out the *dcl-1* gene in a *Δdcl-2* mutant without using the split marker method. *Δdcl-2* protoplasts were transformed with a single gene-deletion DNA cassette generated using overlap extension PCR (Primers listed in Table S1). The DNA construct contained 600 bp of sequence homologous to the upstream region of the *Δdcl-1* gene and 600 bp of downstream homologous sequence fused to the G418 resistance gene under the control of the *Aspergillus nidulans* trpC promoter. Recombination occurred at the homologous arms flanking the resistance gene and the *dcl-1* gene was subsequently replaced by the G418 gene. G418 is an aminoglycosidic antibiotic similar to hygromycin but with no cross-resistance. The G418 resistance gene was amplified from pSCB-TrpC-G418 [15].

2.3. Fungal Transformation

Gene deletion cassettes were transformed into wild-type *S. Sclerotiorum* protoplasts using polyethylene glycol (PEG)-mediated transformation. Protoplasts were prepared as described by Chen et al. [18] with a digestion time of 3 h at RT using the lysing enzyme from *Trichoderma harzianum* (Sigma, St. Louis, MO, USA). PEG-mediated transformation of gene deletion constructs into fungal protoplasts was performed following the protocol described by Rollins et al. [19] with some modifications [20]. Briefly, following PEG transformation, 3 mL of liquid regeneration media (RM) was added to protoplasts and the suspension incubated at 28 °C with shaking (100 rpm, 2–4 h). Molten RM (45 °C) was then added to a final volume of 20 mL and the mixture poured into a petri dish. Plates were grown at 28 °C for 12 h and then overlaid with 5 mL molten RM containing hygromycin for single dicer gene mutants and hygromycin and G418 for the double dicer mutant. Final antibiotic concentrations used for fungal selection were 100 μg/mL for hygromycin and 250 μg/mL for G418. Colonies were transferred to potato dextrose agar (PDA) plates supplemented with the appropriate antibiotic and hyphal-tip transferred at least three times.

To confirm gene deletions, DNA was extracted from transformants and PCR was conducted using primer pairs- F1 and F4, F1 and HY-R, and YG-F and F4 to amplify the target regions (Table S1). PCR amplicons were compared in size to the wild-type gene amplicon. Amplicons of the correct size (indicating successful gene deletion) were sequenced to confirm integration of the marker gene into the correct region. Repeated hyphal tipping and nested PCR (Primers listed in Table S2) were performed to ensure monokaryotic gene deletions in each gene disruption experiment (Figure S1).

2.4. Complementation of dcl-1

For complementation experiments, the Δdcl-1/dcl-2 mutant was transformed with a plasmid (pD-NAT1, Fungal Genetics Stock Center, Manhattan, KS, USA) engineered to contain the full length dcl-1 open reading frame flanked by 2.3 kb of upstream genomic sequence and 1 kb of downstream genomic sequence. The dcl-1 gene and flanking regions were amplified from wtDK3 using primers F1-SacI-Dcl1 and F4-NotI-Dcl1 (Table S1) and inserted into the SacI-NotI site of the vector downstream to the *Aspergillus nidulans* TrpC promoter and *nat1* gene which confers resistance to nourseothricin. Following transformation with the plasmid construct, protoplasts were grown on RM media supplemented with nourseothricin to a final concentration of 200 µg/mL. Transformants were then transferred to PDA plates supplemented with 200 µg/mL nourseothricin and phenotypic analysis was conducted. Constructed plasmids were all transformed into *Escherichia coli* strain DH5α for propagation and plasmid isolation. Constructs were verified using PCR amplification and sequencing prior to protoplast transformation.

2.5. Phenotypic Characterization of Gene Deletion Mutants

Growth assays were conducted on at least 3 replicates each of wtDK3, Δdcl-1, Δdcl-2 and Δdcl-1/dcl-2 cultures. Five-millimeter PDA discs were taken from the edges of actively growing 2-day-old mutant and wild-type cultures and inoculated onto fresh PDA plates. Hyphal diameter was measured 24 h, 48 h and 72 h post inoculation.

2.6. Virulence Assay of Gene Deletion Mutants

Pathogenicity assays were conducted by placing a single 5-mm PDA disc from the edge of an actively growing, 2-day-old culture on the center of a freshly harvested canola leaf or a detached center leaflet (4 to 5 cm long) from the first trifoliate leaf of a soybean or sunflower seedling. At least 3 replicates of the leaves were incubated at 20 ± 1 °C in a growth chamber with a 12 h light-12 h dark photoperiod. Lesion size was calculated 24 h, 48 h and 72 h post inoculation by averaging two perpendicular lesion diameter measurements.

2.7. Transfection of Mutants with In Vitro Transcripts of SsHV2

In vitro transcripts of SsHV2-L were synthesized and transfected into wtDK3 and dicer mutant protoplasts following a published procedure [12]. After >6 transfers, viral infection was confirmed by extraction of total RNA using RNeasy Mini Kit (Qiagen, Hilden, Germany) followed by reverse transcription using Maxima H Minus Reverse Transcriptase (Thermofisher, Waltham, MA, USA) and PCR to amplify a 1.1 kb region corresponding to the viral genome. PCR amplicons were sequenced to confirm identity with the SsHV2-L genome.

2.8. Construction of An Infectious Clone of SsHADV-1 and Transfection of Mutants with SsHADV-1

The 2166 nt genome of SsHADV-1 was chemically synthesized by GeneArt (ThermoFisher Scientific, Waltham, MA, USA) in three segments with the ends flanked by overlapping unique restriction enzyme cutting sites, based on GenBank accession NC_013116.1. The 1-mer genome of SsHADV-1 was reconstructed by ligating three fragments containing restrictions sites SpeI, ApaI, and EagI internal to the viral genome. Using primers 33F and 3R, the viral genome was amplified and cloned into pJET1.2 as a single copy (1-mer) clone. A second copy of the genome was amplified by primers SV2F and 3R′-NotI (Table S1). Both the 1-mer clone and the second copy of the genome were digested with *Spe*I and *Not*I and ligated to form a tandem 1.9-mer clone which was then used for transfection (Figure S2A). Detailed procedure is described in the Supplementary Materials. There were no long concatemers formed because directional cloning with non-complementary sticky ends was performed. Fungal protoplasts (wtDK3) were transfected by PEG-mediated transformation. Infectivity was confirmed by inverse PCR to amplify a 2166 bp fragment (Figure S2B), which indicates that a recombined DNA template was formed. Fungal DNA was extracted and used as a template for rolling circle amplification (RCA) (Illustra Templiphi, GE Health, Little Chalfont, UK) using random primers. The product was then digested with a single cut restriction enzyme and this resulted in a 2166 bp fragment, indicating that no concatemers exist after transfection. The RCA product was also subjected to Sanger sequencing to confirm identity and infectivity. Additionally, after >6 serial transfers to fresh PDA plates, the presence of the replicating virus in fungal hyphae was confirmed once more by PCR amplification using SsHADV-1-specific primers and sequencing. Mutant cultures were infected with SsHADV-1 by extracellular transmission of virus particles from infected wtDK3 growth medium into fungal hyphae. Specifically, plugs were taken from the agar surrounding an SsHADV-1 infected culture of wtDK3 and placed adjacent to plugs taken from the edges of actively growing mutant cultures on fresh PDA plates with corresponding selective antibiotics.

2.9. Preparation of Small RNA Libraries and Sequencing Analysis

Small RNAs were extracted from 4-day-old mycelia using mirVana miRNA Isolation kit (ThermoFisher Scientific) following the manufacturer's protocol. Libraries were prepared using the NEBNext small RNA Library Kit (NEB, Ipswich, MA, USA). The libraries were pooled and sequenced in one lane for 50-nt single-end reads on an Illumina HiSeq4000 at Keck Center, University of Illinois. We sequenced two replicates each of virus-free wtDK3 and *Δdcl-1/dcl-2* as well as five replicates each of wtDK3 infected with SsHV2-L and three replicates of wtDK3 infected with SsHADV-1. Demultiplexed reads were removed of the 3′ adaptors by Trimmomatic [21]. Loci producing sRNAs were identified by ShortStack [22]. The obtained sequences have been deposited in NCBI (the accession will be available during review).

3. Results

3.1. Generation of Disruption Mutants for Dicer Genes

Dicer-like genes in *S. sclerotiorum* were disrupted using the homologous recombination method for gene displacement (Figure 1A) to generate *Δdcl-1*, *Δdcl-2* and *Δdcl-1/dcl-2* mutants directly from wild-type strain DK3 without using a *ΔKu80* strain. Dicer genes were confirmed to be disrupted by extracting DNA from multiple transformants and performing PCR amplification using F1 and F4 primers for initial screening. When the target locus was amplified, wild-type and mutant PCR amplicons differed in size confirming gene deletion (Figure 1B). PCR screening and Sanger sequencing of PCR amplicons confirmed integration of the gene-replacement cassettes into the target region and ruled out ectopic integration of the *hph* gene. Finally, nested PCR was used to rule out heterokaryotic mutation in which both the original dicer genes and disrupted genes occur in different nuclei within fungal hyphae (Table S1). This step was necessary because each transformed protoplast can contain multiple nuclei. Once a monokaryotic mutation was confirmed, further characterization of colony morphology and pathogenicity was carried out.

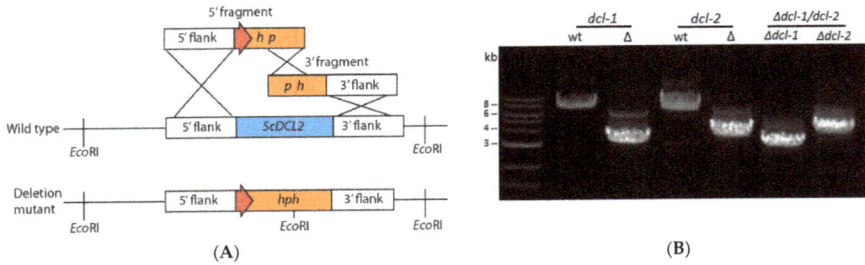

(A)   (B)

Figure 1. (**A**) Generation of deletion mutants for dicer genes in *S. sclerotiorum* using the split-marker gene replacement method (orange: selective marker, ex. *hph*; blue: gene replaced, ex. *dcl-2*; red: TrpC promoter) and (**B**) electrophoresis gel image of PCR amplification to confirm dicer gene disruption using F1–F4 primer pairs. Amplicons of wild-type *dcl-1* and *dcl-2* genes (7.7 kb and 7 kb, respectively) and deletion alleles (3.3 and 3.9 kb) differ in size. Lanes 5 and 6 show deletion alleles (3.1 and 3.9 kb) in the double dicer mutant.

3.2. Effect of Dicer Gene Disruption on S. sclerotiorum Phenotype

We compared the growth rate and colony morphology of dicer mutants to the wild-type strain, wtDK3. Single mutants-Δ*dcl-1* and Δ*dcl-2*- and wtDK3 exhibited similar growth rates, whereas the double Δ*dcl-1*/*dcl-2* disruption mutant exhibited significantly slower growth as indicated by measurements of hyphal diameter ($p < 0.05$) (Figure 2A). No significant difference in phenotype was observed in Δ*dcl-1* or Δ*dcl-2* compared to wtDK3, whereas Δ*dcl-1*/*dcl-2* mutant showed more hyphal branching and feathery colony morphology.

3.3. Effects of Dicer Gene Disruptions on S. sclerotiorum Pathogenicity

To test the pathogenicity of *S. sclerotiorum* dicer mutants, plugs taken from actively growing cultures were used to inoculate detached leaves. Lesion size data collected 24, 48 and 72 h post inoculation showed that there was no difference in the sizes of lesions produced on canola leaves by the single mutants, Δ*dcl-1* or Δ*dcl-2*, compared to wtDK3. However, significantly smaller lesions were produced by the Δ*dcl-1*/*dcl-2* double mutant compared to those produced by wtDK3 ($p < 0.05$) (Figure 2B).

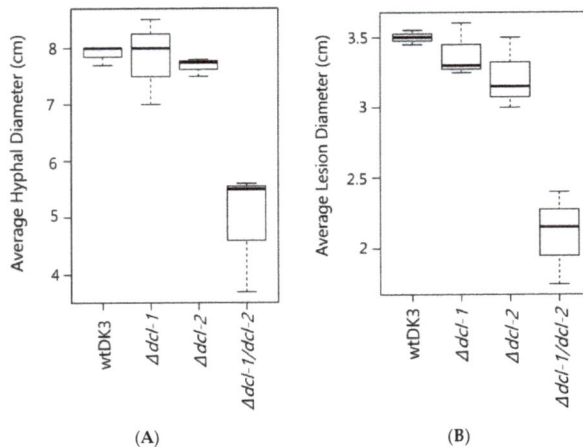

(A)   (B)

Figure 2. (**A**) Average mycelial growth of wild-type *S. sclerotiorum* and dicer gene disruption mutants grown on PDA for four days; and (**B**) lesion diameter measurements 72 hpi comparing wtDK3, Δ*dcl-1*, Δ*dcl-2* and Δ*dcl-1*/*dcl-2* virus-free cultures inoculated on canola leaves.

3.4. Transfection of Dicer Gene Deletion Mutants with SsHV2-L or SsHADV-1 Viruses Consistently Results in Severe Debilitation in the Δdcl-1/dcl-2 Mutant

To examine the effect of viral infection on strains containing deletions of *dcl-1*, *dcl-2* or both genes, mutants were transfected with SsHV2-L or SsHADV-1 via the methods described in the Materials and Methods section. As shown in Figure 3A, the *Δdcl-1* and *Δdcl-2* mutants infected with either mycovirus showed no significant difference in growth or morphology compared to virus-infected wtDK3. In sharp contrast, the *Δdcl-1/dcl-2* mutant showed severe debilitation following virus infection as evidenced by significantly slower growth and hypovirulence on three different crop species (Figure 3B–D). Complementation of *dcl-1* in the double dicer mutant, named as Comp-dcl-1, resulted in growth and phenotype similar to the wild-type strain prior to and following virus infection.

3.5. Infectious Clone of SsHADV-1 Causes Severe Debilitation and Significantly Reduced Virulence in wtDK3 at Lower Temperatures

The SsHADV-1-transformed and serial transferred fungal DNA was extracted to determine the infectivity of the infectious clone. Rolling circle amplification followed by *Xba*I digestion resulted in a 2.2 kb band. Inverse PCR amplification of the viral genome and sequencing confirmed that the synthetic virus is identical to the Chinese strain. Initial viral infection at room temperature (~24 °C) resulted in fairly asymptomatic infection in wtDK3; however, we found that incubation at lower temperatures (~20 °C) in a growth chamber resulted in severe debilitation including significantly slower growth and little to no virulence on inoculated leaves incubated at the same temperature (Figure 3B–D).

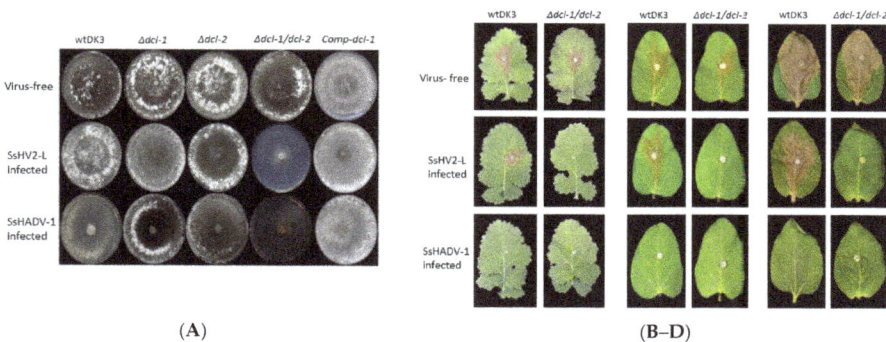

(A)   (B–D)

Figure 3. (**A**) Colony morphology of virus-free and virus-infected wild-type, mutant, and complemented strains: (Top row) virus-free wtDK3, *Δdcl-1*, *Δdcl-2*, *Δdcl-1/dcl-2* and *Comp-dcl-1*. (Middle row) strains infected with hypovirus, SsHV2-L; and (Bottom row) strains infected with SsHADV-1. Cultures were grown for seven days on PDA at room temperature. Virulence assays on: (**B**) detached canola leaves; (**C**) detached soybean leaves; and (**D**) detached sunflower leaves. Plugs were taken from the edge of actively growing wtDK3, *Δdcl-1* (not shown), *Δdcl-2* (not shown) and *Δdcl-1/dcl-2* cultures and inoculated onto detached leaves stored at 20 ± 1 °C. Lesion size was measured 36 h post-inoculation.

3.6. Double Dicer Disruption Mutant Has Reduced 21–24 nt sRNA Accumulation

To examine whether sRNA accumulation is affected by disrupting both dicers, sRNA sequences were profiled by size distribution and 5′ terminal nucleotide in the virus-free *Δdcl-1/dcl-2* mutant and wild-type strain. Although the 5′ terminal nucleotide remained uracil-biased, the size distribution of small RNAs was drastically changed in the double dicer mutant compared to the wild-type strain (Figure 4A,B). Specifically, there was a reduction in the 21–24-nt sRNA fraction in the double mutant compared to the wild-type strain. Notably and similar to *B. cinerea*, sRNA production in *S. sclerotiorum* is not completely eliminated after both dicers are deleted.

3.7. SsHADV-1 and SsHV2-L Are both Processed by Virus-Infected wtDK3

Sequence analysis of the small RNAs produced by either SsHADV-1 or SsHV2-L infected wtDK3 revealed the presence of virus-derived sRNAs (vsiRNAs) within the pool of total small RNAs extracted from these cultures. On average, 14.4% of the total small RNA reads from the SsHV2-L-infected wild-type strain were derived from SsHV2-L, whereas 2.26% of the total small RNA reads from the SsHADV-1 infected wild-type strain were derived from SsHADV-1. Three replicates of each virus infected wild-type strain were analyzed. Surprisingly, only one of the three libraries from SsHADV-1-infected wtDK3 had vsiRNAs. For each barcoded library, 5–10 million reads were obtained and passed QC. The 22-nt sRNAs were the most abundant for both virus-infected wild-type strains (Figure 4C,D) with a preference (>90%) for uracil at the 5′ position. Overall, 77.89% of SsHV2-L derived sRNA aligned to the negative strand, and 22.01% to the positive strand (Figure 4E). Virus-derived small RNAs from all five replicates of SsHV2-L-infected wtDK3 displayed the same even distribution along the viral genome. SsHADV-1 derived sRNA reads aligned non-uniformly to both strands (Figure 4E) with strand biases for the negative strand in the first 350 bases of the coat protein encoding gene and strand biases for the positive strand between nucleotide bases 1000–2200 of the replicase protein encoding gene; overall, 51.6% of the reads aligned to the published positive strand sequence and 48.3% to the negative strand. We found that a significant number of virus-derived sRNAs contained 1-nt terminal mismatches. The majority of SsHADV-1 vsiRNAs contained an A or T at the mismatched 3′-terminus and mismatched A nucleotide at the 5′-terminus. SsHV2-L vsiRNAs contained mismatches primarily at the 3′-terminus involving A and T. Mismatches involving G or C were also found but to a much lower extent (Table 1). SsHV2-L vsiRNAs were also found to contain a high number of internal mismatches at specific positions (Figure S3). For example, the 22-nt long sRNAs have an internal peak of mismatches at the 11th nucleotide.

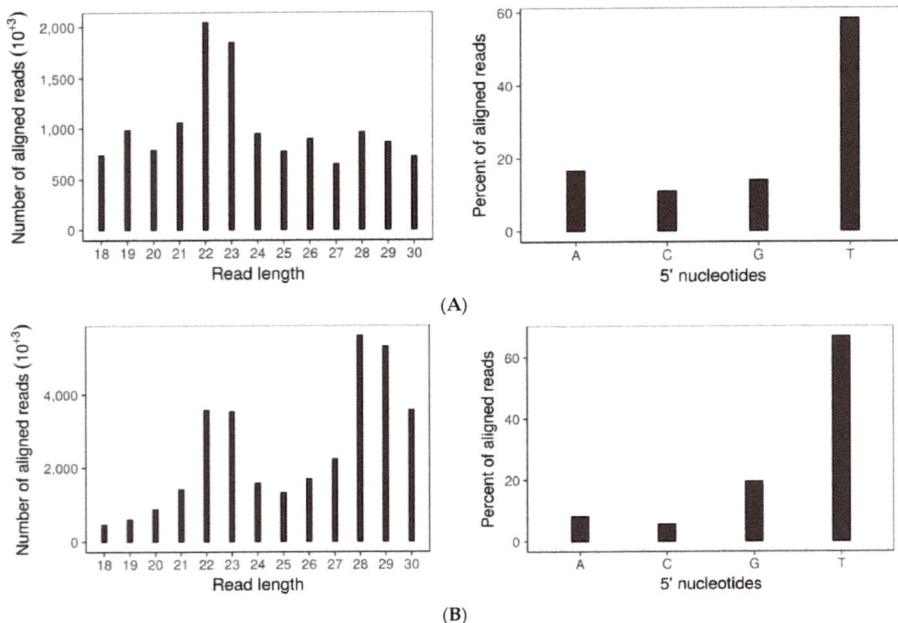

(A)

(B)

Figure 4. *Cont.*

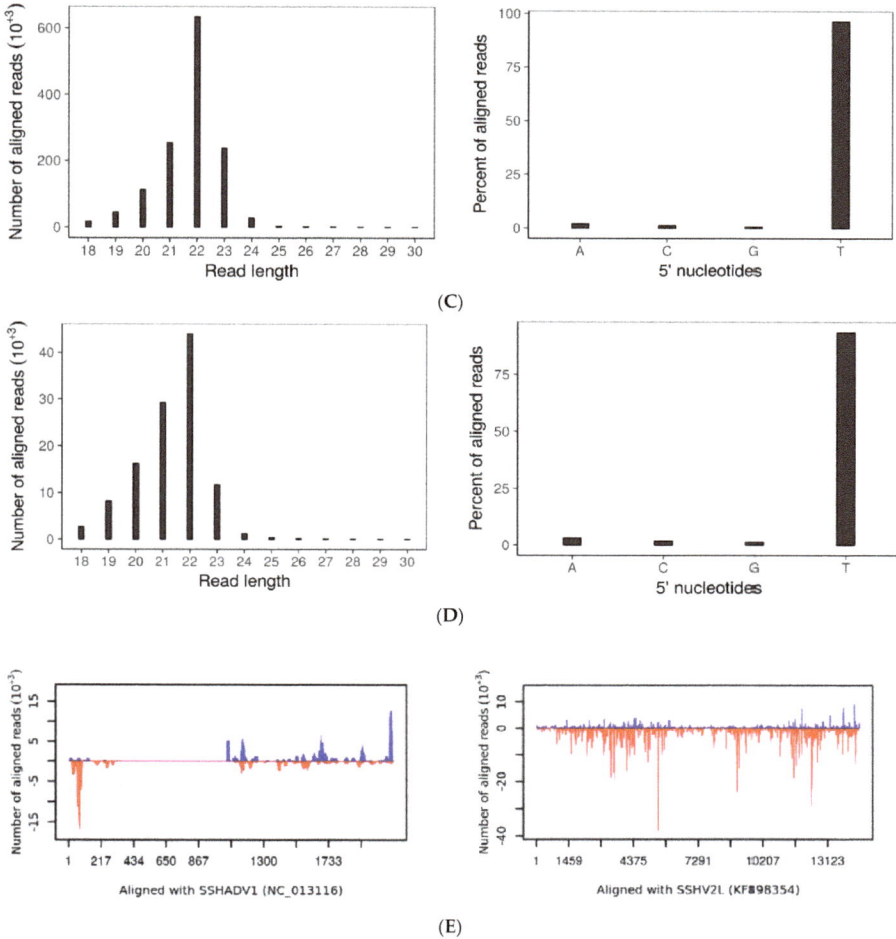

Figure 4. Small RNA: (**A**) Size distribution (left) and frequency of 5′ terminal nucleotides (right) of small RNAs in wtDK3; (**B**) size distribution (left) and frequency of 5′ terminal nucleotides (right) of small RNAs in *Δdcl-1/dcl-2* disruption mutant; (**C**) size distribution (left) and frequency of 5′ terminal nucleotides (right) of small RNAs aligned to SsHV2-L genome; (**D**) size distribution (left) and frequency of 5′ terminal nucleotides (right) of small RNAs aligned to SsHADV-1 genome; and (**E**) distribution of small RNA reads that aligned to the SsHADV-1 genome plus or minus strands (left) and distribution of small RNA reads that aligned to the SsHV2-L genome plus or minus strands (right). Bars above zero indicate alignment to the positive strand, and bars below zero indicate alignment to the negative strand.

Table 1. Percentage of SsHV2-L and SsHADV-1 derived small RNAs containing mismatches relative to viral genomes.

SsHADV-1	5′-terminal mismatch (%)				3′-terminal mismatch (%)			
vsiRNA Sequence length	A	C	G	T	A	C	G	T
18	16.9	1.9	1.1	0.8	18.2	5.0	3.0	14.6
19	4.2	1.1	1.0	2.8	21.0	7.3	3.9	19.6
20	10.3	0.8	0.8	1.1	24.5	4.8	3.2	22.4
21	5.0	0.6	0.8	1.6	27.9	3.2	4.7	22.4
22	26.8	0.8	0.8	1.0	20.2	3.0	2.7	12.3
23	46.1	0.6	0.9	0.7	12.5	2.5	1.5	9.1
24	5.9	1.7	2.0	0.6	28.0	3.4	2.0	24.4
SsHV2-L	**5′-terminal mismatch (%)**				**3′-terminal mismatch (%)**			
vsiRNA Sequence length	A	C	G	T	A	C	G	T
18	1.1	0.4	1.4	0.5	16.6	3.0	6.6	23.1
19	1.2	0.6	1.0	0.3	18.5	3.0	6.1	26.9
20	0.6	0.5	1.1	0.3	21.1	2.6	5.0	26.9
21	0.7	0.5	1.0	0.4	17.1	2.6	5.1	20.6
22	2.3	0.4	1.0	0.4	11.7	3.2	4.5	17.1
23	0.8	0.7	1.7	0.3	14.1	2.5	4.6	19.0
24	0.2	1.3	2.8	0.5	19.6	1.9	5.2	22.0

4. Discussion

We previously compared the three hypovirus strains of SsHV2 and detected inter- and intra-specific recombination near the 5′ end of the genome where a putative virus silencing suppressor is predicted to be located, suggesting the existence of RNA silencing in the host fungus (9). Our study demonstrated that a robust RNA silencing mechanism does exist in the plant pathogenic fungus, *Sclerotinia sclerotiorum*, and established the vital role played by dicer genes in this regulatory pathway. RNA silencing mechanisms in fungi have been said to function primarily in defense against viral nucleic acids, and our results provide additional support for this theory by demonstrating the antiviral function of *S. sclerotiorum* RNA silencing pathways, although we cannot rule out the possibility of other functions as well. Wild-type strains of *S. sclerotiorum* displayed fairly normal phenotype and virulence following virus infection, while RNA-silencing-deficient mutants (specifically a double dicer mutant) displayed significantly slower growth and decreased virulence upon virus infection. Complementation of a single dicer gene in the double dicer mutant reverted viral susceptibility to the wild-type state.

Additionally, our study demonstrated that a ss(+)RNA virus and, notably, a ssDNA virus are not only the triggers but also the targets of RNA silencing in *S. sclerotiorum* based on the production of virus-derived small RNAs (vsiRNAs) in virus-infected wtDK3. Small RNAs are known to influence various cellular functions by altering gene expression at the transcriptional and post-transcriptional level. For this reason, it may be informative to study the impact the accumulation of mycovirus-derived small RNAs may have on *S. sclerotiorum* gene expression since vsiRNAs can encompass a sizeable proportion of total small RNA accumulation in virus-infected strains. In our study, for example, up to 14% of the total small RNA accumulation in SsHV2-L-infected wtDK3 were vsiRNAs. A small number of studies have shown that vsiRNAs may be able to silence certain plant host genes that share an amount of complementarity to them (reviewed in [23]). Furthermore, studies have shown that strand bias during the formation of siRNAs plays a role in determining the functionality of siRNAs [24]. This suggests that vsiRNAs, which displayed strong strand biases in our study, may become incorporated into RISCs in a manner similar to functional endogenous siRNAs and act to silence host fungal genes. Studies involving the pulldown of fungal Argonaute proteins and sequencing of bound small RNAs could be used to test this hypothesis. Finally, additional related studies may involve the introduction of mutations into regions of mycovirus genomes where purported viral suppressors of RNA silencing (VSRs) exist or introducing VSRs into viral genomes that lack them and examining the changes in virus infectivity and in the production of vsiRNAs that occur. Such a study is relevant because it has been shown that viruses that possess VSRs can circumvent RNA silencing processes by various means and limit the production of vsiRNAs [25]. These findings along with others such as ours will provide valuable information on virus–host counter-defense mechanisms.

It is unlikely that the high percentage of virus derived sRNAs that contained terminal mismatches is due to chance or the introduction of errors during the amplification of small RNAs. This is because an obvious pattern of mismatches involving primarily A or T nucleotides at the 5′ and 3′ termini is evident. This suggests that non-random modifications of vsiRNAs may have occurred. A similar pattern of terminal mismatches was also discovered in vsiRNA present in virus-infected *C. parasitica* [26]. One possibility is that mismatches are generated during the production of secondary siRNAs. This would indicate that a significant portion of SsHV2-L and SsHADV-1 derived siRNAs are associated with secondary silencing. The abundance of 22 nt long vsiRNAs found in our study may further support this hypothesis since in plants 22 nt long miRNAs are associated with secondary siRNA production [27].

Only one of the three small RNA libraries from SsHADV-1-infected wtDK3 cultures had a well accumulated small RNA profile. Acute/initial SsHADV-1-infection is marked by severe debilitation, with sectoring growth and an absence of virus-derived small RNA production. This is followed by a strong host immune reaction resulting in the silencing of viral nucleic acids and the remission of acute symptoms. Virus-derived small RNAs become detectable in this latter stage. The two samples with no virus-derived small RNAs detected were possibly obtained from debilitated, sectoring hyphae that had not progressed to symptom remission and hence no vsiRNAs were detectable. A plant geminivirus, Pepper gold mosaic virus, is also associated with a recovery phenotype in plant hosts accompanied by the presence of virus-derived small RNAs [28]. Further quantification of viral titer sector-by-sector to confirm a reduction of the viral DNA titers during the recovery process and a corresponding accumulation of SsHADV-1 derived small RNAs are needed. In the small RNA data from SsHADV-1 infected tissue, hotspots were observed in virus-derived small RNAs from this virus with a 700 bp gap similar to the small RNAs profiled for tomato yellow leaf yellow curl china virus [29]. Opposite strand biases were also observed between the two clusters, possibly because the direction of transcription for the two genes is opposite.

Besides establishing a role for *S. sclerotiorum* dicer genes in antiviral mechanisms, our study also demonstrated that *S. sclerotiorum* dicers contribute to endogenous gene regulation likely through the action of small RNAs generated by these genes. The important roles played by dicer-generated small RNAs are well documented (reviewed in [30]). We found that the deletion of both dicer genes resulted in compromised growth and virulence in the double mutant prior to virus infection. Similar changes were observed in another member of *Sclerotiniaceae*, *Botrytis cinerea* [13], where slower growth and reduced pathogenicity were observed when both dicer genes were disrupted. As in *B. cinerea*, the changes observed in the *S. sclerotiorum* double mutant may be attributable to a significant reduction in small RNA effectors produced by the mutant. Small RNA-seq analysis revealed a reduction in small RNAs 22nt long in the virus-free double dicer mutant compared to the wild-type strain. Notably, production of small RNAs is not completely eliminated upon deletion of both dicer genes (again similar to *B. cinerea* [13]), and this indicates that there may be other dicer-independent pathways that contribute to the generation of sRNAs. By conserved domain search, we found a putative RNaseL gene (GenBank Ss1G_04823), also an RNA-endonuclease-III, which may be responsible for the remaining small RNA processing. RNaseL endonucleases share similarities with yeast Ire1p proteins which are said to be involved in fungal mRNA splicing [31].

The high level of debilitation observed in the double dicer mutant following virus infection was not observed in the virus-infected single dicer mutants. Furthermore, complementation of a single dicer gene was sufficient to restore viral susceptibility to the wild-type state. These findings imply that there is redundancy in the antiviral function of *S. sclerotiorum* dicer genes. Redundancy in dicer antiviral function has not been reported in fungal species; however, a redundancy in dicer function in transgene-induced gene silencing has been found in *Neurospora crassa* [6]. Dicer redundancy in antiviral RNA silencing mechanisms in *S. sclerotiorum* could be validated by small RNA sequence analysis of virus-infected single dicer knockout mutants to demonstrate that the small RNA accumulations (particularly vsiRNAs) are identical to the wild-type strain due to the presence of an intact dicer gene (*dcl-1* or *dcl-2*) in each mutant. This further investigation is outside the scope of this study, however. Once validated, dicer redundancy would then appear to have evolved in a specific lineage of ascomycetes as a conserved anti-invasive nucleic acid mechanism because it is not the case for *Cryphonectria parasitica* [5] or *Colletotrichum higginsianum* [4].

Mycoviruses belonging to the families *Hypoviridae* and *Genomoviridae* are widespread. *S. sclerotiorum* is the host of the sole representative of *Genomoviridae*, SsHADV-1. This viral family is considered part of an emerging group of infectious agents [32] due to its association with other eukaryotes such as vertebrates and invertebrates in addition to fungi. Furthermore, circular ssDNA viruses have polyphyletic origin [33] and SsHADV-1 has been reported to replicate in distant hosts [34]. The unique properties of this virus warrant further studies into its interaction with its host and other organisms. We have demonstrated in our study that SsHADV-1 can be the trigger and target of RNA silencing pathways; however, more studies are needed to help us understand how and when the RNA silencing pathway, which is traditionally triggered by dsRNA molecules, is triggered by DNA viruses. Thus far, one hypothesis that has been put forth for dsDNA viruses is that overlaps in viral transcripts resulting from overlapping or adjacent genes or secondary structures in viral RNA transcripts may serve as the initiators of the RNA silencing response against these viruses [35]. It is unclear how dsRNAs that result in primary siRNA are made in the case of ssDNA viruses but secondary siRNAs are speculated to be made from host-encoded RNA-dependent RNA polymerases and these comprise the majority of siRNAs found in a plant geminivirus (reviewed in [36]).

Overall, the results derived from this study will have broad relevance to efforts to understand the complex interactions between viruses and host RNA silencing pathways. The literature has illustrated the role antiviral RNA silencing mechanisms play in mammalian cells as superimposed on the type I-interferon pathway [37,38]. These interactions may also have implications on developing innovative techniques that utilize viruses for in vivo targeted gene silencing. Using synthetic viruses, disarmed viral nanoparticles have shown high efficiencies in intracellular delivery of gene-targeted therapy using adeno-based vectors in mammals. However, strategies to cross the blood–brain barrier still await to be improved [39]. Although still understudied and rare, single stranded eukaryotic DNA viruses can naturally invade the central nervous system and cause diseases in humans [40,41]. The availability of the SsHADV-1 infectious clone will provide a unique opportunity to understand how animal hosts recognize and defend against foreign single stranded circular DNA.

Supplementary Materials: The following are available online at http://www.mdpi.com/1999-4915/10/4/214/s1, Table S1: Primers used in this study. Table S2: Primers and expected sizes of nested PCR reactions performed to confirm monokaryotic deletion of dicer genes with the corresponding gel image in Figure S1. Figure S1: Electrophoresis gel image to confirm monokaryotic deletion of dicer genes by nested PCR. Lanes as indicated in Table S1. Lane 1–12 from right to left on the upper half of the agarose gel image. Lane 13–18 from right to left on the lower half of the agarose gel image. Ladders are shown on the farthest right and left lanes with corresponding size labelled. Figure S2: Inverse PCR product amplified using SsHADV-1-specific primers to confirm the infectivity of the infectious clone from a >6 times transferred culture. A 2.2 kb band was amplified and confirmed by Sanger sequencing. The product shows the dimer clone has recombined circular template of SsHADV-1 genome and demonstrates the infectivity of the infectious clone assembled from synthetic DNA. Left Lane: 1 kb ladder. Right Lane: 2.2 kb amplicon. Figure S3: Frequency and distribution of mismatches occurring in SsHADV-1 and SsHV2-L-derived sRNAs. A majority of mismatches occur at the 5′ and 3′ termini; however, a significant number of internal mismatches occur at non-terminal positions in SsHV2-L-derived vsiRNAs.

Acknowledgments: This study was supported in part by National Sclerotinia Initiative Grant SA1800330 (to Shin-Yi Lee Marzano) and SDSU startup from USDA Hatch fundSD00H606-16 project number with Accession Number 1009451. The authors also thank Leslie L. Domier of USDA/ARS for initial edits on the manuscript and providing Perl codes for small RNA analyses.

Author Contributions: Pauline Mochama and Shin-Yi Lee Marzano conceived and designed the experiments; Pauline Mochama and Prajakta Jadhav performed the experiments; Achal Neupane analyzed the small RNA data; and Pauline Mochama and Shin-Yi Lee Marzano wrote the paper.

Conflicts of Interest: The authors declare no conflict of interest.

References

1. Baulcombe, D. RNA silencing. *Trends Biochem. Sci.* **2005**, *30*, 290–293. [CrossRef] [PubMed]
2. Waterhouse, P.M.; Wang, M.B.; Lough, T. Gene silencing as an adaptive defence against viruses. *Nature* **2001**, *411*, 834–842. [CrossRef] [PubMed]
3. Baulcombe, D. RNA silencing in plants. *Nature* **2004**, *431*, 356–363. [CrossRef] [PubMed]
4. Campo, S.; Gilbert, K.B.; Carrington, J.C. Small RNA-based antiviral defense in the phytopathogenic fungus Colletotrichum higginsianum. *PLoS Pathog.* **2016**, *12*, e1005640. [CrossRef] [PubMed]
5. Segers, G.C.; Zhang, X.; Deng, F.; Sun, Q.; Nuss, D.L. Evidence that RNA silencing functions as an antiviral defense mechanism in fungi. *Proc. Natl. Acad. Sci. USA* **2007**, *104*, 12902–12906. [CrossRef] [PubMed]
6. Catalanotto, C.; Pallotta, M.; ReFalo, P.; Sachs, M.S.; Vayssie, L.; Macino, G.; Cogoni, C. Redundancy of the two dicer genes in transgene-induced posttranscriptional gene silencing in Neurospora crassa. *Mol. Cell. Biol.* **2004**, *24*, 2536–2545. [CrossRef] [PubMed]
7. Becker, B.; Schmitt, M.J. Yeast Killer Toxin K28: Biology and Unique Strategy of Host Cell Intoxication and Killing. *Toxins* **2017**, *9*, 333. [CrossRef] [PubMed]
8. Laurie, J.D.; Ali, S.; Linning, R.; Mannhaupt, G.; Wong, P.; Güldener, U.; Münsterkötter, M.; Moore, R.; Kahmann, R.; Bakkeren, G.; et al. Genome comparison of barley and maize smut fungi reveals targeted loss of RNA silencing components and species-specific presence of transposable elements. *Plant Cell* **2012**, *24*, 1733–1745. [CrossRef] [PubMed]
9. Filipowicz, W. RNAi: The nuts and bolts of the RISC machine. *Cell* **2005**, *122*, 17–20. [CrossRef] [PubMed]
10. Amselem, J.; Cuomo, C.A.; Van Kan, J.A.; Viaud, M.; Benito, E.P.; Couloux, A.; Coutinho, P.M.; De Vries, R.P.; Dyer, P.S.; et al. Genomic analysis of the necrotrophic fungal pathogens Sclerotinia sclerotiorum and Botrytis cinerea. *PLoS Genet.* **2011**, *7*, e1002230. [CrossRef] [PubMed]
11. Krupovic, M.; Ghabrial, S.A.; Jiang, D.; Varsani, A. Genomoviridae: A new family of widespread single-stranded DNA viruses. *Arch. Virol.* **2016**, *161*, 2633–2643. [CrossRef] [PubMed]
12. Marzano, S.Y.L.; Hobbs, H.A.; Nelson, B.D.; Hartman, G.L.; Eastburn, D.M.; McCoppin, N.K.; Domier, L.L. Transfection of Sclerotinia sclerotiorum with in vitro transcripts of a naturally occurring interspecific recombinant of Sclerotinia sclerotiorum hypovirus 2 significantly reduces virulence of the fungus. *J. Virol.* **2015**, *89*, 5060–5071. [CrossRef] [PubMed]
13. Wang, M.; Weiberg, A.; Lin, F.M.; Thomma, B.P.; Huang, H.D.; Jin, H. Bidirectional cross-kingdom RNAi and fungal uptake of external RNAs confer plant protection. *Nat. Plants* **2016**, *2*, 16151. [CrossRef] [PubMed]
14. Weiberg, A.; Wang, M.; Lin, F.M.; Zhao, H.; Zhang, Z.; Kaloshian, I.; Huang, H.D.; Jin, H. Fungal small RNAs suppress plant immunity by hijacking host RNA interference pathways. *Science* **2013**, *342*, 118–123. [CrossRef] [PubMed]
15. Chang, H.X.; Domier, L.L.; Radwan, O.; Yendrek, C.R.; Hudson, M.E.; Hartman, G.L. Identification of multiple phytotoxins produced by Fusarium virguliforme including a phytotoxic effector (FvNIS1) associated with sudden death syndrome foliar symptoms. *Mol. Plant-Microbe Interact.* **2016**, *29*, 96–108. [CrossRef] [PubMed]
16. Staben, C.; Jensen, B.; Singer, M.; Pollock, J.; Schechtman, M.; Kinsey, J.; Selker, E. Use of a bacterial hygromycin B resistance gene as a dominant selectable marker in Neurospora crassa transformation. *Fungal Genet. Rep.* **1989**, *36*, 79. [CrossRef]
17. Nelson, M.D.; Fitch, D.H. Overlap extension PCR: An efficient method for transgene construction. In *Molecular Methods for Evolutionary Genetics*; Springer: Berlin, Germany, 2012; pp. 459–470.
18. Ge, C.Y.; Duan, Y.B.; Zhou, M.G.; Chen, C.J. A Protoplast Transformation System for Gene Deletion and Complementation in Sclerotinia sclerotiorum. *J. Phytopathol.* **2013**, *161*, 800–806. [CrossRef]

19. Rollins, J.A. Sclerotinia sclerotiorum pac1 Gene Is Required for Sclerotial Development and Virulence. *Mol. Plant-Microbe Interact.* **2003**, *16*, 785–795. [CrossRef] [PubMed]

20. Chung, K.R.; Lee, M.H. Split-Marker-Mediated Transformation and Targeted Gene Disruption in Filamentous Fungi. In *Genetic Transformation Systems in Fungi*; Springer: Berlin, Germany, 2015; Volume 2, pp. 175–180.

21. Bolger, A.M.; Lohse, M.; Usadel, B. Trimmomatic: A flexible trimmer for Illumina sequence data. *Bioinformatics* **2014**, *30*, 2114–2120. [CrossRef] [PubMed]

22. Axtell, M.J. ShortStack: Comprehensive annotation and quantification of small RNA genes. *RNA* **2013**, *19*, 740–751. [CrossRef] [PubMed]

23. Zhang, C.; Wu, Z.; Li, Y.; Wu, J. Biogenesis, function, and applications of virus-derived small RNAs in plants. *Front. Microbiol.* **2015**, *6*, 1237. [CrossRef] [PubMed]

24. Khvorova, A.; Reynolds, A.; Jayasena, S.D. Functional siRNAs and miRNAs exhibit strand bias. *Cell* **2003**, *115*, 209–216. [CrossRef]

25. Csorba, T.; Kontra, L.; Burgyán, J. Viral silencing suppressors: Tools forged to fine-tune host-pathogen coexistence. *Virology* **2015**, *479*, 85–103. [CrossRef] [PubMed]

26. Zhang, X.; Segers, G.C.; Sun, Q.; Deng, F.; Nuss, D.L. Characterization of hypovirus-derived small RNAs generated in the chestnut blight fungus by an inducible DCL-2-dependent pathway. *J. Virol.* **2008**, *82*, 2613–2619. [CrossRef] [PubMed]

27. Shahid, S.; Kim, G.; Johnson, N.R.; Wafula, E.; Wang, F.; Coruh, C.; Bernal-Galeano, V.; Phifer, T.; Westwood, J.H.; Axtell, M.J. MicroRNAs from the parasitic plant Cuscuta campestris target host messenger RNAs. *Nature* **2018**, *553*, 82. [CrossRef] [PubMed]

28. Carrillo-Tripp, J.; Lozoya-Gloria, E.; Rivera-Bustamante, R.F. Symptom remission and specific resistance of pepper plants after infection by *Pepper golden mosaic virus*. *Phytopathology* **2007**, *97*, 51–59. [CrossRef] [PubMed]

29. Yang, X.; Wang, Y.; Guo, W.; Xie, Y.; Xie, Q.; Fan, L.; Zhou, X. Characterization of small interfering RNAs derived from the geminivirus/betasatellite complex using deep sequencing. *PLoS ONE* **2011**, *6*, e16928. [CrossRef] [PubMed]

30. Kurzynska-Kokorniak, A.; Koralewska, N.; Pokornowska, M.; Urbanowicz, A.; Tworak, A.; Mickiewicz, A.; Figlerowicz, M. The many faces of Dicer: The complexity of the mechanisms regulating Dicer gene expression and enzyme activities. *Nucleic Acids Res.* **2015**, *43*, 4365–4380. [CrossRef] [PubMed]

31. Dong, B.; Niwa, M.; Walter, P.; Silverman, R.H. Basis for regulated RNA cleavage by functional analysis of RNase L and Ire1p. *RNA* **2001**, *7*, 361–373. [CrossRef] [PubMed]

32. Halary, S.; Duraisamy, R.; Fancello, L.; Monteil-Bouchard, S.; Jardot, P.; Biagini, P.; Gouriet, F.; Raoult, D.; Desnues, C. Novel single-stranded DNA circular viruses in pericardial fluid of patient with recurrent pericarditis. *Emerg. Infect. Dis.* **2016**, *22*, 1839. [CrossRef] [PubMed]

33. Krupovic, M. Networks of evolutionary interactions underlying the polyphyletic origin of ssDNA viruses. *Curr. Opin. Virol.* **2013**, *3*, 578–586. [CrossRef] [PubMed]

34. Liu, S.; Xie, J.; Cheng, J.; Li, B.; Chen, T.; Fu, Y.; Li, G.; Wang, M.; Jin, H.; Wan, H.; et al. Fungal DNA virus infects a mycophagous insect and utilizes it as a transmission vector. *Proc. Natl. Acad. Sci. USA* **2016**, *113*, 12803–12808. [CrossRef] [PubMed]

35. Jayachandran, B.; Hussain, M.; Asgari, S. RNA interference as a cellular defense mechanism against the DNA virus baculovirus. *J. Virol.* **2012**, *86*, 13729–13734. [CrossRef] [PubMed]

36. Hanley-Bowdoin, L.; Bejarano, E.R.; Robertson, D.; Mansoor, S. Geminiviruses: Masters at redirecting and reprogramming plant processes. *Nat. Rev. Microbiol.* **2013**, *11*, 777. [CrossRef] [PubMed]

37. Li, Y.; Lu, J.; Han, Y.; Fan, X.; Ding, S.W. RNA interference functions as an antiviral immunity mechanism in mammals. *Science* **2013**, *342*, 231–234. [CrossRef] [PubMed]

38. Qiu, Y.; Xu, Y.; Zhang, Y.; Zhou, H.; Deng, Y.Q.; Li, X.F.; Miao, M.; Zhang, Q.; Zhong, B.; Hu, Y. Human virus-derived small RNAs can confer antiviral immunity in mammals. *Immunity* **2017**, *46*, 992–1004. [CrossRef] [PubMed]

39. Hocquemiller, M.; Giersch, L.; Audrain, M.; Parker, S.; Cartier, N. Adeno-associated virus-based gene therapy for CNS diseases. *Hum. Gene Ther.* **2016**, *27*, 478–496. [CrossRef] [PubMed]

40. Zhou, C.; Zhang, S.; Gong, Q.; Hao, A. A novel gemycircularvirus in an unexplained case of child encephalitis. *Virol. J.* **2015**, *12*, 197. [CrossRef] [PubMed]

41. Phan, T.G.; Messacar, K.; Dominguez, S.R.; Da Costa, A.C.; Deng, X.; Delwart, E. A new densovirus in cerebrospinal fluid from a case of anti-NMDA-receptor encephalitis. *Arch. Virol.* **2016**, *161*, 3231–3235. [CrossRef] [PubMed]

Review

Description, Distribution, and Relevance of Viruses of the Forest Pathogen *Gremmeniella abietina*

Leticia Botella [1,*] **and Jarkko Hantula** [2]

[1] Phytophthora Research Centre, Department of Forest Protection and Wildlife Management, Faculty of Forestry and Wood Technology, Mendel University in Brno, Zemědělská 1, 613 00 Brno, Czech Republic
[2] Forest Health and Biodiversity, Natural Resources Institute Finland (Luke), Latokartanonkaari 9, 00790 Helsinki, Finland; jarkko.hantula@luke.fi
* Correspondence: qqbotell@mendelu.cz; Tel.: +420-730-96-1992

Received: 30 October 2018; Accepted: 16 November 2018; Published: 20 November 2018

Abstract: The European race of the ascomycetous species *Gremmeniella abietina* (Lagerberg) Morelet includes causal agents of shoot blight and stem canker of several conifers in Europe and North America, which are known to host a diverse virome. GaRV6 is the latest and sixth mycovirus species reported within *G. abietina*. Before its description, one victorivirus and one gammapartitivirus species were described in biotype A, two mitoviruses in both biotypes A and B and a betaendornavirus in biotype B. Possible phenotypic changes produced by mycoviruses on *G. abietina* mycelial growth have been reported in Spanish mitovirus-free and GaRV6-hosting *G. abietina* isolates, which had higher growth rates at the optimal temperature of 15 °C, but no other major differences have been observed between partitivirus-like dsRNA and dsRNA-free isolates. In this review, we reappraise the diversity of viruses found in *G. abietina* so far, and their relevance in clarifying the taxonomy of *G. abietina*. We also provide evidence for the presence of two new viruses belonging to the families *Fusariviridae* and *Endornaviridae* in Spanish isolates.

Keywords: *Brunchorstia pinea*; conifers; mycovirus; dsRNA; ssRNA; phylogeny; evolution

---

1. Taxonomy of *G. abietina* and Relevance in Forestry

Gremmeniella abietina is a virulent haploid ascomycete responsible for shoot dieback and Scleroderris canker on conifers including spruces, firs, larches, pines, and junipers in North, Central, and South Europe, northeastern North America, and East Asia [1–6].

The taxonomy of *G. abietina* and its relation to forestry is a complex issue, since the taxon is divided into a number of varieties, races, and biotypes. These include two varieties: *G. abietina* var. *abietina* that mainly affects pines, and *G. abietina* var. *balsamea* that attacks firs and spruces [7]. Within *G. abietina* var. *abietina*, three races—Asian, North American, and European—were described based on serological analyses [8]. It has also been proposed that these races would be considered as separate species [9,10]. In the European race, three biotypes have been identified on the basis of symptoms, septa numbers, spore length, and molecular markers; namely, the alpine biotype, biotype A (LTT, large tree type), and biotype B (STT, small tree type) [11–14]. Furthermore, there is a distinctive population of *G. abietina* in Spain that may stem from biotype A of the European race [15,16].

In a taxonomic context, obligate parasites without an extracellular phase, such as mycoviruses, may be considered especially informative because they can only spread through mycelial contacts [17]. Therefore, determining the presence and evolutionary history of mycoviruses in different populations of *G. abietina* will shed light on the origin and pathways of its spread. In the literature, there are a number of examples that support the use of mycoviruses as tracers of the origin and pathways of different plant pathogens, i.e., *Hymenoscyphus fraxineus* [18], *Cryphonectria parasitica* [19] and *Heterobasidion annosum* [20].

The study of mycoviruses may also clarify the level of fungal compatibility between species. There is growing evidence suggesting differences in the fungal incompatibility rate [21], and a number of cases of mycovirus co-specificity in phylogenetically separated fungi have been reported, not only in the laboratory, but also in nature [20,22–29].

2. Occurrence of Viruses in *G. abietina*: Description of Their Genome and Structure, and Phylogenetic Relationships

The presence of putative mycoviral dsRNA was first described in *G. abietina* biotype A by Tuomivirta et al. [30]. They found two independent dsRNA banding patterns in a single *G. abietina* isolate, and later characterised one totivirus (Gremmeniella abietina RNA virus-lone 1, GaRV-L1) and two partitiviruses (Gremmeniella abietina RNA virus multisegmented, GaRV-MS1 and GaRV-MS2), but could not find dsRNA in ascospore isolates [31,32]. Narnaviruses were observed later, and are now known to be present in both biotypes of *G. abietina* [33,34]. Thereafter, biotype B has also been shown to host an endornavirus (Gremmeniella abietina RNA virus XL, GBRV-XL) [35] and, within the Spanish population, a taxonomically uncategorised mycovirus (Gremmeniella abietina RNA virus 6, GaRV6) [36,37].

2.1. Partitiviruses

Three full-length genomes have been characterised in three different strains of *G. abietina*, two in Finnish isolates of biotype A (GaRV-MS1 and 2) [31,32], and one in an isolate of the Spanish population of *G. abietina* (GaRV-MS1-3) [38]. GaMRV-MS1 has its genome divided into three segments, and the largest one contains the ORF (open reading frame) that codes for an RNA-dependent RNA polymerase (RdRp) and has a size of circa (ca.) 1.7 kb, the medium segment (ca. 1.5 kb) codes for a capsid protein (CP), and the smallest one (III) (ca. 1.1 kb) codes for a protein with unknown function (Figure 1A). The comparison of amino acid sequences of the CP, RdRp, and unknown protein revealed that the three full-length sequences described belong to the same species of the genus *Gammapartitivirus* [31,32,38]. Interestingly, GaRV-MS1 appears to have low genetic variability, and it is highly conserved not only in Europe, but also in North America [38]. GaRV-MS1 was detected in 28% of 162 investigated isolates. It primarily occurs in *G. abietina* biotype A (Table 1) but is also present in biotype B in Turkey [29]. When the occurrence of GaRV-MS1 was analysed within each population/biotype, the highest incidence was found in the Spanish population (56% of 50 isolates), followed by the biotype A population in North America (45% of 11 isolates), biotype A in most of Europe (16% of 68 isolates), and biotype B (and the Alpine biotype; in only 6% of 33 isolates). The virus GaMRV-MS1 evolves not only through purifying selection but also, to some extent, via recombination. Strain GaRV-MS1-2 seemed to be a recombinant between the complete CP sequences of GaRV-MS1-1 and GaRV-MS1-3, suggesting that GaRV-MS1-2 or one of its ancestors was a recombinant. Likewise, recombination was identified in the full-length RdRp sequences of GaRV-MS1-3 and 2, with GaRV-MS1-1 indicated to be a recombinant [38].

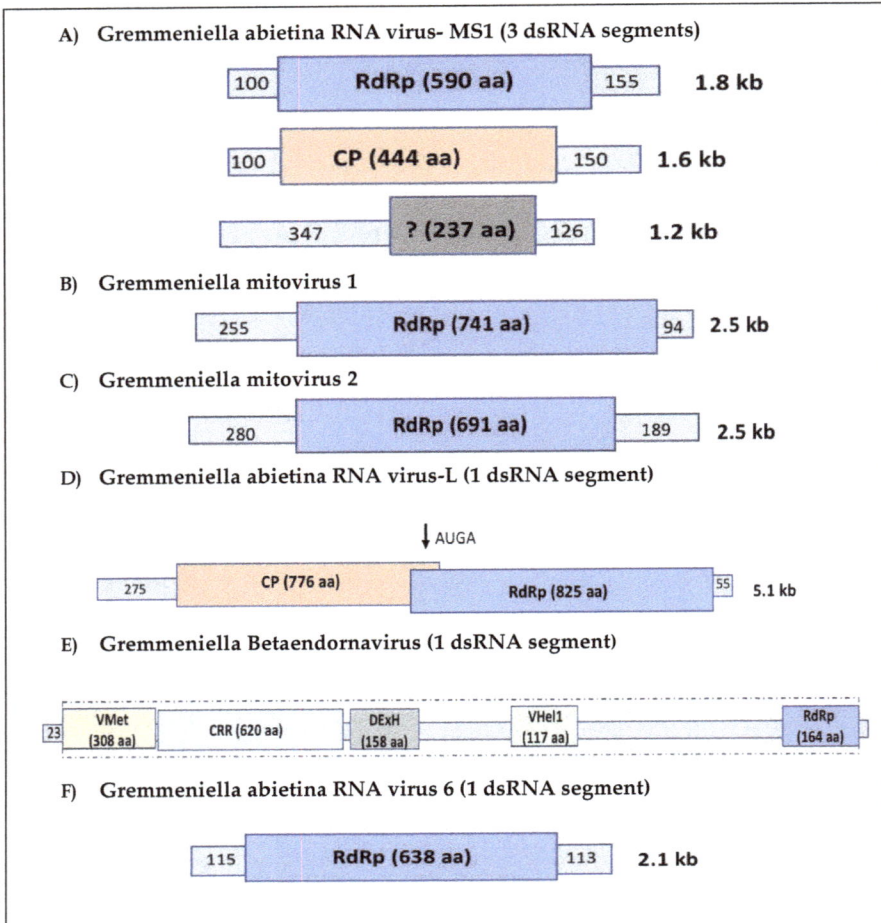

Figure 1. Schematic illustration of the genomic organisation of all fully sequenced *G. abietina* mycoviruses. ORFs and conserved domains are represented by rectangular coloured boxes indicating the coded proteins and the size of their amino acid sequences. They are flanked by 5'- and 3'-UTRs, numbers represent base pairs (bp). The full-length dsRNA segment size is indicated on the right. (**A**) Gremmeniella abietina RNA virus-MS1; (**B**) Gremmeniella mitovirus 1 (previously named Gremmeniella abietina RNA virus-S1); (**C**) Gremmeniella mitovirus 2; (**D**) Gremmeniella abietina RNA virus-L; (**E**) Gremmeniella betaendornavirus 1 (previously Gremmeniella abietina RNA virus-XL) consists of a polyprotein of 3249 and five conserved regions: viral methyltransferase (VMet), cysteine-rich region (CRR), DExH box helicase (DExH), viral RNA helicase 1 (VHel1), and RNA_dep_RNApol2 (RdRp); (**F**) Gremmeniella abietina RNA virus 6.

Table 1. Described viruses in *G. abietina*.

Virus Name	Virus Abbreviation	Full-Length Virus Strains	GenBank Accessions *	Virus Genus	Genome	Fungal Population	Population Studies	Countries Detected	References
Gremmeniella abietina RNA virus multi segmented 1	GaRV-MS1	3	KJ786411-KJ786413, AY089993-5, AY615211-13,	*Gammapartitivirus*	dsRNA	A, B, SP	YES	Canada, USA, Finland, Spain, Montenegro, Italy, Turkey	[31,32,28]
Gremmeniella abietina mitovirus 1	GMV1	3	HE586988, AF534641, AY615209	*Mitovirus*	(+) ssRNA	A, SP	YES	Finland, Spain	[32–34]
Gremmeniella abietina mitovirus 2	GMV2	1	JN654496	*Mitovirus*	(+) ssRNA	B, SP	YES	Finland, Spain	[34]
Gremmeniella abietina RNA virus lone	GaRV-L	2	AF337175, AY615210	*Totivirus*	dsRNA	A	NO	Finland	[31,32]
Gremmeniella abietina RNA virus XL	GaBRV-XL	2	DQ399289-90	*Betaendornavirus*	dsRNA	B	NO	Finland	[35]
Gremmeniella abietina RNA virus 6	GaRV6	1	KJ742567.1	Unassigned †	dsRNA	A, SP	YES	Finland, Canada, Italy, Spain	[36]

† Virus belonging to a new family not defined and submitted in the ICTV yet. * GenBank accession numbers of full-length virus sequences.

We reanalysed the taxonomic statuses of *G. abietina* viruses as the sequence information in GenBank is increasing quickly. This was done simply by using a BLAST search (amino acid sequences) to find the most similar *RdRp* genes, and MAFFT alignment using BLOSUM62 cost matrix to determine identities [39]. The reanalysis was carried out in October 2018. Based on this analysis, the closest relatives of GaRV-MS1 (and other partitiviruses of *G. abietina*) were 20 strains of Pseudogymnoascus destructans partitiviruses (PdPV-pa) [40] and Valsa malicola partitivirus (VmPV; (GenBank accession number AIS37554) with 77–78% identities. The closest strain of PdPV-pa (APG38267.1) with a partial sequence covering amino acid positions 53–362 of GaRV-MS1 RdRp had 78% identity to GaRV-MS1. The VmPV sequence was very short, and covered only amino acid positions 198–331 of GaRV-MS1 RdRP and had 76% identity with it.

2.2. Narnaviruses

Two separate narnavirus populations of the genus *Mitovirus* have been characterised and reported in *G. abietina*: Gremmeniella mitovirus 1 (GMV1) (originally named as GaMRV-S) and Gremmeniella mitovirus 2 (GMV2) (Table 1) [32–34]. They share 94% identity at aa-level, and have a typical mitovirus monopartite genome of around 2.5 kb and GC content of 30% (Figure 1B,C). Using the mitochondrial translation table, both of them code for a single large ORF of ca. 2 kb.

Based on the screening of the 2.5 kb band in 353 isolates of *G. abietina* [34], there was no evidence of mitoviruses in the six Swiss (Alpine biotype), two Turkish (biotype B), and six North American isolates (EU race or biotype A in North America). However, 68 of the 91 Spanish isolates harboured the dsRNA band. Similarly, among the 211 biotype A Finnish isolates analysed, 51 isolates (24%) carried a 2.5 kb dsRNA segment. Only three putative mitoviruses were found among the 37 Finnish biotype B isolates tested (8%).

The population genetic parameters calculated for the two populations suggest that GMV1 is genetically more variable than GMV2. The evolution of both GMV1 and GMV2 is mainly driven by mutation and selection, as no recombination events were detected [34]. Based on the comparison to sequences in the GenBank, GMV1 is most closely related to soybean leaf-associated mitovirus 4 (SlaMV4) [41], soybean leaf-associated mitovirus 2 (SlaMV2) [41], Alternaria arborescens mitovirus 1 (AaMV1) [42], and Alternaria brassicicola mitovirus 1 (AbMV1) [43]. The highest similarity was observed to the partial sequence of SSlaMV4, which was aligned with amino acid positions 1–529 of GMV1 RdRp with 44% identity. For the full sequences of SlaMV2, AaMV1, and AbMV1, RdRp had 43%, 42% and 42% identities with that of GMV1, respectively.

GMV1 is only observed in the Finnish biotype A and Spanish strains, whereas GMV2 infects the Finnish biotype B and the Spanish population. Thereby, the Spanish population of *G. abietina* harbours mitovirus strains that, in Finland, occur separately in biotype A and B strains. Therefore, the Spanish population is the first one hosting distantly related mycoviruses of a single genus in one population of *G. abietina*. This may suggest that horizontal transmission of viruses could have occurred between biotype B and the Spanish population (A type origin) in Spain, although biotype B has never been observed there. Furthermore, GMV2 has been observed only in one of the four studied localities in Spain, suggesting a certain local differentiation of virus populations among the Spanish *G. abietina* [34].

2.3. Totiviruses

GaRV-L1 and GaRV-L2 are putative members of the same species belonging to the genus *Victorivirus*. They were sequenced from two Finnish isolates belonging to biotype A [31]. Their genome lengths are ca. 5 kb, and show 90% overall identity (RdRp 98%). GaRV-L1 and 2 contain two large partially overlapping ORFs (Figure 1D). The first ORF starts at approximately nucleotide 270 from the 5′ end of the coding strand. Starting nucleotides for the second ORF were located in positions located 2.6 kb from the 5′ end in both cases. The protein encoded by the second ORF contained all eight conserved motifs of RdRps of viruses infecting lower eukaryotes [44]. Based on BLAST analysis using the RdRp amino acid sequence, the closest relatives of these viruses are Penicillium aurantiogriseum totivirus 1 (PaTV1) [45], Penicillium digitatum virus 1 (PdV1) [46], and Aspergillus mycovirus 178 (AMV178) [47]. PaTV1 and PdV1 covered the full RdRp sequence of GaRV-L1 with identities of 60% and 59%, whereas AMV178 covered positions 53–825 of GaRV-L1 RdRp sequence with an identity of 60%. The first ORF in both isolates coded for a putative CP, as BLAST searches indicated high similarity with analogous proteins of the viruses described above.

Although there are no population studies available on GaRV-L, it should be noted that these viruses have only been observed in biotype A [31].

2.4. Endornaviruses

The virus with the largest genome found to date is GaBRV-XL, described in two Finnish isolates of *G. abietina* B type (Table 1) [35]. It is a linear, monopartite dsRNA virus of ca. 11 kb (Figure 1E), and belongs to the genus *Betaendornavirus*. The GaBRV-XL genomic structure has five conserved sites, and encodes for a putative 3249 aa polyprotein with four regions showing high similarity to putative viral methyltransferases, DExH box helicases, RNA helicase 1 of viruses, and RNA-dependent RNA polymerases. The closest relatives of GaBRV-XL, according to the BLAST search with RdRp motif of the polyprotein, are Discula destructiva virus 3 (DdV3) [48] and several endornaviruses from *Sclerotinia sclerotiorum* (including SsEV2-A) [49]. The DdV3 sequence covered only amino acid positions 107–165 in GaBRV-XL sequence with 72% identity, whereas SsEV2-A sequence covered the complete RdRp motif with 69% identity. In order to follow the International Committee on Taxonomy of Viruses (ICTV) rules, we propose renaming this virus as Gremmeniella betaendornavirus 1 (GBEV1).

2.5. Gremmeniella abietina RNA Virus 6

GaRV6 consists of a polymerase segment of ca. 2.1 kb with 54.7% GC content (Figure 1F) [36]. It seems to be part of an unclassified mycovirus group that is likely to be proposed as a novel virus family [37]. Its members are hosted by important plant pathogens, such as *Fusarium graminearum* [50], *Rhizoctonia solani* [51], *C. parasitica* (GenBank accession numbers KC549809 and KC549810), *H. annosum* [20], and the endophyte *Curvularia protuberata* [52]. The closest relatives of GaRV6 based on BLAST search using *RdRp* gene are Fusarium graminearum dsRNA mycovirus 5 (FgV5) [53] and Fusarium graminearum dsRNA mycovirus-4 (FgV4) [50]. The FgV-5 and FgV4 sequences covered amino acid positions 95–638 and 60–638 of GaRV6 RdRp with identities of 48% and 46%, respectively. This virus group is closely related to the *Partitiviridae* family [20,37,50]. Most of these viruses have bipartite genomes with sizes resembling those of partitiviruses (1700 to 2400 bp). The larger segment encodes the RdRp, whereas the smaller segment appears to have one or two ORFs coding proteins with unknown functions. However, the smaller genome segment of Ustilaginoidea virens RNA virus 4 (UvRV4) [54] and Heterobasidion RNA virus 6 (HetRV6) [20] have not been detected.

GaRV6 primarily appeared in the Spanish population of *G. abietina*, where its genetic diversity was minimal, despite its relatively high abundance (46% of 50 isolates screened) [36]. GaRV6 has also been detected by RT-qPCR in three other isolates belonging to biotype A in Italy, Canada, and Finland [55].

2.6. Multiple Virus Infections and Evidence of the Existence of Novel Viruses

G. abietina commonly hosts more than one virus in a single isolate [30–32,36]. In biotype A, up to five dsRNAs are found in some isolates [30], in B type three [35] and, in the Spanish population, there are up to eight different dsRNA-banding patterns described, some of which include up to eight bands (Figure 2), suggesting the existence of other *Gremmeniella* viruses not reported yet. Nothing is known about the interactions of multiple viruses co-inhabiting the same mycelia, nor of their effects on the host phenotype.

Here, we provide further information about the exceptional richness of the virus community hosted by Spanish isolates of *G. abietina* noted by Botella et al. [36]. Two dsRNA bands of putative viruses were detected in several isolates, including P3-7 and 06P (Figure 2) which were further analysed as previously (Supplementary Data). These isolates of *G. abietina* have been included in the previously mentioned studies about GaRV-MS1, GMV1 and GaRV6 [34,36,38].

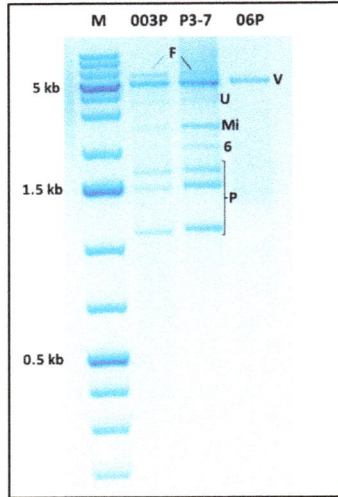

Figure 2. dsRNA banding pattern of the isolates 003P, P3-7, and 06P of Gremmeniella abietina after DNase I and S1 nuclease treatment. M, DNA marker (GeneRuler 1 kb Plus DNA Ladder, 75–20,000 bp, Thermo Scientific); **F**, dsRNAs corresponding to a putative fusarivirus; **V**, dsRNA corresponding to a putative victorivirus, **U**, unknown band, **Mi**, dsRNA corresponding to Gremmeniella mitovirus 1; **6**, dsRNA band corresponding to GaRV6; **P**, dsRNA corresponding to Gremmeniella abietina RNA virus MS1.

2.6.1. The ca. 6 kb dsRNA Band

A partial 2309 bp sequence of the ca. 6 kb dsRNA fragment (GenBank accession number LR031261; Figure 2) from isolate P3-7 was determined as previously described (Supplementary Data) [21,23,27,29]. The comparison in BLASTX (Table 2), resulted in the highest similarity (65%) with the RdRp of Macrophomina phaseolina single-stranded RNA virus 1 (MpSRV1) followed by Neofusicoccum luteum fusarivirus 1 (62%). These viruses are unclassified ssRNA viruses belonging to Fusariviridae, which is a recently proposed family not yet officially accepted by the ICTV. The sequence included a 332 bp conserved domain with an E-value of 3.34×10^{-3}, belonging to a hypothetical ATP-dependent RNA helicase HrpB of *Haemophilus influenzae* (Accession number PRK11664).

Table 2. BLASTX comparison of a 2228 bp RdRp sequence obtained from cloning 6 kb dsRNA of P3-7 *G. abietina* isolate.

GenBank BLASTX Database	Sequence ID	Family	Identities	Matching Region [1]
Macrophomina phaseolina single-stranded RNA virus 1	ALD89094.1	*Fusariviridae* *	217/335 (65%)	512 to 845
Neofusicoccum luteum fusarivirus 1	ARO52688.1	*Fusariviridae* *	205/333 (62%)	527 to 856
Penicillium aurantiogriseum fusarivirus 1	YP_009182154.1	*Fusariviridae* *	198/328 (60%)	511 to 838
Penicillium roqueforti ssRNA mycovirus 1	YP_009052456.1	*Fusariviridae* *	177/313 (57%)	500 to 812
Fusarium graminearum dsRNA mycovirus-1	YP_223920.2	*Fusariviridae* *	179/334 (54%)	530 to 863
Sodiomyces alkalinus fusarivirus 1	ATP75827.1	*Fusariviridae* *	183/326 (56%)	445 to 770
Rosellinia necatrix fusarivirus 1	YP_009047147.1	*Fusariviridae* *	80/329 (55%)	522 to 850
Pleospora typhicola fusarivirus 1	YP_009182158.1	*Fusariviridae* *	178/329 (54%)	521 to 849
Fusarium poae fusarivirus 1	YP_009272906.1	*Fusariviridae* *	172/326 (53%)	496 to 818
Agaricus bisporus virus 11	AQM49938.1	*Fusariviridae* *	151/314 (48%)	592 to 901

* Proposed family not official in the ICTV yet; [1] Matching region in the aa sequence of the database sequences.

An ORF with a length of 987 bp, coding for a protein of 328 aa, was detected. This sequence was obtained from a contig of 3 overlapping clones (682, 743 and 438 bps) and 5 reverse transcription (RT) PCR products. Specific primers (Table S1) were designed based on the first contig achieved by cloning the 6 kb dsRNA band (Supplementary Data). The aa sequence had 61% identity with the MpSRV1 RdRp gene. We propose the name Gremmeniella fusarivirus 1 (GFV1) for this mycovirus.

A similar banding pattern has also been found in other isolates of *G. abietina* [56], such as 003P (Figure 2). However, the frequency of GFV1 in Spanish or other *G. abietina* populations has not been determined, with more specific methods such as reverse transcription (RT) PCR.

2.6.2. The ca. 5 kb dsRNA Band

The 1852 bp and 1261 bp sequences (accession numbers LR030278 and LR030279) were determined from a 43-clone contig from a ca. 5 kb dsRNA fragment of Spanish isolate 06P (Figure 2). The BLASTX analyses with the two nucleotide sequences showed that they code for RdRp and CP proteins of a putative new virus of the genus *Victorivirus*. The BLASTX comparison of the partial RdRp sequence had the highest similarity (78%) with Sclerotinia nivalis victorivirus 1, followed by Aspergillus foetidus slow virus 1 (52%) (Table 3), but was distantly related to the previously known victoriviruses, GaRV-MS1 and GaRV-MS2 (39% and 38% identities in Blastx, respectively), from *G. abietina* bictype A. The partial RdRp fragment contains an ORF of 1557 bp that codes for a 512 aa protein.

Table 3. BLASTX comparison of a 1852 bp RdRp sequence obtained from cloning 5 kb dsRNA of 06P *G. abietina* isolate.

GenBank BLASTX database	Sequence ID	Family	Identities	Matching Region [1]
Sclerotinia nivalis victorivirus 1	YP_009259368.1	*Totiviridae*	456/587 (78%)	247 to 833
Aspergillus foetidus slow virus 1	YP_009508249.1	*Totiviridae*	306/584 (52%)	254 to 837
Beauveria bassiana victorivirus 1	AMQ11131.1	*Totiviridae*	302/585 (52%)	258 to 841
Sphaeropsis sapinea RNA virus	NP_047558.1	*Totiviridae*	314/587 (53%)	254 to 837
Rosellinia necatrix victorivirus 1	BAM36400.1	*Totiviridae*	397/587 (67%	107 to 692
Beauveria bassiana victorivirus NZL/1980	YP_009032633.1	*Totiviridae*	301/585 (51%)	258 to 841
Soybean-associated double-stranded RNA virus 1	ALM62239.1	*Totiviridae*	311/587 (53%)	255 to 840
Ustilaginoidea virens RNA virus 1	AGO04407.1	*Totiviridae*	302/584 (52%)	243 to 825
Bipolaris maydis victorivirus 1	AXB26764.1	*Totiviridae*	289/588 (49%)	253 to 835
Botryotinia fuckeliana totivirus 1	YP_001109580.1	*Totiviridae*	291/581 (50%)	266 to 836

[1] Matching region in the aa sequence of the database sequences.

In the case of the partial CP fragment, the consensus sequence derived from a seven-clone contig of 1261 bp also showed the highest similarity (81%) with Sclerotinia nivalis victorivirus 1, followed by Tolypocladium cylindrosporum virus 1 (68%) (Table 4). According to the 60% maximum identity criterion of the ICTV 9th report (2011) for the genus *Victorivirus*, this virus should be considered a strain of Sclerotinia nivalis victorivirus 1 because it shows 78% RdRp and 81% CP similarity to Sclerotinia nivalis victorivirus 1. Therefore, we designate the virus as Sclerotinia nivalis victorivirus 1—strain 1 from *Gremmeniella abietina* (SnVV1-Ga1)

A similar dsRNA banding pattern has also been found in other isolates of Spanish *G. abietina* [56], such as 003P and P3-7 (Figure 2), but its presence among *G. abietina* isolates has not been determined with specific methods.

Table 4. BLASTX comparison of a 1261 nt capsid protein (CP) sequence obtained from cloning 5 kb dsRNA of 06P *G. abietina* isolate.

GenBank BLASTX Database	Sequence ID	Family	Identities	Matching Region [1]
Sclerotinia nivalis victorivirus 1	YP_009259367.1	*Totiviridae*	293/363 (81%)	1 to 363
Tolypocladium cylindrosporum virus 1	YP_004089629.1	*Totiviridae*	247/362 (68%)	1 to 362
Beauveria bassiana victorivirus NZL/1980	YP_009032632.1	*Totiviridae*	237/361 (66%)	1 to 361
Helminthosporium victoriae virus 190S	NP_619669.2	*Totiviridae*	240/363 (66%)	1 to 363
Bipolaris maydis victorivirus 1	AXB26763.1	*Totiviridae*	240/363 (66%)	1 to 363
Botryotinia fuckeliana totivirus 1	YP_001109579.1	*Totiviridae*	227/363 (63%)	1 to 363
Aspergillus foetidus slow virus 1	YP_009508248.1	*Totiviridae*	221/360 (61%)	1 to 360
Fusarium asiaticum victorivirus 1	AYD49681.1	*Totiviridae*	225/364 (62%)	1 to 363
Ustilaginoidea virens RNA virus L	YP_009094184.1	*Totiviridae*	223/363 (61%)	1 to 363
Rosellinia necatrix victorivirus 1	BAM36399.1	*Totiviridae*	227/362 (63%)	1 to 361
Sclerotinia sclerotiorum victorivirus 1 (partial)	AWY10937.1	*Totiviridae*	208/263 (79%)	1 to 233

[1] Matching region in the aa sequence of the database sequences.

3. Mycoviruses and Their Clarifying Message on the Complex Diversity of *G. abietina*

One of the fundamental questions in evolutionary biology is the degree to which the diversification of parasites and/or symbionts is associated with the diversification of their hosts [57]. Mycoviruses with RNA genomes produce mostly cryptic infections, replicate only in the fungal cytoplasm, and are transferred only through anastomosis and/or via sporal dispersal [58]. Such mechanisms would indicate that evolutionary patterns of mycoviruses can be expected to follow those of their hosts [59,60]. The diversification of the European race of *G. abietina* into different biotypes, and possible species [9], is mostly the outcome of geographical separation and ecological adaptation. Furthermore, its virus communities provide insights about this history of dispersion, with a special relevancy to the establishment of new populations in Spain and Turkey.

The European race of *G. abietina* is known to host only one species of gammapartitivirus (GaRV-MS1) [38]. In northern Europe, it has only been found in biotype A isolates, but it occurs also in Turkish isolates, which belong to biotype B (Table 1) [16,34]. On the other hand, the lack of GaRV-MS1 in biotype B in northern Europe and its occurrence in Turkey suggests that the virus transmission must have occurred between biotypes A and B during their evolutionary history. This virus transmission may have happened prior to the introduction of *G. abietina* to Turkey or, alternatively, in Turkey with an unrecognised biotype A population. In the first scenario, biotype B isolates arrived in Turkey with GaRV-MS1 viruses and, in the second scenario, they would have received the viruses in Turkey. A third possibility is that GaRV-MS1 viruses occurred in both biotypes during their evolutionary history, but were later lost in B-type with the exception of the Turkish population. The presence of GaRV-MS1 in the European race (biotype A) in North America, introduced from Europe in 1977 [61], indicates that the virus travelled with its host, and has not changed much over these years.

The virus community inhabiting the Spanish *G. abietina* population supports its origin as biotype A [16] because (i) recombination events have been detected between three strains of GaRV-MS1 present in Finland and Spain and (ii) GaRV6 primarily appears in Spain, but was also found in single isolates of biotype A from Canada, Italy and Finland, and (iii) the Spanish population also harbours both mitovirus species GMV1 and GMV2, associated with biotype A and B strains in northern Europe, respectively.

The occurrence of GMV2 in both biotype B and the Spanish population, together with its extremely low genetic variability, would be in accordance with a recent host switch and subsequent adaptation to a new host [34]. Natural interspecies transmission of viruses has been shown for *Cryphonectria* sp. and *Cryphonectria parasitica* [22], *Sclerotinia homoeocarpa* and *Ophiostoma ulmi* [23], *O. ulmi* and *Ophiostoma novo-ulmi* [24], and within species of the genera *Heterobasidion* [20,25,26], *Sclerotinia* [27,28], and *Rhizoctonia* [29]. However, biotype B of *G. abietina* has never been shown to exist in Spain and, therefore, the timing and place of this possible host jump remains open. It could have occurred in the past, or even today, in the mountains of northern Spain, where the snow cover in wintertime is enough to fulfil its need [12,13,62]. Further light on this issue would be obtained by a more detailed study on the presence of biotype B in possible candidate areas.

No evidence is available for the presence of mitoviruses in the Swiss (Alpine biotype), Turkish (biotype B), and North American isolates (biotype A). Therefore, the Spanish population of *G. abietina* is the only one hosting the two different mitoviruses, which underlines its unique position within the taxonomy of *G. abietina* var. *abietina*, extending also to its mycovirus community.

A certain level of specialisation is detected in Finland [31,32,35], where both *G. abietina* biotypes (or species) cohabit. In Finland, viruses occurring in biotype A and biotype B diverge (Table 1). The Finnish biotype A is known to host four viruses (GaRV-MS1, GMV1, GaRV6, and GaRV-L) while, within the *G. abietina* biotype B, two different viruses (GMV2 and GBEV1) have been reported [35].

Interestingly, we report, here, that the Spanish population hosts a fusarivirus, GFV1, and also an unrelated victorivirus, SnVV1, the presence of which, in other populations of *G. abietina*, has not been tested directly, but the lack of corresponding dsRNA banding patterns in Finnish populations suggests its unlikeliness. Therefore, it can also be hypothesised that these viruses represent transmission from some distantly related fungi, such as *Sclerotinia nivalis*, to the Spanish population of *G. abietina*. This possibility is supported by the emerging evidence of virus transmissions over wide taxonomic distances known from other fungi [60,63–65] and, therefore, should be tested in the future.

Altogether, phylogenetic and prevalence studies of the distinct viruses provide evidence that support separation of biotypes A and B from each other, as well as on the origin of separate populations of *G. abietina* in North America, Spain, and Turkey.

4. Consequences of Viral Infections on *G. abietina*

Two pathogenicity tests with *G. abietina* biotype A isolates harbouring GaRV-L and GaRV-MS dsRNA patterns were performed, but no firm conclusion could be drawn on the effect of dsRNA viruses on *G. abietina* pathogenicity towards *Pinus sylvestris* [32]. All isolates were pathogenic to host plants, but both dsRNA-containing and dsRNA-free isolates were found among the most pathogenic isolates, and there was no difference between the pathogenicity of isolates containing GaRV-L and GaRV-MS patterns. However, it should be noted that no pairs of isogenic dsRNA-containing and dsRNA-free isolates were available for the pathogenicity tests. Additionally, the inoculated trees were genetically different, and the pathogenicity test setup could only measure how mycelium is able to grow in phloem after inoculation [32].

Romeralo et al. [66] noted that the existence of GMV1 appeared to have a substantial increasing effect on the growth rate of *G. abietina* at its optimum growing temperature, 15 °C. Whether this increase in mycelial growth is connected to the virulence of their hosts has not been tested. In contrast, GaRV6 occurrence had a negative impact on the growth rate of *G. abietina* on artificial media [45].

Based on the results described above, at least some viruses may have effects on the phenotype of *G. abietina*. However, conclusions about the real and specific role of each type of these viruses cannot be fully determined, since multiple virus infections are common in this conifer pathogen [30–32,35,36,55,56]. For example, isolate P3-7, has a large viral dsRNA load (Figure 2). It hosts GaRV6, GaRV-MS1, and GMV1, as well as both novel species of *Fusarivirus* and *Victorivirus* reported in this review, and had the lowest growth rate in the experiment by Botella et al. [55]. Conversely, in the same study, isolate 06P had the highest growth rate in the three temperatures tested, and it only hosts the possible victorivirus reported here. Thus, it is very necessary to carry out experiments, including pairs of cured and single virus infected isogenic *G. abietina* isolates; unfortunately, up to now, virus removal trials with higher temperatures or cycloheximide have failed with this pathogen [55]. Furthermore, it has not been possible to infect viruses of *G. abietina* in dual culture experiments used for *C. parasitica* since the 1960s (and, thereafter, for many other fungi) and, therefore, artificial transmission methods should also be developed for *G. abietina* viruses.

5. Conclusions

G. abietina hosts a distinct and abundant community of viruses belonging to different genera and families of mycoviruses. Its diversification and spread in different countries has also affected the divergence of its virus community. Based on a single study, it appears that multiple virus infections may affect the phenotype of the host, but the effects of single viruses remain obscure, and should be determined by in vivo and in vitro experiments using isogenic pairs of single-infected and virus-free isolates.

Supplementary Materials: The following is available online at http://www.mdpi.com/1999-4915/10/11/654/s1, Table S1: Primers used in this study for the determination of the virus sequences.

Author Contributions: L.B. and J.H. contributed to the writing, editing and content of this manuscript.

Funding: This work was supported by "Indicators of tree vitality" (CZ.1.07/2.3.00/20.0265), funded by the OP Education for Competitiveness, the European Social Fund and the Czech Ministry of Education, Youth and Sport. And, Project Phytophthora Research Centre Reg. No. CZ.02.1.01/0.0/0.0/15_003/0000453, co-financed by the European Regional Development Fund.

Acknowledgments: L.B. would like to thank her children Daniel and Alex Dvořák Botella that they sleep well at night, and she could find the time to write this manuscript during her maternity leave.

Conflicts of Interest: The authors declare no conflict of interest.

References

1. Donaubauer, E. Distribution and hosts of *Scleroderris lagerbergii* in Europe and North America. *For. Pathol.* **1972**, *2*, 6–11. [CrossRef]
2. Yokota, S.; Uozumi, T.; Matsuzaki, S. Scleroderris canker of Todo-Fir in Hokkaido, Northern Japan I. Present status of damage, and features of infected plantations. *For. Pathol.* **1974**, *4*, 65–74. [CrossRef]
3. Morelet, M. La maladie à *Brunchorstia*. *Eur. J. For. Pathol.* **1980**, *10*, 268–277. [CrossRef]
4. Barklund, P.; Rowe, J. *Gremmeniella abietina* (*Scleroderris lagerbergii*), a primary parasite in a Norway spruce die-back. *For. Pathol.* **1981**, *11*, 97–108. [CrossRef]
5. Kaitera, J.; Seitamäki, L.; Jalkanen, R. Morphological and ecological variation of *Gremmeniella abietina* var. *abietina* in *Pinus sylvestris*, *Pinus contorta* and *Picea abies* sapling stands in northern Finland and the Kola Peninsula. *Scand. J. For. Res.* **2000**, *15*, 13–19. [CrossRef]
6. Santamaria, O.; Alves-Santos, F.M.; Diez, J.J. Genetic characterization of *Gremmeniella abietina* var. *abietina* isolates from Spain. *Plant Pathol.* **2005**, *54*, 331–338. [CrossRef]
7. Petrini, O.; Toti, L.; Petrini, L.E.; Heiniger, U. *Gremmeniella abietina* and *G. laricina* in Europe: characterization and identification of isolates and laboratory strains by soluble protein electrophoresis. *Can. J. Bot.* **1990**, *68*, 2629–2635. [CrossRef]
8. Dorworth, C.E.; Krywienczyk, J. Comparisons among isolates of *Gremmeniella abietina* by means of growth rate, conidia measurement, and immunogenic reaction. *Can. J. Bot.* **1975**, *53*, 2506–2525. [CrossRef]
9. Uotila, A.; Hantula, J.; Vaatanen, A.K.; Hamelin, R.C. Hybridization between two biotypes of *Gremmeniella abietina* var. *abietina* in artificial pairings. *For. Pathol.* **2000**, *30*, 211–219. [CrossRef]
10. Hamelin, R.C.; Rail, J. Phylogeny of *Gremmeniella* spp. based on sequences of the 5.8S rDNA and internal transcribed spacer region. *Can. J. Bot.* **1997**, *75*, 693–698. [CrossRef]
11. Uotila, A. Physiological and morphological variation among Finnish *Gremmeniella abietina* isolates. *Commun. Inst. For. Fenn.* **1983**, *119*, 12.
12. Hamelin, R.C.; Lecours, N.; Hansson, P.; Hellgren, M.; Laflamme, G. Genetic differentiation within the European race of *Gremmeniella abietina*. *Mycol. Res.* **1996**, *100*, 49–56. [CrossRef]
13. Hellgren, M.; Högberg, N. Ecotypic variation of *Gremmeniella abietina* in northern Europe: Disease patterns reflected by DNA variation. *Can. J. Bot.* **1995**, *73*, 1531–1539. [CrossRef]
14. Hantula, J.; Muller, M. Variation within *Gremmeniella abietina* in Finland and other countries as determined by Random Amplified Microsatellites (RAMS). *Mycol. Res.* **1997**, *101*, 169–175. [CrossRef]
15. Santamaria, O.; Pajares, J.A.; Diez, J.J. First report of *Gremmeniella abietina* on *Pinus halepensis* in Spain. *Plant Pathol.* **2003**, *52*, 425. [CrossRef]

16. Botella, L.; Tuomivirta, T.T.; Kaitera, J.; Carrasco Navarro, V.; Diez, J.J.; Hantula, J. Spanish population of *Gremmeniella abietina* is genetically unique but related to type A in Europe. *Fungal Biol.* **2010**, *114*, 778–789. [CrossRef] [PubMed]

17. Ghabrial, S.A.; Suzuki, N. Viruses of Plant Pathogenic Fungi. *Annu. Rev. Phytopathol.* **2009**, *47*, 353–384. [CrossRef] [PubMed]

18. Schoebel, C.N.; Botella, L.; Lygis, V.; Rigling, D. Population genetic analysis of a parasitic mycovirus to infer the invasion history of its fungal host. *Mol. Ecol.* **2017**, *26*, 2482–2497. [CrossRef] [PubMed]

19. Bryner, S.F.; Rigling, D.; Brunner, P.C. Invasion history and demographic pattern of Cryphonectria hypovirus 1 across European populations of the chestnut blight fungus. *Ecol. Evol.* **2012**, *2*, 3227–3241. [CrossRef] [PubMed]

20. Vainio, E.J.; Hyder, R.; Aday, G.; Hansen, E.; Piri, T.; Doğmuş-Lehtijärvi, T.; Lehtijärvi, A.; Korhonen, K.; Hantula, J. Population structure of a novel putative mycovirus infecting the conifer root-rot fungus *Heterobasidion annosum* sensu lato. *Virology* **2012**, *422*, 366–376. [CrossRef] [PubMed]

21. Milgroom, M.G.; Lipari, S.E.; Ennos, R.A.; Liu, Y.-C. Estimation of the outcrossing rate in the chestnut blight fungus, *Cryphonectria parasitica*. *Heredity* **1993**, *70*, 385–892. [CrossRef]

22. Liu, Y.-C.; Linder-Basso, D.; Hillman, B.I.; Kaneko, S.; Milgroom, M.G. Evidence for interspecies transmission of viruses in natural populations of filamentous fungi in the genus *Cryphonectria*. *Mol. Ecol.* **2003**, *12*, 1619–1628. [CrossRef] [PubMed]

23. Deng, F.; Xu, R.; Boland, G.J. Hypovirulence-associated double-stranded RNA from *Sclerotinia homoeocarpa* is conspecific with *Ophiostoma novo-ulmi* Mitovirus 3a-Ld. *Phytopathology* **2003**, *93*, 1407–1414. [CrossRef] [PubMed]

24. Buck, K.W.; Brasier, C.M.; Paoletti, M.; Crawford, L.J. Virus transmission and gene flow between two species of the Dutch elm disease fungi, *Ophiostoma ulmi* and *O. novo-ulmi*: Deleterious viruses as selective. *Br. Ecol. Soc.* **2003**, *15*, 26–45.

25. Vainio, E.J.; Hakanpää, J.; Dai, Y.-C.; Hansen, E.; Korhonen, K.; Hantula, J. Species of *Heterobasidion* host a diverse pool of partitiviruses with global distribution and interspecies transmission. *Fungal Biol.* **2011**, *115*, 1234–1243. [CrossRef] [PubMed]

26. Kashif, M.; Hyder, R.; De Vega Perez, D.; Hantula, J.; Vainio, E.J. Heterobasidion wood decay fungi host diverse and globally distributed viruses related to Helicobasidium mompa partitivirus V70. *Virus Res.* **2015**, *195*, 119–123. [CrossRef] [PubMed]

27. Melzer, M.S.; Ikeda, S.S.; Boland, G.J. Interspecific transmission of double-stranded RNA and hypovirulence from *Sclerotinia sclerotiorum* to *S. minor*. *Phytopathology* **2002**, *92*, 780–784. [CrossRef] [PubMed]

28. Boland, G.J. Fungal viruses, hypovirulence, and biological control of *Sclerotinia* species. *Can. J. Plant Pathol.* **2004**, *26*, 6–18. [CrossRef]

29. Charlton, N.D.; Carbone, I.; Tavantzis, S.M.; Cubeta, M.A. Phylogenetic relatedness of the M2 double-stranded RNA in *Rhizoctonia* fungi. *Mycologia* **2008**, *100*, 555–564. [CrossRef] [PubMed]

30. Tuomivirta, T.T.; Uotila, A.; Hantula, J. Two independent double-stranded RNA patterns occur in the Finnish *Gremmeniella abietina* var. *abietina* type A. *For. Pathol.* **2002**, *32*, 197–205. [CrossRef]

31. Tuomivirta, T.T.; Hantula, J. Two unrelated double-stranded RNA molecule patterns in *Gremmeniella abietina* type A code for putative viruses of the families *Totiviridae* and *Partitiviridae*. *Arch. Virol.* **2003**, *148*, 2293–2305. [CrossRef] [PubMed]

32. Tuomivirta, T.T.; Hantula, J. Three unrelated viruses occur in a single isolate of *Gremmeniella abietina* var. *abietina* type A. *Virus Res.* **2005**, *110*, 31–39. [CrossRef] [PubMed]

33. Tuomivirta, T.T.; Hantula, J. Gremmeniella abietina mitochondrial RNA virus S1 is phylogenetically related to the members of the genus *Mitovirus*. *Arch. Virol.* **2003**, *148*, 2429–2436. [CrossRef] [PubMed]

34. Botella, L.; Tuomivirta, T.T.; Vervuurt, S.; Diez, J.J.; Hantula, J. Occurrence of two different species of mitoviruses in the European race of *Gremmeniella abietina* var. *abietina*, both hosted by the genetically unique Spanish population. *Fungal Biol.* **2012**, *116*, 872–882. [CrossRef] [PubMed]

35. Tuomivirta, T.T.; Kaitera, J.; Hantula, J. A novel putative virus of *Gremmeniella abietina* type B (Ascomycota: Helotiaceae) has a composite genome with endornavirus affinities. *J. Gen. Virol.* **2009**, *90*, 2299–2305. [CrossRef] [PubMed]

36. Botella, L.; Vainio, E.J.; Hantula, J.; Diez, J.J.; Jankovsky, L. Description and prevalence of a putative novel mycovirus within the conifer pathogen *Gremmeniella abietina*. *Arch. Virol.* **2015**, *160*, 1967–1975. [CrossRef] [PubMed]

37. Nibert, M.L.; Ghabrial, S.A.; Maiss, E.; Lesker, T.; Vainio, E.J.; Jiang, D.; Suzuki, N. Taxonomic reorganization of family *Partitiviridae* and other recent progress in partitivirus research. *Virus Res.* **2014**, *188*, 128–141. [CrossRef] [PubMed]

38. Botella, L.; Tuomivirta, T.T.; Hantula, J.; Diez, J.J.; Jankovsky, L. The European race of *Gremmeniella abietina* hosts a single species of gammapartitivirus showing a global distribution and possible recombinant events in its history. *Fungal Biol.* **2015**, *119*, 125–135. [CrossRef] [PubMed]

39. Kuraku, S.; Zmasek, C.M.; Nishimura, O.; Katoh, K. aLeaves facilitates on-demand exploration of metazoan gene family trees on MAFFT sequence alignment server with enhanced interactivity. *Nucleic Acids Res.* **2013**, *41*, W22–W28. [CrossRef] [PubMed]

40. Thapa, V.; Turner, G.G.; Hafenstein, S.; Overton, B.E.; Vanderwolf, K.J.; Roossinck, M.J. Using a novel partitivirus in *Pseudogymnoascus destructans* to understand the epidemiology of White-Nose Syndrome. *PLoS Pathog.* **2016**, *12*, e1006076. [CrossRef] [PubMed]

41. Marzano, S.-Y.L.; Domier, L.L. Novel mycoviruses discovered from metatranscriptomics survey of soybean phyllosphere phytobiomes. *Virus Res.* **2016**, *213*, 332–342. [CrossRef] [PubMed]

42. Komatsu, K.; Katayama, Y.; Omatsu, T.; Mizutani, T.; Fukuhara, T.; Kodama, M.; Arie, T.; Teraoka, T.; Moriyama, H. Genome sequence of a novel mitovirus identified in the phytopathogenic fungus *Alternaria arborescens*. *Arch. Virol.* **2016**, *161*, 2627–2631. [CrossRef] [PubMed]

43. Chen, Y.; Shang, H.H.; Yang, H.Q.; Da Gao, B.; Zhong, J. A mitovirus isolated from the phytopathogenic fungus *Alternaria brassicicola*. *Arch. Virol.* **2017**, *162*, 2869–2874. [CrossRef] [PubMed]

44. Bruenn, J.A. A closely related group of RNA-dependent RNA polymerases from double-stranded RNA viruses. *Nucleic Acids Res.* **1993**, *21*, 5667–5669. [CrossRef] [PubMed]

45. Nerva, L.; Ciuffo, M.; Vallino, M.; Margaria, P.; Varese, G.C.; Gnavi, G.; Turina, M. Multiple approaches for the detection and characterization of viral and plasmid symbionts from a collection of marine fungi. *Virus Res.* **2016**, *219*, 22–38. [CrossRef] [PubMed]

46. Niu, Y.; Zhang, T.; Zhu, Y.; Yuan, Y.; Wang, S.; Liu, J.; Liu, D. Isolation and characterization of a novel mycovirus from *Penicillium digitatum*. *Virology* **2016**, *494*, 15–22. [CrossRef] [PubMed]

47. Hammond, T.M.; Andrewski, M.D.; Roossinck, M.J.; Keller, N.P. Aspergillus mycoviruses are targets and suppressors of RNA silencing. *Eukaryot. Cell* **2008**, *7*, 350–357. [CrossRef] [PubMed]

48. Rong, R.; Rao, S.; Scott, S.W.; Carner, G.R.; Tainter, F.H. Complete sequence of the genome of two dsRNA viruses from *Discula destructiva*. *Virus Res.* **2002**, *90*, 217–224. [CrossRef]

49. Khalifa, M.E.; Pearson, M.N. Molecular characterisation of an endornavirus infecting the phytopathogen *Sclerotinia sclerotiorum*. *Virus Res.* **2014**, *189*, 303–309. [CrossRef] [PubMed]

50. Yu, J.; Kwon, S.-J.; Lee, K.-M.; Son, M.; Kim, K.-H. Complete nucleotide sequence of double-stranded RNA viruses from *Fusarium graminearum* strain DK3. *Arch. Virol.* **2009**, *154*, 1855–1858. [CrossRef] [PubMed]

51. Zheng, L.; Liu, H.; Zhang, M.; Cao, X.; Zhou, E. The complete genomic sequence of a novel mycovirus from *Rhizoctonia solani* AG-1 IA strain B275. *Arch. Virol.* **2013**, *158*, 1609–1612. [CrossRef] [PubMed]

52. Marquez, L.M.; Redman, R.S.; Rodriguez, R.; Stout, R.G.; Rodriguez, R.J.; Roossinck, M. A Virus in a fungus in a plant: Three-way symbiosis required for thermal tolerance. *Science* **2007**, *315*, 513–515. [CrossRef] [PubMed]

53. Wang, L.; Wang, S.; Yang, X.; Zeng, H.; Qiu, D.; Guo, L. The complete genome sequence of a double-stranded RNA mycovirus from *Fusarium graminearum* strain HN1. *Arch. Virol.* **2017**, *162*, 2119–2124. [CrossRef] [PubMed]

54. Jiang, Y.; Zhang, T.; Luo, C.; Jiang, D.; Li, G.; Li, Q.; Hsiang, T.; Huang, J. Prevalence and diversity of mycoviruses infecting the plant pathogen *Ustilaginoidea virens*. *Virus Res.* **2015**, *195*, 47–56. [CrossRef] [PubMed]

55. Botella, L.; Dvořák, M.; Capretti, P.; Luchi, N. Effect of temperature on GaRV6 accumulation and its fungal host, the conifer pathogen *Gremmeniella abietina*. *For. Pathol.* **2017**, *47*, e12291. [CrossRef]

56. Leticia Botella, L.; Tuomivirta, T.T.; Hantula, J.; Diez, J.J. Presence of viral dsRNA molecules in the Spanish population of *Gremmeniella abietina*. *J. Agric. Ext. Rural Dev.* **2012**, *4*, 211–213. [CrossRef]

57. Göker, M.; Scheuner, C.; Klenk, H.-P.; Stielow, J.B.; Menzel, W. Codivergence of mycoviruses with their hosts. *PLoS ONE* **2011**, *6*, e22252. [CrossRef] [PubMed]
58. Pearson, M.N.; Beever, R.E.; Boine, B.; Arthur, K. Mycoviruses of filamentous fungi and their relevance to plant pathology. *Mol. Plant Pathol.* **2009**, *10*, 115–128. [CrossRef] [PubMed]
59. Varga, J.; Vágvölgyi, C.; Tóth, B. Recent advances in mycovirus research. *Acta Micrcbiol. Immunol. Hung.* **2003**, *50*, 77–94. [CrossRef] [PubMed]
60. Linder-Basso, D.; Dynek, J.N.; Hillman, B.I. Genome analysis of *Cryphonectria hypovirus 4*, the most common hypovirus species in North America. *Virology* **2005**, *337*, 192–203. [CrossRef] [PubMec]
61. Dorworth, C.E.; Krywienczyk, J.; Skilling, D.D. New York isolates of *Gremmeniella abietina* (*Scleroderris lagerbergii*) identical in immunogenic reaction to European isolates [*Pinus*]. *Plant Dis. Rep.* **1977**, *61*, 887–890.
62. Marosy, M.; Patton, R.F.; Upper, C.D. A conductive day concept to explain the effect of low temperature on the development of Scleroderris shoot blight. *Ecol. Epidemiol.* **1989**, *79*, 1293–1301.
63. Yaegashi, H.; Kanematsu, S. Natural infection of the soil-borne fungus *Rosellinia necatrix* with novel mycoviruses under greenhouse conditions. *Virus Res.* **2016**, *219*, 83–91. [CrossRef] [PubMed]
64. Yaegashi, H.; Nakamura, H.; Sawahata, T.; Sasaki, A.; Iwanami, Y.; Ito, T.; Kanematsu, S. Appearance of mycovirus-like double-stranded RNAs in the white root rot fungus, *Rosellinia necatrix*. in an apple orchard. *FEMS Microbiol. Ecol.* **2013**, *83*, 49–62. [CrossRef] [PubMed]
65. Vainio, E.J.; Pennanen, T.; Rajala, T.; Hantula, J. Occurrence of similar mycoviruses in pathogenic, saprotrophic and mycorrhizal fungi inhabiting the same forest stand. *FEMS Microbiol. Ecol.* **2017**, *93*, fix003. [CrossRef] [PubMed]
66. Romeralo Tapia, C.; Botella, L.; Santamaría, O.; Diez, J. Effect of putative mitoviruses on in vitro growth of *Gremmeniella abietina* isolates under different laboratory conditions. *For. Syst.* **2012**, *21*, 515. [CrossRef]

Article

Genetic and Phenotypic Characterization of Cryphonectria hypovirus 1 from Eurasian Georgia

Daniel Rigling [1,*], Nora Borst [1], Carolina Cornejo [1], Archil Supatashvili [2] and Simone Prospero [1]

[1] WSL Swiss Federal Research Institute, Zürcherstrasse 111, 8903 Birmensdorf, Switzerland;
    nora@breuert.eu (N.B.); carolina.cornejo@wsl.ch (C.C.); simone.prospero@wsl.ch (S.P.)
[2] Vasil Gulisashvili Forestry Institute, Agricultural University of Georgia, 0186 Tbilisi, Georgia;
    archil.tbilisi@hotmail.com
* Correspondence: daniel.rigling@wsl.ch; Tel.: +41-44-739-2415

Received: 19 October 2018; Accepted: 30 November 2018; Published: 3 December 2018

Abstract: *Cryphonectria hypovirus 1* (CHV-1) infects the chestnut blight fungus *Cryphonectria parasitica* and acts as a biological control agent against this harmful tree disease. In this study, we screened the recently characterized *C. parasitica* population in Eurasian Georgia for the presence of CHV-1. We found 62 CHV-1 infected *C. parasitica* isolates (9.3%) among a total of 664 isolates sampled in 14 locations across Georgia. The prevalence of CHV-1 at the different locations ranged from 0% in the eastern part of the country to 29% in the western part. Sequencing of two specific regions of the viral genome one each in ORFA and ORFB revealed a unique CHV-1 subtype in Georgia. This subtype has a recombinant pattern combining the ORFA region from the subtype F2 and the ORFB region from subtype D. All 62 viral strains belonged to this Georgian CHV-1 subtype (subtype G). The CHV-1 subtype G strongly reduced the parasitic growth of *C. parasitica* isolates from Georgia, with a more severe effect on the European genepool compared to the Georgian genepool. The CHV-1 subtype detected in Georgia provides a valuable candidate for biological control applications in the Caucasus region.

Keywords: mycovirus; populations study; *Cryphonectria parasitica*; chestnut blight; *Castanea sativa*; biological control

1. Introduction

Mycoviruses in plant pathogenic fungi have attracted interest because of their potential to be used as biological control agents against plant diseases [1]. One of the best studied examples is *Cryphonectria hypovirus 1* (CHV-1), which infects the chestnut blight fungus *Cryphonectria parasitica* [2,3]. This fungus is native to east Asia and has been introduced into North America and Europe [4]. After introduction of the pathogen, both continents experienced severe chestnut blight epidemics, which in eastern North America eliminated the American chestnut (*Castanea dentata*) as a dominant tree species [4]. The epidemic in Europe on the native European chestnut (*Castanea sativa*) initially took a very similar course, but after a certain time, recovery of many chestnut stands was observed [5]. This recovery has been attributed to hypovirulence, which refers to the viral disease caused by CHV-1 in *C. parasitica* [6]. The hypovirus induces a hypovirulent phenotype characterized by reduced virulence and sporulation capacity of the infected *C. parasitica* strains [7]. CHV-1 has an RNA genome, which contains two open reading frames (ORFs A and B). Both ORFs encode polyproteins, which are autocatalytically processed. The viral polymerase and helicase genes are located in the larger ORFB of the viral genome [2].

CHV-1 is located in the cytoplasm of the fungus and can be horizontally transmitted between fungal individuals through hyphal anastomosis [8]. From hypovirus-infected strains, CHV-1 may be vertically transmitted into asexual spores, which carry the hypovirus to new hosts [7,9]. The discovery of hypovirulence has led to the development of a biological control method against chestnut blight based on treatments of chestnut blight cankers with hypovirus-infected *C. parasitica* strains [10]. Hypovirulence is now a widespread phenomenon in many chestnut growing areas of Europe, either naturally or after biological control treatments [5,11,12].

Different CHV-1 subtypes or lineages have been identified in the European *C. parasitica* population [13,14]. Their distribution reflects the different introduction events of the fungal host. The Italian CHV-1 subtype (subtype I) is the most widespread and is linked to the introduction of *C. parasitica* into Italy in the 1930s [13]. Together with its fungal host, this subtype has spread across a large part of the chestnut growing areas in central, southern and southeastern Europe, including western Turkey [15–17]. Additional subtypes, designated F1, F2, and D/E, have mainly been found in western Europe (France, Spain and Germany) [14,18,19]. Interestingly, CHV-1 strains related to subtype F2 have also been identified in the eastern Black Sea region of Turkey [17]. Some subtypes apparently originated from recombination events [14,20]. The CHV-1 subtypes differ in their effects on the fungal host, with some variation within subtypes [7,21]. Strains of the Italian subtype typically have a mild effect on their hosts while subtypes F1 and F2 have a more severe impact [16,22,23].

Besides CHV-1, several other mycoviruses have been detected in *C. parasitica*, including not only additional members in the genus hypovirus (CHV-2, CHV-3, and CHV-4) but also viruses in other genera [24]. Some of these viruses (e.g., CHV-2, CHV-3) do also affect growth and fitness of *C. parasitica*. CHV-1, however, is most relevant for biological control of chestnut blight and to date the only mycovirus reported in *C. parasitica* from Europe.

The *C. parasitica* populations in the Caucasus region, i.e., in the most eastern distribution range of European chestnut, have only recently been genetically characterized. Microsatellite analysis indicated that most of the *C. parasitica* isolates from Georgia and Azerbaijan belong to a genepool, which largely differ from the genepools previously identified in Europe [25,26]. This finding points to an independent introduction of the fungus in the Caucasus area. However, *C. parasitica* isolates belonging to the European genepool were also identified in the western part of Georgia. These isolates were closely related to those previously described in neighboring Turkey, suggesting that they migrated from this country into Georgia.

The objectives of this study were to determine (1) the prevalence of CHV-1 in the *C. parasitica* population in Georgia, (2) the relatedness of the local CHV-1 strains to known CHV-1 subtypes from Europe, and (3) the phenotypic effects of the CHV-1 strains on fungal isolates from the European and Georgian genepool of *C. parasitica*.

2. Materials and Methods

2.1. Fungal Isolates

One set of isolates of *C. parasitica* was obtained from a previous study on the population structure of the chestnut blight fungus in Georgia [25]. Five additional populations (Shemoqmedi, Korbouli, Kumistavi, Satsable, Pshaveli) were sampled for this study in 2012 resulting in a total of 664 *C. parasitica* isolates (Table 1). All isolates were screened for the white culture morphology (Figure 1), which typically is associated with an infection by CHV-1 [7,16]. To assess the culture morphology, the isolates were grown on potato dextrose agar (PDA, Difco™, Becton Dickinson, Sparks, MD, USA) as described previously [27]. All isolates exhibiting a white or intermediate (between white and orange) culture morphology were tested for the presence of CHV-1 by RT-PCR and sequencing.

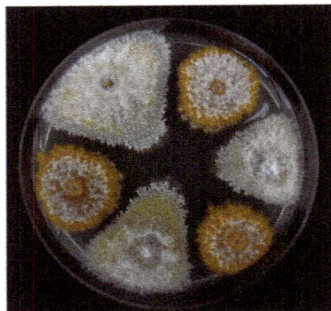

Figure 1. Culture morphology of CHV-1-free (orange) and CHV-1 infected (white) *C. parasitica* isolates from Georgia on PDA.

2.2. RNA Extraction and RT-PCR

Isolates were grown on PDA plates overlaid with cellophane sheets for at least 6 days at room temperature in the dark. Mycelia reaching a diameter of approx. 6 cm were harvested, lyophilized, and ground to a fine powder in a mixer mill (MM 300 from Retsch, Haan, Germany). Total RNA was extracted from approx. 20 mg mycelial powder using the Norgen Plant/Fungi RNA Purification kit (Norgen Biotek Corp., Thorold, ON, Canada). Complementary DNA (cDNA) was synthesized from total RNA with random hexamer primers using the Maxima First Strand cDNA Synthesis kit from Fermentas (Thermo Fisher, Waltham, MA, USA).

2.3. Sequencing of CHV-1

Two different regions, one in ORFA and one in ORFB, were chosen for sequencing. The ORFA region corresponds to the positions 1471–2165 in the nucleotide sequence CHV-1/Euro7 [22] and was amplified using the primers described by Bryner, Rigling and Brunner [15]. The ORFB region was slightly modified after Feau et al. [14] and corresponds to the position 6264–6978. The forward primer ORFB-12aF (5′-AGACCTCAATCGGGTCTCCCT-3′) and the reverse primer ORFB-12aR (5′-TTCAACCACACGACGAGTTCG-3′) were used for PCR amplification. Sanger sequencing was performed using the same primers and the BigDye Terminator v3.1 Cycle Sequencing kit (Applied Biosystems, Foster City, CA, USA). Sequences were assembled and edited with DNADynamo (Blue Tractor Software Ltd., North Wales, UK). The GenBank accession numbers of all sequences are listed in Table S1.

2.4. Phylogenetic Analysis of CHV-1

Sequences were aligned using the web service Clustal Omega provided by the European Bioinformatics Institute (EMBL-EBI). Reference sequences for each CHV-1 subtype were generated as above or obtained from GenBank and are also listed in Table S1. Phylogenetic trees were reconstructed by maximum likelihood (ML) using the software PhyML 3.0 [28] on the ATGC platform (www.atgc-montpellier.fr). PhyML treats gaps systematically as unknown characters. Two random trees were used as starting trees corresponding to two random starts in order to decrease the chance of becoming trapped in a local maximum of the likelihood function. The starting trees were estimated using the BioNJ algorithm. Support for each internal branch of the phylogeny was calculated using the nonparametric bootstrap (B) option with 1000 pseudoreplicates, which estimates a phylogeny for each replicate. The online execution of PhyML includes the Smart Model Selection [29], which selects the model that best fit data under the Akaike Information Criterion and uses the selection direct in phylogenetic reconstruction.

Single locus trees were used to detect conflicting phylogenetic signal between the ORFA and ORFB region. For this purpose, clades of individual gene trees were examined for well-supported ($\geq$70% of 1000 replicates) conflict between ML phylogenies. Following Vijaykrishna et al. [30], the change of the topological position of individuals was interpreted as a sign of reticulate evolutionary history. Unrooted phylogenies, including branch lengths and bootstrap support values, were graphically represented with SplitsTree4 (v.4.14.4; [31]).

2.5. Phenotypic Effects of the Georgian CHV-1 Subtype on C. parasitica

The effect of the Georgian CHV-1 subtype on the growth of the infected *C. parasitica* strains was assessed by conducting two different experiments in vitro similar as described by Bryner and Rigling [23]. The experiment on PDA plates was considered to reflect the saprophytic growth while that on dormant chestnut stems the parasitic growth of the fungus. For the experiments, 9–10 hypovirus-infected isolates with white culture morphology of each of the three Georgian genetic clusters of *C. parasitica* (CpGeo20 cluster, CpGeo75 cluster and CpGeo97 cluster) were randomly chosen among the 62 hypovirus-infected isolates identified in this study (Table S1). The CpGeo20 and CpGeo75 clusters represent the Georgian *C. parasitica* genepool, whereas the CpGeo97 cluster is related to isolates in vc type EU-1, which are present in Turkey and belong to the European genepool. To compare the effects of the Georgian and the Italian CHV-1 subtype, 10 Turkish isolates belonging to the vc type EU-1 and infected by the Italian CHV-1 subtype [21] were included in these experiments. From each virus-infected isolate, an isogenic virus-free isolate was obtained through single conidial isolation as described by Bryner and Rigling [21], resulting in 40 pairs of virus-infected and virus-free isolates. Virus-free isolates were distinguished from virus-infected isolates by their orange culture morphology (Figure 1).

2.5.1. Growth on PDA Plates

For this experiment, three mycelial plugs (6 mm in diameter) originating from the growing margin of a 5-day-old pure culture on PDA of each hypovirus-infected isolate and its corresponding isogenic hypovirus-free isolate were placed each in the center of a PDA plate. The plates were sealed with Parafilm and incubated at 20 °C in a climate chamber under a 14L:10D photoperiod. Forty-eight hours after inoculation, the size of each culture was assessed the first time by measuring with a millimeter ruler the two cardinal diameters through two orthogonal axes previously drawn on the bottom of each plate. Measurements were repeated every 24 h for the next four days. Since the shape of the growing cultures were not perfect circles, the geometric mean diameter of an ellipse was calculated.

2.5.2. Growth on Dormant Chestnut Stems

The same fungal isolates used for the previous experiment were inoculated on dormant chestnut (*Castanea sativa*) stems (5–10 cm in diameter, 50 cm in length) that were cut in December from three different sprout clusters in southern Switzerland (Ticino). Prior to inoculation, both ends of the stems were sealed with paraffin. For inoculation, four circular wounds (6 mm in diameter) reaching the cambium were made with a cork borer along the axis of each stem. Wounds were arranged 7 cm from each end of the stem and 12 cm apart from each other. Each wound was filled with one mycelial plug (6 mm in diameter) originating from the margin of a pure culture previously grown on PDA for 5 days in the dark. To prevent desiccation, wounds were subsequently sealed with masking tape. The inoculated stems were distributed horizontally on plastic grids located 5 cm above the bottom of 12 plastic containers (57 cm × 37 cm × 13 cm) so that each container hosted stems of all three sprout clusters. Hypovirus-free and hypovirus-infected isolates were assigned in pairs to chestnut stems. For each of the 40 pairs of isolates, three replicated inoculations were performed which were randomly distributed across containers and stems to avoid any influence of these factors on fungal growth. Containers were filled with demineralized water to a depth of 2 cm, closed with nontransparent lids and incubated at a constant temperature of 20 °C. After 28 days, the sizes of the lesions that developed at the inoculation points were determined using a millimeter ruler. Since these lesions had an elliptic shape, both the longitudinal and transversal diameters were measured. Thereafter, the geometric mean diameter was calculated.

2.6. Data Analysis

The CHV-1 virulence was quantified as the difference between the mean performances of the hypovirus-infected isolates and those of the corresponding isogenic hypovirus-free isolates and were given as a proportion (%) of the performances of the hypovirus-free isolate. A negative value indicated a higher performance of hypovirus-free isolates, whereas a positive value showed a higher performance of hypovirus-infected isolates. In the growth experiment on PDA, the performance of the *C. parasitica* isolates was estimated as the mean radial growth rate during the phase of linear growth [23]. To determine this phase, linear regression implemented in Microsoft Excel for Mac 2011 (Version 14.3.9) was performed (criterion: $R^2 > 0.98$) on the geometric mean diameters of the cultures at 72, 96 and 120 h after inoculation. In the experiment with the dormant chestnut stems, the performance of the inoculated isolates corresponded to the geometric mean diameter of the lesion 28 days after inoculation.

All statistical analyses were performed in SPSS, version 22.0 (IBM® SPSS® Statistics). Mean performances of isolates were tested for significant differences ($p < 0.05$) by conducting a Tukey's honestly significant difference test (Tukey's HSD) in conjunction with a one-way analysis of variance (One Way ANOVA). This test identifies any difference between two means that is greater than the expected standard error. Model assumptions were checked with a Tukey-Anscombe plot and a Q-Q plot in R 3.1.2. Residuals were normally distributed and displayed constant error variances. To compare the effect of the Georgian and Italian CHV-1 subtypes on *C. parasitica* isolates from the European gene pool (CpGeo97 cluster from Georgia and EU-1 isolates from Turkey), an independent-samples *t*-test ($p < 0.01$) was performed. To test for a linear relationship between the virus effect on the fungal growth on dormant stems and on PDA, the Pearson's correlation was calculated for all tested isolates.

3. Results

3.1. Prevalence and Genetic Characterization of CHV-1

Among the 664 *C. parasitica* isolates from Georgia, 62 isolates were tested positive for CHV-1 (Table 1). Of these 62 hypovirus-infected isolates, 58 exhibited a white and 4 an intermediate (between white and orange) culture morphology.

Table 1. Prevalence of CHV-1 in the *Cryphonectria parasitica* populations sampled in Georgia.

Population [1]	Region	N C.p. [2]	N CHV-1 [3]	% CHV-1
Keda (Ked)	Adjara	48	0	0
Shemoqmedi (She)	Adjara	47	14	29.8
Tkilnari (Tkhi)	Adjara	48	5	10.4
Gezruli (Gez)	Imereti	45	8	17.8
Korbouli (Kor)	Imereti	44	2	4.5
Kumistavi (Kum)	Imereti	50	12	24.0
Satsable (Sab)	Imereti	47	11	23.4
Satsire (Sat)	Imereti	71	3	4.2
Tskalthashua (Tska)	Imereti	46	4	8.7
Mukhuri (Muk)	Samegrelo-Zemo Svaneti	48	1	2.1
Taleri (Tal)	Samegrelo-Zemo Svaneti	43	2	4.7
Axalsopeli (Axa)	Kakheti	29	0	0
Khecili (Khe)	Kakheti	49	0	0
Pshaveli (Psh)	Kakheti	49	0	0
Total		664	62	9.3

[1] Abbreviation of each population in brackets. [2] Number of *C. parasitica* isolates per population. [3] Number of *C. parasitica* isolates tested positive for CHV-1.

The prevalence of CHV-1 ranged from 0% to 29.8% with an average of 9.3%. CHV-1 was detected in three regions in western Georgia (Adjara, Imereti, Samegrelo-Zemo Svaneti), but not in the region Kakheti (populations of Axa, Khe and Psh), which is located in the eastern part of the country (Figure 2).

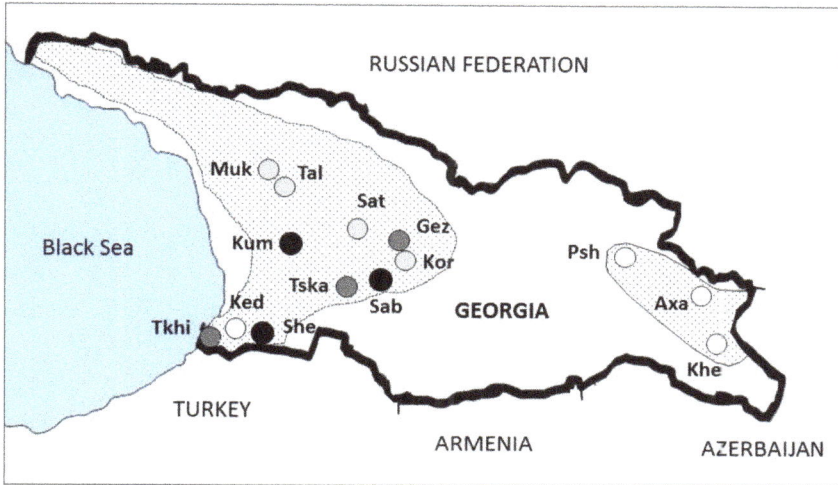

Figure 2. Prevalence and geographical distribution of CHV-1 in the *C. parasitica* populations in Georgia.

We obtained ORFA sequences from all 62 CHV-1 strains detected in Georgia. One strain failed to amplify the ORFB region resulting in 61 sequences for this region. In the ORFA region (695 nt), 100 sites were polymorphic, which corresponds to an average of 0.143 substitutions per site. Among the 62 sequences of the ORFA region, 52 haplotypes were identified. In the ORFB region (715 nt), 95 sites were polymorphic giving an average of 0.133 substitutions per site. A total of 54 haplotypes were found among the 61 sequences of the ORFB region. Pairwise identities among the ORFA sequences ranged from 97.3% to 100% and among the ORFB sequences from 96.6% to 100%.

3.2. Phylogenetic Analysis

Including the reference sequences, the ORFA alignment contained 71 sequences and the ORFB alignment 70 sequences. Individual trees for both ORFs obtained from maximum likelihood analyses in PhyML are shown in the Figure 3. In both trees, all CHV-1 strains from Georgia clustered together in one group. In ORFA, this group was clearly separated from all other CHV-1 subtypes (D, E, F, F2, and I) with maximal support. A clear separation from the subtypes F1, F2, and I was also evident for ORFB. In this region, the CHV-1 strains from Georgia were closely related to the reference strains in subtypes D and E.

Comparison of both trees showed gene tree discordance, which occurs when phylogenies obtained from individual genes differ among themselves. This is clearly evident for the Georgian CHV-1 strains, which are most closely related to CHV-1 subtype F2 in ORFA but to subtype D in ORFB. Such contrasting tree topographies can be caused by genetic recombination that can occur in host cells co-infected with different virus strains [14]. Forcing incongruent phylogenies into a single species-tree analysis may result in discordant molecular convergence, which does not reflect any biological property of the sites where it occurs, being solely the result of the technical bias [32]. Therefore, the two ORF regions were not analyzed in a concatenated dataset. Since all CHV-1 strains from Georgia formed a distinct group in both ORFs, we propose a new subtype designation for this group, namely the Georgian CHV-1 subtype (CHV-1 subtype G). Subtype G is most closely related to the European lineage A2B2, i.e., A2 in ORFA and B2 in ORFB, described by Feau et al. [14].

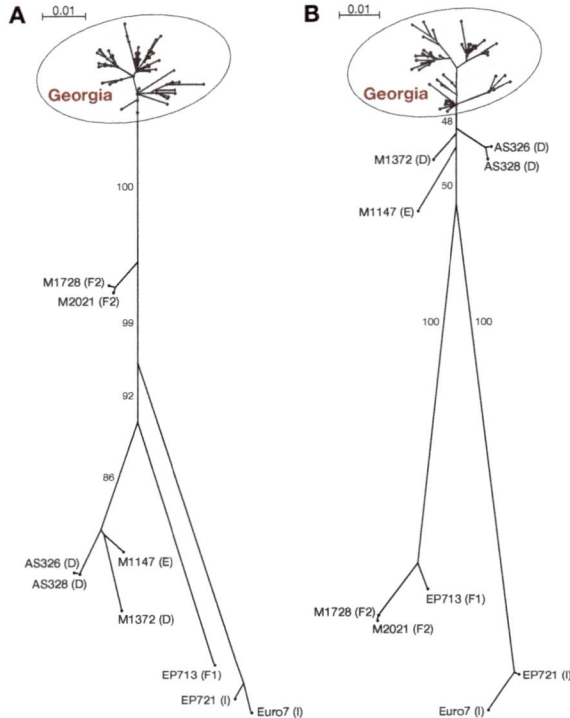

Figure 3. Individual gene trees of CHV-1 strains from Georgia and reference strains generated under maximum likelihood criterion (PhyML) using the GTR + gamma substitution model. (**A**) Gene tree of 71 ORFA sequences of CHV-1 (695 nt, 211 (30.4%) parsimony informative sites). (**B**) Gene tree of 70 ORFB sequences of CHV-1 (715 nt, 210 (29.4%) parsimony informative sites). Trees are unrooted. The scale bar represents the number of substitutions per site for a unit branch length. Numbers beside branches represent bootstraps support (%) of 1.000 bootstrap replicates. For the sake of simplicity, values of terminal branches are not shown and all CHV-1 strains from Georgia are indicated by a grey circle. Samples ID of the CHV-1 reference subtypes (D, E, F, F2 and I) are written beside nodes with the subtype denomination in brackets.

3.3. Effect of the Georgian CHV-1 Subtype on Growth of C. parasitica

The Georgian CHV-1 subtype had a different effect on the growth of its fungal host on dormant chestnut stems (parasitic growth) and on PDA medium (saprophytic growth) (Figure 4). On dormant chestnut stems, bark lesions induced by hypovirus-infected fungal isolates were significantly ($p = 0.017$) smaller than those caused by hypovirus-free isolates, independently from the genetic background of the fungal host. The mean hypovirus effect on lesion size across all isolates ranged from -22.8% (She27) to -92.7% (Tkhi48), with a mean value of $-59.1\% \pm 3.8$ ($\pm$ SE; Table S2). The strongest hypovirus effect was observed on the *C. parasitica* cluster CpGeo97 ($-74.6\% \pm 5.9$) (Figure 4A). The isolates belonging to the clusters CpGeo20 and CpGeo75 experienced a milder effect ($-48.3\% \pm 4.4$ and $-58.6\% \pm 6.7$, respectively). Hypovirus-induced growth reduction differences were, however, only significant ($p < 0.05$) between cluster CpGeo20 and cluster CpGeo97. On PDA, the presence of the hypovirus did not significantly ($p = 0.44$) affect the growth rate of the infected *C. parasitica* isolates. A hypovirus infection resulted in a slight growth stimulation ($+7.6\% \pm 3.4$) and no significant differences were detected among the different genetic clusters of the fungus (cluster CpGeo97: $+1.5\% \pm 6.2$; cluster CpGeo20: $+8.1\% \pm 4.2$; cluster CpGeo75: $+12.5\% \pm 7$; Figure 4B).

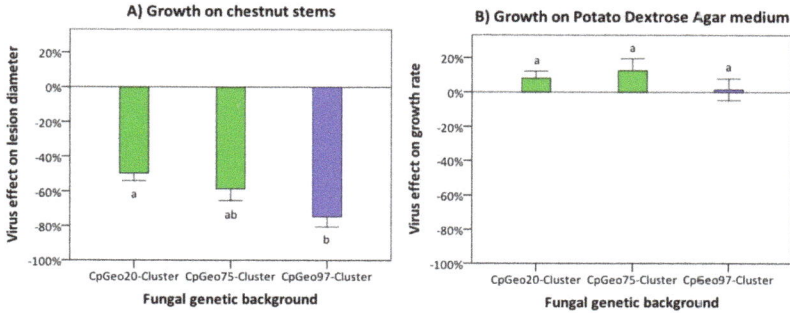

Figure 4. Effect of the Georgian CHV-1 subtype on the growth of its fungal host on dormant chestnut stems (**A**) and on PDA medium (**B**). For analyses, *C. parasitica* isolates were grouped according to their genetic background into the three genetic clusters CpGeo20 ($n = 10$), CpGeo75 ($n = 10$), and CpGeo97 ($n = 9$). Clusters CpGeo20 and CpGeo75 belong to the Georgian gene pool of *C. parasitica*, whereas cluster CpGeo97 to the European gene pool (for details on fungal genotypes see Prospero et al. [25]). The virus effect was quantified as the difference in performance between the hypovirus-infected and its isogenic hypovirus-free isolate as percentage of the performance of the hypovirus-free isolate. Error bars represent standard errors. Values with different letters differ significantly ($p < 0.05$).

3.4. Comparison of the Georgian and Italian CHV-1 Subtypes

The effect of the Georgian and Italian CHV-1 subtype on growth of the fungal host was compared in *C. parasitica* isolates belonging to the CpGeo97 cluster (European genepool). This cluster is represented by isolates from Georgia infected by the Georgian subtype ($n = 9$) and by isolates from Turkey infected by the Italian subtype ($n = 10$). Both subtypes reduced the fungal growth on dormant chestnut stem (Figure 5A), but the inhibitory effect of the Georgian subtype ($-74.6\% \pm 5.9$) was significantly higher ($p < 0.01$) than that of the Italian subtype ($-12.5\% \pm 6.9$). On PDA medium, both the Georgian and the Italian CHV-1 subtype slightly stimulated the growth of infected *C. parasitica* isolates (Figure 5B). The difference between the two subtypes was minor (Georgian subtype: $1.5\% \pm 6.2$; Italian subtype: $15.5\% \pm 3.3$) and statistically not significant ($p > 0.05$).

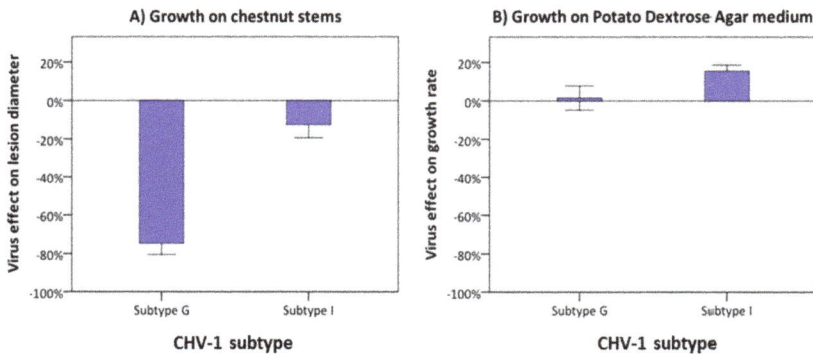

Figure 5. Effect of the Georgian (G) and the Italian (I) CHV-1 subtypes on the growth of *C. parasitica* isolates from the European gene pool on dormant chestnut stems (**A**) and on PDA (**B**). The virus effect was quantified as the difference in performance between the hypovirus-infected and its isogenic hypovirus-free isolate as percentage of the performance of the hypovirus-free isolate. Error bars represent the standard errors. Values with different letters differ significantly ($p < 0.05$).

4. Discussion

As in Europe, chestnut forests in the Caucasus are suffering from the introduced chestnut blight fungus, *Cryphonectria parasitica*. In previous studies, a unique *C. parasitica* genepool was identified in this region, which is not related to the genepools occurring in Europe [25,26]. In the present study, we demonstrate that the Georgian *C. parasitica* population harbors a unique *Cryphonectria hypovirus* population. All hypovirus strains detected formed distinct phylogenetic groups in both regions analyzed. Following the CHV-1 subtype designation used for European hypoviruses ([13,33], we propose the new hypovirus group to be named the Georgian CHV-1 subtype (subtype G).

In relation to the previously described CHV-1 subtypes from Europe, conflicting phylogenetic signals between the ORFA and ORB regions were obtained for the Georgian CHV-1 strains. In ORFA the analyzed strains were most closely related to the CHV-1 subtype F2 and in ORFB to the subtypes D/E. Finding gene tree discordance in viral populations is not surprising, as phylogenetic incongruence has been often observed within viral strains [34,35]. This phenomenon may be the consequence of viral recombination, which occurs when two different virus strains co-infect a single cell. The viral RNA-dependent-RNA-polymerase can dissociate from the first genome and continue replication by using the second distinct genome as template [30,36]. The result is a novel mosaic-like genome with regions from different sources. Recombination events have been previously described in CHV-1 and very likely have contributed to the evolution of CHV-1 subtypes in Europe [14,20]. The study of Feau et al. [14] showed that two recombinant CHV-1 lineages (subtype F1 and I) were more frequent than the putative parental strains suggesting an increased invasiveness of the recombinants.

Mlinarec et al. [20] recently provided full genome sequences of all European subtypes including one CHV-1 strain from Georgia. Recombinant analysis suggested that the strain from Georgia has a recombinant pattern between CHV-1 subtype I (minor parent) and the subtypes F1, D, E (major parent). Our study indicates that these putative parental strains do not occur in Georgia and suggests that the recombination event took place outside of this area. Western Europe (France, Spain) could be the center of origin of the Georgian subtype since all putative parental CHV-1 subtypes have been found in that area [14,19,37]. However, there is no indication of any relationship between the *C. parasitica* population in western Europe and Georgia [25]. To our knowledge there are also no records for any movement of chestnut plant material between the two areas, which could have spread hypovirus-infected *C. parasitica* isolates. Alternatively, and more likely, the Georgian subtype has evolved somewhere in Asia, i.e., in the native range of *C. parasitica* and CHV-1. *C. parasitica* isolates harboring such recombinant CHV-1 strains were then introduced into Georgia and founded both the fungal and virus population in this country. To our knowledge artificial application of hypovirus-infected *C. parasitica* strains to control chestnut blight have never been performed in the Caucasus region including eastern Turkey. Thus, the unique CHV-1 subtype identified in Georgia was very likely introduced there together with its fungal host from Asia. The recent finding of related CHV-1 strains in Eastern Turkey [17] suggests that CHV-1 subtype G is slowly migrating westwards.

Our study revealed distinct hypovirus x fungus interactions for parasitic growth of the fungus on dormant chestnut stems, but not for saprophytic growth on PDA plates. On chestnut stems, the Georgian CHV-1 subtype was found to have a different effect on the growth of *C. parasitica* depending on the genetic background of the fungal host strains. This was most evident when comparing the Georgian CpGeo20 cluster with the European CpGeo97 cluster, which experienced a significant greater growth reduction than the former cluster. It has been hypothesized that mycoviruses will adapt to its fungal hosts and evolve towards milder strains, in order not to severely reduce the fitness of the hosts [21,38]. This hypothesis is based on the fact that most mycoviruses, including CHV-1, do not have an extracellular phase and fully depend on their hosts for survival and spread [1,2]. Along this line, our results suggest that the Georgian CHV-1 subtype is more adapted to the Georgian *C. parasitica* genepool than to the European genepool. This assumption is consistent with the scenario that the Georgian CHV-1 subtype has been present for a longer time in the Georgian genepool and only recently infected the European genepool that had migrated from Turkey into Georgia. Similarly, the Georgian CHV-1 subtype had a more severe effect on the European genepool than the Italian subtype, which has shared a long invasion history with the European *C. parasitica* population [15].

Hypovirus-mediated biological control of chestnut blight has been successful in many areas in Europe, either naturally or after artificial application [5]. The basis of biocontrol is provided by the ability of CHV-1 to reduce the parasitic growth (virulence) of infected *C. parasitica* isolates. An additional important characteristic of hypovirulence is the transmissibility of CHV-1 among fungal individuals. This horizontal hypovirus transmission is controlled by the vegetative incompatibility system in *C. parasitica* [8]. If barriers imposed by this system are not too high, CHV-1 is able to infect a large proportion of a fungal population [39]. In Europe, where hypovirulence is successful it is not uncommon to find a hypovirus prevalence of more than 50% [3]. In Georgia, we found three locations with a CHV-1 prevalence between 20% and 30%. In most other locations, the prevalence was much lower with zero detection in the eastern part of the country. Even at a relatively low prevalence, CHV-1 might still mitigate the chestnut blight severity in Georgia at least to a certain degree and particularly in the western part of the country. Our phenotypic analysis supports this assumption by demonstrating that the Georgian CHV-1 subtype strongly reduces the parasitic growth of all three genetic clusters of *C. parasitica* identified in Georgia. It is interesting to note that Georgian CHV-1 subtype did not reduce the saprophytic growth of *C. parasitica* as assessed on PDA plates. This is a typical characteristic of many CHV-1 strains, particularly those of the Italian subtype [16]. The saprophytic ability of hypovirus-infected *C. parasitica* to growth and sporulate on dead chestnut stems is thought to be important for the success of natural hypovirulence in Europe [9] and probably also plays a role in the natural chestnut forests in Georgia.

In eastern Georgia, where CHV-1 was not found, biological control applications are highly feasible using CHV-1 strains from the western part of the country. The genetic diversity of *C. parasitica* is low in the eastern part with a single dominant genotype (CpGeo20) present [25]. Therefore, there are almost no vegetative incompatibility (*vic*) barriers, which should favor the spread of an artificially applied hypovirus. Hypovirulence application in the western part of the country is more problematic since the local *C. parasitica* population is genetically highly diverse as determined by microsatellite markers [25]. Vegetative incompatibility so far has not been studied in Georgia. Studies in other areas, however, showed that *vic* genotype diversity is highly correlated with microsatellite diversity [26,40,41]. Therefore, we can assume that *vic* barriers are strong in western Georgia and are probably responsible for the low prevalence of CHV-1 in local *C. parasitica* populations.

In our study, we focused on CHV-1, which is the most efficient biological control agent against chestnut blight and to date the only mycovirus reported in *C. parasitica* from Europe. In North America and Asia, several other mycoviruses have been found in the chestnut blight fungus [24]. It is likely that *Cryphonectria* mycoviruses other than CHV-1 also occur in Europa including Georgia, but screening for the presence of such viruses remains to be done.

5. Conclusions

Our study reveals the occurrence of a unique CHV-1 subtype in the *C. parasitica* population in Georgia. This finding is consistent with a previous study showing that there is a unique fungal genepool in Georgia, which is not related to other *C. parasitica* genepools in Europe. The Georgian CHV-1 subtype belongs to the more severe hypoviruses as it strongly reduced the parasitic growth of *C. parasitica*. Although this CHV-1 subtype overall has a strong phenotypic effect on *C. parasitica*, it seems to be more adapted to the Georgian than to the European genepool of *C. parasitica*. This suggests a longer interaction history with the Georgian than with the European genepool, which probably had migrated only recently from Turkey into Georgia. The CHV-1 subtype in Georgia provides a valuable biological control agent for hypovirulence applications in the Caucasus region.

Supplementary Materials: The following are available online at http://www.mdpi.com/1999-4915/10/12/687/s1, Table S1: List of CHV-1 strains used in this study and GenBank accession numbers of the CHV-1 sequences, Table S2: Effect of the Georgian CHV-1 subtype on the growth of the infected *Cryphonectria parasitica* isolates on PDA and on dormant chestnut stems.

Author Contributions: D.R., S.P. and N.B. conceived and designed the experiments; N.B. performed the experiments; N.B., D.R. and C.C. analyzed the data; A.S. contributed samples; D.R., N.B., C.C. and S.P. wrote the paper.

Funding: This research was funded by the Swiss National Science Foundation (Scopes project IZ73Z0-152525).

Acknowledgments: We thank Paata Torchinava and Bedzina Tavadze for helping with the field sampling in Georgia and Esther Jung and Hélène Blauenstein for helping with fungal isolations and virus sequencing.

Conflicts of Interest: The authors declare no conflict of interest.

References

1. Ghabrial, S.A.; Suzuki, N. Viruses of plant pathogenic fungi. *Annu. Rev. Phytopathol.* **2009**, *47*, 353–384. [CrossRef] [PubMed]

2. Nuss, D.L. Hypovirulence: Mycoviruses at the fungal-plant interface. *Nat. Rev. Microbiol.* **2005**, *3*, 632–642. [CrossRef] [PubMed]

3. Rigling, D.; Prospero, S. *Cryphonectria parasitica*, the causal agent of chestnut blight: Invasion history, population biology and disease control. *Mol. Plant Pathol.* **2018**, *19*, 7–20. [CrossRef] [PubMed]

4. Anagnostakis, S.L. Chestnut blight—The classical problem of an introduced pathogen. *Mycologia* **1987**, *79*, 23–37. [CrossRef]

5. Heiniger, U.; Rigling, D. Biological control of chestnut blight in Europe. *Annu. Rev. Phytopathol.* **1994**, *32*, 581–599. [CrossRef]

6. Choi, G.H.; Nuss, D.L. Hypovirulence of chestnut blight fungus conferred by an infectious viral cDNA. *Science* **1992**, *257*, 800–803. [CrossRef]

7. Peever, T.L.; Liu, Y.C.; Cortesi, P.; Milgroom, M.G. Variation in tolerance and virulence in the chestnut blight fungus-hypovirus interaction. *Appl. Environ. Microb.* **2000**, *66*, 4863–4869. [CrossRef]

8. Cortesi, P.; McCulloch, C.E.; Song, H.Y.; Lin, H.Q.; Milgroom, M.G. Genetic control of horizontal virus transmission in the chestnut blight fungus, *Cryphonectria parasitica*. *Genetics* **2001**, *159*, 107–118.

9. Prospero, S.; Conedera, M.; Heiniger, U.; Rigling, D. Saprophytic activity and sporulation of *Cryphonectria parasitica* on dead chestnut wood in forests with naturally established hypovirulence. *Phytopathology* **2006**, *96*, 1337–1344. [CrossRef]

10. Grente, M.J. Les formes hypovirulentes d'*Endothia parasitica* et les espoirs de lutte contre le chancre du châtaignier. *Académie D'agriculture de France* **1965**, *51*, 1033–1036.

11. Diamandis, S. Management of chestnut blight in Greece using hypovirulence and silvicultural interventions. *Forests* **2018**, *9*, 492. [CrossRef]

12. Bryner, S.F.; Prospero, S.; Rigling, D. Dynamics of Cryphonectria hypovirus infection in chestnut blight cankers. *Phytopathology* **2014**, *104*, 918–925. [CrossRef] [PubMed]

13. Gobbin, D.; Hoegger, P.J.; Heiniger, U.; Rigling, D. Sequence variation and evolution of *Cryphonectria hypovirus 1* (CHV-1) in Europe. *Virus Res.* **2003**, *97*, 39–46. [CrossRef]

14. Feau, N.; Dutech, C.; Brusini, J.; Rigling, D.; Robin, C. Multiple introductions and recombination in *Cryphonectria hypovirus 1*: Perspective for a sustainable biological control of chestnut blight. *Evol. Appl.* **2014**, *7*, 580–596. [CrossRef] [PubMed]

15. Bryner, S.F.; Rigling, D.; Brunner, P.C. Invasion history and demographic pattern of *Cryphonectria hypovirus 1* across European populations of the chestnut blight fungus. *Ecol. Evol.* **2012**, *2*, 3227–3241. [CrossRef] [PubMed]

16. Robin, C.; Lanz, S.; Soutrenon, A.; Rigling, D. Dominance of natural over released biological control agents of the chestnut blight fungus *Cryphonectria parasitica* in South-Eastern France is associated with fitness-related traits. *Biol. Control* **2010**, *53*, 55–61. [CrossRef]

17. Akilli, S.; Serce, C.U.; Katircioglu, Y.Z.; Maden, S.; Rigling, D. Characterization of hypovirulent isolates of the chestnut blight fungus, *Cryphonectria parasitica* from the Marmara and Black Sea regions of Turkey. *Eur. J. Plant Pathol.* **2013**, *135*, 323–334. [CrossRef]

18. Peters, F.S.; Busskamp, J.; Prospero, S.; Rigling, D.; Metzler, B. Genetic diversification of the chestnut blight fungus *Cryphonectria parasitica* and its associated hypovirus in Germany. *Fungal Biol.* **2014**, *118*, 193–210. [CrossRef]

19. Trapiello, E.; Rigling, D.; Gonzalez, A.J. Occurrence of hypovirus-infected *Cryphonectria parasitica* isolates in northern Spain: An encouraging situation for biological control of chestnut blight in Asturian forests. *Eur. J. Plant Pathol.* **2017**, *149*, 503–514. [CrossRef]

20. Mlinarec, J.; Nuskern, L.; Jezic, M.; Rigling, D.; Curkovic-Perica, M. Molecular evolution and invasion pattern of *Cryphonectria hypovirus 1* in Europe: Mutation rate, and selection pressure differ between genome domains. *Virology* **2018**, *514*, 156–164. [CrossRef]

21. Bryner, S.F.; Rigling, D. Hypovirus virulence and vegetative incompatibility in populations of the chestnut blight fungus. *Phytopathology* **2012**, *102*, 1161–1167. [CrossRef] [PubMed]

22. Chen, B.S.; Nuss, D.L. Infectious cDNA clone of hypovirus CHV1-Euro7: A comparative virology approach to investigate virus-mediated hypovirulence of the chestnut blight fungus *Cryphonectria parasitica*. *J. Virol.* **1999**, *73*, 985–992. [PubMed]

23. Bryner, S.F.; Rigling, D. Temperature-dependent genotype-by-genotype interaction between a pathogenic fungus and its hyperparasitic virus. *Am. Nat.* **2011**, *177*, 65–74. [CrossRef]

24. Hillman, B.I.; Suzuki, N. Viruses of the chestnut blight fungus, *Cryphonectria parasitica* *Adv. Virus Res.* **2004**, *63*, 423–472.

25. Prospero, S.; Lutz, A.; Tavadze, B.; Supatashvili, A.; Rigling, D. Discovery of a new gene pool and a high genetic diversity of the chestnut blight fungus *Cryphonectria parasitica* in Caucasian Georgia. *Infect. Genet. Evol.* **2013**, *20*, 131–139. [CrossRef] [PubMed]

26. Aghayeva, D.N.; Rigling, D.; Prospero, S. Low genetic diversity but frequent sexual reproduction of the chestnut blight fungus *Cryphonectria parasitica* in Azerbaijan. *For. Pathol.* **2017**, *47*, e12357. [CrossRef]

27. Bissegger, M.; Rigling, D.; Heiniger, U. Population structure and disease development of *Cryphonectria parasitica* in European chestnut forests in the presence of natural hypovirulence. *Phytopathology* **1997**, *87*, 50–59. [CrossRef] [PubMed]

28. Guindon, S.; Dufayard, J.F.; Lefort, V.; Anisimova, M.; Hordijk, W.; Gascuel, O. New algorithms and methods to estimate maximum-likelihood phylogenies: Assessing the performance of PhyML 3.0. *Syst. Biol.* **2010**, *59*, 307–321. [CrossRef]

29. Lefort, V.; Longueville, J.E.; Gascuel, O. SMS: Smart model selection in PhyML. *Mol. Biol. Evol.* **2017**, *34*, 2422–2424. [CrossRef]

30. Vijaykrishna, D.; Mukerji, R.; Smith, G.J.D. RNA virus reassortment: An evolutionary mechanism for host jumps and immune evasion. *PLoS Pathog.* **2015**, *11*, e1004902. [CrossRef]

31. Huson, D.H.; Bryant, D. Application of phylogenetic networks in evolutionary studies. *Mol. Biol. Evol.* **2006**, *23*, 254–267. [CrossRef] [PubMed]

32. Mendes, F.K.; Hahn, Y.; Hahn, M.W. Gene tree discordance can generate patterns of diminishing convergence over time. *Mol. Biol. Evol.* **2016**, *33*, 3299–3307. [CrossRef] [PubMed]

33. Allemann, C.; Hoegger, P.; Heiniger, U.; Rigling, D. Genetic variation of *Cryphonectria* hypoviruses (CHV1) in Europe, assessed using restriction fragment length polymorphism (RFLP) markers. *Mol. Ecol.* **1999**, *8*, 843–854. [CrossRef]

34. Wani, S.A.; Sahu, A.R.; Mishra, B.P.; Kumar, A.; Priya, G.B.; Padhy, A.; Sahoo, A.P.; Tiwari, A.K.; Gandham, R.K.; Singh, R.K. Whole genome sequence analysis of viruses; moving beyond single/partial gene based phylogenies in context of epidemiology and genetic evolution. *Adv. Anim. Vet. Sci.* **2015**, *3*, 435–443. [CrossRef]

35. Kostaki, E.G.; Karamitros, T.; Stefanou, G.; Mamais, I.; Angelis, K.; Hatzakis, A.; Kramvis, A.; Paraskevis, D. Unravelling the history of hepatitis B virus genotypes A and D infection using a full-genome phylogenetic and phylogeographic approach. *eLife* **2018**, *7*, e36709. [CrossRef] [PubMed]

36. Simon-Loriere, E.; Holmes, E.C. Why do RNA viruses recombine? *Nat. Rev. Microbiol.* **2011**, *9*, 617–626. [CrossRef] [PubMed]

37. Zamora, P.; Martin, A.B.; Rigling, D.; Diez, J.J. Diversity of *Cryphonectria parasitica* in western Spain and identification of hypovirus-infected isolates. *For. Pathol.* **2012**, *42*, 412–419. [CrossRef]

38. Bryner, S.F.; Rigling, D. Virulence not only costs but also benefits the transmission of a fungal virus. *Evolution* **2012**, *66*, 2540–2550. [CrossRef]

39. Brusini, J.; Robin, C. Mycovirus transmission revisited by in situ pairings of vegetatively incompatible isolates of *Cryphonectria parasitica*. *J. Virol. Methods* **2013**, *187*, 435–442. [CrossRef]

40. Jezic, M.; Krstin, L.; Rigling, D.; Curkovic-Perica, M. High diversity in populations of the introduced plant pathogen, *Cryphonectria parasitica*, due to encounters between genetically divergent genotypes. *Mol. Ecol.* **2012**, *21*, 87–99. [CrossRef]

41. Prospero, S.; Rigling, D. Invasion genetics of the chestnut blight fungus *Cryphonectria parasitica* in Switzerland. *Phytopathology* **2012**, *102*, 73–82. [CrossRef] [PubMed]

viruses

MDPI

Article

Detection of a Conspecific Mycovirus in Two Closely Related Native and Introduced Fungal Hosts and Evidence for Interspecific Virus Transmission

Corine N. Schoebel *, Simone Prospero, Andrin Gross and Daniel Rigling

Swiss Federal Institute for Forest, Snow and Landscape Research, WSL, Zuercherstrasse 111, 8903 Birmensdorf, Switzerland; Simone.Prospero@wsl.ch (S.P.); andrin.gross@wsl.ch (A.G.); Daniel.rigling@wsl.ch (D.R.)
* Correspondence: corine.schoebel@wsl.ch; Tel.: +41-44-739-25-28

Received: 24 October 2018; Accepted: 10 November 2018; Published: 13 November 2018

Abstract: *Hymenoscyphus albidus* is a native fungus in Europe where it behaves as a harmless decomposer of leaves of common ash. Its close relative *Hymenoscyphus fraxineus* was introduced into Europe from Asia and currently threatens ash (*Fraxinus* sp.) stands all across the continent causing ash dieback. *H. fraxineus* isolates from Europe were previously shown to harbor a mycovirus named *Hymenoscyphus fraxineus Mitovirus 1* (HfMV1). In the present study, we describe a conspecific mycovirus that we detected in *H. albidus*. HfMV1 was consistently identified in *H. albidus* isolates (mean prevalence: 49.3%) which were collected in the sampling areas before the arrival of ash dieback. HfMV1 strains in both fungal hosts contain a single ORF of identical length (717 AA) for which a mean pairwise identity of 94.5% was revealed. The occurrence of a conspecific mitovirus in *H. albidus* and *H. fraxineus* is most likely the result of parallel virus evolution in the two fungal hosts. HfMV1 sequences from *H. albidus* showed a higher nucleotide diversity and a higher number of mutations compared to those from *H. fraxineus*, probably due to a bottleneck caused by the introduction of *H. fraxineus* in Europe. Our data also points to multiple interspecific virus transfers from *H. albidus* to *H. fraxineus*, which could have contributed to the intraspecific virus diversity found in *H. fraxineus*.

Keywords: *Chalara fraxinea*; *Hymenoscyphus pseudoalbidus*; ash dieback; Narnaviridae; evolution; invasive species; horizontal virus transmission

---

1. Introduction

In the past years, species invasions have become an important topic in the scientific literature due to increasing awareness of the consequences of intensified international trade of plant material [1–3]. In the case of introduced (exotic) plant pathogens, we frequently observe invasions by species that have a harmless close relative in the invaded area and which are often more virulent due to the lack of co-evolution with the novel host. This can have catastrophic impacts on forest ecosystems (e.g., [4,5]). So far, most studies have focused on the invasive species itself, while little attention has been paid to possible interactions with closely related native species that may occupy similar ecological niches and share evolutionary histories. The duration of this interaction is often limited because the invasive species outcompetes its native sister species (e.g., *Cryphonectria parasitica* vs. *C. radicalis* in Europe; [6]). Nevertheless, during this short timeframe, gene flow between native and introduced species may occur, resulting in varying levels of gene introgression between the two interacting taxa. The occurrence of hybridization between allopatric species following attained sympatry because of anthropogenic introduction has been reported for example in the basidiomycete genus *Heterobasidion* [7] and in the oomycete genus *Phytophthora* [8].

In the present study we focus on the interaction of *Hymenoscyphus fraxineus* and *H. albidus*, which belong to the family Helotiaceae and are closely related [9]. The first species is native to Asia and in Europe is an invasive pathogen causing ash dieback (ADB), mainly on common (*Fraxinus excelsior*) and narrow-leaved ash (*Fraxinus angustifolia*) [10]. The latter species is a native endophyte of common ash in Europe (e.g., [11]). Due to its severe ecological consequences, there has lately been a great interest in the biology, ecology, and population genetics of *H. fraxineus* (e.g., [12–16]). However, little is known about *H. albidus* [17,18], as well as its interaction with *H. fraxineus*. Field reports indicate that since the onset of the ADB epidemic in Europe, the frequency of detection of *H. albidus* has decreased strongly, with the species being at risk of local extinction [19,20]. Both species show a similar life cycle and, more importantly, occupy the same sporulation niche, i.e., rachis and petioles of last year's ash leaves in the litter [17,18,20]. The introduced *H. fraxineus* is most likely outcompeting its native relative because of its superior ability to cause leaf infections and its massive production of airborne ascospores [21]. After the first detection of the disease in the 1990s in Eastern Europe [22], *H. fraxineus* is now abundant in most ash stands all across mainland Europe and the UK [15,23,24].

Fungal viruses (mycoviruses) are widespread in all major groups of fungi [25] and a mycovirus, *Hymenoscyphus fraxineus Mitovirus 1* (HfMV1) was recently also discovered in *H. fraxineus* [26]. HfMV1 is a putative member of the genus Mitovirus (family *Narnaviridae*) that are only present in fungi. Mitoviruses are commonly found in fungi and represent the simplest mycoviruses with an unencapsidated single strand RNA genome of approximately 2.5 kb. The genome contains one single open reading frame (ORF), encoding the RNA-dependent RNA polymerase (RdRp). As many other RNA viruses, mitoviruses are mutating at high rates [27]. Mitoviruses are located and translated in the mitochondria [28,29]. It has previously been shown that some mycoviruses may be transferred between related fungal species e.g., in the genera *Cryphonectria*, *Heterobasidion*, *Sclerotinia*, as well as between fungal taxa e.g., *Sclerotinia homoeocarpa* and *Ophiostoma novo-ulmi* [30–33]. In many of these studies, evidence for interspecific virus transmission was based on the occurrence of conspecific viruses in different fungal host species.

In the present study, we tested the occurrence of a conspecific mitovirus in *H. albidus* and *H. fraxineus* using the RdRP gene as a marker for population genetic analyses. Specifically, we (i) determined the presence and prevalence of HfMV1 sequences in *H. albidus* isolates, which were collected in Switzerland and France before the arrival and spread of the invasive *H. fraxineus*; (ii) determined the phylogenetic relationship of the viral sequences detected in *H. albidus* and *H. fraxineus*; and (iii) compared the genetic diversity of the viruses in *H. albidus* and *H. fraxineus* populations from the same geographic regions, to determine the possible origin of the conspecific virus in the two closely related host species.

2. Materials and Methods

2.1. Isolates Used in the Study

Overall, 67 isolates of *H. albidus* from France and Switzerland and 221 HfMV1-positive isolates of *H. fraxineus* from Europe (mainly Switzerland) were analyzed (Table 1). All *H. albidus* isolates were recovered from colonized petioles of common ash. The *H. fraxineus* isolates were also obtained from common ash, either from petioles or bark lesions (for details see [14,26]). All *H. albidus* isolates were identified by ITS sequencing using ITS1 and ITS4 primers [34]. The *H. fraxineus* isolates were identified either by ITS sequencing or microsatellite genotyping [12]

Table 1. Overview of the *Hymenoscyphus albidus* and *H. fraxineus* isolates included in the present study.

Host	Region (Abbreviation)	Canton (s)/Region	Number of Samples Screened	Number of Samples with HfMV1	Sampling Year (s)	Reference for Fungal Isolates
H. albidus						
Switzerland	South-Western Switzerland (SW-CH)	VS, VD	67	34	2008, 2009, 2012	This study
	Southern Switzerland (S-CH)	TI	13	13	2009, 2010, 2012	[35]; this study
	Northern Switzerland (N-CH)	GL, ZH	33	12	2009	[35]
France	North-Western France (NW-F)	Bretagne	14	5	2012	[17]; this study
H. fraxineus						
Switzerland	South-Western Switzerland (SW-CH)	VS, VD	**	221	2013, 2016	[13]; this study
	Southern-Switzerland * (S-CH)	TI, GR	**	41	2014, 2013, 2016	[15]; this study
	Northern-Switzerland (N-CH)	SG, SZ, UR	**	90	2013, 2014	[13]; this study

* Including two HfMV1 positive isolates from Northern Italy; ** HfMV1 positive isolates were selected, no data on number initially screened.

2.2. Fungal Cultivation and RNA Extraction

For each sample, a small piece of growing pure culture was transferred to a new cellophane covered malt agar plate [13]. These were incubated for 5 weeks in the dark at room temperature. Thereafter, the mycelia were harvested, lyophilized and then milled in a MixerMill (MM33, Retsch GmbH, Haan, NRW, Germany). For each sample, approximately 20 mg of the ground mycelium were used for RNA extraction, using the PureLink Pro 96 total RNA Purification Kit (Invitrogen, Carlsbad, CA, USA), following the manufacturer's instructions.

2.3. Screening for HfMV1 and Viral Sequencing

All *H. albidus* and *H. fraxineus* isolates were screened for the presence of HfMV1 using a specific PCR assay [26]. Complementary DNA (cDNA) was produced with the Maxima First Strand cDNA Synthesis Kit (Life Technologies, Carlsbad, CA, USA) according to the manufacturer's protocol. For sequencing, PCR products were purified using Illustra ExoProStar (GE Healthcare Life Sciences, Pittsburgh, PA, USA) according to the instructions. Sanger sequencing of the PCR products was done on both strands using the same primers as for PCR. In addition, for 40 isolates (8 for *H. albidus*, 32 for *H. fraxineus*) the full length viral RdRP gene (2151 bp) was sequenced in both directions as described in [26]. To account for mutations at primers sites in some of the isolates, additional primers had to be designed to obtain the complete ORF (see Table S2).

2.4. Phylogenetic Analyses

The maximum likelihood (ML) phylogenetic analysis was conducted for the nucleotide alignment. The alignment quality was first double-checked using the MACSE aligner [36]. A total of 40 separate runs each with 10,000 replicates were conducted using the 'rapid bootstrap analysis and tree search' algorithm, resulting in a best ML tree with branch lengths and bootstrap support values. Two separate analyses were performed, (i) for the full length RdRP gene (2151 bp, N = 40) and (ii) for the partial RdRP gene (495 bp, N = 255). For the partial RdRP gene, three distinct data partitions with joint branch length optimization in RAxML v8.2.4 [37] implementing the GTRGAMMA model of nucleotide substitution were used.

The Bayesian analysis was conducted for each of the two datasets separately, using MRBAYES v3.2.6 [38]. Three independent runs, each consisting of 100 million generations with a burn-in of 80 million and a sampling frequency of 1000 were performed.

The resulting data was visualized using the software TREEGRAPH v2 [39] and FIGTREE v1.4.2 (http://tree.bio.edac.uk/software/figtree/).

2.5. Genetic Diversity, Differentiation and Evolution

DNASP v5 [40] was used to determine the nucleotide diversity estimated by the average number of differences per site between two sequences (π), the number of haplotypes observed (h), the haplotype diversity (Hd), the number of segregating sites (S), the number of total mutations (η), and the average number of differences (K) per country/population were calculated for the two datasets 'partial ORF' and 'full ORF'. Pairwise identity percentages for the nucleotide and amino acid sequences were calculated using Multiple Sequence Comparison by Log- Expectation (MUSCLE; http://www.ebi.ac.uk/Tools/msa/muscle).

3. Results

In the present study, we describe the first fungal virus detected in *H. albidus*. HfMV1, which was previously described for *H. fraxineus*, could now also be detected in its close relative, *H. albidus*. Although vast areas of Switzerland were sampled, from 2013 onward *H. albidus* could not be isolated any longer.

3.1. Prevalence of HfMV1 in H. albidus

All *H. albidus* isolates analysed originated from Swiss and French regions where *H. fraxineus* had not been officially reported at the time of sampling (Figure 1). By using a HfMV1-specific PCR assay, we detected HfMV1-sequences in 33 out of 67 *H. albidus* isolates (49.3%). More precisely, 28 out of 53 Swiss isolates (52.8%) and 5 out of 14 French isolates (35.7%) harbored HfMV1

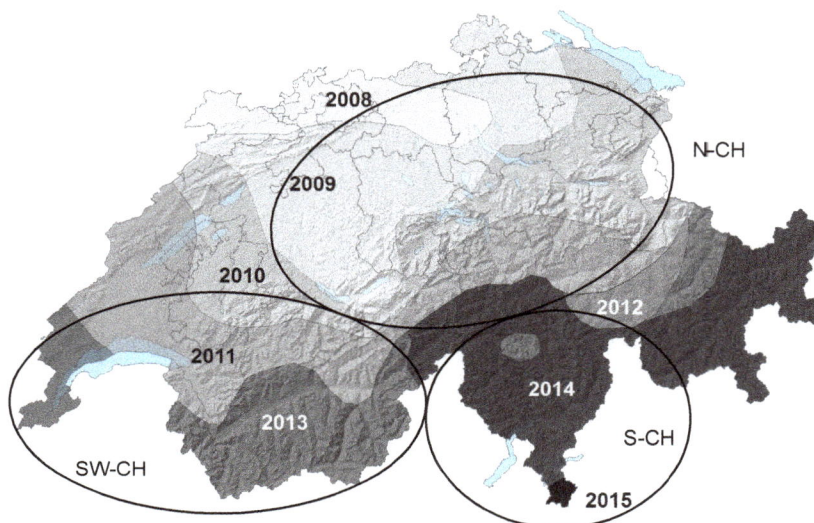

Figure 1. Geographic origin of the Swiss *Hymenoscyphus albidus* and *H. fraxineus* samples analyzed in this study. Grey tones display the year of first report of ash dieback. Light grey lines display Swiss canton borders. Abbreviations: S-CH: Southern Switzerland, SW-CH: South-Western Switzerland, N-CH: Northern Switzerland. Modified after [41].

3.2. Pairwise Sequence Identity

All virus sequences obtained from *H. albidus* and *H. fraxineus* showed high similarities and were of identical length for both the full and partially sequenced RdRP gene. Average pairwise sequence identity (PWI) was 95% (range between 93% to 97%), when comparing the full ORF amino acid sequences (717 AA) of HfMV1 originating from the two fungal host species. For the nucleotide sequences (NT) the average PWI was 91% (90% to 93%). When considering only HfMV1 sequences from *H. albidus*, the PWI value was on average 97% (92% for NT), whereas for HfMV1 sequences from *H. fraxineus*, the PWI value was 98% (96% for NT). According to the International Committee on Taxonomy of Viruses, a PWI < 40 indicates different species, whereas PWI > 90 is indicative that the compared sequences belong to the same virus species [42]. Therefore, we can conclude that all viral sequences analyzed in this study belong to the same species, which was previously described as HfMV1 [26].

3.3. Phylogenetic Relationships of HfMV1 across Two Different Host Species

For the full RdRP gene of HfMV1 (2151 bp), three well supported groups of sequences were detected (posterior probability, PP = 1; Figure 2). RAxML analysis evidenced very similar three groups, with >90% bootstrap support (Figure S1). In accordance with [15], these groups were called HfMV1 group 1, HfMV1 group 2 and HfMV1 *H. albidus* group. All viral sequences obtained from *H. albidus* belonged to a separate group, which consisted of long-branched sequences. HfMV1 group 2 also included long-branched sequences, all from Switzerland. HfMV1 group 1, on the other hand, was characterized by short branches with little differentiation and low support values. The sequences of the German and Polish isolates, which were initially used for HfMV1 primer design [26], came to reside in this group (Hf_DE_C436 and Hf_PL_C428 in Figure 2). The full RdRP sequences from *H. fraxineus* were allocated both to HfMV1 group 1 (11 isolates, light grey in Figure 2) and HfMV1 group 2 (19 isolates, black in Figure 2). All 8 full RdRP sequences obtained from *H. albidus* formed the *H. albidus* group. In addition, also two full RdRP sequences of *H. fraxineus* isolates obtained from necrotic lesions cluster here (arrows and blue font in Figure 2).

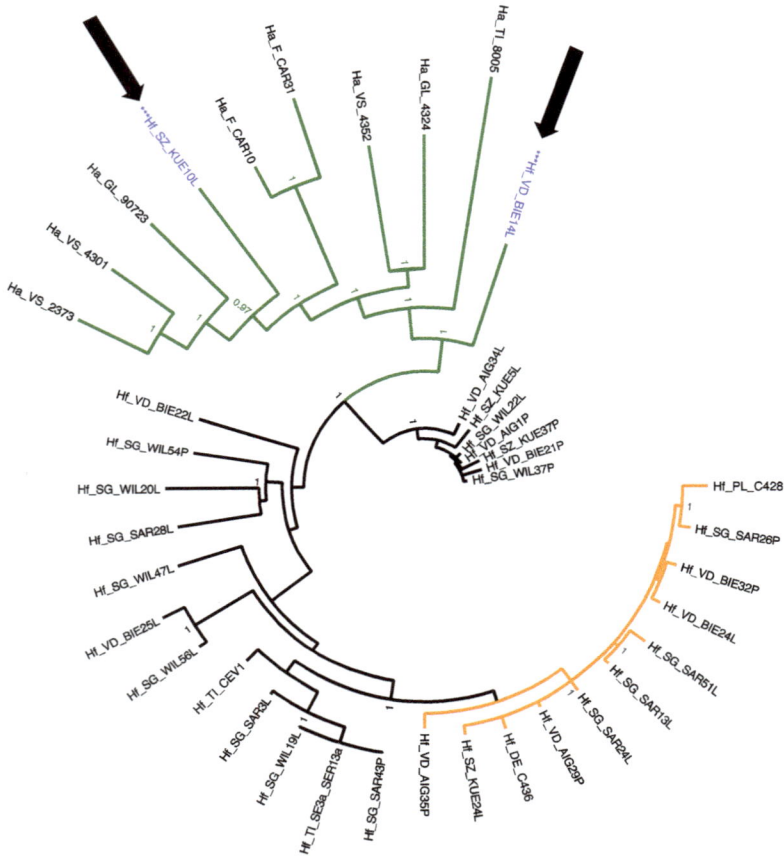

Figure 2. Phylogenetic tree resulting from the MrBayes analysis conducted with 40 (8 *H. albidus*, 32 *H. fraxineus*) full sequences (2151 bp) of the RdRP gene of the mitovirus HfMV1. Green color depicts HfMV1 *H. albidus* group isolates. Black color depicts HfMV1 group 2 isolates and orange color HfMV1 group 1 isolates. Blue color and *** marks *H. fraxineus* isolates within the *H. albidus* group. Hf characterizes viral sequences from *H. fraxineus*, Ha from *H. albidus* isolates. For abbreviation see Table 1. Posterior probabilities are shown at the respective branch if they were >0.95. Arrows indicate intermixing of *H. albidus* and *H. fraxineus* HfMV1 sequences.

For the partial RdRP gene of HfMV1 (495 bp), two major groups of sequences (PP = 1, bootstrap value 61%) were detected (Figure 3). In accordance with the full RdRP gene sequences and [14] these groups were again called HfMV1 group 1 and HfMV1 group 2. The *H. albidus* sequences formed three distinct clusters, which were all closer related to HfMV1 group 2 than to HfMV1 group 1 (Figure 3).

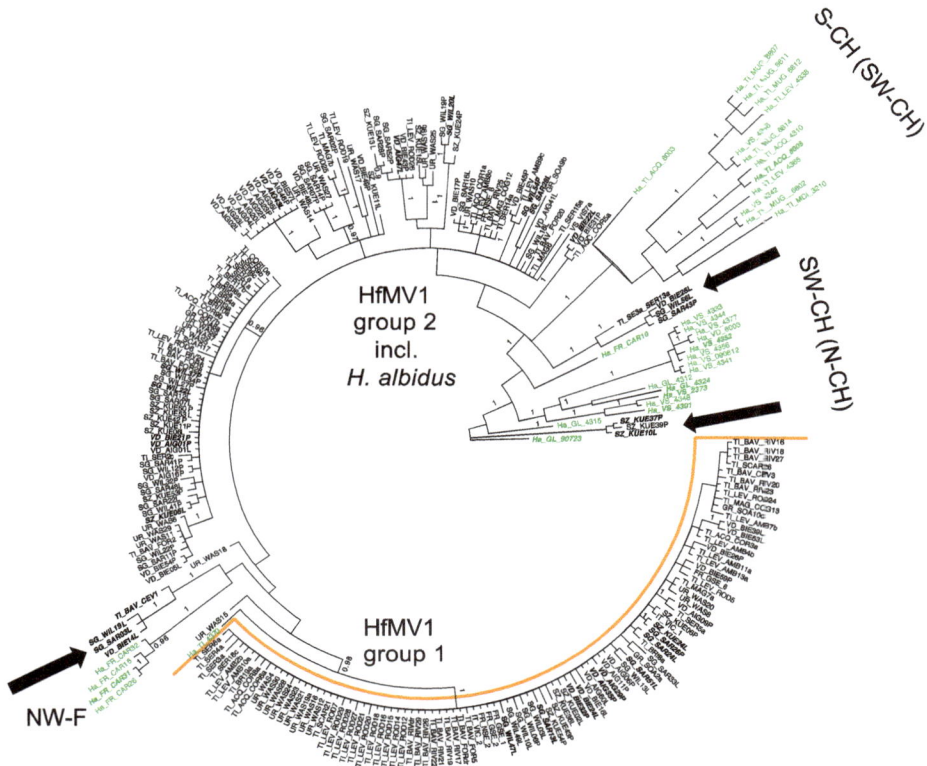

Figure 3. MrBayes analysis conducted with 255 sequences of the partial RdRP gene (495 bp) of the mitovirus HfMV1. Green color depicts *H. albidus* isolates. Posterior probabilities are shown at the respective branch if they were >0.95. Each sample is named with the sampling year, followed by the sampling locality (for abbreviation see Table 1) and sample name. Arrows indicate intermixing of *H. albidus* and *H. fraxineus* HfMV1 sequences. Samples marked in bold and italics are included in the full-length analysis.

No structure could be detected in the overall HfMV1 population (i.e., combining *H. albidus* and *H. fraxineus*), neither geographically nor chronologically, using area or sampling year as priors. Nonetheless, within HfMV1 sequences of *H. albidus*, a certain geographic clustering was visible, with French (NW-F) isolates grouping together most distinctly (Figure 3). The 12 HfMV1 sequences from Southern Switzerland, together with two sequences from South-Western Switzerland (SW-CH) formed the second most distinct cluster (Figure 3 and Figure S2). Most sequences from SW-CH also grouped together in a separate cluster. Noteworthy, 12 *H. fraxineus* HfMV1 sequences grouped within the three *H. albidus* clusters (indicated by black arrows in Figure 3).

3.4. Population Genetic Parameters

3.4.1. Full RdRP Gene

Given that all full-length RdRP sequences were different, the haplotype diversity (Hd) was one in both *H. fraxineus* and *H. albidus* (Table 2). The nucleotide diversity estimated by the average number of differences per site between two sequences (π) was 0.03 across all HfMV1 sequences from both host species for the full ORF (2151 bp; N = 40; Table 2). Among HfMV1 sequences from

H. albidus, π was 0.077, and 0.021 for *H. fraxineus* sequences. The average number of nucleotide differences K, was 52 across all sequences from both species and ranged between 38 in *H. fraxineus* and 163 in *H. albidus*. The total number of mutations (η) was 380 for the entire population (*H. fraxineus* and *H. albidus* combined) and was lower in the Swiss *H. fraxineus* (238) than in the Swiss *H. albidus* population (403; Table 2).

Table 2. Population genetic parameters for the full length RdRP gene sequences (2151 bp) of HfMV1 recovered from *Hymenoscyphus albidus* and *H. fraxineus*.

Population	N	Sequence Length	Net Sites	S	η	h	Hd	π	K
H. fraxineus	32	2151	1791	219	239	30	1.0	0.021	38
H. albidus	8	2151	2122	400	464	8	1.0	0.077	163
H. fraxineus CH	30	2151	1851	218	238	28	1.0	0.021	39
H. albidus CH	6	2151	2129	360	403	6	1.0	0.076	163
H. fraxineus & *H. albidus* total	40	2151	1774	328	380	38	1.0	0.029	52

Abbreviations: N, number of isolates; Sequence length, entire sequence length in bp; Net sites, sequence length in analysis; S, number of segregating sites; η, total number of mutations occurred in that population; h, number of haplotypes; Hd, haplotype diversity; π, nucleotide diversity estimated by the average number of differences per site between two sequences; K, average number of nucleotide differences; CH, only samples from Switzerland (30 out of 32 in *H. fraxineus* and 6 out of 8 in *H. albidus*).

3.4.2. Partial RdRP Gene

For the partial RdRP sequences (495 bp; N = 255), π was on average 0.02 (0.08 for *H. albidus* and 0.007 for *H. fraxineus*; Table 3). Hd was on average 0.84 and ranged between 0.90 and 1.00 in *H. albidus* and between 0.7 and 0.99 in *H. fraxineus*. K, the average number of nucleotide differences, was 5 across all sequences from both species and ranged between 3 in *H. fraxineus* and 30 in *H. albidus*, all from Southern Switzerland (S-CH; Figure 1). The total number of mutations (η) was 89 for the entire population (*H. fraxineus* and *H. albidus* combined) and lower in Swiss *H. fraxineus* (58) than in Swiss *H. albidus* populations (122; Table 3).

Table 3. Population genetic parameters for the partial RdRP gene sequences (495 bp) of HfMV1 recovered from *Hymenoscyphus albidus* and *H. fraxineus*.

Population	N	Sequence Length	Net Sites	S	η	h	Hd	π	K
H. fraxineus all	221	495	348	49	57	48	0.79	0.007	2.35
H. fraxineus CH	214	495	349	50	58	49	0.80	0.007	2.41
H. fraxineus S-CH *	90	495	359	33	37	28	0.80	0.008	2.98
H. fraxineus SW-CH	41	495	485	89	102	27	0.96	0.039	19.21
H. fraxineus N-CH	85	495	458	73	81	53	0.96	0.019	8.72
H. fraxineus NW-F	5	495	488	16	16	3	0.70	0.014	6.60
H. albidus all	34	495	426	99	129	28	0.99	0.080	34.02
H. albidus CH	29	495	427	94	122	24	0.99	0.078	33.39
H. albidus S-CH	12	495	443	79	94	12	1.00	0.067	29.74
H. albidus SW-CH	13	495	467	76	88	10	0.96	0.052	24.03
H. albidus NW-F	5	495	494	34	34	4	0.90	0.026	14.40
H. fraxineus & *H. albidus* total	255	495	348	71	89	73	0.84	0.015	5.33

Only populations ≥5 samples were considered. Abbreviations: N, number of isolates; Sequence length, entire sequence length in bp; Net sites, sequence length in analysis; S, number of segregating sites; η, total number of mutations occurred in that population; h, number of haplotypes; Hd, haplotype diversity; π, nucleotide diversity estimated by the average number of differences per site between two sequences; K, average number of nucleotide differences; CH: Switzerland, S-CH: Southern Switzerland, SW-CH: South-Western Switzerland, N-CH: Northern Switzerland, NW-F: North-Western France, * including 2 isolates from Northern Italy.

4. Discussion

The main objective of this study was to determine the presence and prevalence of the mitovirus HfMV1 [26] in the leaf endophyte *H. albidus* and to use the viral RdRP gene for population genetic analyses. Furthermore, we aimed to compare the genetic diversity of HfMV1 in *H. albidus*, which is presumably native to Europe, to that in the closely related and introduced fungus *H. fraxineus*, the causal agent of ash dieback (ADB). Both fungi occupy the same ecological niche, colonizing and infecting ash leaves and forming fruiting bodies on leaf petioles in the litter. While *H. fraxineus* is very abundant, *H. albidus* is now difficult to find and possibly locally extinct [19,20]. Our study supports this observation, as from 2013 onwards we could only isolate *H. fraxineus* even in areas where *H. albidus* had previously been officially reported. It is generally assumed that this is the result of a displacement of *H. albidus* due to the competition for the sporulation niche with its relative *H. fraxineus*. As *H. fraxineus* acts as a primary leaf pathogen and produces massive airborne spore clouds (see also [21]), it is outcompeting the native *H. albidus*.

Genetic analyses confirmed that the mitovirus detected in *H. albidus* effectively corresponds to the mitovirus HfMV1 previously described in the invasive pathogen *H. fraxineus*. First, the viral RdRP gene has an identical ORF length in both fungi. Second, the full RdRP gene sequences obtained from *H. albidus* and *H. fraxineus* showed high pairwise identities (94.5% for the RdRP amino acid sequences, 91.1% for nucleotide sequences). As this finding indicates that HfMV1 is conspecific in *H. fraxineus* and *H. albidus*, we propose the same virus name, HfMV1, for both fungal species. This is the first description of a fungal virus in the leaf endophyte *H. albidus*.

The overall prevalence of HfMV1 in *H. albidus* is lower (49.3%) than previously described for *H. fraxineus* isolates from the European mainland and Great Britain (78.7% and 67%, respectively; [15,24]). This difference could be due to the fact that we analyzed far more *H. fraxineus* compared to *H. albidus* isolates. Furthermore, there could be a geographic difference, as the *H. albidus* isolates in this study were obtained mainly from Switzerland (and France), whereas the *H. fraxineus* isolates were collected throughout Europe (and Great Britain). The lower prevalence of HfMV1 in *H. albidus* could also be due to the PCR assay used for virus screening, which presumably did not detect all HfMV1 variants in this species. Nonetheless, the prevalence of HfMV1 in both *Hymenoscyphus* species is in accordance with the ranges observed for mitoviruses in other fungal species e.g., in *Gremmeniella* [43].

All *H. albidus* isolates analyzed in this study, including those infected by HfMV1, were collected in Switzerland and North-Western France prior to the arrival of ADB (Figure 1; [35]). Therefore, we can assume that HfMV1 must have been present in *H. albidus* before the introduction of *H. fraxineus* into Europe. This assumption is supported by the fact that *H. albidus* viruses form a distinct, well supported phylogenetic group when considering the full viral RdRP gene (Figure 2). The *H. albidus* viruses also exhibit a higher genetic diversity than the *H. fraxineus* viruses. Moreover, clustering by geographic region is evident for HfMV1 strains from *H. albidus*. Specifically, sequences from Switzerland are clearly separated from those from North-Western France. Within Switzerland, sequences from Southern Switzerland are quite distinct and only partially intermix with HfMV1 sequences from South-Western Switzerland. Such spatial genetic patterns are expected due to local differentiation of HfMV1 in native populations of *H. albidus*. In contrast, we did not detect any geographic structuring for HfMV1 sequences obtained from *H. fraxineus*, which is in accordance with previous studies for the fungus and the virus [12–15]. In summary, these findings suggest that HfMV1 is also native to Europe where it occurs as a genetically distinct group in *H. albidus*.

The presence of a conspecific mitovirus in *H. fraxineus* and *H. albidus* could be the result of a parallel virus evolution from a common ancestor. Mitoviruses are the most widespread fungal viruses and HfMV1only encodes one single gene. Hence, possibilities for mutations are limited and similar HfMV1 variants may form independently (in parallel) in different fungal hosts, particularly if the host species are closely related. Since mating between the two fungal species, which also possess different mating systems (homothallic in *H. albidus* vs. heterothallic in *H. fraxineus*), has never been observed [44], transfer of HfMV1 from *H. albidus* to *H. fraxineus* through interspecific hybridization does not seem to be a plausible hypothesis.

The detection of HfMV1 in *H. albidus* sheds new light on previous research, where only viral sequences obtained from *H. fraxineus* were investigated [14]. Based on the discovery of two HfMV1 groups in Europe, it was hypothesized that two divergent HfMV1 strains were introduced together with two *H. fraxineus* strains. The two HfMV1 groups in *H. fraxineus* were also detected in the present study for both the full and partial viral RdRP gene sequences. Although the *H. albidus* viruses are well separated from these two groups, they seem to be more closely related to HfMV1 group 2 than to group 1 (Figures 2 and 3). Most interestingly, there are several HfMV1 sequences from *H. fraxineus*, which cluster together with *H. albidus* viruses, i.e., these viruses are closer related to those in *H. albidus* than to those in *H. fraxineus* (arrows in Figures 2 and 3). This pattern is consistent with the hypothesis of an interspecies virus transmission from the native *H. albidus* to the invasive *H. fraxineus*. In particular, there are 2 full sequences of *H. fraxineus* isolated from necrotic lesions that cluster with the *H. albidus* group (Figure 2). Our results suggest at least three cross-species transmission events of HfMV1. Further interspecies virus transfers cannot be ruled out particularly within HfMV1 group 2, which could explain the higher genetic diversity observed in this group compared to group 1 [15]. In plant pathogenic fungi, cross-species transmission of mycoviruses has been previously described in different genera, e.g., *Cryphonectria hypovirus 1* (family Hypoviridae) in the genus *Cryphonectria* [30] and the *Heterobasidion RNA virus 1* (HetRV1; family Partitiviridae) in the genus *Heterobasidion* [31]. For HetRV1, Vainio et al. [31] detected high nucleotide level similarity (98%) between HetRV1 obtained from taxonomically distant *H. parviporum* and *H. australe*, thus suggesting a recent HetRV1 transmission in nature. Furthermore, Deng et al. [33] reported a mitovirus to be conspecific in *Sclerotinia homoeocarpa*, the causal agent of Dollar spot, and *Ophiostoma novo-ulmi*, the causal agent of Dutch elm disease, with 92.4% (nucleotide) and 95.1% (amino acid) sequence identities between viral strains and [45] detected 95% identity of BcMV1 and *Ophiostoma novo-ulmi mitovirus 3b*.

In contrast to plant and animal pathogenic viruses, natural vectors are largely unknown in fungal viruses, which typically have no extracellular phase [25]. In vitro experiments (e.g., [30,32]) have shown that mycoviruses are transmitted horizontally between fungal species via hyphal anastomosis. Recent genomic studies revealed that the genomes of *H. fraxineus* and *H. albidus* are highly similar [46] and 75% of the *H. albidus* reads could be mapped to the *H. fraxineus* reference genome [16]. Hence, hyphal anastomosis and transfer of genetic elements between both fungal species without the final hybridization event cannot be excluded. The low detection rate of such events (i.e., intermixing of HfMV1 sequences from *H. fraxineus* with sequence clusters of HfMV1 from *H. albidus*) in this study might be simply due to the fact that it is still too early in time to detect them more frequently. Moreover, it has to be kept in mind, that most likely we did not capture the full picture and all transmission events, as only a small number of *H. albidus* isolates from Switzerland and France were investigated in the present study. Future sampling of *H. fraxineus* in areas where divergent HfMV1 strains were found in *H. albidus* (e.g., Southern Switzerland or North-Western France), as well as the inclusion of additional *H. albidus* samples from other countries could reveal additional evidence of interspecies transmission of HfMV1.

5. Conclusions

Our study shows that the mitovirus HfMV1 that was previously identified in the ash dieback pathogen *H. fraxineus* [26] is also occurring in the congeneric species *H. albidus*. Sampling history together with phylogenetic and population genetic analyses suggest that HfMV1 is a native mycovirus in the European *H. albidus* population. The occurrence of a conspecific mitovirus in *H. albidus* and *H. fraxineus* is most likely the result of parallel virus evolution in the two fungal hosts. In addition, our study provides evidence for interspecies virus transmission from *H. albidus* to *H. fraxineus*, which contributed to the viral diversity observed in the invasive *H. fraxineus* population in Europe.

Supplementary Materials: The following are available online at http://www.mdpi.com/1999-4915/10/11/628/s1, **Figure S1:** Phylogenetic tree resulting from the RaxML analysis conducted with 40 (8 *H. albidus*, 32 *H. fraxineus*) full sequences (2151 bp) of the RdRP gene of the mitovirus HfMV1. Green color depicts HfMV1 *H. albidus* group isolates. Black color depicts HfMV1 group 2 isolates and grey color HfMV1 group isolates. Blue color and *** mark *H. fraxineus* isolates within the *H. albidus* group. Hf characterizes viral sequences from *H. fraxineus*, Ha from *H. albidus* isolates. For abbreviation see Table 1. Bootstrap values are shown if >70. Arrows indicate intermixing of *H. albidus* and *H. fraxineus* HfMV1 sequences. **Figure S2:** RAxML analysis conducted with 255 sequences (495 bp)

Viruses **2018**, *10*, 628

of the partial RdRP gene of the mitovirus HfMV1. Each sample is named with the sampling year, followed by the sampling locality (for abbreviation see Table 1) and sample name. Bootstrap values are show if ≥60. Samples marked in bold and italics are included in the full-length analysis. **Table S1:** Details of the isolates in the study with sample locations, years and GenBank accession numbers. **Table S2:** List of additional primers used to sequence the full open reading frame of *Hymenoscyphus fraxineus mitovirus 1* and 5′ and 3′ flanking regions of HfMV1. Tm = annealing temperature in degrees Celsius.

Author Contributions: Conceptualization, C.N.S., D.R. and S.P.; Methodology, C.N.S., D.R. and S.P.; Formal Analysis, C.N.S.; Investigation, C.N.S.; Resources, D.R. and A.G.; Data Curation, C.N.S.; Writing—Original Draft Preparation, C.N.S.; Writing—Review & Editing, C.N.S., D.R., A.G. and S.P.; Visualization, C.N.S.; Funding Acquisition, D.R.

Funding: This research was funded by the Lithuanian-Swiss cooperation program to reduce economic and social disparities within the enlarged European Union (project grant agreement no. CH-3-SMM-C1/12).

Acknowledgments: We would like to thank O. Holdenrieder, V. Queloz, and Th. Kirisits for providing *H. albidus* cultures. We are very thankful to E. Jung, S. Pfister, O. Zuberbühler, K. Moosbrugger, J. Schwarz, U. Oggenfuss and C. Cornejo for much appreciated help in the laboratory. We are grateful to the members of the COST Action FP1103 (FRAXBACK) for initiating sample exchange and to Lea Stauber for careful proofreading of the manuscript.

Conflicts of Interest: The authors declare no conflict of interest.

References

1. Westphal, M.I.; Browne, M.; MacKinnon, K.; Noble, I. The link between international trade and the global distribution of invasive alien species. *Biol. Invasions* **2008**, *10*, 391–398. [CrossRef]

2. Santini, A.; Ghelardini, L.; De Pace, C.; Desprez-Loustau, M.L.; Capretti, P.; Chandelier, A.; Cech, T.; Chira, D.; Diamandis, S.; Gaitniekis, T.; et al. Biogeographical patterns and determinants of invasion by forest pathogens in Europe. *New Phytol.* **2013**, *197*, 238–250. [CrossRef] [PubMed]

3. Hantula, J.; Mu, M.M.; Uusivuori, J. International plant trade associated risks: Laissez-faire or novel solutions. *Environ. Sci. Policy* **2014**, *37*, 158–160. [CrossRef]

4. Grünwald, N.J.; Garbelotto, M.; Goss, E.M.; Heunges, K.; Prospero, S. Emergence of the sudden oak death pathogen *Phytophthora ramorum. Trends Microbiol.* **2012**, *20*, 131–138. [CrossRef] [PubMed]

5. Rigling, D.; Prospero, S. *Cryphonectria parasitica*, the causal agent of chestnut blight: Invasion history, population biology and disease control. *Mol. Plant Pathol.* **2018**, *19*, 7–12. [CrossRef] [PubMed]

6. Hoegger, P.J.; Rigling, D.; Holdenrieder, O.; Heiniger, U. *Cryphonectria radicalis*: Rediscovery of a lost fungus. *Mycologia* **2002**, *94*, 105–115. [CrossRef] [PubMed]

7. Gonthier, P.; Nicolotti, G.; Linzer, R.; Guglielmo, F.; Garbelotto, M. Invasion of European pine stands by a North American forest pathogen and its hybridization with a native interfertile taxon. *Mol. Ecol.* **2007**, *16*, 1389–1400. [CrossRef] [PubMed]

8. Érsek, T.; Nagy, Z.A. Species hybrids in the genus *Phytophthora* with emphasis on the alder pathogen Phytophthora alni: A review. *Eur. J. Plant Pathol.* **2008**, *122*, 31–39. [CrossRef]

9. Gross, A.; Han, J.G. *Hymenoscyphus fraxineus* and two new *Hymenoscyphus* species identified in Korea. *Mycol. Prog.* **2015**, *14*, 1. [CrossRef]

10. Gross, A.; Zaffarano, P.L.; Duo, A.; Grünig, C.R. Reproductive mode and life-cycle of the ash dieback pathogen *Hymenoscyphus pseudoalbidus. Fungal Genet. Biol.* **2012**, *49*, 977–986. [CrossRef] [PubMed]

11. Baral, H.O.; Bemmann, M. *Hymenoscyphus fraxineus* vs. *Hymenoscyphus albidus*—A comparative light microscopic study on the causal agent of European ash dieback and related foliicolous, stroma-forming species. *Mycology* **2014**, *5*, 228–290. [PubMed]

12. Bengtsson, S.B.K.; Vasaitis, R.; Kirisits, T.; Solheim, H.; Stenlid, J. Population structure of *Hymenoscyphus pseudoalbidus* and its genetic relationship to *Hymenoscyphus albidus. Fungal Ecol.* **2012**, *5*, 147–153. [CrossRef]

13. Burokiene, D.; Prospero, S.; Jung, E.; Marciulyniene, D.; Moosbrugger, K.; Norkute, G.; Rigling, D.; Lygis, V.; Schoebel, C.N. Genetic population structure of the invasive ash dieback pathogen *Hymenoscyphus fraxineus* in its expanding range. *Biol. Invasions* **2015**, *17*, 2743–2756. [CrossRef]

14. Gross, A.; Hosoya, T.; Queloz, V. Population structure of the invasive forest pathogen *Hymenoscyphus pseudoalbidus. Mol. Ecol.* **2014**, *23*, 2943–2960. [CrossRef] [PubMed]

15. Schoebel, C.N.; Botella, L.; Lygis, V.; Rigling, D. Population genetic analysis of a parasitic mycovirus to infer the invasion history of its fungal host. *Mol. Ecol.* **2017**, *26*, 2482–2497. [CrossRef] [PubMed]

16. Sonstebo, J.H.; Smith, A.V.; Adamson, K.; Drenkhan, R.; Solheim, H.; Hietala, A. Genome-wide population diversity in *Hymenoscyphus fraxineus* points to an eastern Russian origin of European Ash dieback. *bioRxiv* **2017**, 154492. [CrossRef]

17. Kirisits, T.; Dämpfle, L.; Kräutler, K. *Hymenoscyphus albidus* is not associated with an anamorphic stage and displays slower growth than *Hymenoscyphus pseudoalbidus* on agar media. *For. Pathol.* **2013**, *43*, 386–389.

18. Brasier, C.; King, K.; Kirisits, T.; Orton, E.; Webber, J. High frequency of vegetative incompatibility combined with haploid selfing in the native European ash foliage coloniser *Hymenoscyphus albidus*. *Fungal Ecol.* **2017**, *28*, 11–24. [CrossRef]

19. McKinney, L.V.; Thomsen, I.M.; Kjær, D.; Bengtsson, S.B.K.; Nielsen, L.R. Rapid invasion by an aggressive pathogenic fungus (*Hymenoscyphus pseudoalbidus*) replaces a native decomposer (*Hymenoscyphus albidus*): A case of local cryptic extinction? *Fungal Ecol.* **2012**, *5*, 663–669. [CrossRef]

20. King, K.M.; Webber, J.F. Development of a multiplex PCR assay to discriminate native *Hymenoscyphus albidus* and introduced *Hymenoscyphus fraxineus* in Britain and assess their distribution. *Fungal Ecol.* **2016**, *23*, 79–85. [CrossRef]

21. Hietala, A.M.; Timmermann, V.; Barja, I.; Solheim, H. The invasive ash dieback pathogen *Hymenoscyphus pseudoalbidus* exerts maximal infection pressure prior to the onset of host leaf senescence. *Fungal Ecol.* **2013**, *6*, 302–308. [CrossRef]

22. Kowalski, T.; Holdenrieder, O. Pathogenicity of *Chalara fraxinea*. *For. Pathol.* **2009**, *39*, 1–7. [CrossRef]

23. Pautasso, M.; Aas, G.; Queloz, V.; Holdenrieder, O. European ash (*Fraxinus excelsior*) dieback—A conservation biology challenge. *Biol. Conserv.* **2013**, *158*, 37–49. [CrossRef]

24. Orton, E.S.; Brasier, C.M.; Bilham, L.J.; Bansal, A.; Webber, J.F.; Brown, J.K.M. Population structure of the ash dieback pathogen, *Hymenoscyphus fraxineus*, in relation to its mode of arrival in the UK. *Plant Pathol.* **2018**, *67*, 255–264. [CrossRef] [PubMed]

25. Ghabrial, S.A.; Suzuki, N. Viruses of plant pathogenic fungi. *Annu. Rev. Phytopathol.* **2009**, *47*, 353–384. [CrossRef] [PubMed]

26. Schoebel, C.N.; Zoller, S.; Rigling, D. Detection and genetic characterization of a novel mycovirus in *Hymenoscyphus fraxineus*, the causal agent of ash dieback. *Infect. Genet. Evol.* **2014**, *28*, 78–86. [CrossRef] [PubMed]

27. Li, W.H. *Molecular Evolution*; Sinauer Associates, Inc.: Sunderland, MA, USA, 1997.

28. Polashock, J.J.; Hillman, B.I. A small mitochondrial double-stranded (ds) RNA element associated with a hypovirulent strain of the chestnut blight fungus and ancestrally related to yeast cytoplasmic T and W dsRNAs. *Proc. Natl. Acad. Sci. USA* **1994**, *91*, 8680–8684. [CrossRef] [PubMed]

29. Polashock, J.; Bedker, P.; Hillman, B. Movement of a small mitochondrial double-stranded RNA element of *Cryphonectria parasitica*: Ascospore inheritance and implications for mitochondrial recombination. *Mol. Gen. Genet.* **1997**, *256*, 566–571. [CrossRef] [PubMed]

30. Liu, Y.C.; Linder-Basso, D.; Hillman, B.I.; Kaneko, S.; Milgroom, M.G. Evidence for interspecies transmission of viruses in natural populations of filamentous fungi in the genus *Cryphonectria*. *Mol. Ecol.* **2003**, *12*, 1619–1628. [CrossRef] [PubMed]

31. Vainio, E.J.; Hakanpää, J.; Dai, Y.C.; Hansen, E.; Korhonen, K.; Hantula, J. Species of *Heterobasidion* host a diverse pool of partitiviruses with global distribution and interspecies transmission. *Fungal Biol.* **2011**, *115*, 1234–1243. [CrossRef] [PubMed]

32. Melzer, M.; Ikeda, S.S.; Boland, G.J. Interspecific transmission of double-stranded RNA and hypovirulence from *Sclerotinia sclerotiorum* to *S. minor*. *Phytopathology* **2002**, *92*, 780–784. [CrossRef] [PubMed]

33. Deng, F.; Xu, R.; Boland, G. Hypovirulence-associated double-stranded RNA from *Sclerotinia homoeocarpa* is conspecific with *Ophiostoma novo-ulmi* mitovirus 3a-Ld. *Phytopathology* **2003**, *93*, 1407–1414. [CrossRef] [PubMed]

34. White, T.; Bruns, T.; Lee, S.; Taylor, J. Amplification and direct sequencing of fungal ribosomal RNA genes for phylogenetics. In *PCR Protocols: A Guide to Methods and Applications*; Innis, M., Gelfand, D., Sninsky, J., White, T., Eds.; Academic Press: New York, NY, USA, 1990; pp. 315–322. [CrossRef]

35. Queloz, V.; Grünig, C.R.; Berndt, R.; Berndt, R.; Kowalski, T.; Sieber, T.N.; Holdenrieder, O. Cryptic speciation in *Hymenoscyphus albidus*. *For. Pathol.* **2011**, *41*, 133–142. [CrossRef]

36. Ranwez, V.; Harispe, S.; Delsuc, F.; Douzery, E.J.P. MACSE: Multiple Alignment of Coding Sequences accounting for frameshifts and stop codons. *PLoS ONE* **2011**, *6*, e22594. [CrossRef] [PubMed]

37. Stamatakis, A. RAxML-VI-HPC: Maximum likelihood-based phylogenetic analyses with thousands of taxa and mixed models. *Bioinformatics* **2006**, *22*, 2688–2690. [CrossRef] [PubMed]

38. Ronquist, F.; Teslenko, M.; Van der Mark, P.; Ayres, D.L.; Darling, A.; Höhna, S.; Larget, B.; Liu, L.; Suchard, M.A.; Huelsenbeck, J.P. MrBayes 3.2: Efficient Bayesian phylogenetic inference and model choice across a large model space. *Syst. Biol.* **2012**, *61*, 539–542. [CrossRef] [PubMed]

39. Stöver, B.C.; Müller, K.F. TreeGraph 2: Combining and visualizing evidence from different phylogenetic analyses. *BMC Bioinform.* **2010**, *11*, 7. [CrossRef] [PubMed]

40. Librado, P.; Rozas, J. DnaSP v5: A software for comprehensive analysis of DNA polymorphism data. *Bioinformatics* **2009**, *25*, 1451–1452. [CrossRef] [PubMed]

41. Rigling, D.; Hilfiker, S.; Schöbel, C.; Meier, F.; Engesser, R.; Scheidegger, C.; Stofer, S.; Senn-Irlet, B.; Queloz, V. *Das Eschentriebsterben: Biologie, Krankheitssymptome und Handlungsempfehlungen*; Merkbl. Publisher: Birmensdorf, Switzerland, 2016.

42. Buck, K.W.; Esteban, R.; Hillman, B.I. Narnaviridae. In *Virus Taxonomy: Eighth Report of the International Committee on Taxonomy of Viruses*, 1st ed.; Fauquet, C.M., Mayo, M.A., Maniloff, J., Desselberger, U., Ball, L.A., Eds.; Elsevier Academic Press: San Diego, CA, USA, 2005; pp. 751–756.

43. Botella, L.; Tuomivirta, T.T.; Vervuurt, S.; Diez, J.J.; Hantula, J. Occurrence of two different species of mitoviruses in the European race of *Gremmeniella abietina* var. *abietina*, both hosted by the genetically unique Spanish population. *Fungal Biol.* **2012**, *116*, 872–882. [PubMed]

44. Wey, T.; Schlegel, M.; Stroheker, S.; Gross, A. MAT–gene structure and mating behavior of *Hymenoscyphus fraxineus* and *Hymenoscyphus albidus*. *Fungal Genet. Biol.* **2016**, *87*, 54–63. [CrossRef] [PubMed]

45. Wu, M.; Zhang, L.; Li, G.; Jiang, D.; Ghabrial, S.A. Genome characterization of a debilitation-associated mitovirus infecting the phytopathogenic fungus *Botrytis cinereal*. *Virology* **2010**, *406*, 117–126. [CrossRef] [PubMed]

46. Stenlid, J.; Elfstand, M.; Cleary, M.; Ihrmark, K.; Karlsson, M.; Davydenko, K.; Brandström Durling, M. Genomes of *Hymenoscyphus fraxineus* and *Hymenoscyphus albidus* Encode Surprisingly Large Cell Wall Degrading Potential, Balancing Saprotrophic and Necrotrophic Signatures. *Baltic For.* **2017**, *23*, 89–106.

viruses

MDPI

Communication

Co-Infection with Three Mycoviruses Stimulates Growth of a *Monilinia fructicola* Isolate on Nutrient Medium, but Does Not Induce Hypervirulence in a Natural Host

Thao T. Tran, Hua Li, Duy Q. Nguyen, Michael G. K. Jones and Stephen J. Wylie *

Plant Biotechnology Research Group-Virology, Western Australian State Agricultural Biotechnology Centre, School of Veterinary and Life Sciences, Murdoch University, Perth 6150, Australia; t.tran@murdoch.edu.au (T.T.T.); perthmuzi@yahoo.com (H.L.); q.nguyen@murdoch.edu.au (D.Q.N.); m.jones@murdoch.edu.au (M.G.K.J.)
* Correspondence: s.wylie@murdoch.edu.au; Tel.: +61-893-606-600

Received: 12 November 2018; Accepted: 18 January 2019; Published: 21 January 2019

Abstract: *Monilinia fructicola* and *Monilinia laxa* are the most destructive fungal species infecting stone fruit (*Prunus* species). High-throughput cDNA sequencing of *M. laxa* and *M. fructicola* isolates collected from stone fruit orchards revealed that 14% of isolates were infected with one or more of three mycoviruses: Sclerotinia sclerotiorum hypovirus 2 (SsHV2, genus *Hypovirus*), Fusarium poae virus 1 (FPV1, genus *Betapartitivirus*), and Botrytis virus F (BVF, genus *Mycoflexivirus*). Isolate M196 of *M. fructicola* was co-infected with all three viruses, and this isolate was studied further. Several methods were applied to cure M196 of one or more mycoviruses. Of these treatments, hyphal tip culture either alone or in combination with antibiotic treatment generated isogenic lines free of one or more mycoviruses. When isogenic fungal lines were cultured on nutrient agar medium in vitro, the triple mycovirus-infected parent isolate M196 grew 10% faster than any of the virus-cured isogenic lines. BVF had a slight inhibitory effect on growth, and FPV1 did not influence growth. Surprisingly, after inoculation to fruits of sweet cherry, there were no significance differences in disease progression between isogenic lines, suggesting that these mycoviruses did not influence the virulence of *M. fructicola* on a natural host.

Keywords: brown rot; stone fruit; *Prunus*; mycovirus; hypervirulence; hypovirulence; isogenic

---

1. Introduction

Mycoviruses are viruses that infect fungi. They have been identified from all major fungal phyla, namely the Zygomycota, Chytridiomycota, Ascomycota, and Basidiomycota [1,2]. Since mycoviruses were first described [3], the partial or complete genomes of more than 250 mycoviruses have been sequenced [4]. Fungi can be multiply infected with closely-related and distantly-related viruses [5]. Most mycoviruses have double-stranded (ds) or single-stranded (ss) RNA genomes, and some groups do not encode a coat protein.

The influence that mycoviruses have on the ecology of their hosts is not well studied. Some mycoviruses reduce the ability of the fungal host to cause disease in plants. These are known as hypovirulent mycoviruses, and they have potential as biological control agents. The most well-known are Cryphonectria hypoviruses 1 and 2 (CHV1, CHV2), which significantly decreased the virulence of the fungus *Cryphonectria parasitica*, the causal agent of chestnut blight [6,7]. Other hypovirulent mycoviruses reduce the pathogenicity of white mold fungus *Sclerotinia sclerotiorum* [8,9] and white root rot fungus *Rosellinia necatrix* [10]. In contrast, several mycoviruses are associated with hypervirulence, described as a higher level of virulence or sporulation in their fungal hosts [11–13]. Other mycoviruses are reported to be associated with latent infections [8,14–16].

Monilinia fructicola and *M. laxa* were first recorded in Western Australian stone fruit production regions in 1997 [17,18], and they have since spread to all other stone fruit production regions in the state [19]. They incur costs in control (fungicides, gathering and destroying mummified fruit) and crop losses. Very little is known about mycoviruses that infect *Monilinia* species. Tsai, et al. [20] identified seven virus-like double-stranded RNA species in 36 of 49 *M. fructicola* isolates infecting nectarine and peach orchards in New Zealand. Although not characterized genetically, the authors described virus-like particles resembling those of partitiviruses, totiviruses, tobraviruses and furoviruses. They identified no differences in host growth rates between isolates with and without virus infection. In this current study, we identified three mycoviruses infecting a collection of *M. fructicola* and *M. laxa* isolates, and undertook to determine if these viruses influenced growth rates of infected fungal cultures in vitro, and the virulence of the pathogen on a natural host.

2. Materials and Methods

2.1. Fungus Collection and Isolation

Eighteen *M. laxa* and ten *M. fructicola* isolates were collected from symptomatic flowers, twig cankers and fruits from stone fruit orchards in Western Australia (Table S1). Conidia were collected in a drop of water, then spread on 1% water agar media to separate single spores using a microscope. Individual spores were transferred to V8 agar medium (V8 juice 200 mL, distilled water 800 mL, agar 15 g) and incubated in the dark at 22 °C. After a week, a 5 × 5 mm square of agar containing actively-growing hyphae was excised from the edge of the mycelium and transferred to V8 liquid medium, and placed on a shaker at 100 rpm in the dark at 22 °C. After about a week, DNA and RNA were extracted for further studies.

2.2. DNA and RNA Extraction

Nucleic acids were extracted from 100 mg of mycelium, which was frozen in liquid nitrogen and ground to a fine powder using a mortar and pestle. The powdered mycelium was transferred into a 1.5 mL centrifuge tube containing 450 μL extraction buffer (0.1 M NaCl, 50 mM Tris (hydroxymethyl) aminomethane (Tris) pH8.0, 0.5 mM Ethylenediaminetetraacetic acid (EDTA) pH8.0, 1% Sodium dodecyl sulfate (SDS), 1% Polyvinylpolypyrrolidone (PVPP) and 450 μL phenol-chloroform (50:50) saturated with Tris-EDTA buffer (pH 8.0). The mixture was homogenized before being centrifuged at room temperature for 2 min. After that, 400 μl of the aqueous phase was transferred to a new tube containing 400 μl of phenol-chloroform, then mixed and centrifuged for 2 min. Then, 300 μL of the aqueous phase containing nucleic acids was removed to a new tube containing 58 μL of absolute ethanol to which 200 mg of cellulose powder (CF11, Whatman) was added and mixed. After centrifuging at high speed for 1 min, the material was separated into two phases, the DNA-containing supernatant and the pellet containing RNA. The supernatant (250 μL) was transferred to a fresh tube and 20 μL 3M NaOAC pH5.2 and 625 μL absolute ethanol was added to precipitate DNA. The pellet containing RNA was washed three times with 750 μL application buffer (0.1 mM NaCl, 50 mM Tris pH8.0, 0.5 mM EDTA pH8.0, absolute ethanol). The RNA pellet was eluted in 450 μL elution buffer (0.1 mM NaCl, 50 mM Tris pH8.0, 0.5 mM EDTA pH8.0). The RNA solution (450 μL) was removed and precipitated after incubation at −20 °C in a new tube containing 1 mL absolute ethanol and 45 μL 3M NaOAC pH5.2 for RNA collection.

DNA was used to identify the fungal species based on the comparison of internal-transcribed spacer (ITS) region sequences of ribosomal RNA genes, as previously described [19]. RNA was used for virus identification in *M. fructicola* isolate M196 by high-throughput sequencing.

2.3. cDNA Synthesis, PCR Amplification and Library Preparation for High-Throughput Sequencing

cDNA synthesis was carried out in 20 μL volume containing the following components: 4 μL 5× GoScript™ Buffer; 2 μL 0.1 mM DDT; 1 μL 10 mM deoxynucleotides; 1 μL GoScript™ reverse transcriptase (Promega Corporation, Sydney, Australia); 1 μL 10 mM Tris EDTA; 8 μL water; 1 μL RNA, and 1 μL random primer adaptor (5'-CGTACAGTTAGCAGGCNNNNNNNNNNNNN-3', where N is any nucleotide, annealed to the complement of the adaptor sequence added to cDNA molecules). The mixture was incubated at 25 °C for 10 min; 42 °C for 60 min; and 72 °C for 15 min. cDNA was amplified by PCR using primer 5'-CGTACAGTTAGCAGGC-3', which annealed to the complement of the adaptor sequence added to cDNA molecules, in 20 μL volume of 10 μL 2× GoTaq®Green Master Mix (Promega Corporation, Sydney, Australia); 4 μL cDNA products; 5 μL water; 1 μL barcode primer. The reaction was carried out with an initial cycle of 5 min at 94 °C; 40 cycles of 10 min at 94 °C, 20 s at 45°C and 30 s at 72 °C, an extension cycle of 5 min at 72 °C, and a final cycle of 5 min at 37 °C. The library was purified, quantified, and paired-end sequenced on an Illumina MiSeq platform at the Australian Genome Research Facility.

2.4. Sequence Analysis

The sequences obtained were analyzed as previously described [21]. Briefly, in CLC Genomics Workbench (Qiagen, Sydney) the reads were trimmed using a quality score of 0.05 and of ambiguous bases, then of primer sequences and indices after assigning them to source bins. *De novo* assembly was carried out on the trimmed reads to form contigs >300 nt in length. Assembly parameters were word size of 40 and bubble size of 50. Contigs were compared to sequences lodged in GenBank (National Center for Biotechnology Information, NCBI) databases (https://blast.ncbi.nlm.nih.gov) using Blastn and Blastx [22] to identify virus-like sequences. Overlapping contigs were joined together when possible, and missing sequences were determined after designing specific primers to span the gaps (Table S2). Depth of sequencing was determined by mapping raw reads back to consensus sequences using the 'map to reference' function in Geneious v9.1.7 (Biomatters, Auckland, New Zealand) [23]. Annotation of the genomes was done manually in Geneious v9.1.7 after comparison with related sequences, and amino acid sequences were deduced from the nucleotide sequences of open reading frames (ORF).

Specific primers (Table S3) were designed from consensus sequences and used to confirm the presence of these three mycoviruses in other fungal isolates. When an amplicon was detected, both strands were sequenced using the Sanger method to confirm presence of the virus.

2.5. Generation of Isogenic Fungal Lines Free of Mycoviruses

Four methods were tested alone or in combination to eliminate viruses from fungal isolates:

(1). Cold treatment. Fungal mycelium was stored in 30% glycerol at −80 °C for two years before being recovered on V8 liquid medium.

(2). Temperature shock. Fungal mycelium was stored in 30% glycerol at −80 °C for two years was heated to 30 °C for 30 s, incubated in liquid nitrogen for 45 s, then heated again to 30 °C for 45 s. Mycelium was then recovered on V8 liquid medium.

(3). Hyphal tipping. Hyphal tips were harvested from the edges of rapidly-growing colonies, and sub-cultured to a fresh water agar plate.

(4). Antibiotic treatment. Antibiotics were added singly to water agar when the autoclaved media had cooled to approximately 50 °C. Antibiotic concentrations used were 12.5 mg/L cycloheximide, 100 mg/L kanamycin, and 250 mg/L streptomycin. Hyphal tips were inoculated to plates and cultures incubated in the dark at 22 °C for six days before hyphal tips were harvested from it and the process repeated. Each line was treated this way five times before it was tested for the presence of all three mycoviruses by RT-PCR using species-specific primers. Species-specific primers were designed from high-throughput sequences (Table S3).

Virus-free lines were maintained in culture for up to six months before their virus-free status was reconfirmed by RT-PCR with species-specific primers. These virus-free lines were used for subsequent growth rate and virulence experiments.

2.6. Virulence on Cherry

Isolate M196 (triple virus-infected) and isogenic M196 lines free of at least one virus (M196-1, M196-4, M196-6) (Table 1) were grown on V8 agar medium without antibiotics at 22 °C for six days before a 2 × 2 mm square of mycelium was harvested from the margin of the colony. This was placed on the surface of a washed and dry cherry fruit (*Prunus avium*) of cultivar Bing. Inoculated fruits were incubated in 100% humidity in the dark at room temperature (17–20 °C) for seven days. Each fungal isolate was inoculated to 36 cherry fruits; each fruit was treated as a replication. At the end of the incubation period, the widest extent of the fungal lesion was measured using a compass and ruler. The entire experiment was repeated twice.

Table 1. Presence (+) or absence (−) of mycoviruses from isogenic lines of *M. fructicola* M196 after treatment. The antibiotic treatments were combined with hyphal tipping.

Mycoviruses [a]	−80 °C Storage	Temperature Shock	Hyphal Tipping	Cycloheximide	Kanamycin	Streptomycin
SsHV2	+	+	−	−	−	−
FpV1	+	+	−	+	+	+
BVF	+	+	−	+	−	−
Number of lines obtained [b]	6	1	1	2	1	1
Name of line	M196	−	M196-1	M196-4	M196-6	−

[a] SsHV2, Sclerotinia sclerotiorum hypovirus 2; FpV1, Fusarium poae virus 1; BVF, Botrytis virus F. [b] Number of isogenic lines showing the given pattern out of six treated lines sub-cultured from parent isolate M196. [c] Name and source of line used in subsequent experiments.

2.7. Mycelial Growth In Vitro

Four virus-free and virus-infected isogenic lines (M196, M196-1, M196-4, M196-6) were grown on V8 agar plates at 22 °C for six days. Each isolate was inoculated into 36 V8 plates (three replications × 12 plates per replication). After six days, the maximum diameter of each colony was measured and differences in morphology recorded. The entire experiment was repeated twice.

To determine the differences between treatments, a one-way analysis of variance (ANOVA) was done at a significance level of 0.05 using SPSS version 24.

3. Results

3.1. Sequencing Analysis and Virus Assays

A MiSeq sequencing run created 13,264,058 reads of 100 nt. Barcode (index) sequences were used to assign reads to samples. After screening for quality and trimming the barcode and primer sequences, *de novo* assembly resulted in 37,146 contigs within the size ranges of 300 to 5385 nucleotides. Blastn and Blastx analysis of the contigs revealed 20 that shared nucleotide or amino acid identities with viral sequences. Where two or more contigs mapped to the same virus, gaps in the sequence were filled by RT-PCR so that the sequences of complete or almost complete genomes were determined (Table S2). Three previously-identified mycoviruses infected *M. fructicola* M196 (Table S1). RT-PCR assays using species-specific primers (Table S3) with appropriate controls were used to confirm the presence of the viruses and to check for their presence in isogenic lines over time. Isogenic lines of isolate M196 lacking one, two or three viruses were maintained in culture for 26 weeks before in vitro and *in planta* growth rates were measured.

3.1.1. Sclerotinia sclerotiorum Hypovirus 2

A contiguous virus-like sequence of 13,535 nucleotides was obtained from *M. fructicola* isolate M196. The nucleotide and deduced amino acid sequences shared the greatest identities with the replicase gene of three previously described isolates of Sclerotinia sclerotiorum hypovirus 2 (SsHV2) (genus *Hypovirus*, family *Hypoviridae*) classified as a ssRNA virus. Alignment of the new sequence revealed that it shared 85–89% pairwise nucleotide and 93–95% amino acid identities with SsHV2 isolates 5427 from New Zealand (KF525367) [24], isolate SsHV2 from the USA (KF898354) [25], and isolate SX247 from China (KJ561218) [13]. The demarcation criterion for hypovirus species is less than 50% pairwise nucleotide identity over the complete genome [26]. Although identity with Rosellinia necatrix hypovirus 2 is 54%, identities with other SsHV2 isolates is far higher, and so we propose this to be a member of species *Sclerotinia sclerotiorum hypovirus 2*. It is designated Sclerotinia sclerotiorum hypovirus 2 isolate Monilinia-TNS. SsHV2 was identified only in *Monilinia* M196.

The sequence of SsHV2-Monilinia-TNS is estimated to represent 93% of the complete genome. The SsHV2-Monilinia-TNS genome comprises one large ORF, which is incomplete at the 5′end. The ORF extended from nucleotide 1–13,205 where it was terminated by an opal stop codon (UGA). The ORF is followed by a 3′ untranslated region (UTR) of 333 nucleotides, present from nucleotides 13,206 to 13,535. The conserved RdRp core motifs V and VI (S/TG x3 T x3 NS/T x22 GDD) (where x is any amino acid residue) [27] was present as TG x3 T x3 DS x38 GDD. No poly(A) tail region was detected. The SsHV-2 sequence was assigned GenBank accession MH665657.

3.1.2. Fusarium poae Virus 1

Two contigs representing the complete genomic segments of a bipartite virus were identified from *M. fructicola* M196. The sequence shared the highest sequence identity with published isolates of Fusarium poae virus 1 (FpV1) (genus *Partitivirus*, family *Partitiviridae*), a double-stranded RNA virus. One segment (RNA1) encodes the replicase (RdRp) and the other (RNA2) encodes the coat protein. Comparison of the 2100 nucleotide sequence of RNA1 revealed that it shared 90% pairwise identity with RNA1 of two FpV1 isolates: A11 from Slovakia (AF047013) and 240374 from Japan (LC150606). The deduced amino acid sequence of the RdRp segment shared 92% identity with the RdRps of these two isolates. The ORF of RNA1 extended from nucleotide 54-2084 where it was terminated by an opal stop codon, encoding an RdRp-like protein of 689 amino acid residues with an estimated molecular weight of 79.9 kDa. The conserved core RdRp motifs V and VI [28] were present as SG x3 T x3 DS x29 GDD. The 5′UTR extended from nucleotide 1 to 53 and the 3′UTR from nucleotide 2085 to 2160.

The RNA2 segment, encoding the coat protein (CP) gene, was 2093 nucleotides in length. The single ORF encoding a protein deduced to be 654 amino acid residues in length, with an estimated mass of 72.6 kDa. Surprisingly, a partial RdRp-like motif (T x2 DS x27 GDD) was present with the CP ORF. The deduced amino acid sequence of ORF1 of RNA2 shared greatest identity (85–86%) with CPs of the two other FpV1 isolates, and the nucleotide identity was 83–84% with them. The 5′ UTR extended from 1 to 102 nt, and the 3′ UTR from 2017 to 2093 nt.

The levels of identity of this new virus with isolates of FpV1 were above the species demarcation of partitiviruses recommended by the ICTV [29] of <40% amino acid identity between the RdRps, and so the isolate is proposed here as a member of species *Fusarium poae virus 1*. The isolate from M196 was designated Fusarium poae virus 1 isolate Monilinia-TNS. FpV1-Monilinia-TNS RNA1 was assigned GenBank accession MH665658, and FpV1-Monilinia-TNS RNA2 assigned GenBank accession MH665659.

3.1.3. Botrytis Virus F

Contigs representing partial genomic sequences of a virus were detected from *M. fructicola* M196. The deduced amino acid sequences of fragments each shared 90–95% identity to those of an isolate of Botrytis virus F (BVF) (accession NP068550) (genus *Mycoflexivirus*, family *Gammaflexiviridae*), a ssRNA virus. Together, the three fragments were estimated to represent about 30% of the BVF complete genome, designated Botrytis virus F isolate Monilinia-TNS. One of the fragments held the core RdRp motifs V and VI as SG x3 T x3 NT x21 GDD. The three contigs of BVF-Monilinia-TNS were assigned GenBank accessions MH665660, MH665661, and MH665662.

3.2. The Presence of Mycoviruses in Other Monilinia Isolates

Eighteen *M. laxa* isolates and nine *M. fructicola* isolates were screened using SsHV2, FpV1 and BVF-specific primers (Table S3). Three *M. laxa* isolates harbored one or more of these viruses, including M82 (Fusarium poae virus 1), M84 (Fusarium poae virus 1 and Botrytis virus F), and M140 (Fusarium poae virus 1) (Table S1), but none of the other nine *M. fructicola* isolates tested held these viruses.

3.3. Elimination of Mycoviruses

The cold treatment and temperature shock methods failed to eliminate any mycoviruses, and the temperature shock method resulted in the death of most cultures (Table 1). On the other hand, hyphal tipping with and without antibiotics was effective at curing *M. fructicola* M196 of one or more of the three mycoviruses. It is unclear whether any of the antibiotics played a significant role in eliminating the viruses because tipping alone without antibiotics generated isogenic line M196-1 that was free of all three mycoviruses. Treatment with 12.5 mL/L cycloheximide eliminated SsHV2 from two isogenic lines, one of which, M196-4, was used in subsequent experiments. Treatment with kanamycin (100 mg/L), or with 250 mg/L streptomycin, eliminated both SsHV2 and BVF, but not the partitivirus FpV1 (Table 2). All isogenic lines were assayed by RT-PCR using virus-specific primers (Table S3, Figure S1) before and after subsequent experiments, and these tests confirmed the maintenance of their virus status.

Table 2. Summary of colony diameter on V8 medium and lesion diameter on cherry of isogenic *M. fructicola* lines.

Line	Treatment [a]	Viruses present [b]	Colony Diameter on V8 Medium			Lesion Diameter on Cherry Fruit		
			Range (mm)	Mean (mm)	Variance	Range (mm)	Mean (mm)	Variance
M196	None	SsHV2; FpV1; BVF	68–80	73.8	11.8	3–21	8.7	26.5
M196-1	Hyphal tipping	None	56–70	66.4	12	3–22	9.6	33.7
M196-4	Cycloheximide	FpV1; BVF	55–70	62.9	16.9	3–21	8.9	29.5
M196-6	Kanamycin	FpV1	60–71	66.0	11.5	3–25	10.9	33.4

[a] Antibiotic treatments were combined with hyphal tipping; [b] SsHV2, Sclerotinia sclerotiorum hypovirus 2; FpV1, Fusarium poae virus 1; BVF, Botrytis virus F.

3.4. Mycoviruses Influenced Growth In Vitro

Triple-infected parent isolate M196 grew significantly faster ($p < 0.05$) than lines lacking one or more viruses. After six days in culture, the mean diameter of colonies of isolate M196 was 73.8 mm, while virus-cured lines M196-1, M196-4, and M196-6 were less: 66.4, 62.9, and 66.0 mm, respectively (Table 2).

Where SsHV2 was absent but FpV1 and BVF were present (M196-4), colony growth was suppressed, suggesting that SsHV2 enhanced growth. There was no significant difference ($P > 0.05$) in growth between virus-free line M196-1 and FpV1-infected M196-6, indicating that FpV1 had no influence on growth. Where FpV1 and BVF co-occurred (M196-4), mycelial growth was suppressed compared to virus-free line M196-1. Although FpV1 alone had no apparent effect on growth, in combination with BVF it suppressed growth. This result could be interpreted as BVF alone suppressing growth and FpV1 remaining latent or as a synergistic suppressive influence of both viruses (Tables 2–4, Figure S2).

Table 3. *p*-values of paired comparisons between *M. fructicola* isogenic lines of colony diameter on V8 medium and lesion diameter on cherry fruits.

Line [a]	V8 Medium			Cherry Fruit		
	M196	M196-1	M196-4	M196	M196-1	M196-4
M196-1	2.20e^{-13}	-		0.5	-	
M196-4	4.30e^{-19}	1.50e^{-04}	-	0.8	0.6	-
M196-6	1.60e^{-14}	0.6	0.0007	0.09	0.4	0.1

p-value < 0.05 indicates a significant difference between means. [a] Isolate M196 contains SsHV2-Monilinia-TNS, FpV1-Monilinia-TNS, BVF-Monilinia-TNS, isogenic M196-1 line is virus-free, M196-4 line contains FpV1-Monilinia-TNS, BVF-Monilinia-TNS, M196-6-Monilinia-TNS line carries FpV1-Monilinia-TNS.

Table 4. Possible effects on mycelial growth by each mycovirus.

Virus	V8 Medium	Cherry Fruit
SsHV2-Monilinia-TNS	Increase mycelial growth	No effect on lesion
FpV1-Monilinia-TNS	No effect on mycelial growth	No effect on lesion
BVF-Monilinia-TNS	Decreases mycelial growth	No effect on lesion

3.5. Influence of Mycoviruses on Virulence of M. fructicola

When cherry fruits were inoculated with the four isogenic lines of M196, there was no significant difference ($p > 0.05$) in the diameters of the resulting lesions. Although mean lesion size was lowest for M196 (8.7 mm), and virus-cured lines grew faster (9.6, 8.9, and 10.9 mm for M196-1, M196-4, and M196-6, respectively) (Table 3), they were not significantly different ($p > 0.05$) (Table 3). Mock inoculated fruits used as controls never became infected with *Monilinia* or other pathogens during the course of the experiment.

4. Discussion

The *M. fructicola* and *M. laxa* isolates tested were not widely infected with mycoviruses. Of the 18 *M. laxa* isolates and 10 *M. fructicola* isolates tested, only three *M. laxa* and one *M. fructicola* isolate were infected with one or more of three mycoviruses. This overall infection rate of 14% is far below that reported in the only other study of mycoviruses from *Monilinia* [20]. That study reported 76% of *M. fructicola* isolates tested were infected with at least one mycovirus. This report was based on RNA profiles and visualization of virus-like particles, not by sequence analysis [20]. Visualization of RNA species by electrophoresis or virus particles by TEM may not be sensitive enough to detect very low-titer viruses or those lacking a coat protein, and it provides only clues to identity based on genome size and/or virion shape and size. A shotgun sequencing approach, as was used here, should be capable of detecting all RNA-based viruses present, and the resulting sequence provide evidence for taxonomic placement.

The genome sequences of SsHV2 and FpV1 obtained using this approach were complete or almost complete, while that of BVF was partial. The incomplete genome may be a function of the low titer of BVF relative to the other two viruses, but it is possibly a function of differential RNA extraction efficiencies of the extraction procedure (cellulose-based) used.

Monilinia fructicola isolate M196 was infected with three mycoviruses, all of which were originally identified from other host genera, and none had previously been identified from Australia. Their presence in Australia probably reflects anthropogenic international translocations of mycovirus-infected *M. fructicola* isolates infecting *Prunus* fruit and germplasm, although the other known hosts of these mycoviruses, *Sclerotinia sclerotiorum*, *Fusarium poae* and *Botrytis cinerea*, all occur in Australia and may carry these viruses. Notably, *M. fructicola* and other known fungal hosts of the three viruses identified all have international distribution, they are all serious plant pathogens, and they are all members of the family Sclerotiniaceae [30].

Until now, SsHV2 has been identified in *S. sclerotiorum* in China, New Zealand and the USA [8,25], but not from *Monilinia* and not from Australia. Two groups have shown SsHV2 induces hypovirulence in its host. A study with isogenic lines of *S. sclerotiorum* infected with SsHV2 and an endornavirus showed that the presence of these mycoviruses reduced mycelial pigmentation and sclerotia formation in vitro. SsHV2 induced hypovirulence of *S. sclerotiorum* on lettuce and soybean, delaying production of sclerotia and reducing their numbers, but its influences on mycelial growth rate or virulence were not recorded [25]. Hu et al. [13] reported SsHV2 was associated with hypovirulence of *S. sclerotiorum* on canola (*Brassica napus*) in China. In contrast, our studies with SsHV2 infecting a *M. fructicola* line (co-infected with with FpV-1 and BVF) did not indicate hypovirulence. Our studies suggest SsHV2 enhanced *Monilinia* mycelial growth in vitro, it did not visibly change pigmentation or conidia production, and it did not appear to influence virulence on cherry. Unfortunately, we were unable to generate a line containing only SsHV2 to confirm these findings.

This is the first report of the betapartitvirus FpV1 being identified in a host beyond *Fusarium poae*, and its first report from Australia [29]. In *F. poae* [31] and here in *M. fructicola*, FpV1 does not seem to induce abnormal morphology or changes to virulence, and this is a common observance for partitiviruses generally [1]. Known exceptions are Aspergillus fumigatus partitivirus-1 (AfuPV-1) which induced abnormal aconidial sectors and a light pigmentation phenotype in *Aspergillus fumigatus* [28], and Sclerotinia sclerotiorum partitivirus-1 (SsPV1) that induced a hypovirulence phenotype in *Sclerotinia sclerotiorum* after damaging cell organelles [8].

Botrytis virus F is the sole species in genus *Mycoflexivirus*, family *Gammaflexiviridae*, a group closely aligned to other fungus-infecting viruses in the *Deltaflexiviridae* and to plant-infecting viruses in the *Alphaflexiviridae* and *Betaflexiviridae* [32]. BVF was the first virus identified from *Botrytis cinerea*, on strawberries in New Zealand [33]. It was also identified from a fungus associated with grapes in South Africa [34]. Our results suggest that BVF slightly reduced *M. fructicola* growth in vitro. That BVF may negatively impact growth of *M. fructicola* is of potential interest, given that both *B. cinerea* and *M. fructicola* are important plant pathogens, together affecting over 200 plant species [35,36]. The possibility that BVF induces hypovirulence should be studied further.

Investigation of the influence of mycoviruses on the growth of *M. fructicola* is not a trivial exercise, involving 'curing' parent fungal isolates of viruses to obtain isogenic lines [37]. Culturing fungi with cycloheximide is used widely to eliminate RNA mycoviruses [38]; however, this approach is not always successful. Incubation of *Aspergillus niger* cultures infected with a number of virus-like particles on cycloheximide failed to eliminate any of the viruses [39,40]. It was unclear if the antibiotics used in our experiments had any influence on virus elimination because hyphal tip culture alone without antibiotics effectively cured line M196-1 of all three viruses.

A limitation of our experiment is that conclusions of hypervirulence are based on linear colony growth measurements, not on measurements of total biomass, metabolism, or fecundity. It is unclear how the triple-infection by viruses might stimulate *M. fructicola* growth in vitro, but not affect virulence on fruit. In another study, growth of a chryso-like-virus-infected strain of *Alternaria alternate* in vitro was severely restricted, while virulence against fruit was enhanced [41]. The agar plate and the fruit are very different environments. On the plate, the complex cellular interactions between the fungal cells and plant cells is lacking [42–44]. We can only speculate that infection by one or more of these viruses, probably SsHV1, provided a means by which *M. fructicola* could metabolize nutrients more efficiently on the plate than could the virus-free isogenic line, but the mechanism for this is unknown.

Supplementary Materials: The following are available online at http://www.mdpi.com/1999-4915/11/1/89/s1, Figure S1: Presence and absence of mycoviruses in isogenic fungal lines treated to remove mycoviruses; Figure S2. Comparison of typical plates of four isogenic lines of *M. fructicola* isolate M196 inoculated on V8 media after 5 days incubation in the dark at 25 °C; Table S1. *Monilinia* isolates from Western Australia and their mycovirus-infection status; Table S2. Primers used to fill the gaps of the virus sequences; Table S3: Species-specfic primers used to reconfirm the presences of mycoviruses in fungal hosts.

Author Contributions: Conceptualization, T.T.T., H.L. and S.J.W.; methodology, T.T.T. and D.Q.N.; software, T.T.T.; validation, S.J.W., and H.L.; formal analysis, T.T.T.; investigation, T.T.T., D.Q.N. and S.J.W.; resources, S.J.W. and M.G.K.J.; writing—original draft preparation, T.T.T.; writing—review and editing, S.J.W. and H.L.; supervision, S.J.W., H.L. and M.G.K.J.

Funding: This research received no external funding.

Acknowledgments: T.T.T. and D.Q.N. each received a scholarship provided jointly by Vietnam International Education Development (VIED) and Murdoch University.

Conflicts of Interest: The authors declare no conflict of interest.

References

1.   Ghabrial, S.A.; Suzuki, N. Viruses of plant pathogenic fungi. *Annu. Rev. Phytopathol.* **2009**, *47*, 353–384. [CrossRef] [PubMed]

2.   Pearson, M.N.; Beever, R.E.; Boine, B.; Arthur, K. Mycoviruses of filamentous fungi and their relevance to plant pathology. *Mol. Plant Pathol.* **2009**, *10*, 115–128. [CrossRef] [PubMed]

3.   Hollings, M. Viruses associated with a die-back disease of cultivated mushroom. *Nature* **1962**, *196*, 962–965. [CrossRef]

4.   Xie, J.; Jiang, D. New insights into mycoviruses and exploration for the biological control of crop fungal diseases. *Annu. Rev. Phytopathol.* **2014**, *52*, 45–68. [CrossRef] [PubMed]

5.   Ong, J.W.; Li, H.; Sivasithamparam, K.; Dixon, K.W.; Jones, M.G.; Wylie, S.J. The challenges of using high-throughput sequencing to track multiple bipartite mycoviruses of wild orchid-fungus partnerships over consecutive years. *Virology* **2017**, *510*, 297–304. [CrossRef] [PubMed]

6.   Nuss, D.L. Biological control of chestnut blight: An example of virus-mediated attenuation of fungal pathogenesis. *Microbiol. Rev.* **1992**, *56*, 561–576. [PubMed]

7.   Myteberi, I.F.; Lushaj, A.B.; Keča, N.; Lushaj, A.B.; Lushaj, B.M. Diversity of *Cryphonectria parasitica*, hypovirulence, and possibilities for biocontrol of chestnut canker in Albania. *Int. J. Microb. Res. Rev.* **2013**, *1*, 11–21.

8.   Xie, J.; Xiao, X.; Fu, Y.; Liu, H.; Cheng, J.; Ghabrial, S.A.; Li, G.; Jiang, D. A novel mycovirus closely related to hypoviruses that infects the plant pathogenic fungus *Sclerotinia sclerotiorum. Virology* **2011**, *418*, 49–56. [CrossRef]

9.   Liu, L.; Wang, Q.; Cheng, J.; Fu, Y.; Jiang, D.; Xie, J. Molecular characterization of a bipartite double-stranded RNA virus and its satellite-like RNA co-infecting the phytopathogenic fungus *Sclerotinia sclerotiorum. Front. Microbiol.* **2015**, *6*, 406. [CrossRef]

10.   Kanematsu, S.; Shimizu, T.; Salaipeth, L.; Yaegashi, H.; Sasaki, A.; Ito, T.; Suzuki, N. Genome rearrangement of a mycovirus *Rosellinia necatrix* megabirnavirus 1 affecting its ability to attenuate virulence of the host fungus. *Virology* **2014**, *450*, 308–315. [CrossRef]

11.   Ahn, I.P.; Lee, Y.H. A viral double-stranded RNA up regulates the fungal virulence of *Nectria radicicola. Mol. Plant Microbe Interact.* **2001**, *14*, 496–507. [CrossRef] [PubMed]

12.   Lee, K.M.; Cho, W.K.; Yu, J.; Son, M.; Choi, H.; Min, K.; Lee, Y.W.; Kim, K.H. A comparison of transcriptional patterns and mycological phenotypes following infection of *fusarium graminearum* by four mycoviruses. *PLoS ONE* **2014**, *9*, e100989. [CrossRef] [PubMed]

13.   Hu, Z.; Wu, S.; Cheng, J.; Fu, Y.; Jiang, D.; Xie, J. Molecular characterization of two positive-strand RNA viruses co-infecting a hypovirulent strain of *Sclerotinia sclerotiorum. Virology* **2014**, *464*, 450–459. [CrossRef] [PubMed]

14.   Yaegashi, H.; Kanematsu, S.; Ito, T. Molecular characterization of a new hypovirus infecting a phytopathogenic fungus, Valsa ceratosperma. *Virus Res.* **2012**, *165*, 143–150. [CrossRef] [PubMed]

15.   Wang, S.; Kondo, H.; Liu, L.; Guo, L.; Qiu, D. A novel virus in the family Hypoviridae from the plant pathogenic fungus *Fusarium graminearum. Virus Res.* **2013**, *174*, 69–77. [CrossRef] [PubMed]

16.   Koloniuk, I.; El-Habbak, M.H.; Petrzik, K.; Ghabrial, S.A. Complete genome sequence of a novel hypovirus infecting *Phomopsis longicolla. Arch. Virol.* **2014**, *159*, 1861–1863. [CrossRef]

17.   House, M. *Plant Diseases Act 1914*; Government Gazette, Western Australia Government Printer; State Law Publisher: Perth, Australia, 1997; Volume 235, p. 7507.

18. AMRiN. Occurrence Record: WAC—WAC9462 Monilinia. Department of Agriculture and Food—Western Australia. 2015. Available online: http://amrin.ala.org.au/occurrences/e656aaee-e0c4-4330-ab2deb15023f876a;jsessionid=4F6F18291AFA044E7FCB9C2AFFF8A7DC (accessed on 21 January 2019).

19. Tran, T.T.; Li, H.; Nguyen, D.Q.; Sivasithamparam, K.; Jones, M.G.K.; Wylie, S.J. Spatial distribution of *Monilinia fructicola* and *M. laxa* in stone fruit production areas in Western Australia. *Aust. Plant Pathol.* **2017**, *46*, 339–349. [CrossRef]

20. Tsai, P.F.; Pearson, M.N.; Beever, R.E. Mycoviruses in *Monilinia fructicola*. *Mycol. Res.* **2004**, *108*, 907–912. [CrossRef]

21. Wylie, S.J.; Li, H.; Jones, M.G. Yellow tailflower mild mottle virus: A new tobamovirus described from *Anthocercis littorea* (Solanaceae) in Western Australia. *Arch. Virol.* **2014**, *159*, 791–795. [CrossRef]

22. Altschul, S.F.; Gish, W.; Miller, W.; Myers, E.W.; Lipman, D.J. Basic local alignment search tool. *J. Mol. Biol.* **1990**, *215*, 403–410. [CrossRef]

23. Kearse, M.; Moir, R.; Wilson, A.; Stones-Havas, S.; Cheung, M.; Sturrock, S.; Buxton, S.; Cooper, A.; Markowitz, S.; Duran, C.; et al. Geneious basic: An integrated and extendable desktop software platform for the organization and analysis of sequence data. *Bioinformatics* **2012**, *28*, 1647–1649. [CrossRef] [PubMed]

24. Khalifa, M.E.; Pearson, M.N. Characterisation of a novel hypovirus from *Sclerotinia sclerotiorum* potentially representing a new genus within the Hypoviridae. *Virology* **2014**, *464*, 441–449. [CrossRef] [PubMed]

25. Marzano, S.Y.L.; Hobbs, H.A.; Nelson, B.D.; Hartman, G.L.; Eastburn, D.M.; McCoppin, N.K.; Domier, L.L. Transfection of *Sclerotinia sclerotiorum* with in vitro transcripts of a naturally occurring interspecific recombinant of Sclerotinia sclerotiorum hypovirus 2 significantly reduces virulence of the fungus. *J. Virol.* **2015**. [CrossRef] [PubMed]

26. King, M.Q.; Adams, M.J.; Carstens, E.B.; Lefkowitz, E.J. *Virus Taxonomy*, 9th ed.; Elsevier Academic Press: Amsterdam, the Netherlands, 2012.

27. Koonin, E.V.; Dolja, V.V.; Morris, T.J. Evolution and taxonomy of positive-strand RNA viruses: Implications of comparative analysis of amino acid sequences. *Crit. Rev. Biochem. Mol.* **1993**, *28*, 375–430. [CrossRef] [PubMed]

28. Bhatti, M.F.; Jamal, A.; Petrou, M.A.; Cairns, T.C.; Bignell, E.M.; Coutts, R.H. The effects of RNA mycoviruses on growth and murine virulence of *Aspergillus fumigatus*. *Fungal Genet. Biol.* **2011**, *48*, 1071–1075. [CrossRef]

29. Akinsanmi, O.A.; Mitter, V.; Simpfendorfer, S.; Backhouse, D.; Chakraborty, S. Identity and pathogenicity of *Fusarium* spp. isolated from wheat fields in Queensland and northern New South Wales. *Aust. J. Agr. Res.* **2004**, *55*, 97–107. [CrossRef]

30. Cannon, P.F.; Kirk, P.M. *Fungal Families of the World*; CABI: Wallingford, UK, 2017; pp. 327–328.

31. Cho, W.K.; Lee, K.M.; Yu, J.; Son, M.; Kim, K.H. Insight into mycoviruses infecting *Fusarium* species. *Adv. Virus Res.* **2013**, *86*, 273–288. [PubMed]

32. Carstens, E.B. Ratification vote on taxonomic proposals to the International Committee on Taxonomy of Viruses (2009). *Arch. Virol.* **2010**, *155*, 133–146. [CrossRef] [PubMed]

33. Howitt, R.L.; Beever, R.E.; Pearson, M.N.; Forster, R.L. Genome characterization of *Botrytis* virus F, a flexuous rod-shaped mycovirus resembling plant 'potex-like' viruses. *J. Gen. Virol.* **2001**, *82*, 67–78. [CrossRef]

34. Al Rwahnih, M.; Daubert, S.; Urbez-Torres, J.R.; Cordero, F.; Rowhani, A. Deep sequencing evidence from single grapevine plants reveals a virome dominated by mycoviruses. *Arch. Virol.* **2011**, *156*, 397–403. [CrossRef]

35. Holb, I.J. Brown rot blossom blight of pome and stone fruits: Symptom, disease cycle, host resistance, and biological control. *Int. J. Hortic. Sci.* **2008**, *14*, 15–21.

36. Williamson, B.; Tudzynski, B.; Tudzynski, P.; van Kan, J.A. *Botrytis cinerea*: The cause of grey mould disease. *Mol. Plant Pathol.* **2007**, *8*, 561–580. [CrossRef] [PubMed]

37. Tran, T.T.; Li, H.; Nguyen, D.Q.; Jones, M.G.K.; Sivasithamparam, K.; Wylie, S.J. *Monilinia fructicola* and *Monilinia laxa* isolates from stone fruit orchards sprayed with fungicides displayed a broader range of responses to fungicides than those from unsprayed orchards. *Eur. J. Plant Pathol.* **2018**, *1*, 1–15. [CrossRef]

38. Fink, G.R.; Styles, C.A. Curing of a killer factor in *Saccharomyces cerevisiae*. *Proc. Natl. Acad. Sci. USA* **1972**, *69*, 2846–2849. [CrossRef] [PubMed]

39. Varga, J.; Kevei, F.; Vágvölgyi, C.; Vriesema, A.; Croft, J.H. Double-stranded RNA mycoviruses in section *Nigri* of the *Aspergillus* genus. *Can. J. Microb.* **1994**, *40*, 325–329. [CrossRef]

40. Aoki, N.; Moriyama, H.; Kodama, M.; Arie, T.; Teraoka, T.; Fukuhara, T. A novel mycovirus associated with four double-stranded RNAs affects host fungal growth in *Alternaria alternata*. *Virus Res.* **2009**, *140*, 179–187. [CrossRef] [PubMed]
41. Govrin, E.M.; Levine, A. The hypersensitive response facilitates plant infection by the necrotrophic pathogen *Botrytis cinerea*. *Curr. Biol.* **2000**, *10*, 751–757. [CrossRef]
42. Berger, S.; Papadopoulos, M.; Schreiber, U.; Kaiser, W.; Roitsch, T. Complex regulation of gene expression, photosynthesis and sugar levels by pathogen infection in tomato. *Physiol. Plantarum* **2004**, *122*, 419–428. [CrossRef]
43. Van Kan, J.A. Licensed to kill: The lifestyle of a necrotrophic plant pathogen. *Trends Plant Sci.* **2006**, *11*, 247–253. [CrossRef]
44. Wang, D.; Pajerowska-Mukhtar, K.; Culler, A.H.; Dong, X. Salicylic acid inhibits pathogen growth in plants through repression of the auxin signaling pathway. *Curr. Biol.* **2007**, *17*, 1784–1790. [CrossRef]

MDPI
St. Alban-Anlage 66
4052 Basel
Switzerland
Tel. +41 61 683 77 34
Fax +41 61 302 89 18
www.mdpi.com

Viruses Editorial Office
E-mail: viruses@mdpi.com
www.mdpi.com/journal/viruses

www.ingramcontent.com/pod-product-compliance
Lightning Source LLC
Chambersburg PA
CBHW051711210326
41597CB00032B/5447